AF601918

Transformation of Indian Agriculture through Innovative Technologies

The Editors

Dr Shahid Ahamad presently working as Chief Scientist/Associate Director Research at Directorate of Research in Sher-E-Kashmir University of Agricultural Sciences & Technology of Jammu (Chatha) J.&K. Earlier he served as Jr. Scientist (Plant Pathology), Sr. Scientist/Programme Coordinator in KVKs and Dy. Director Research. He is also working as Nodal Officer, PME Cell. His research interests are plant diseases management, maize pathology, organic crop production and research management. He Conferred Fellow of Society of Plant Protection Sciences (2005) IARI, New Delhi & Fellow of Indian Phytopathological Society in 2015. He wins many awards viz. Young Achiever Awards-2011; Distinguished Service Awards (2012); Scientist of the Year Award (2013) and Eminence Scientist Awards (2018) for his outstanding contribution in the field of Plant Pathology. He served many Scientific Societies in different capacity viz. Councilor (2012-15) Society of Plant Protection Sciences, Division of Nematology, I.A.R.I., New Delhi; Zonal President of Indian Phytopathological Society (NZ) for 2017. He is also working as editorial members in different society of national and international repute. As an academician, He has published more than 250 publications in national and international reputed journals/magazines/ News papers etc. including 10 books in different aspects of agriculture/ Plant Pathology. He is having 20 years of experience in research, extension and teaching activities in the field of Plant Pathology in general and agriculture in particular.

Prof. Jag Paul Sharma is presently working as Director Research at Sher-e-Kashmir Univers ity of Agricultural Sciences and Technology of Jammu. Earlier, he has served as Senior Scientist (Plant Breeding), Leh; Senior Scientist (Pulses), Samba; Professor & Head, Division of Vegetable Science and Floriculture; Associate Director Research and Dean, Faculty of Agriculture, SKUAST-J, Chatha. As a plant breeder, developed 12 varieties in different crops viz 03 in wheat, 02 in barley and 01 each in chickpea, knolkhol, methi, beet palak, okra, coriander, broccoli and basmati, whereas, 04 strains of different vegetable crops are in advance stage of testing. As an academician, he contributed 09 books in the field of agriculture including one text book and published more than 100 research papers in journals of national and international repute. His research interests are organic crop production, crop hybridization, plant tissue culture, seed production and research management. He envisions India as emerging hub of quality agricultural products of global competence and strongly believes in scientific cooperation in sharing of knowledge on international issues such as food and nutritional security, climate change, agro-ecological sustainability, natural resource management and social security. He has been the co-convenor of Vth Dean's Committee of ICAR on Agricultural Education. He is also a member of Project Review Committee on CODER, ASAR of Department of Science & Technology, Govt. of India, State Biodiversity Board of Govt. of J&K.

Transformation of Indian Agriculture through Innovative Technologies

– *Editors* –

Shahid Ahamad

Chief Scientist/Associate Director Research

Jag Paul Sharma

Director Research

Directorate of Research
Sher-e-Kashmir University of Agricultural Sciences and Technology of Jammu,
Main Campus Chatha, Jammu – 180 009 (J&K)

2019

Daya Publishing House®

A Division of

Astral International Pvt. Ltd.

New Delhi–110 002

ISBN: 9789390384655 (HB)

Published by : **Daya Publishing House®**
A Division of
Astral International Pvt. Ltd.
– ISO 9001:2015 Certified Company –
4736/23, Ansari Road, Darya Ganj
New Delhi-110 002
Ph. 011-43549197, 23278134
E-mail: info@astralint.com
Website: www.astralint.com

Digitally Printed at : **Replika Press Pvt. Ltd.**

Foreword

Doubling farmers' income by 2022 is the mission of Government of India to achieve the mission. Indian agriculture awaits transformation through innovative technological interventions. Government of India has taken a number of important policy decisions and initiated measures to boost agriculture sector, such as instituting soil health cards, more crop per drop through efficient irrigation, strengthening government procurement policy on pulses, introducing neem-coated urea, building more assets under MGNREGA, expanding crop insurance for farmers, building a common agricultural e-market via e-NAM and allied sectors etc.

Advanced agricultural approaches are also required for sustainable solutions to the huge national problem of 'hidden hunger' and it may be achieved by proper crop management practices, value addition and promotion of specialized crops. The book provides wider coverage on innovative social, cultural and technical aspects, such as transformation of rural India through agricultural technologies, modern fruit production technologies, quality maize for augmenting income of the farmers, entrepreneurship development in beekeeping, maize hybrids technology, pulse production technology, impact of climate change on agriculture, indigenous technical knowledge in weather forecasting, direct-seeded rice, diversification of agriculture, nanotechnology in plant disease management, women empowerment and self help groups (SHGs), empowerment of rural youths for rural transformation, integrated pest management in vegetables, saffron diseases and their management, role of markets and market-led extension in doubling the farmers' income etc have been covered in this book.

Attempts made by **Dr. Shahid Ahamad, Chief Scientist /Associate Director Research** and **Dr. Jag Paul Sharma, Director Research**, Directorate of Research, SKUAST-Jammu in the form of a book Volume I entitled "**Transformation of Indian Agriculture through Innovative Technologies**" are worth appreciation. I feel that this volume shall find use with the researchers, policy makers and students, other stake-holders in Indian agriculture. I congratulate the authors for their painstaking efforts in bringing out this publication.

(Prof. Pradeep K. Sharma)

Prof. Pradeep K. Sharma

Vice Chancellor

SKUAST, Jammu (J & K)

Preface

A recent survey by the Economist revealed that the world population has increased by 90 per cent in the past 40 years while food production has increased only by 25 per cent per head. With an additional 1.5 billion mouth to feed by 2020, farmers worldwide have to produce 39 per cent more. The new millennium promises excitement and hope for the future by new advancement in agriculture research. Doubling farmers' income by 2022 is the mission of Government of India to achieve the mission. Indian agriculture awaits transformation through innovative technological interventions. Government of India has taken a number of important policy decisions and initiated measures to boost agriculture sector, such as instituting soil health cards, more crop per drop through efficient irrigation, strengthening government procurement policy on pulses, introducing neem-coated urea, building more assets under MGNREGA, expanding crop insurance for farmers, building a common agricultural e-market via e-NAM and allied sectors etc.

Advanced agricultural approaches are also required for sustainable solutions to the huge national problem of 'hidden hunger' and it may be achieved by proper crop management practices, value addition and promotion of specialized crops. The book provides wider coverage on innovative social, cultural and technical aspects, such as transformation of rural India through agricultural technologies, modern fruit production technologies, quality maize for augmenting income of the farmers, entrepreneurship development in beekeeping, maize hybrids technology, pulse production technology, impact of climate change on agriculture, indigenous technical knowledge in weather forecasting, direct-seeded rice, diversification of agriculture, nanotechnology in plant disease management, women empowerment and self help groups (SHGs), empowerment of rural youths for rural transformation, integrated pest management in vegetables, saffron diseases and their management, role of markets and market-led extension in doubling the farmers' income etc have been covered in this book.

There has been a long felt need for a book exclusively on the "Transformation of Indian Agriculture through Innovative Technologies". An effort has been made to collect information on various aspects of Agriculture viz Plant Pathology, Entomology, Agri. Economics, Agronomy, Agri. Extension,Horticulture,Vegetable Science etc. Chapters in this book have been contributed by noted agricultural Scientists of India. I am grateful and indebted to all the learned galaxy of contributors who have spent their considerable time in contributing the chapters

A deep sense of respect and gratitude to Prof. Pradeep K. Sharma, Hon'ble Vice Chancellor, SKUAST-JAMMU, Dr. Deepak Kher,DPM, Dr. R. K. Arora, ADE & I/c KVKs; Dr. V.K.Razdan, Dr.Anil Gupta, Dr.R.K.Salgotra, Dr.Banarsi Lal for their technical guidance and help in bringing out this book. It is our privilege to have foreword written of these book by Prof. Pradeep K. Sharma, Hon'ble Vice Chancellor, SK University of Agricultural Sciences and Technology of Jammu . This is added feather to our effort and we are thankful to him for the same.

Last but not the least, I must acknowledge and oblige the M/S Astral International Publishing House, New Delhi who has been always considerate and positive to our contentions and efforts. I look forward that this book will be made available to teachers, Scientists, students and other stake holders at reasonable and economic price.

Dr. Shahid Ahamad

Dr. Jag Paul Sharma

Contents

List of Contributors

Akas Sharma
Assistant Professor
Division of Fruit Science
Chatha, SKUAST-Jammu – 208 009, J&K

Ali Anwar
Professor & Head,
Division of Plant Pathology
SKUAST-K, Shalimar – 190 025, J&K

Alka Kushwaha
Assistant Professor
Department of Botany,
D.A.V. College, Kanpur – 208 001, U.P.

Anil Bhushan
Assistant Professor,
Regional Agricultural Research Station,
Rajouri, SKUAST-Jammu – 208 009, J&K

Banarsi Lal
Krishi Vigyan Kendra, Reasi
SKUAST-Jammu, J&K

Anil Kumar
AICRP-Weed Management,
Division of Agronomy,
SKUAST-Jammu – 208 009, J&K

Anil Gupta
Division of Plant Pathology,
Faculty of Agriculture,
SKUAST-Jammu, J&K

Arif Hussain
Division of Plant Pathology
SKUAST-K, Shalimar – 190 025, J&K

Balkaran Singh Sandhu
Krishi Vigyan Kendra,
Sri Muktsar Sahib – 152 026, Punjab

Ghulam Hassan
Division of Plant Pathology
SKUAST-K, Shalimar – 190 025, J&K

B.C. Sharma
Division of Agronomy, FOA,
SKUAST-Jammu – 208 009, J&K

Hafeez Ahmad
Professor,
Division of Entomology,
SKUAST-Jammu – 208 009, J&K

Bharat Singh Ghanghas
Department of Agricultural Extension,
CCSHAU, Hisar – 125 004, Haryana

Jag Paul Sharma
Director Research,
SKUAST-Jammu – 208 009, J&K

Bharti Deshmukh
Assistant Professor,
Firozpur Agricultural College,
PAU, Punjab

Jyoti Kachroo
Division of Agricultural Economics and Agribusiness Management
SKUAST-Jammu – 208 009, J&K

Brij Nandan
Pulse Research Sub Station Samba
SKUAST-Jammu, J&K

Kiran Kour
Advanced Centre for Horticulture Research,
Udheywalla, Talab Tillo, Jammu

D.B. Abrol
Division of Entomology
SKUAST-Jammu – 208 009, J&K

Kumari Sharda
Krishi Vigyan Kendra,
Banka, Bihar

Devidner Sharma
Division of Entomology
SKUAST-Jammu – 208 009, J&K

Mahender Singh
Agromet Section, Division of Agronomy,
FOA, SKUAST-Jammu – 208 009, J&K

Dharmendra Kumar
Krishi Vigyan Kendra,
Banka, Bihar

Mahesh Kumar
Division of Entomology
SKUAST-Jammu – 208 009, J&K

Mohammad Najeeb Mughal
Division of Plant Pathology
SKUAST-K, Shalimar – 190 025, J&K

Pawan Kumar Sharma
KVK, Kathua,
SKUAST-Jammu, J&K

Moni Gupta
Division of Biochemistry,
Faculty of Agriculture,
SKUAST-Jammu – 208 009, J&K

Rajesh Kumar
Regional Agricultural Research Station,
Rajouri,
SKUAST-Jammu, J&K

Mudasir Hassan
Division of Plant Pathology
SKUAST-K, Shalimar – 190 025, J&K

Ranbir Singh
Division of Plant Pathology,
Faculty of Agriculture,
SKUAST-Jammu – 208 009, J&K

Nirmaljit Singh Dhaliwal
Krishi Vigyan Kendra,
Sri Muktsar Sahib – 152 026, Punjab

R. Puniya
AICRP-Weed Management,
Division of Agronomy,
SKUAST-Jammu – 208 009, J&K

Parshant Bakshi
Advanced Centre for Horticulture Research,
Udheywalla, Talab Tillo, Jammu, J&K

Ravinder Singh Sudan
Senior Scientist (PBG) and Incharge,
AICRP Maize Centre
SKUAST-Jammu, Udhampur, J&K

Sachin Gupta
Division of Plant Pathology,
Faculty of Agriculture,
SKUAST-Jammu – 208 009, J&K

Sunita Rani
Division of Plant Pathology,
Faculty of Agriculture,
SKUAST-Jammu – 208 009, J&K

S.P. Singh
Division of Agricultural Economics and Agribusiness Management,
SKUAST-Jammu – 208 009, J&K

Udit Narain
Department of Plant Pathology,
C.S.A. University of Agriculture and Technology
Kanpur – 208 002, Uttar Pradesh

Sudhakar Dwivedi
Division of Agricultural Economics and Agribusiness Management,
SKUAST-Jammu – 208 009, J&K

Vandana Krishna
Department of Botany,
D.A.V. College, Dehradun, Uttarakhand

Veena Sharma
Agromet Section,
Division of Agronomy, FOA,
SKUAST-Jammu – 208 009, J&K

Vikas Sharma
Regional Agricultural Research Station,
Rajouri
SKUAST-Jammu, J&K

Suheel Ahmad Ganai
Division of Entomology
SKUAST-Jammu – 208 009, J&K

V.K. Ambardar
Division of Plant Pathology
SKUAST-K, Shalimar – 190 025, J&K

Vishal Mahajan
KVK, Kathua,
SKUAST-Jammu – 208 009, J&K

Transformation of Indian Agriculture through Innovative Technologies *Pages* **1–9**
Editor: **Dr. Shahid Ahamad & Dr. Jag Paul Sharma**
Published by: **ASTRAL INTERNATIONAL PVT. LTD., NEW DELHI**

1 Applying Seasonal Climate Prediction to Agricultural Production

Veena Sharma and Mahender Singh

Weather

The success or failure of agriculture crop production is mainly determined by the weather parameters. Agricultural productivity largely depends upon weather. The effects of meteorological conditions are most pronounced on high yielding varieties of crop with increased sensitivities to environmental conditions, requiring maximum optimization of water, air, thermal and nutritional conditions. The biological potential of the plants manifests itself best in favorable conditions and is severely reduced when conditions are adverse. This results in large fluctuations in annual crop yields. Weather manifests its influence on agricultural operations and farm production through its effects on soil and plant growth. Out of the total annual crop losses, a substantial portion is because of aberrant weather. The loss could be minimized by making adjustment with coming weather through timely and accurate weather forecasting. While farmers are often flexible in dealing with weather and year-to-year variability, there is nevertheless a high degree of adaptation to the local climate in the form of established infrastructure, local farming practice and individual experience. Climate change can therefore be expected to impact on agriculture, potentially threatening established aspects of farming systems but also providing opportunities for improvements.

Need of the Forecast

Weather varies with space and time, hence, its forecast is desirable for effective planning and management of agricultural practices and to minimize the farm losses.

Agricultural operations can be advanced or delayed with the help of advanced weather forecast from three to ten days. Using information on the effect of weather and climatic factors on agricultural productivity in an educated manner can not only reduce damage, but can also make it possible to obtain additional yield without significant financial outlays. Agriculturally relevant forecast is not only useful for efficient management of farm inputs but also leads to precise impact assessment (Gadgil, 1989).The complete avoidance of all farm losses due to weather factor is not possible but it can be minimized to some extent by making adjustments through timely and accurate information of weather forecast. The development of response strategy (Everingham *et al.*, 2002) help farmers realize the potential benefits of using weather-based agro-meteorological information in minimizing the losses due to adverse weather conditions, thereby improving yield, quantity and quality of agricultural productions. For effective planning and management of agricultural practices such as selection of cultivar, sowing, need-based application of fertilizer, pesticides, insecticides, efficient irrigation and harvest, weather forecasts in all temporal ranges are desirable. Weather forecast in short and medium ranges greatly contribute towards making short-term adjustments in daily agricultural operations which minimize losses resulting from adverse weather conditions and improve yield and quantity and quality of agricultural productions.

From a farmer's perspective, the forecast value increases if the weather and climate forecasts are capable of influencing their decisions on key farm management operations (Gadgil *et al.*, 2002) as wind speed and direction and cloud cover play an important role during infestation of pest and disease. Thus, it becomes essential to relate with the requirements of farmers(Hansen*, 2002), understand their needs and give the forecast in appropriate spatial and temporal range (Hansen**, 2002). This ultimately helps in increasing the reliability of the forecast and thus in better adoption of the weather-based advisory. These weather-based agro-advisories have been helping the farming community to take advantage of prognosticated weather conditions and thereby form a response strategy.

Issuance of Forecast

Weather forecasts are given by IMD-Pune. Location-specific 5-days weather forecast for six parameters, *viz.* rainfall, cloud cover, wind direction and speed, and minimum and maximum temperature is issued twice a week (Tuesday and Friday). The State Agromet Centres (SAMC) at IMD's forecasting offices located at state capitals, provide district level weather forecast to AMFUs biweekly (Tuesday and Friday), who will in turn issue crop and area specific advisories to farming community. Location-specific weather forecasts for six parameters, *viz.* rainfall, cloud cover, wind direction and speed, and minimum and maximum temperature are obtained by AMFUs twice a week.

Agro-advisory (Three Tier Agromet Advisory System)

The forecasts are converted into farm-level advisories by AMFUs and disseminated to the farmers in vernacular language through mass media. The forecast, in quantitative terms, is issued for six parameters, *viz.* rainfall, cloud cover, wind direction and speed, and minimum and maximum temperature.

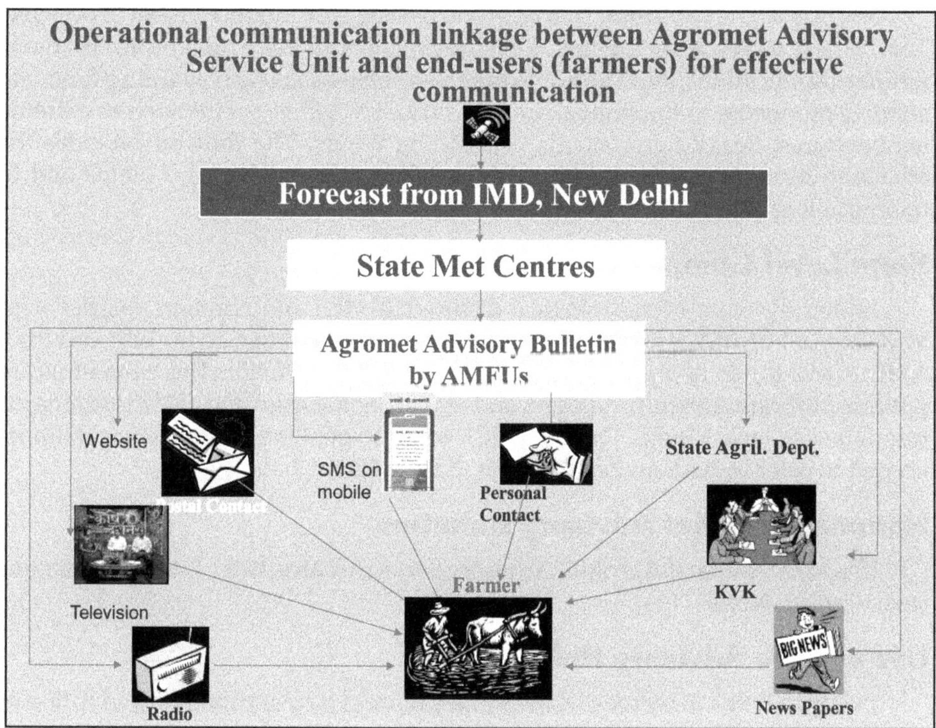

Figure 1.1. The Complete Flow Diagram of the Agro Advisory.

District AgroMet Advisories Bulletins

Issued by AMFUs and bulletin basically contains information on current and past week weather, crop information and weather based advisories. The temporal range of the forecast is five days. The Bulletin contains the forecast on six parameters *viz.* rainfall, cloud cover, wind direction and speed, and minimum and maximum temperature along with their likely impact on crops. The AMFUs are located within SAU headquarters, their regional research stations and ICAR institutes. AMFUs receive weather forecast from SAMC on bi-weekly basis (Tuesday and Friday).The AMFUs constitute an advisory board consisting of an inter-disciplinary group of agricultural and extension specialists, such as, Plant Pathologists, Soil Scientists, Entomologists, Horticulturists, Agronomists *etc.*, discuss the forecast of their zone to formulate the agro-advisory. The main stress is given to the preparation of advisories. These advisories contain location specific and crop specific farm level advisories. So Advisory content varies with location, season, weather, crop condition, and local management practices. Advisories are prepared in local language containing description of prevailing weather, soil and crop condition, and suggestions for taking appropriate measures to minimize the loss and also optimize input in the form of irrigation, fertilizer or pesticides.

AMFUs also take output from crop and pest disease models wherever possible. This helps to increase the timeliness of spraying operations, irrigation applications, fertilizer applications, *etc.* The advisories also serve as an early warning function, alerting producers to the implications of various weather events such as extreme temperatures, heavy rains, floods, and strong winds. The Bulletin provides the advice to farmers on the measures to be taken to maintain crop yields and to minimize weather-induced losses.

State Level Composite AAS Bulletins

Issued by State Meteorological Centre (SAMC) and contains district wise advisories. The SAMCs prepare state composite bulletin based on the bulletins from AMFUs and the bulletin from State Agricultural Department. These bulletins are provided to State Agricultural Dept. and other functionaries and also disseminated through state level media. These SAMCs will also generate other technical inputs needed to cater to AAS and co-ordinate the services at state level.

National AgroMet Advisory Bulletins

Issued by National AgroMet Advisory Service Centre, IMD, Pune and contains state wise advisories.

Use of Agro Advisory Bulletin

Bulletins helps in increasing awareness among farmers about the adoption of weather based advisories and their positive impacts. Weather forecast and weather based agromet advisories help in increasing the economic benefit to the farmers by suggesting them the suitable management practices according to the weather conditions.

The activities of AAS are aimed at lending support to agriculturists and planners by giving advance information relating to weather conditions for planning day to day agricultural operations, crop-yield forecast based on meteorological data, information relating to progress of monsoon, commencement of sowing rains, conducting studies on evapo-transpiration to know crop water demand and on the effect of weather on plant pests and diseases

Dissemination

Once the weather based AAS bulletins are prepared by AMFUs these AAS bulletins are disseminated to the farmers of the region through mass media, such as T.V., All India Radio, Newspapers in vernacular language and also through Internet and personal contact with the progressive farmers through extension workers. The bulletin in printed form/public notice is also disseminated to the contact farmers in several villages by telephone/fax, post, poster and hand delivery. Media plays a crucial role in the dissemination of the AAS and can be taken as the nodal agency for effective outreach to the end user that is the farmer. Public awareness programmes in 'kisanmelas' are regularly held by agricultural universities to educate farmers on the usage of farm advisories.

Among these, radio and television remained the most popular and easily accessible media through which 60–70 per cent farmers could get information on weather forecast. Literate farmers comprising about 20 per cent could read the detailed weather-based agroadvisory bulletin in newspapers and take corrective action. On several occasions, they contact agricultural extension workers and take their advice for various operations. It is concluded from the study that almost 90 per cent of the farmers gave more importance to forecast of rainfall and temperature over wind speed, wind direction and cloud cover (Maini and Rathore, 2011).

Feedback Mechanism

Periodic feedback on worthiness of forecast and usefulness of advisories is also obtained. This feedback is obtained weekly, monthly and also annually. Feedback from selected farmers and Research and Development under different SAU's are being documented on whether they have adjusted their day-to-day farming operations in response to the advice laid in AAS and also on their additional requirements. About 60–80 per cent of the AAS farmers are keen to have weekly cumulative rainfall forecast for carrying out their various farm operations. About 20 per cent of the AAS farmers are interested to obtain the seasonal forecast of rainfall in order to plan their sowing and harvesting operations in advance. Almost all AAS farmers opined that the weather bulletin is received by them on time and 60–70 per cent feel that the rainfall forecast and the temperature trends are reliable. Almost 90 per cent of the farmers gave more importance to forecast of rainfall and temperature over wind speed, wind direction and cloud cover (Maini and Rathore, 2011).

Annual review meetings are conducted for the evaluation of the use of AAS and weather information. These meetings are held at different SAUs on rotation. All nodal officers from the AMFU's and scientists from IMD and ICAR participate in this meet for the assessment of the performance of AMFU's. Also, further possible ways for improvement in the existing system are discussed.

Verification of Location Specific Forecast Issued to AAS Units

In order to evaluate the skill of the forecasts issued, the rigorous verification of the forecast is done as a routine. The verification of weather forecasts is done for four seasons *viz.*, Pre- monsoon (March to May), monsoon (June-September),Post – monsoon (October to December) and winter (January and February) as per the guidelines of NCMRWF (Anon., 1999). For evaluation of usability of forecast of quantitative precipitation and temperature an error structure is formulated. The forecast of rainfall, cloud cover, temperature, wind speed and direction is verified by calculating the error structure. Initially, the error structure is used to categorize the forecast given as correct, usable or unusable based on the per cent deviation in the forecast values as compared to observed values as per the guidelines of National Centre for Medium Range Weather Forecasting (NCMRWF) (Anon., 1999). The correct and usable cases are summed up and the combined values indicate the per cent usability of the forecasts of various parameters to the total events occurred in respective parameter. The Ratio Score (RS) is also worked out for rainfall forecast and this score varies from 0 to 1, with 1 indicating perfect forecast (Anon., 1999). In addition, Hanssen and Kuipers Score (H.K. Score) is also worked out for rainfall forecast which indicated the skill of the forecast given (Anon., 1999). It ranges between -1 to +1, with 0 indicating no skill. The skill of maximum temperature (Tx) and minimum temperature (Tn) is given in terms of correlation (CC) and RMSE.

Economic Impact

User Requirement

The types of economic decision which require agro- meteorological products can be categorized according to three time scales:

- ✰ Long-term planning for agricultural development (rational allocation ofland, choice of crops, selection of species and varieties)
- ✰ Medium-term planning for the next season (choice of farming area, crop varieties, *etc.*)
- ✰ Short-term decisions regarding imminent farming operations (choice of optimal sowing and harvest dates, dates and quantities for fertilization, dates and quantities for irrigation, *etc.*).

Each type of decision requires the appropriate meteorological information.

In the first of the three categories listed above, this will involve basic climatological data and long-term forecasts.

In the second case, it will involve seasonal forecasts, monthly forecasts, and various agrometeorological forecasts on moisture availability, yield *etc.*

In the third case, it will involve short-term forecasts, medium-term forecasts, and special recommendations for crop-growth.

Service Requirement

Internal Perspectives

- ✰ **To establish the worthiness of the service**: Economic impact has to becarried out in order to know its potential benefits.
- ✰ **Service credibility**: Credibility is always closely linked to forecast verification. Hence, economic impact studies need to be carried out to establish credibility in the eyes of the potential users if optimum benefits are to be derived from the marketing of the service.
- ✰ **Service accountability or justification**: Assessment of the service helps justifying the costs and the ongoing need and existence of such a service.

External Perspectives

- ✰ By quantifying the benefits of this service one comes to know the needs of the users, their level of satisfaction and their further expectation. Consequently the progressive user provides a positive feedback and increased response of progressive users drive the service. The outcome includes better services over time, services with better utility and most likely with better perceived accuracy. Secondly, through these interactive educational initiatives, policy makers and other clients become sensitized to and better informed about the value of these services, which results in improved decision making.

National Perspective

- ✰ On the national scale, more knowledgeable decision-making leads to improved practices and attitudes, enhanced productivity, a more nationally relevant economic society and more socially acceptable practices.

Global Scale

On the global scale, more knowledgeable decision-making leads to improved practices and attitudes, enhanced productivity, a more internationally relevant economic society and more socially acceptable practices.

Benefits or expectations from these studies

- ✰ Will give an insight into forecasting skill and reach of the service and also its economic value in terms of money,
- ✰ Will help in taking better decision. Application of these methods for assessing economic and social benefits can produce information leading to the efficient production and supply of services,
- ✰ Will help in cultivar selection, their dates of sowing/planting/ transplanting, dates of intercultural operations, dates of harvesting and also performing postharvest operations,
- ✰ Will give site-specific forecast information and corresponding advisories that will help maximize output and avert crop damage or loss. The service will also help growers anticipate and plan for chemical applications, irrigation scheduling, disease and pest outbreaks and many more weather related agriculture-specific operations,
- ✰ Will give agromet advisories that will increase profits by consistently delivering actionable weather information, analysis and decision support for farming situations such as:
- ❖ To manage pests through forecast of relative humidity, temperature and wind:
  - ❖ Progressive water management through rainfall forecasts,
  - ❖ To protect crop from thermal stress through forecasting of extreme temperature conditions.

Above all, along with many other situations agro advisory will help increase the crop protection, protect resources and preserve the environment.

Economic Impact Studies

To know the impact of forecast and advisories given from Agromet Advisory Service Unit, regular estimation of economic benefit/loss on account of adoption of the agro advisory issued by the AMFUs compared with non AAS farmers is done.

Two key impact indicators are used: (a) ability of the forecast to influence decision of farmers and adopt appropriate measures to reduce weather-induced losses, and (b) economic and other benefits accruing to farmers due to change in their farm management decisions due to weather forecast. The economic benefits are either in terms of cost saving or reduction in yield losses. Estimation of these benefits involves getting information on use of weather forecasts in decision making, changes in farmer's decisions, changes in cultivars, date of sowing, inputs use, yields, *etc.* These changes are computed using a 'with' and 'without' approach. Difference in input use, cost and crop yield is tested for their statistical significance.

For this purpose, a field survey of the study area is conducted and feedback from various farmers is collected and summarized by recording the yield of crops from two situations *viz.*, recommended practices with agromet advisory and recommended practices without agromet advisory. The per cent gain in income from different crops by the AAS farmers is to the tune of 5.3 to 25.0 per cent over non AAS farmers (Manjappa, K. and Yeledalli, 2013).

Conclusion

The agro advisory is an innovative inter-departmental extension service, with a goal to deliver weather-wise management of agriculture. It may be concluded that agro advisory helps in bringing out substantial awareness among farmers about weather. Their timely availability and quality will help them in decision making. It also helps in encouraging the adoption and use of modern agricultural production technologies and practices, in promoting weather-based irrigation management, pest/disease management, *etc.* The application of agromet advisory bulletin, based on currentand forecasted weather is a useful tool for enhancing the production and income. AAS farmers receive weather forecast based agro-advisories, including optimum use of inputs for different farm operations. Due to judicious and timely utilization of inputs, production cost for the AAS farmers reduced. The increased yield level and reduced cost of cultivation leads to increased net returns.

References

Anonymous, 1999, Verification of medium range weather forecast. In: "Guide for Agrometeorological Advisory Service". (Eds. S.V. Singh, L.S. Rathore and H.K.N. Trivedi). pp.73-93. (DST, NCMRWF, New Delhi).

Everingham, Y. L., Muchow, R. C., Stone, R. C., Inman-Bamber, G., Singels, A. and Bezuidenhout, C. N. 2002. Enhanced risk management and decision-making capability across the sugarcane industry value chain based on seasonal climate forecasts. *Agric. Syst.*, 74(3), 459–477.

Gadgil, S., 1989, Monsoon variability and its relationship with agricultural strategies. Paper presented at Int. symp. Climate variability and food security in developing countries. Held at New Delhi during February 5-7, 1987. pp. 249–267.

Gadgil, S., Seshagiri Rao, P. R. and Narahari, K. 2002. Use of climate information for farm-level decision-making: rainfed groundnut in southern India. *Agric. Syst.*, 74(3), 431–457.

Hansen*, J. W. 2002. Realising the potential benefits of climate prediction to agriculture: issues, approaches, challenges. *Agric. Syst.*, 74(3), 309–330.

Hansen**, J. W. 2002. Applying seasonal climate prediction to agricultural production. *Agric. Syst.*, 74(3), 305–307.

Maini Parvinder and Rathore, L. S.2011. Economic impact assessment of the Agro-meteorological Advisory Service of India. *Current Science*. 101(10): 25.

Manjappa, K. and. Yeledalli, S. B. 2013. Validation and assessment of economic impact of agro advisories issued based on medium range weather forecast for Uttara Kannada district of Karnataka Agromet Advisory Service Unit Agricultural Research Station (Paddy), Sirsi – 581 401, India, *Karnataka J. Agric. Sci.*,26 (1): (36-39).

Transformation of Indian Agriculture through Innovative Technologies *Pages* **11–21**
Editor: **Dr. Shahid Ahamad & Dr. Jag Paul Sharma**
Published by: **ASTRAL INTERNATIONAL PVT. LTD., NEW DELHI**

2 Minimize Weather Risk in Agricultural Planning and Management through Agromet Advisory Services in Rural Areas

Mahender Singh, Bharat Singh Ghanghas, Veena Sharma and B.C. Sharma

Importance of Agromet Advisory Services

Weather and climate plays a major role before and during the cropping season. The management of weather and climate risks in agriculture has become an important issue due to climate change. The Intergovernmental Panel on Climate Change (IPCC) has highlighted multiple climate risks for agriculture and food security as well as the potential of improved weather and climate early warning systems to assist farmers. The impact of the agricultural sector is wide-ranging and extends to economic growth, food security, poverty reduction, livelihoods, rural development and the environment. In India small scale farmers constitute the major part of the supply base and improvements in productivity and food production will need to come from them. Moreover, the poorest half of the population benefits significantly more from agricultural growth than growth in other sectors of the economy (UN, 2008; World Bank, 2007). However, to do so, they need to be able to compete in tradition (urban markets) and emerging institutional markets from the globalized agri-business system that is providing food and fiber to growing markets. Most small scale farmers supply the price-driven wet markets, but have difficulties accessing the value-driven institutional markets. To compete and remain competitive, small scale must increase volume, quality and consistency of supply as well as productivity.

Weather information for agricultural operations shall be a tailored product that can be effectively used in crop planning and its management. A comprehensive weather based farm advisory is an interpretation of how the weather parameters, in future and present will affect crops, livestock and farm operations and, suggests actions to be taken. Anderson (2007) defines the terms agricultural extension and advisory services as "the entire set of organizations that support and facilitate people engaged in agricultural production to solve problems and to obtain information, skills and technologies to improve their livelihoods. The Agro-Met Advisory Services (AAS) will be more effective if they are given in simple and local language that farmers can understand it. In order to make the Agro-Met Advisory Services more successful and continuous process, it is to be supported with: (a) agro meteorological database, (b) crop conditions, (c) real time weather, research results on crop-weather relationships, and (d) skilled manpower in multidisciplinary resources and users interface. Wise use of weather and climate information can help to make better-informed policy, institutional and community decisions that reduce related risks and enhance opportunities, improve the efficient use of limited resources and increase crop, livestock and fisheries production. Weather based agro-advisory is one such adaptation strategy in which India has already shown considerable progress.

However, this needs to be propagated throughout the region for betterment of the farming community in the region. As a policy maker, educating people and sensitizing them to understand the farming sector is important to avoid inaccurate, inappropriate policies.

The Agromet Advisory Services of the India Meteorological Department (IMD) in the Ministry of Earth Sciences is a small step in this direction, aimed at "weatherproofing" farm production. I strongly believe that if recommendations emerged out of the consultation meeting are adopted by the member states, there are enormous opportunities to minimize the hazardous effect of climatic parameters and increase productivity in the region through agro-met advisory services.

Agro Advisory Services and Agricultural Development

Indian agriculture is passing through a critical phase as the rate of increase in crop production is barely keeping pace with the increase in population rates. Despite technological advances, such as improved varieties, genetically modified organisms, and irrigation systems, weather is still a key factor in agricultural productivity, as well as soil properties and natural communities. The effect of climate on agriculture is related to variability in local climates rather than in global climate patterns. Impact of climatic parameters on agricultural production and minimizing crop productivity losses through weather forecast and advisory service.

Extension of agromet advisory services also has an important role to play in helping the research establishment tailor technology to the agro ecological and resource circumstances of farmers. Extension thus has a dual function in bridging blocked channels between scientists and farmers: it facilitates both the adoption of technology and the adaptation of technology to local conditions. The first involves translating information from the store of knowledge and from new research to farmers, and the second by helping to articulate for research systems the problems and constraints faced by farmers. Moreover, it has increasingly been recognized in recent years that important innovations, for example, those relevant in natural resource management, are developed by farmers themselves rather than from agricultural research stations. Agricultural advisory services can play an important role in promoting the spread of farmer-based innovations. These several interactions among research, extension, education and farmers are well articulated in a world view described as agricultural knowledge and information systems (AKIS), which can serve as a useful organizing principle for discussions of policy relevant to agricultural advisory services (Anderson 2003, FAO/WB 2000). Adoption of innovations by farmers is inevitably affected by many factors. In general, farmers will adopt a particular technology if it usefully suits their socioeconomic and agro-ecological circumstances. The availability of improved technology, access to "modern" inputs and resources, and profitability at an acceptable level of risk are among the critical factors in the adoption process. Adoption can be influenced by educating farmers about improved varieties, cropping techniques, optimal input use, prices and market conditions, more efficient methods of production management, storage, nutrition, *etc.*

Scenario of Agro-meteorological Advisory Service in India

India Meteorological Department (IMD) started weather services for farmers in the year 1945. It was broadcast by All India Radio in the form of Farmer's Weather Bulletin (FWB). Subsequently, in the year 1976; IMD started Agro-Meteorological agricultural Advisory Service (AAS) from its State meteorological Centers, in collaboration with Agriculture Departments of the respective State governments. Though these services are being regularly provided by IMD for the past many years, the demand of the farming community could not be fully met due to certain drawbacks in the system. In view of that IMD launched Integrated Agromet Service in the country for 2007 in collaboration with different organizations/institutes.

At present district-level weather forecast is provided to 130 Agro-met Field Units (AMFUs) located at the State Agriculture Universities (SAUs), institutes of Indian Council of Agriculture Research (ICAR), IITs, *etc.*

There is need to further improvement in these services particularly through preparing the weather forecast at a level smaller than a district to block than to village level, extend the temporal range of weather forecast and also aggressive extension, outreach and agro-met advisory dissemination system. In order to operate at block level, there is a strong need to set an operational unit at District level. Hence, it is proposed to set up District Agro-met Units (DAMUs) in the country. These stations may be collocated with the existing Krishi Vigyan Kendras (KVKs), which are operating through State Agriculture Universities, ICAR Institutions and NGOs *etc.* and are funded and technically guided by ICAR.

Agromet Advisory Services in Minimize Weather Risk in Agricultural Planning and Management

The sources of weather and climate-related risks in agriculture are numerous and diverse: limited water resources, drought, desertification, land degradation, erosion, hail, flooding, early frosts and many more. Effective weather and climate information and advisory services can inform the decision-making of farmers and improve their management of related agricultural risks. Such services can help develop sustainable and economically viable agricultural systems, improve production and quality, reduce losses and risks, decrease costs, increase efficiency in the use of water, labour and energy, conserve natural resources, and decrease pollution by agricultural chemicals or other agents that contribute to the degradation of the environment. Thus, the importance of the Agromet Advisory Services that have now been established at district levels in India.

These Services meet the real-time needs of farmers and contribute to weather-based crop/livestock management strategies and operations dedicated to enhancing crop production and food security. They can make a tremendous difference in agricultural production by assisting farmers in taking the advantage of benevolent weather and in minimizing the adverse impact of malevolent weather. Adoption can be influenced by educating farmers about improved varieties, cropping techniques, optimal input use, prices and market conditions, more efficient methods of production management, storage, nutrition, *etc.*

Some of the adaptive measures communicated through the Agromet Advisory bulletins are:

- Adjustment of planting dates to minimize the effect of temperature increase-induced spikelet sterility to reduce yield instability, by avoiding having the flowering period to coincide with the hottest period.
- Changing the cropping calendar to take advantage of the wet period and to avoid extreme weather events (*e.g.*, cyclones and storms) during the growing season.
- Cultivation of crop varieties those are resistant to lodging (*e.g.* short rice cultivars) which withstands strong winds during the sensitive stage of crop growth.
- Development of cultivars resistant to climate change; adopting new farm techniques that respond to the management of crops under stressful conditions, plant pests and disease
- Shifts on sowing date of crops for more effective use of the soil moisture content.
- Moving forward the dates of crop sowing in a crop rotation calendar and farmers to plant a second crop that could even be vegetable with a short growth period.
- With increased evapotranspiration, orientation toward a shift from conventional crops to types of agriculture that are not vulnerable to evapotranspiration

- Cultivation of heat resistant crop varieties by utilizing genetic resources that may be better adapted to warmer and drier conditions.
- Growing of suitable cultivars (to counteract compression of crop development), increasing crop intensities (*i.e.*, the number of successive crops produced per unit area per year), or planting different types of crops.

The agromet advisory bulletins contain possible risk mitigation measures for the major crops and livestock. Based on the weather forecast, a group of interdisciplinary scientists and agromet scientists at AMFU prepare district-level agromet advisory bulletins. These bulletins are sent to the farmers and other stakeholders of the corresponding district.

Scope for the Agromet Advisory Service in India

The Agro-meteorological Advisory Service (AAS) is a mechanism to apply relevant meteorological information to help the farmer to use it for improving agricultural production.

The main emphasis is to collect and organize climate/weather, soil and crop information, and to amalgamate them with weather forecast to assist farmers in taking management decisions.

1. District level information with Agro Climatic and National Agricultural Research Project **(NARP) nomenclature** of zone distributions;
2. Dissemination of information on Forecasted Weather by IMD;
3. Agro-Met Advisory (SMS based advisories to be adopted);
4. District Level advisories available at present to be used for dissemination;
5. A strong Feedback mechanism/Help Desk to be developed so as to capture the observations of the farmers and improve upon the Advisories accordingly;
6. To contact organizations having expertise in conversions of Roman to Vernacular Text. IIT Kanpur has done considerable work for language conversions. CDAC may also be contacted.
7. Technology for SMS dissemination to be explored keeping in mind how to send text in 160 characters, text to voice conversions *etc.*;
8. IMD and related sources to transfer advisories and related information to NIC to be ported on portal;
9. Advisory Bulletins to farmers and stakeholders regarding weather sensitive agricultural operations to mitigate weather based risk on crop cultivation;
10. Advisories on Forecasted Weather, Variety Selection, Field Selection, Sowing Time, Crop Health and Agro-met Advisory to be provided by experts at the State and National Level;
11. SMS Alerts on
    a. Weather Forecast

   b. Impact on Crop,
   c. Impact on livestock
   d. Impact on fisheries
12. Grievance management through multiple service delivery channels to the farmers;
13. Study of Existing Automation of Meteorological Bulletin done by Pune on for Disseminating the following:
   a. Forecast based on IMD 5 Day Medium Range data
   b. Early Warning System (EWS) on natural calamities
   c. Dissemination over SMS/e-mail/Web (Govt. to Farmer and Govt. to Citizen)
   d. Actual Weather Summary based on Daily Weather Data
14. Dissemination of Agro-Met Advisories
   a. Based on the derived Meteorological Forecast
   b. Content from Domain Experts at Central/State/AMFU and KVKs, with Early Warning System (EWS) on Pest/Disease infestation.
   c. Dissemination over SMS/e-mail/Web (Govt. to Farmer)

Mechanism of Agromet Advisory Services

Today, IMD is implementing operational agrometeorological schemes across the country under a five-tier structure:

- ✰ Top-level policy planning body in Delhi
- ✰ Execution by the National Agromet Service headquarters in Pune
- ✰ Coordination and monitoring by State Agromet Centres
- ✰ Definition of the agro-meteorological zone
- ✰ District or local level extension and training for input management advisory service

This structure includes State Agricultural Universities, Institutes of Indian Council of Agricultural Research and Indian Institutes of Technology. Without it, the district Agromet Advisory Services would not be sustainable.

There are a few bottlenecks in the existing services like optimum observations, seamless weather forecast, manpower and permanency of staff in AMFUs, real-time information flow particularly crop and pest/disease information, establishment of connectivity, outreach/extension mechanism, R&D support for Agro-meteorology *etc.* Dissemination of right information at right time to each and every farmer of the country is a challenging job.

Dissemination Mechanism of Agromet Advisory Service

There are 130 district level Agromet Field Unit for prepare Agromet advisory in collaboration with State Department of Agriculture on Tuesday and Friday. The Agromet advisory is disseminated through All India Radio (AIR), Print Media,

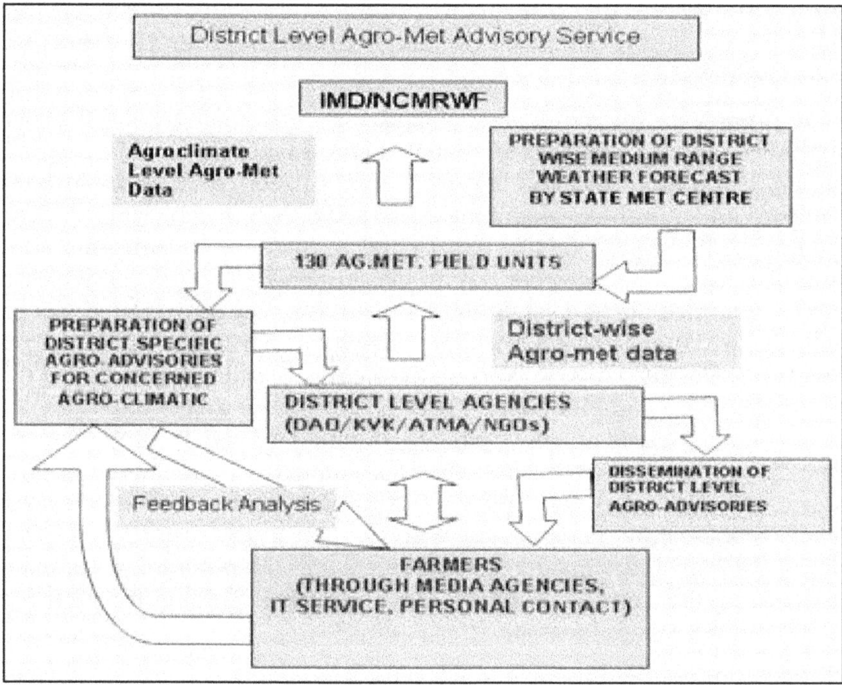

Figure 2.1. Flow Chart of the Mechanism of District Agromet Advisory Services.

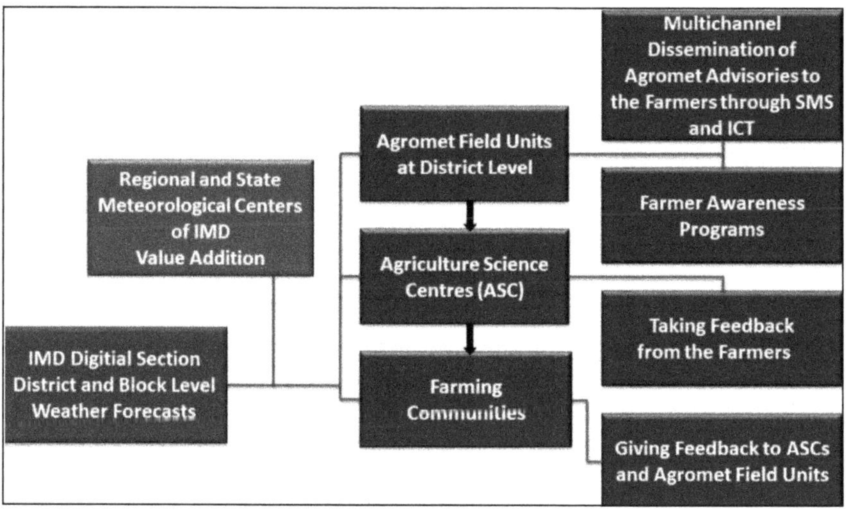

Figure 2.2. Agromet Advisory Services Organizational Structure and their Function.

Doordarshan, Website and SMS. Inclusion of these services will cover large fraction of farmers in the country to get the benefit. Moreover, Agro met Division; IMD is establishing linkages with the State level ICTs like Kisan Kerala, e-livestock. ICT for agriculture knowledge management needs to be establishment, *etc.* All FM channels of AIR and now casting centers of Doordarshan under Prasar Bharati may also be included.

The proposed projects for dissemination of Agro-met advisory bulletin are through IFFCO KIsan Sanchar Ltd., Reutor Market Light (RML), MahaAgri, Vritti Solutions, Handygo, Common Service Centers (CSC), and National Bank for Agriculture and Rural Development (NABARD), MS Swami Nathan Research Foundation (MSSRF) *etc.* IFFCO has started to disseminate agro-met advisory through five voice messages per day to the farmers in free of cost, which cover area of their immediate interest. Initially in 16 states they have started this project namely Delhi, Uttar Pradesh, Punjab, Haryana, Rajasthan, Madhya Pradesh, Odisha, West Bengal, Gujarat, Karnataka, Kerala, Tamilnadu, Andhra Pradesh, Bihar, Maharashtra and Himachal Pradesh. Handygo is a Telecom Value Added Services company based at New Delhi and maintains a very healthy relation with leading mobile operators in the industry. The company is disseminating agro-met advisories through IVR (Interactive Voice Response) system to five states (Punjab, Haryana, Maharashtra, Gujarat and West Bengal) in the country. Its rural IVR service on 55678 by is designed to serve Farmers through providing information on weather, crops, livestock and fisheries, Govt. and Bank offerings for rural, Health, Family Planning, Mandi Rates *etc.*

Reuters India offers its mobile Farming Information Services to farmers on crops, weather and commodity prices, in the States of Maharashtra and Punjab, through its Scheme - Reuters Market Light. Airtel in collaboration with IFFCO (IKSL), in the State of Andhra Pradesh, through its Grameena Mobile Kranti scheme, offers its mobile services, to strengthen the cooperative movement in rural India by empowering the farmers through information. Nokia too has launched its agricultural information services through its mobile services in India. Some of the best applications for the farmers, presently available in India are the AGMARKNET Portal (http: //agmarknet.nic.in), Seed NET Portal (http: //seednet.gov.in), National Horticulture Mission Portal (http: //nhm.nic.in) which is yet to provide information dissemination through Mobile telephones.

Feedback Mechanism of Agromet Advisory Service

Periodic feedback on worthiness of forecast and usefulness of advisories is also obtained. This feedback is obtained weekly, monthly and also annually. Feedback from selected farmers and Research and Development under different SAU's are being documented on whether they have adjusted their day-to-day farming operations in response to the advice laid in AAS and also on their additional requirements. Regular feedback from farmers, State Agricultural Departments and Agricultural Universities/ICAR and other related Institutes would be collected and processed for further improvement of services. Feedback information would also be collected from Regional and Narrowcasting stations of DD, AIR, and FM channels, KVKs, ATMA, CSCs, NGOs, VRCs and VKCs and also through Kisan Melas. The delivery of the content is in Regional Languages. A feedback mechanism in which farmers can let us know that how relevant the information provided by IMD has also been set up.

Impact of Agro-met Services

The agro-met advisory services through various channels have resulted in significant increases in farm productivity, resulting in increased availability of food and higher income generation. They have helped the farmers not only in increasing their productions but also reducing their losses due to changing weather patterns and others problems.

The economic benefit of the agro-met services runs in crores. The Ministry Earth Sciences had engaged National Council of Applied Economic Research (NCAER) to carry out a comprehensive study on "Impact Assessment and Economic benefits of Weather and Marine Services." This study was carried out during September and October 2010 and restricted to main end users *i.e.* Farmers for Agro meteorological Advisory Services. The field study was carried out in 12 states and 1 Union territory It was revealed that economic profit estimates can vary between Rs. 50,000 Crore (where 24 per cent farmers receive weather information) to 211,000 Crore (where all farmers receive weather information). This shows that its economic returns depend on the proportion of farmers receiving information.

That is why there is need to ensure that all farmers, small, marginal and big, are able to benefit from the Agro-met Advisory Services. This will no doubt go a long way in not only increasing production but also raising the income of farmers and reducing their losses and ultimately distress. But for this infrastructure to match the demand will have to be created and that will indeed be a big challenge for the government as well as other stakeholders.

Communication Gaps between Research, Advisory Services, and Farmers of Rural Areas

In the traditional research context, agricultural scientists tend to overlook situations at the farm level. Their research projects are often oriented at producing publications rather than solving concrete on-farm problems. Producers on the other hand expect immediate answers to local problems, and are not concerned with experimental details or the goals and objectives of the scientists. Many linkage problems between major institutional actors are caused by a lack of coordinated planning, poor communication between linkage partners, and absence of follow through with actual linkage resource planning or implementation. In addition, there is typically little or no involvement at all of representative farmers or their organizations. A lack of appropriate communication structures, methodologies and tools results in poor identification of farmers' needs and priorities, inappropriate research programs, poor or irrelevant extension information and technologies and finally, low farmers' take-up of technology innovations. From the agricultural and rural development perspective communication is considered as a social process designed to bring together agricultural technicians and farmers in a two-way process where people are both senders and receivers of information and cocreators of knowledge.

The information/knowledge needs on agriculture, livestock, innovation, markets and alternative employment are quite location and group specific. Remote areas are particularly disadvantaged and may need continued subsidized support in communication infrastructure and services. It may therefore be necessary for governments to provide incentives for information providers (research, farmers with local knowledge, market information providers, government on regulations) to adequately communicate information to rural areas.

Rural communication is an interactive process in which information, knowledge and skills, relevant for development are exchanged between farmers, extension/ advisory services, information providers and research either personally or through media such as radio, print and more recently the new "Information and Communication Technologies" (ICTs).

New Challenges

For sustainable Climate Smart Agriculture (CSA), it is essential to converges other sectors like energy and water that are essential for capitalizing on potential synergies with agriculture, livestock, forestry and fisheries. CSA also help in enhancing over all prosperity by developing villages in to "Climate Smart Villages" that are Weather Smart, Water Smart, Carbon Smart, Nitrogen Smart, Energy Smart and Knowledge Smart. The CSA provides appropriate climate change solutions through government – private sector intervention in policies and financial investments. Agromet Advisory Services is an example of such alignment which at present is accessible to over 2.5 million farmers in India and Weather Based Crop Insurance adopted more than 9 million farmers. These interventions will help in breaking barriers for adoption, especially among farmers. This is just the beginning and more number of interventions shall be staged in the near future. Climate smart agriculture is not a new agricultural system or a set of practices, but is a new approach, a way to guide the needed changes of agricultural systems, given the necessity to jointly address food security and climate change. While it is wonderful to popularize CSA, it's equally important to see whether these ideas can be put into practice. The need of the hour is to quickly implement the happenings and success stories in our numerous Agricultural Labs to the Lands of our Farmers. CSA, however, is not a generic agricultural technology or practice that can be universally applied, but it is location specific and knowledge intensive approach which requires site-specific assessments to identify suitable agricultural technologies and practices. However, the good news is that, climate analogues (finding future climate for potential adaptation) already exist and it is estimated that 70 per cent of expected future climatic conditions already exist somewhere location on the earth. It is essential to identify those analogues and adopt them for environmental sustainability. The most worrying factor for our Farmers is the lack of enthusiasm amongst their sons that will render the Lands almost ruder-less. Farmers are in fact today desperate as they are unable to get adequate pricing for their crops despite a good harvest.The high cost of inputs like Chemicals, Fertilizers, Seeds and Labour and low price for their final produce are forcing farmers to sell their land or send their young sons to Cities and Towns in search of new pastures

Conclusion: The Way Forward

We are convinced that, with better and higher resolution forecasts in future, the farmers will definitely be very much pro-active to weather and weather based farming to enhance their capability towards climate smart and, therefore, climate resilient agriculture. This approach of efficient use of ICT in taking the weather based advisories to the doorstep of the farmers is the first of its kind in India, and up scaling it will benefit the farmer many fold. Farmers are willing to pay for advisory services as long as they recognize financial benefits and profits from the services. Due to judicious and timely utilization of inputs, production cost for the AAS farmers reduced. The increased yield level and reduced cost of cultivation leads to increased net returns.

The total eradication of agricultural development problems can be achieved through extension service approach if the role of Agromet Advisory Service is properly conceived and effectively administered. New approaches to ICT for agricultural development and natural resource management exist to overcome extension challenges and new concepts are emerging for anticipation, learning and problem solving between the key players. The availability of extension agents and the ability to exchange and validate new techniques and/or productions to introduce at the farm level by researchers and specialists are highly desired.

REFERENCES

Anderson, J.R. (2007), 'Agricultural Advisory Services', Background Paper for the World Development Report 2008. http: //siteresources.worldbank.org/INTWDR2008/Resources/27950871191427986785/Anderson_AdvisoryServices.pdf

Anderson, J.R. and Feder, G. (2003), 'Rural Extension Services', World Bank Policy Research Working Paper 2976, World Bank, Washington D.C. http: //econ.worldbank.org/files/24374_wps2976.pdf

FAO (2003), 'Statement at the Ministerial Conference of the WTO - Circulated by H.E. Mr

Hartwig de Haen, Assistant Director-General', Fifth Session, Cancun, 10-14 September, Doc No: WT/MIN(03)/ST/61.

FAO/WB (2000), *Agricultural Knowledge and Information System for Rural Development (AKIS/RD): Strategic Vision and Guiding Principles*, Food and Agriculture Organization of the United Nations and the AKIS Thematic Group of the World Bank, Rome.

United Nations (2008), 'Trends in Sustainable Development – Agriculture, rural development, land, desertification and drought', Department of Economic and Social Affairs of the United Nations, New York. http: //www.un.org/esa/sustdev/publications/trends2008/fullreport.pdf

Transformation of Indian Agriculture through Innovative Technologies *Pages* 23–44
Editor: **Dr. Shahid Ahamad & Dr. Jag Paul Sharma**
Published by: **ASTRAL INTERNATIONAL PVT. LTD., NEW DELHI**

3 Importance of Indigenous Technical Knowledge in Weather Forecasting

Dharmendra Kumar and Kumari Sharda

Indigenous knowledge is systemic bodyof knowledge acquired by local people through the accumulation of experiences, informal experiments and intimate understanding of the environment in a given culture. Indigenous knowledge systems are dynamic, changing through indigenous mechanism of creativity and innovativeness as well as through contact with other local and international knowledge systems.

Weather is certainly the most important factor determining the success or the failure of agricultural enterprises. It manifests itself through its effects on soil, plant growth as well as on every phase of animal growth and development. Also crop and animal diseases are greatly influenced by weather. In all, weather accounts for approximately three fourth of the annual loss in farm production both directly and indirectly.However, the crop losses can be reduced substantially by affecting adjustments through timely and accurate weather forecasts. Such weather forecast support and also provides guidelines for long range or seasonal planning and selection of crops best suited to the anticipated climatic conditions. At present we have many improved technologies for making weather forecasts as well as for their dissemination. Previously when there was no such technology available, farmers used to plan and practice based on their prediction on many natural, cultural and social phenomena. Some of these are discussed below [4]:

Local communities over the centuries have devised weather-related customs that, they say, help bring in copious showers.If a bleak monsoon is forecast, people in parts of India's hinterland gather together after the morning meal without washing their hands and pray for rain with their hands held up to the sky.In another community, farmers keep frogs in an earthen pot covered by a piece of new cloth dyed in turmeric. They continuously pour water over it and shout "frog in water" while moving around the streets. They also pour water over other people of the village and on themselves – all this so that rains don't fail. Of late there have been failures in rain forecast. In the last ten to fifteen years or so dramatic changes have occurred in patterns of the various seasons. Monsoon rains have become unpredictable. Sometimes it arrives early, followed by a dry spell.Then you have heavy rains at the end of the season, leading to floods and damage to standing crops. These changes have confused even birds and animals and they are trying to adapt themselves to the changing weather.

Rain forecasting based on the Hindu almanac is a common practice among farmers. "TattaSanketam" or basket indicator is one such custom. A little boy is asked to lift a basket of jowar on his head in which a tumbler is placed. While doing so the tumbler falls. The planetary position in the direction where the tumbler falls is indicative of rain.Between mid-December and mid-January, the farmers in Andhra Pradesh's Anantapur town go through a ritual every year. A priest collects grains from every farmer in the village. He carries this to Bhadrachalam town nearby to seek divine blessings, and returns the energised grains with a promise of a good crop. Weather forecasts for agriculture can be grouped into short range forecast (up to 48 hours), medium range forecast (3-10 days) and long range forecast (one week to entire season). Each plays an important role in farm operations and planning of agricultural activities. Most important efforts since time immemorial have been on rain making and weather fore casting.

Table 3.1. Difference between Weather and Climate

Weather	*Climate*
Refers to physical state of atmosphere at a given time in terms of temperature, rainfall, relative humidity and sunshine hours and so on.	Refers to average state of weather (Required at least 30 years of data to find out the climatic normals)
Decides animal behaviour and thereby milk productivity	Decides geographical distribution of animals and integrated farming systems (Eg. Tropical, sub-tropical, temperate and polar animals)
Day to day livestock farming is dependent on weather	Livestock farming is dependent on climate
Animal pest and disease incidence is dependent on weather	Hot spot areas of animal pest and disease incidence and their geographical distribution can be delineated based on climate
Extremes like floods and droughts and cold and heat waves depend on weather	Extreme weather prone areas can be delineated based on climate. It decides sea level rise.

Seasons in India

1. Summer- March to May (Temperature)
2. Monsoon-June to September (Rainfall)

3. Post monsoon- October to November (Rainfall)
4. Winter- December to February (Temperature)

Weather Prediction

I) **Short range**: Upto 72 hours
   a) Now casting: 0 - 2 hours
   b) Very short range: 0 –12 hours
      Main users: Aviation and Marine agencies

II) **Medium range**: Beyond 3 days up to 10 days.
   Main users: Farmers

III) **Long range**: Beyond 10 days; a few weeks to a month or a season

Table 3.2. Intensity of Precipitation of Rain

Rainfall Amount (in mm)	*Description Term Used*
0.0	No rain
0.1 to 2.4	Very light rain
2.5 to 7.5	Light rain
7.6 to 35.5	Moderate rain
35.6 to 64.4	Rather heavy rain
64.5 to 124.4	Heavy rain
124.5 to 244.4	Very Heavy rain
Exceeding 244.4	Extremely very heavy

Table 3.3. Spatial Distribution

Isolated/at one or two places	25 per cent or less of the area
Scattered/at a few places	Between 26 per cent and 50 per cent of the total area
Fairly widespread/at many places	Between 51 per cent and 75 per cent of the total area
Widespread/at most places	More than 75 per cent of total area

Short Range ITK (Indigenous Technical Knowledge) [2]

☆ There will be heavy rain pour, if the rainbow appears on the eastern sky at the time of raining. If it is on the western sky, there will be no rain.

☆ When crab comes to the bund, it may rain.

☆ Red clouds in the east sky, it may rain in the next day.

☆ Frogs crocking in chorus then it is followed by rain.

☆ When dragon flies, fly down(low), there will be rainfall within a day or two.

- Increased mosquito bite predicts rain.
- Dense fog in the early morning indicates no rain.
- If there is an accumulation of clouds in the South-East direction in a layered form accompanied by winds blowing from the southern direction then it is claimed Rain.
- If there is a swelling on the lower portion of the camel's legs then rainfall is predicted by the farmers. The swellings are probably caused due to higher relative humidity.

Medium Range ITK

- The closer circle to the moon, the nearer is the shower and vice versa.
- If a snail climbs certain trees, there will be no rain.

Long Range ITK

- If the Tatiharibird (Lapwing) lays eggs on the higher portion of the lake bunds or on the top of any structure, the coming season carries heavy rainfall. If the same bird lays eggs on the lower side of the tank or on the floor of the water body structure, drought is anticipated.
- Further it is also believed that if a single egg is laid by the Lap wing bird, then there will be rainfall only for one month out of four months of the rainy season. If two eggs are laid then rainfall will occur for two months and similarly four eggs indicate there will be rainfall during all the four months of the rainy season.
- If the clouds thunder on the first day of mid April, there will be no rain for 72 days.
- If crow cries during night and fox howls during the day, then there would be severe drought.
- If the "Tillbohara" (Dragon fly), which appears generally in the rainy season, are observed to swarm in a large group over a water surface (Pond) then dry weather is predicted but if they swarm over open dry lands or fields then early rainfall is predicted by the farmers.
- If the "Khejri" tree bears good fruit in a particular year then farmers predict good rainfall during the next rainy season and vice versa less rain is predicted in the event of a poor fruit crop.

Social and Cultural Beliefs

Many cultural, social and religious beliefs and activities superstitious pertaining to the prediction of future weather prevail since generations. From time immemorial farmers have predicted the weather on the basis of these beliefs/activities. The following are some examples from the western Himalayan region.

- If the first 10-15 of the month "Jeth" (May-June) are very hot then good rainfall/monsoon is predicted during the ensuing rainy season. This results probably from the low pressure zone in north-west India that is generated due to the high temperatures.

- ✰ The Soolini Mela (Festival) is organised in Solan, during the month of June every year. People of this area firmly believe that rainfall will occur on the very day of the festival or one day before or day after the festival
- ✰ It is also believed, that when grey coloured clouds descend below the hill tops then they definitely cause rainfall.
- ✰ If the Chakkala-Belan, (rolling pin and board), used in the Kitchen, show moisture on them then within few days rainfall is expected.
- ✰ In villages elderly farmers usually carry a small bag for "Tambaku" (Tobacco) for *Hukka* (Smoking device). When this bag shows more moisture in the *Tambakku* then farmers predict rainfall within one or two days. **[6]**

Some Folklore Regarding Weather Forecasting

The folk-lore of the popular poet Gag and his wife Bhahdari, who lived during the 17th century, regarding weather forecasting are still very popular in northern India. Some are given as under:

- ✰ "When strong eastern winds blow continuously then it is estimated that the rainy season has come"
- ✰ When days are very hot and there is dew at night, then according to Gag, there are very limited chances of rainfall.
- ✰ When cloudy days are accompanied by clear nights and the eastern winds blow somewhat strongly, according to Gag no rainfall is predicted. This is accompanied by shortage of water in ponds, rivers *etc.* Consequently clothes are washed using water from wells.
- ✰ When rainbow is formed in the direction of Bengal then there will be rainfall, if not by the evening then definitely by next morning. During the rainy season, if a cloud appears on Friday and Saturday, rainfall is predicted either for Sunday or Monday.

Predicting Weather with Animals

Is it true that animals can sense things that we humans sometimes can't? Many animals predict the arrival of summer or rain with their changing behaviour, because they sense that the days are getting longer or hotter. Animals are very attuned to day length and the seasons and other cues from nature that help them in many ways[10]. Below is the list of some wonder animals that can tell you the weather that's about to change:

Cows – [11] and [7]

According to legend, when cows sense bad weather, they become restless and antsy and begin to swat flies with their tails or lie down in the pasture to save a dry spot.

- ☆ Scientists at the Universities of Arizona and Northern Missouri proved that cows lie down when it is cold and stand when it is hot for long hours.
- ☆ Oxen sniff the air and *bull leads the cows to pasture, expect rain; if the cows precede the bull, the weather will be uncertain.*
- ☆ When horses and cattle stretch out their necks and sniff the air, it will rain.
- ☆ When cattle lie down in the pasture, it indicates early rain.

Sheep and Goat

Not so well known signs are goats flapping ears and sheep huddling together. Goats flap ears when the humidity increases as they feel uneasy. Flapping tends to cool their ears. Sheep huddle for a similar reason. **[4.].** When sheep gather in a huddle, tomorrow will have a puddle." Although this rhyme is cute, the weather that comes with it isn't. It's believed that you can expect a storm when these animals crowd together and shield each other. [7]

Sheep turn into the wind and if sheep ascend hills and scatter, expect clear weather. The huddling of goats is said to be an indicator of approaching rains also when Sheep move in group.

- ☆ If there is a swelling on the lower portion of the camel's legs then rainfall is predicted by the farmers. The swellings are probably caused due to higher relative humidity.

Groundhogs

In America, the most popular animal that is believed to be able to predict the weather is the groundhog. If see his shadow, it means there will be six more weeks of winter; if not, we can look forward to an early spring [7].

Squirrels

Squirrels aren't very good weather forecasters. Squirrels have been associated with winter weather lore for their nut-gathering abilities ahead of wintry weather. The bushiness of squirrels' tails has also been believed to be a harbinger of a bad winter.

Frogs

This one is the easiest to judge. Frogs generally live under rocks. When there is a change in the atmospheric pressure the amount of air available to them under the rocks becomes less and they come out and croak. This is seen as a sure sign of rain **[9].**

- ☆ Loud and slimy amphibians, the frogs, are said to croak even longer and louder than usual when bad weather is on the horizon. When you hear their volume increase, you can assume a storm is brewing., its reliability is high.
- ☆ Frogs in the well making continuous sound [17].

Common Frog (*Rana temporaria*)

- ☆ If the frog croaks in the water body in the afternoon until sunset, rain will be coming soon even during spring and winter.[5]
- ☆ **Crab** making bigger hole in the channels. [17]

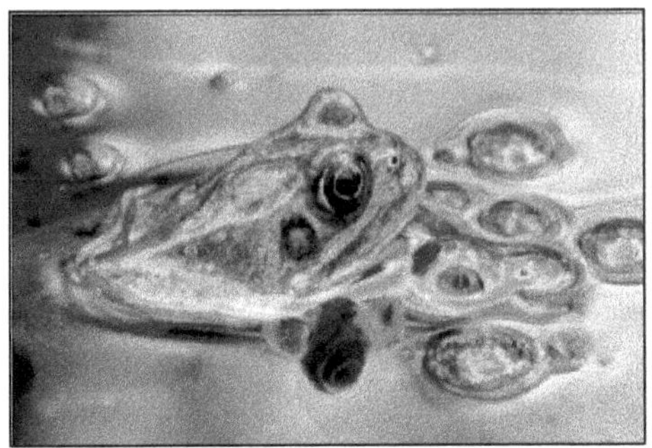

- ☆ Appearance of increasing number of geckos on the walls and ceilings
- ☆ Emergence of fish swimming near the surface of ponds, alert farmers about coming rains.[3]
- ☆ **Swine** are restless.When pigs gather leaves and straw in all, expect a cold winter. [4]
- ☆ When rabbits are fat in October and November, expect a long, cold winter.

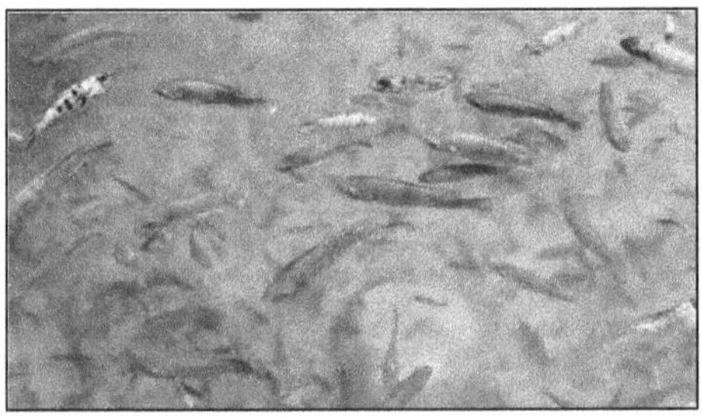

- **Dog and Cat**: Expect rain when dogs eat grass, cats purr and wash sneeze.
- If the mole digs its hole 2½ feet deep, expect severe weather; if two feet deep, not so severe; if one foot deep, a mild winter.
- **Bats**flying late in the evening indicates fair weather.
- If the groundhog sees its shadow on Candlemas Day (February 2), six more weeks of winter remain.
- *Wolves always howl more before a storm.*The otherwise slow **red hairy caterpillar** makes a dash for shelter before rainfall while foxes howl at dawn and dusk, **owls** hooting.

Indigenous Knowledge in Zimbabwe [18]

- Presence of millipedes, frogs means onset of rain
- Rock rabbits crying in morning and evenings means hot season begins
- Presence of reptiles more means hot season begins

Predicting Weather with Birds

Depending on how low the bird's fly, how bad the weather is going to be, can be gauged. If birds are flying high, the weather is clear. But if they're flying closer to the ground, the air pressure of a storm system is causing them pain at higher altitudes. Most birds have a special middle-ear receptor called the vital organ, which can sense incredibly small changes in barometric pressure. So if the activity at feeders suddenly becomes much more intense and they start flying close to the ground, be sure that a storm may be approaching [7] AND [9].

- If crows fly in pairs, expect fine weather; a crow flying alone is a sign of foul weather.
- *Hawks flying high means a clear sky. When they fly low, prepare for blow.*Birds tend to stop flying and take refuge at the coast if a storm is coming. They'll also fly low to avoid the discomfort of the falling air pressure.
- When seagulls fly inland, expect a storm. When fowls roost in daytime, expect rain.

- Petrels gathering under the stern of a ship indicates bad weather.
- Birds tend to get very quiet before a big storm. If you've ever been walking in the woods before a storm, the natural world is eerily silent! Birds also sing if the weather is improving.
- Birds singing in the rain indicates fair weather approaching. The whiteness of a goose's breastbone indicates the kind of winter: A red of dark-spotted bone means a cold and stormy winter; few or light-colored spots mean a mild winter.
- Partridges drumming in the fall means a mild and open winter.
- When domestic geese walk east and fly west, expect cold weather.
- If birds in the autumn grow tame, the winter will be too cold for game.
- When the rooster goes crowing to bed, he will rise with watery head.
- When the swallow's nest is high, the summer is very dry. When the swallow buildeth low, you can safely reap and sow.
- It is believed that on a hot summer day the cry of the bird called "Nialu" for water brings rainfall.
- During the rainy season farmers observe the "Matilari" bird (House swift) and they predict heavy rainfall if the bird flies high in the sky.
- If the Maina bird bathes in the water it indicates that there will be rainfall within one or two days.
- During long hot days in summer if the cry of theapiha bird is heard then people believe that God will quench her thirst and there will be rainfall after one or two days.

- ☆ A group of sparrows frolicking in the sand indicates that there will be rainfall that day or the next day and if they are observed to be playing in water then it is believed that the weather will be dry for some days to come.
- ☆ If the "Jonks" (Leechs) are immobile/stationary at the water surface (Pond) then dry weather is predicted but if they move rapidly in the upward and downward direction in water then rainfall is predicted.
- ☆ If the "Tatihari" bird (Lapwing) lays her eggson the higher portion of the field then heavy rainfall is predicted during the coming rainy season but if the eggs are laid in the lower portion of the field then a drought is predicted. These birds never construct a nest but lay their eggs on bare soil.
- ☆ Further it is also believed that if a single egg is laid, then there will be rainfall only for one month out of four months of the rainy season. If two eggs are laid then rainfall will occur for two months and similarly four eggs indicate there will be rainfall during all the four months of the rainy season.
- ☆ If the "Tillbohara" (Dragon fly), which appears generally in the rainy season, are observed to swarm in a large group over a water surface (Pond) then dry weather is predicted but if they swarm over open dry lands or fields then early rainfall is predicted by the farmers**[6].**
- ☆ Dragon flies and black sparrows flying in a group few meters above ground [17]

Poultry

- ☆ Poultry sit in a place for long Reliability is medium and used by women and herders as a supportive indicator[17]

Domestic Chicken

- ☆ If they search for food even during rain, then rain may last the whole day
- ☆ If they stop searching food and take shelter means the rain is expected to cease soon. Mizoram, North-east India. [5]

Eagles

- ☆ More number of eagles and birds fly on the top from west to east
- ☆ As a supportive indicator for good rains during monsoon season. [17]

Lapwing Bird

- ☆ In Australia If lapwing bird lay eggs on upper part of field good rain is expected and poor or no rain if lay egg on lower part of field. [18]

Bamboo Patridge (*Bambusico lafytchii*)

- ☆ If male roar frequently in spring and summer in the morning, rain expected
- ☆ If rain in the morning and roaring of the bird that time indicates rain will soon stop for that day [5].

End of precipitation is also important for farmers as its approach

- ☆ Rain will stop in a relatively short period once sparrows start flying faster again
- ☆ Also, swallows flying low near rice fields indicate to farmers to go back to fields despite appearance of unstoppable rains.[3]

Indigenous Knowledge in Zimbabwe [16]

Presence of storks	Onset of rain	

Guinea fowls laying eggs	Onset of summer
Swallows flying at low altitudes	Rain to fall immediately
Birds singing and flying high	Stable dry weather
Birds seeking shelter in day	Cloudy and humid weather

Indigenous Knowledge in Tanzania [15]

Occurrence of cattle egrets	Onset of rain and good rainy season
Singing of coucal in early morning	Imminent rainfall and good rainy season
Swallow tailed bee eater flying all over	Heavy rain coming
Pangolins	Indicator in rainfall prediction

Indigenous Knowledge in Gujarat [13]

Sparrows bathing in dust	Good rain	
Peacocks cry frequently	Rain within a day or two	
Crows cry in night and foxes during day	Severe drought	
Lapwing birds lay eggs during night	Heavy rains	
Snake climbs up on trees	Drought	

Indigenous Knowledge in Andhra Pradesh [18]

Poultry inserting feathers in soil	Short range	
Biting nature of housefly	Short range	
Parabolic flight of open bill stork	Short range	

Squeaking of owls	Short range	
Migration of parakeets in north-south direction	Short range	

Predicting Weather with Plants

Entomologist Dr D Jagadeeswar Reddy says, " There are plants too that predict rain. Plant phenology or the science of plant appearances is sometimes accurate to the date. While the Golden shower tree reaches flowering peak exactly 45 days before rains start, a neem tree in full bloom means ample rain. The village headman examines the length of the fruit called Flame of the Forest. If the seed at the bottom has matured, it will rain well at the beginning of the season. If the middle seed is matured, the middle part of the season will see good rain. And a good top seed is an indicator of end-of-season rain. Uniformly developed seeds bring uniform rain throughout. These and many other such thumb rules drive rain prediction in rural Andhra Pradesh [4]. Dried appearance of neem tree in summer Indicates heavy drought. [1]

Ficus

Flowering and generation of new leaves indicates near rainfall onset[17]

Peach

- ☆ Flowering pattern of peach species, If peach or plum flowers grow from the basal region to the terminal in flowering season, it is predicted a good rainy season. [5]
- ☆ A significant flowering of mango tree indicates a potential drought season. [15]
- ☆ When adventitious root of banyan tree (*Ficus* sp.) start sprouting, rain in 2-4 days.
- ☆ When babul tree (*Acacia nilotica*) flowers, rain in 10-15 days.

Indigenous Knowledge South India

- ☆ When neem kernels ripen and start falling, rain after 10-15 days
- ☆ In castor (*Ricinus* sp.) and ber (*Ziziphus nummularia*) when bud start sprouting, rain within 10-15 days

Indigenous Knowledge in Gujarat [13]

Mahuda (*Madhuca latifolia*)	Good foliage	Good monsoon
Bamboo species	Good foliage	Drought
Ber (*Zizyphus mauritiana*)	Heavy flush of fruit	Average monsoon
Darbha grass (*Eragrostis cynosuroides*)	Good foliage	Good monsoon
Pipal (*Ficus relegiosa*)	Good foliage	Adequate rain
Neem (*Azadirachta indica*)	Heavy flush of fruit	Drought

Predicing Weather with Sun and Moon

Visible Spectrum around the Sun and the Moon

People predicted weather after observing the visible spectrum around the sun or moon. If the spectrum around the sun had a greater diameter than that around the moon, they predicted rainfall after a day or two. Some people based their weather prediction on the nature of the solar halo. All the photometers are a luminous phenomenon produced by the reflection, refraction, diffraction or interference of light from the *sun* or moon. The visible spectrum of light around the sun or moon is called halo, or carona according to its distance from the sun or moon. If the distance is more, then it is called the halo phenomenon, which is caused by a layer of thin veil of cirrus clouds *i.e.* non rain bearing clouds. But if the distance is less, it is called corona phenomena produced by somewhat dense clouds which may cause rainfall. The accuracy of this indigenous observation can be as high as 50 per cent.

Cloud and Wind Direction

If there is an accumulation of clouds in the South-East direction in a layered form accompanied by winds blowing from the southern direction then it is claimed that there will be rainfall within a day or two**[6]**.

Weather Proverbs and Prognostics about Rain and Clouds

- ☆ Unusual clearness in the atmosphere, with distant objects seen distinctly, indicates rain.
- ☆ Red sky at night, sailor's delight. Red sky at morning, sailor takes warning.

- ☆ Evening red and morning gray are sure signs of a fine day. Evening gray and morning red put on your hat or you'll wet your head.
- ☆ If it rains before seven, it will clear before eleven.
- ☆ Rain from the south prevents the drought, but rain from the west is always best.
- ☆ Anvil-shaped clouds bring on a gale.
- ☆ A cloud with a round top and flat base carries rainfall on its face.
- ☆ When small clouds join and thicken, expect rain.
- ☆ Black clouds in the north in winter indicates approaching snow.
- ☆ A curdly sky will not leave the earth long dry.
- ☆ If you see clouds going crosswind, there is a storm in the air.
- ☆ Hen scarts and filly tails make lofty ships wear low sails.
- ☆ Clouds floating low enough to cast shadows on the ground are usually followed by rain.
- ☆ Mackerel sky, mackerel sky, never long wet, never long dry.
- ☆ If three nights dewless there be, 'twill rain, you're sure to see.
- ☆ If heavy dew soon dries, expect fine weather; if it lingers on the grass, expect rain in 24 hours.
- ☆ With dew before midnight, the next day sure will be bright.
- ☆ If you wet your feet with dew in the morning, you may keep them dry for the rest of the day.
- ☆ The higher the clouds, the finer the weather.
- ☆ If you spot wispy, thin clouds up where jet airplanes fly, expect a spell of pleasant weather.

- ☆ Keep an eye, however, on the smaller puff clouds (cumulus), especially if it's in the morning or early afternoon. If the rounded tops of these clouds, which have flat bases, grow higher than the one cloud's width, then there's a chance of a thunderstorm forming.

Clear Moon, Frost Soon

- ☆ When the night sky is clear, Earth's surface cools rapidly–there is no cloud cover to keep the heat in. If the night is clear enough to see the Moon and the temperature drops enough, frost will form. Expect a chilly morning!

When Clouds Appear like Towers, the Earth is Refreshed by Frequent Showers

When you spy large, white clouds that look like cauliflower or castles in the sky, there is probably lots of dynamic weather going on inside. Innocent clouds look like billowy cotton, not towers. If the clouds start to swell and take on a gray tint, they're probably turn into thunderstorms. Watch out!

Rainbow in the Morning gives you Fair Warning

A rainbow in the morning indicates that a shower is west of us and we will probably get it.

Ring around the Moon? Rain Real Soon

A ring around the moon usually indicates an advancing warm front, which means precipitation. Under those conditions, high, thin clouds get lower and thicker as they pass over the moon. Ice crystals are reflected by the moon's light, causing a halo to appear.

Rain Foretold, Long Last. Short Notice, Soon will Pass.

If you find yourself toting an umbrella around for days "just in case," rain will stick around for several hours when it finally comes. The gray overcast dominating the horizon means a large area is affected. Conversely, if you get caught in a surprise shower, it's likely to be short-lived.

Red Sky at Night, Sailors Delight. Red Sky in Morning, Sailors take Warning.

- ☆ A reddish sunset means that the air is dusty and dry. Since weather in North American latitudes usually moves from west to east, a red sky at sunset means dry weather–good for sailing–is moving east. Conversely, a reddish sunrise means that dry air from the west has already passed over us on their way easy, clearing the way for a storm to move in.

☆ If the colour of the clouds is similar to the colour of the wings of the Titar bird (Partridge) *i.e.* grey or black-grey and strong eastern winds are also blowing then assured rainfall is predicted by the farmers. The clouds of a colour similar to that of the said bird are rain bearing clouds *i.e.* of cumulonimbus type [8].

REFERENCES

1. Anandraja N., Rathakrishnan T.,Ramasubramanian M., Saravanan P. and Suganthi N S.(2008). Indigenous weather and forecast practices of coimbatore district farmers of Tamilnadu. *Indian Journal of Traditional Knowledge,* 7(4): 630-633.
2. BalasubramanianT.N.http: //nptel.ac.in/courses/126104005/LectureNotes/Week-4_04-Traditional per cent 20knowledge per cent 20on per cent 20weather per cent 20forecast per cent 20and per cent 20their per cent 20validity.pdf
3. Barayazarra and Puri, (2011). Smelling the monsoon: Senses and traditional weather forecasting knowledge among the Kenyahbadeng farmers of Sarawak, Malaysia. *Indian Journal of Traditional Knowledge.* 10(1): 21-30.
4. Curious customshttp: //www.aljazeera.com/indepth/features/2013/12/predicting-rain-indian-style-201312191248555584479.html
5. Chinlampianga, M.(2011).Traditional knowledge, weather prediction and bioindicators: A case study in Mizoram, Northeastern India. *IJTK* 10(1): 207-211.
6. file: ///G: /climate per cent 20rain/Chapter per cent 208 per cent 20- per cent 20Weather per cent 20forecasting.html
7. https: //weather.com/science/news/animals-predicting-weather-20130201#/1
8. http: //www.accuweather.com/en/weather-news/animals-insects-winter-forecast/19137666
9. http: //www.aljazeera.com/indepth/features/2013/12/predicting-rain-indian-style-201312191248555584479.html
10. http: //www.almanac.com/content/weather-proverbs-and-prognostics-animal
11. http: //www.skymetweather.com/content/lifestyle-and-culture/6-animals-that-can-help-predict-weather/
12. https: //weather.com/science/news/animals-predicting-weather-20130201#/1

13. Kanani, 2006.Indigenous knowledge Gujarat()Kanani, P.R. (2006). Testing of traditional methods of weather forecasting in Gujarat usingthe participatory Approach". En Nirmaladevi. A.V.; Balasubramanian, T.D.: Balasubramanian, A.V.: Nirmala Devi, T.D. (ed.). Traditional knowledge systems of India and Sri Lanka. Papers presented at the COMPAS Asian regional workshop in traditional knowledge systems and their current relevance and applications, 3-5 July 2006, Bangalore. Chennai.

14. Kijazi, A.L., Chang'a, L.B., Liwenga, E.T., Kanemba, A. and Nindi, S.J. (2013). The use of indigenous knowledge in weather and climate prediction in Mahenge and Ismani wards, Tanzania. *Journal of Geography and Regional Planning*, 6(7): 274-280.

15. PareekA. andTrivedi, P.C. (2011). Cultural values and indigenous knowledge of climate change and disaster prediction in Rajasthan, India. *Indian Journal of Traditional Knowledge*. 10(1): 183-189.

16. Risiro, Joshua; Mashoko, Dominic; Doreen; Tshuma, T.; Rurinda, Elias (2012). Indigenous Knowledge in Zimbabwe.Weather Forecasting and Indigenous Knowledge Systems in Chimanimani District of Manicaland, Zimbabwe. Academic journal article *Journal of Emerging Trends in Educational Research and Policy Studies*

17. Shankar Ravi.K., PochaiahMarathy, Murthy V.R.K. and Ramakrishna Y.S. (2008). Indigenous Rain forecasting in Andhrapradesh.Central Research Institute for Dryland Agriculture, Santoshnagar, Hyderabad-500059.

18. Shoko, K., and Shoko, N. (2013). Indigenous Weather Forecasting Systems: A Case Study of the Abiotic Weather Forecasting Indicators for Wards 12 and 13 in Mberengwa District Zimbabwe. *Asian Social Science*, 9(5), 285-297. http: // dx.doi.org/10.5539/ass.v9n5p285

Transformation of Indian Agriculture through Innovative Technologies *Pages* **45–56**
Editor: **Dr. Shahid Ahamad & Dr. Jag Paul Sharma**
Published by: **ASTRAL INTERNATIONAL PVT. LTD., NEW DELHI**

4 Direct Seeded Rice: A Approach toward Resource Conservation in South-Western Zone of Punjab

Balkaran Singh Sandhu and Nirmaljit Singh Dhaliwal

Introduction

Rice-wheat cropping system occupies about 13.5 million ha area in Indo-Gangetic Plains. In India 10 m ha area occupied under this cropping system and which feeds about 20 per cent of the world population (Saharawat *et al.*, 2010). Rice is a major source of food for more than half of the world population. In India, it is a staple food crop. It occupies 43.95million ha with a total production of 106.54 million tonnes of rice (Anon, 2014). In Punjab during 2014-15, rice is a major *kharif* crop on an area of about 2.85 million hectares with a total production of 11.26million tones of rice (Anon, 2015). In recent years, the major emphasis has been on alternative resource conservation technologies for both rice and wheat crops to reduce the cost of cultivation and to enhance the profit margin of farmers (Singh *et al.*, 2006). The economy of farmers of Punjab is largely dependent on Rice-Wheat system, so diverting them form rice cultivation is very daunting task. So, we have to select some other production techniques which can sustain the rice production as well as are eco-friendly and resource conservative. Lot of Studies shows that in Punjab rice can successfully grown by dry seeded into unpuddled soils. With direct seeding of rice, seed is sown and sprouted directly into field, eliminating the laborious process of planting the seedlings. Direct Seeded Rice is a plough towards a new set of principles based on minimal soil disturbance, management of crop residues and innovative cropping systems is the good option of farming under rice-wheat cropping system.DSR is a technique based on minimum soil disturbance and is a good option of farming under rice-wheat cropping system (Singh *et al.*, 2012) It is the technology which has low labour cost, water saving, low soil degradation, energy efficient and eco-friendly characteristics (Chauhan *et al.*, 2006; Chauhan *et al.*, 2012, Kumar and Ladha, 2011; Mahajan *et al.*, 2009; Sudhir-Yadav *et al.*, 2010). DSR is fast replacing transplanted rice in areas where good drainage and irrigation facilities available (Joshi, *et al.*, 2013). DSR is very good technique for growing rice crop in those area where water and labour shortage are major problems.So, in order to protect natural resources especially water, there is need to replace puddled transplanted rice with the DSR. Direct seeding of rice is possible provided there is a good crop establishment as well as adequate weed control methods are available to keep the crop free from weeds (Rao *et al.*, 2007; Mishra and Singh, 2012), however in absence of proper weed control, rice yields are reduced drastically in DSR.

Drivers of the Shift from Puddled Transplanting to Direct Seeding of Rice

Water Shortage

Over exploitation of ground water is major constrain to the sustainability of traditional system of the puddled transplanted rice in north-west India due to higher losses of water through puddling, surface evaporation and percolation. Agriculture's share of freshwater is likely to be decline by 8–10 per cent because of increasing the competition from urban and industrial sectors. The groundwater depletion at the rate of 4.0 cm per year in Punjab and Haryana (Yadav *et al.*, 2010). Transplant rice requires continuous flooding in and has relatively need high water input and consumed about 20-40 per cent of the total water required for growing crop (Farooq *et al.*, 2011,Walia *et al.*, 2011). Therefore, the need of hour is to develop alternative systems that require less water (Kumar *et al.*, 2015; Saharawat *et al.*, 2010).

Labour Shortage

In Punjab and Haryana, agriculture is heavily depended on migrant labour from Uttar Pradesh and Bihar state. Now a days the flow of labour decreased due to implementation of the national rural employment guarantee act 2007, in which 100 days compulsory work given by those states. Secondly, the agriculture sector labour diverted to non-agriculture sector due to rapid economic growth. In Punjab, there is restriction for paddy transplanting before 15 June. Due to this the demand of labour for transplant paddy after 15 June increased, which further increase the cost of transplanting. All these factors raise the cost of cultivation and delay the planting of crop.

Soil Health Issue

Traditional rice-wheat system degrade thesoil and water resources and threatening thesustainability of system. It destroys soil structure and adversely affects soil productivity. After puddling and it make adverse effects on the soil environment for the succeeding wheat and other upland crops. Puddling results in a complete breakdown of soil aggregates, destruction of macropores, and formation of a hard pan at shallow depth, which causes water logging.

Early Maturity

According to research done by PAU, that delay in 1 week in optimum time reduces the wheat yield upto 1.5 q/acre. For increasing wheat yield timely planting will must. DSR crop mature 7-10 days earlier as compared to puddled transplant crop. It allows more effective growth period to paddy crop within the same duration. A physiological shock to crop due to uprooting and harmonizing during re-establishment after the transplanting is clearly avoided, which in turn helps to timely sowing of wheat crop.

Ecofriendly

Traditionally paddy cultivation done by puddling the soil water. Puddling emits large quantity of methane (CH4), which is one of the major green house gas (GHGs) contributing to global warming. Whereas, in direct seeding rice, their no need of puddled the soil. So DSR is more ecofriendly from the puddled transplant rice.

Economics

Due to shortage of migratory labor in the state transplantation cost has been increase to Rs 2200 to 2500 per acre. In DSR, there is no transplanting which saves lot of money. Transplanting is done under saturated condition which requires high energy input and more irrigated water is needed. Pudddling generally cost excess of Rs 1500 per acre.

Production Techniques for Direct Seeded Rice

Laser Land Leveling and Field Preparation

Leveling of the field is very important to improve water management in direct seeded rice. Proper depth for placement of seed is must for achieving good crop establishment. The fields with high and low-lying areas often poor crop establishment under direct sowing due to uneven depth of seedling and uneven distribution of irrigation water. Due to this the laser land leveling should be done at least a month before sowing. After this the field should be irrigated to identify uneven areas in the field which can be taken care through fine leveling again. This irrigation is also stimulated weed and previous crop seed to germinate, which can killed before sowing the crop. For obtaining good crop of direct sowing rice, field should be ploughed twice with disc harrow followed by two cultivations with cultivator and one planking to prepare fine seed bed for raising good crop of rice.

Selection of Soil

Sowing of rice/basmati crop by direct seeded should be done only in medium to heavy textured soils. Cultivation of direct sowing under lighter soil faced the Iron deficiency. Do not cultivate DSR in high alkaline soil or soil having poor underground water. In such type of conditions, the DSR crop faced poor crop establishment and which further decrease the yield to very large extend. Every year change the DSR field due to this the problem of different weeds and volunteer rice decreases.

Time of Sowing

The sowing time is very crucial factor in determining the yield and quality of rice/basmati under direct sowing. Early sowing increase the water requirement due to high evaporation demand whereas late sowing results in uneven crop establishment by flooding due to rain. Sowing of rice varieties can be done during first fortnight of June while for Basmati sowing can be done during second fortnight of June. Make ensure that the plants have established before the onset of monsoon. Sowing after pre-seeding irrigation or applying irrigation just after seeding is very helpful to support germination and early growth when pre-monsoon showers are insufficient. Vigorous early growth of rice crop before the arrival of inundating monsoon rains reduces seeding mortality due to submergence and, by hastening crop development, makes it easier to ensure timely planting of succeeding crops after rice harvest.

Seed Rate and Method of Sowing

For obtaining good yield under direct seeded rice the optimum plant density is necessary. Crop density higher than the optimum can cause an increase in the proportion of sterile tillers, attack of foliar diseases and lodging of the crop. However, crop density below the optimum reduces resource utilization and leads to lower yields. Seed rate of 20-25 kg per hectare should be used for rice and basmati varieties. However the Seeding depth is also play a vital role in uniform germination of rice seedlings in direct sowing. The seeding depth should be 2-3 cm. Placement of seeds too deep or shallow reduces seedling emergence. Only use DSR seed-cum-fertilizer drill of 20 cm row spacing.Placement of Seeds too deep or shallow adversely affects the dynamics of seedlingemergence. Seeding depth can be adjusted with the depth control wheels fitted to most drills/planters. Before planting, it is also advisable to ensure that the wheels and seed tynes are leveled. Planking after seeding can also create better seed-soil contact, which is especially important when seeding is done after pre-sowing irrigation in the absence of rain.

Seed Priming and Treatment

DSR is typically sown at shallow depth (< 2.5 cm) and dry soil conditions are commonly the main constraint to rapid establishment of a good crop stand. In such situations, pre-hydration of seeds for 10-12 hours (seed priming) can improve germination. Primedseeds are subsequently dried in shade to decrease moisture content, which facilitates the proper functioning of seed metering mechanisms during planting. Priming induces a wide range of biochemical changes in the seed, the products of which persist following desiccation and are expressed quickly once the seeds again absorb water. Thus priming accelerates seed germination and crop emergence. Priming can also be used to treat the rice seed with bavistin or streptocycline to eliminate or reduce certain seed and soil borne diseases. 20-

25 kg seed of rice/basmati can treat with 2.5g of streptocycline + 50g of bavistin (carbendazim) for 8-10 hours to reduce seed and soil borne disease.

Varieties

Some fast growing paddy/basmati varieties are most suitable under direct seeding rice.Short duration variety PR 115 is very suitable under direct seeding while basmati varieties Pusa basmati 1121 and Pusa basmati 1509 are very suitable for direct sowing.

Nutrition

For direct seeded coarse rice apply 150 kg nitrogen (325 kg urea) per hectare in 3 equal splits at 2,5 and 9 weeks after sowing. For direct seeded basmati apply 60 kg nitrogen (135kg urea) per hectare in 3 equal splits at 3,6 and 9 weeks after sowing. Apply P and K only when soil test shows deficiency of these nutrients. Under scarcity of water, chlorosis among seedlings appears in the youngest leaf about three weeks after transplanting. These symptoms are due to deficient with iron and for this apply 2-3 foliar application of 1 per cent (1 kg Ferrous sulphate dissolved in 100 liter of water) of ferrous sulphate. DSR crop also shows the zinc deficiency symptoms appear 2-3 weeks after transplanting. The lower leaves become rusty brown near the base and ultimately dry up. The seedlings with zinc deficiency remain stunted and tillerless. To control this apply 62.5 kg/ha of zinc sulphateheptahydrate (21 per cent) per acre or 40 kg of zinc sulphate monohydrate (33 per cent) ha at puddling in case previous crop in field had shown the symptoms of zinc deficiency. Where the deficiency is noticed in the growing crop, apply this quantity of zinc sulphate.

Irrigation Water Management

The physics, chemistry and biology of direct dry seeded differ considerably from puddled soil mainly because of the differences in water regimes and the practice of puddling. Whereas puddled fields show up soil cracking behavior very early if adequate water is not maintained in the field. In DSR soils generally do not crack as substantially up on drying. Hence, for irrigation scheduling, the common rule applied in TPR of irrigating at the first sign of hairline cracks may not be suitable for DSR. From the perspective of water stress, rice does not require continuous submergence across growing season but any attempt to reduce irrigation should not be at the cost of yield penalty. Therefore, water stress must be avoided 2 weeks after seedling emergence at tiller, panicle initiation and flowering growth stages. Direct sowing provides an excellent opportunity to enhance water productivity because the fields no longer require continuous flooding. Direct sowing can be seeded in moist seed bed or in dry soil immediately followed by irrigation. In heavy textured soils, sowing should be done with pre-sowing irrigation. In medium textured soils, both moist and dry seed bed approaches are adopted. If the soil is dry at the time of sowing, then apply irrigation immediately after sowing. To fulfill the water requirement of the crop, apply irrigation at 5-7 days interval depending upon the soil type. The interval may be adjusted with rainfall. Stop irrigation 10 days before harvesting.

Weed Management

For getting good yield from direct seeded rice the effective weed management must be needed. Weeds are a main concern for high productivity of the direct seeding. Effective weed management in DSR depends on several factors, including the timeliness of the control operations during the early crop growth stages and, in some cases, good control in preceding crops. Integrated approaches to weed management combine multiple tactics and knowledge of site specific field conditions are essential to increase the efficacy and sustainability of weed control. The focus of weed management should ideally combine agronomic practices that increase crop competitiveness with the judicious use of chemical and other methods of direct weed control.

Cultural Methods

A. Stale seed bed: In this technique, by applying light irrigation weed seed germination and then emerged seedlings are killed by using a non-selective herbicide before sowing. It has observed that this technique reduces weed population by 53 per cent.

B. Good crop establishment: uniform establishment of healthy rice seedlings increases crop competitive ability which further suppresses weed growth.

C. mulching and cover crops: Mulch provides a barrier to emerging weeds. Also decomposing residues released some allele-chemicals which have inhibitory effects on germination of some weed species.

Chemical Weed Control

Selection of herbicides for use in direct sowing depends on the weed flora present in the different field (Table 4.1). Single herbicides have strength and weakness, *e.g.* Bispyribac is very good control over grasses but can't control Leptochloa. The control efficacy of different herbicides is also contingent on use of proper spray techniques. Weed can germinate in flushes during early growing stage. DSR field contains both aerobic and anaerobic grasses, sedges and broad leaf weeds.

Table 4.1. Different Weed Flora Present in DSR

	Grassy Weed	*Sedges*	*Broadleaf Weeds*
Rice weeds	*Echinochlo acrusgalli*	*Cyperus difformis*	*Ammania baccifera*
	Echinochlo acolonum	*Cyperus iria*	*Caesulia axillaris*
	Leptochlo achinensis	*Cyperus compressus*	*Eclipta alba*
	Paspalum distichum	*Fimbristylis miliacea*	*Monocharia vaginalis*
	Ischaemum rugosum		*Sphenochlea zeylanica*
Non-rice weeds	*Dactyloctenium aegyptium*	*Cyperus rotundus*	*Amaranthus viridis*
	Digitaria ciliaris		*Commelina* sp.
	Eleusine indica		*Digera arvensis*
	Eragrostis sp.		*Euphorbia hirta*
			Phylanthus niruri
			Trianthema portulacastrum

Most of cotton growing area of Sri Muktsar sahib district of Punjab shifted toward DSR. Due to this, aerobic weed especially *Cyperusrotundus*is rapidly becoming a major weed at initial stage of DSR. There are different types of herbicides present for controlling weeds in DSR which are written as follows:

A. Pre-Plant Herbicides

For pre-plant herbicides to be effective. Light irrigation must be applied for facilitates weed seed germination and emergence also better uptake of pre-plant herbicides. These germinate weeds can killed by timely use of glyphosate, paraquat, or mechanically by 1-2 shallow ploughing.

- ✰ Glyphosate should be applied 5 to 7 days before seeding. It is a non-selective and systemic herbicide that controls all annual and perennial weeds. Best results are obtained when all weeds are in active growth stages. Glyphosate is absorbed by foliage and rapidly translocated throughout in the plant. It is inactivated immediately when contact with soil and has no residual activity. Glyphosate controls all weeds including Cynadondactylon and Cyperusrotundus. Use glyphosate 1.0 kg a.i./ha mixed with 500 L water ha^{-1}.
- ✰ Paraquat is a non-systemic contact herbicide. It is used in areas infested with annual weeds at the rate of 0.3 kg a.i. mixed with 400 L water ha^{-1}. It is contact-type herbicide and is not translocated belowground storage organs of the perennial weeds. If a large number of perennial weeds are present in the field, then use of glyphosate is a better choice.

B. Pre-Emergence Herbicides

These are generally used before emergence of weeds and are applied immediately after the sowing of the crops within 2 days.

- ✰ Pendimethalin (Stomp) is applied @ 750 g a.i./ha for managing weeds in DSR. However, it should be properly applied in a clod free soils. High soil moisture is a increases the efficacy of herbicide. This herbicide controls most of annual grasses and broadleaf weeds
- ✰ Pyrazosulfuron is applied @ 15 g a.i./ha for the control of grasses, broad leaf and sedges. High soil moisture is a adequate for higher efficacy of this chemical

C. Post-Emergence Herbicides

These are used to control emerged weeds present in rice field. These are selective herbicides and for better efficacy should be sprayed with flat fan nozzles in 400-500 L water ha^{-1}. Following options can be used depending on types of weeds:

- ✰ Bispyribac (Nominee gold) @ 25 g a.i./ha at 15-25 DAS is effective in controlling all three types of weed flora (grasses, broadleaved and sedges). Apply this herbicide when the crop is infested with swank and paddy mothas. However, this herbicide is poor in controlling over *Leptochloa* spp., *Eragrostis, Dactyloctenium aegyptium* and *Cyperus rotundus.*

- Fenoxaprop (Rice star) @ 67 g a.i./ha at 25 DAS is effective against non rice weeds (Table 4.1) including Leptochloa and *D. aegyptium*. Fenoxaprop can also control *Cynodon dactylon* in rice. Apply this herbicide, when the filed is infested with leptochloa, Madhana and takri grass. However, this herbicide is poor in controlling over swank and paddy mothas.
- Azimsulfuron (Segment) @ 20 g a.i./ha and ethoxysulfuron (Sun rice) @18 g a.i./ha) is effective in controlling broadleaf weeds and sedges including *Cyperus rotundus* but is weak on grasses. Therefore, for mixed weed flora, it should be tank mixed with other herbicide which control grasses effectively.

Guidelines for Effective Use of Herbicide

- Use flat fan nozzle with multiple nozzle booms for spraying. With single nozzle boom (if multiple nozzle boom is not available), cut type nozzle should be used.
- Select right herbicide at right dose and time depending on types of weed flora present in the field.
- Rotate herbicides to avoid development of resistance in weed biotypes against any single herbicide.
- For pre-emergence herbicide, apply when there is sufficient soil moisture.
- It is essential that clean water is used for making spray solution.

Management of Volunteer Seed

As the adoption of DSR increases the volunteer seed is likely to became a problem especially if the fields are sown with different rice/basmati variety to the one grown previously. These plants become a competitor for next rice crop and reduce its yield and quality. Volunteer seed can controlled by growing short duration pulses and green manure crops after the harvesting of *rabi* crops, kills the volunteer seed by ploughing during seed-bed preparation. These seed can also be controlled by applying pre-sowing irrigation to encourage germination of these seeds which may controlled by applying paraquat or glyphosate to kill all green vegetation.Do not use the seed of DSR crop for next year. Purchase the seed from recognize seed source.

DSR Cultivation in Sri Muktsar Sahib District of Punjab

During *Kharif* 2014, the direct seeded rice/basmati was cultivated on 1,12,000 hectare in the Punjab state. Out of this 28,000 ha (Figure 4.1) was only in Sri Muktsar sahib district which consists the 25 per cent of total area of Punjab. In *kharif* 2015 area under DSR was reduces to 22,000 and 20,000 ha (Figure 4.1) respectively during 2015 and 2016 in the district. Most of area under direct sowing in Sri Muktsar Sahib district was cultivated under basmati. Area under basmati crop reduces in 2015 and 2016 as compared to 2014 due to fall in basmati price. Krishi Vigyan Kendra, Sri Muktsar sahib has done well affords to promote Direct Seeded Rice(DSR) since *kharif* 2010. Earlier, transplanting of basmati depends upon availability of migratory

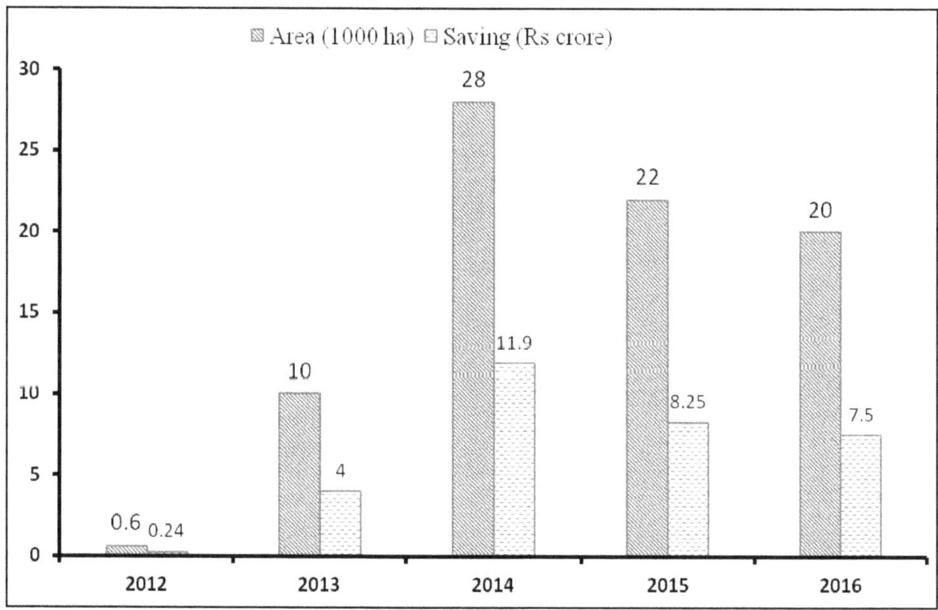

Figure 4.1. Area and Savings under DSR in Muktsar District of Punjab during different Years.

labour. Due to shortage of migratory labor in the state transplantation cost has been increase to Rs 2200 to 2500 per acre. However, the farmer using more ground water for cultivation of Basmati crop. This results in a steady decline in the depth of underground water in the state.Now a days major concern for researcher as well as for farmers are decline of underground water, saving of energy for pumping of groundwater, increasing running cost of tractor for pudling and other operations and availability of labour during transplanting. To overcome these problems Punjab Agricultural University, Ludhiana has recommended direct seeded rice technology for cultivation in Punjab.

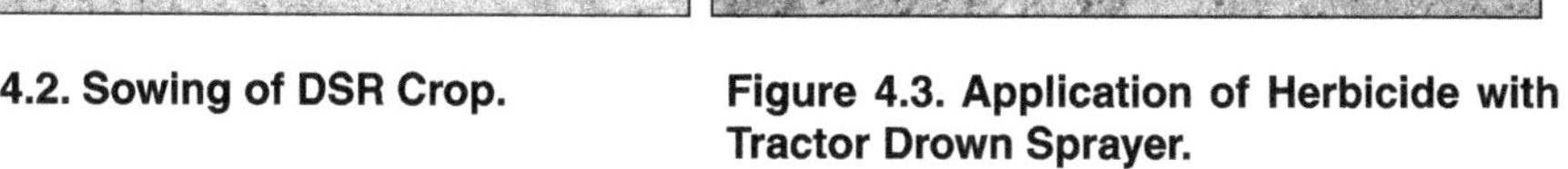

Figure 4.2. Sowing of DSR Crop.

Figure 4.3. Application of Herbicide with Tractor Drown Sprayer.

Figure 4.4. DSR Field.

Figure 4.5. Application of Fertilizer with Tractor Drown Modified Sprayer.

Future Developments

In Muktsar District DSR technique is adopted by many of large land holding farmers. With the help of KVK, Muktsar, they modified their old cotton sprayer according to direct seeded rice crop and spray pre-emergence herbicide, post-emergence herbicide, fertilizer, insecticide spray. This tractor drown sprayer is more accurate and efficient. Pre-emergence herbicideapplication is more uniform and farmer may use upto 500-700 liter of water by adjusting its gears. This type of sprayer is time saving and economics specially among large holding farmers.

REFERENCES

Anonymous (2014). Area and production of rice in India. *httpl/www.indiastat.com.*

Anonymous (2015). *Package of practices for crops of Punjab - Kharif 2015.* Punjab Agricultural University, Ludhiana, p. 1.

Chauhan, B.S., Gupta, R.K., Ladha, J.K., Singh, S., Singh, R.J., Jat, M.L., Saharawat, Y., Singh, V.P., Singh, S.S., Gill, M.S., Alam, M., Mujeeb, H., Singh, U.P., Mann, R., Pathak, H., Singh, B.S., Bhattacharya, P. and Malik, R.K. (2006) *Production technology for direct seeded rice. Rice wheat consortium for the Indo-Gangetic Plains.* Technical Bulletin 8; New Delhi, India. p.16.

Chauhan, B.S., Mahajan, G., Sardana, V., Timsina, J. and Jat, M.L. (2012) Productivity and sustainability of the rice-wheat cropping system in the Indo-Gangetic Plains of the Indian subcontinent: problems, opportunities, and strategies. *Adv. Agronomy.* 117: 315-369.

Farooq M, Siddique K H M, Rehman H, Aziz T, Dong J L, Wahid A (2011). Rice direct seeding: Experiences, challenges and opportunities. *Soil Tillage Res* 111: 87–98.

Joshi, E., Dinesh Kumar, D., Lal, B., Nepalia, V., Gautam, P., and Vyas, A.K. (2013). Management of direct seeded rice for enhanced resource - use efficiency. *Plant Knowledge Journal* 2(3): 119-134.

Kumar A, Kumar S, Dahiya K, Kumar S and Kumar M (2015) Productivity and economics of direct seeded rice (*Oryza sativa* L.) *Journal of Applied and Natural Science* 7 (1): 410–416.

Kumar V, Ladha J.K (2011) Direct seeded rice: Recent development and future research needs. AdvAgron. 111: 297-413.

Mahajan, G., Chauhan, B.S. and Johnson, D.E. (2009) Weed management in aerobic rice in *Northwestern* Indo-Gangetic Plains. *J. Crop Improv.* 23(4): 366-382.

Saharawat Y S, Singh B, Malik R K, Ladha J K, Gathala M, Jat M L, Kumar V (2010). Evaluation of alternative tillage and crop establishment methods in a rice–wheat rotation in North Western IGP. *Field Crops Res.* **116**: 260–267.

Singh N, Singh B, Rai A B, Dubcy A K and Rai A (2012). Impact of Direct Seeded Rice (DSR) For Resource Conservation. *Indian Res J Ext Edu* Special Issue **2**: 6-9.

Singh S, Bhushan L, Ladha J K, Gupta R K, Rao A N, Sivaprasad B (2006). Weed management in dry-seeded rice (*Oryza sativa*) cultivated in the furrow-irrigated raised-bed planting system. *Crop Protection,* **25**: 487–495.

Sudhir-Yadav, Gill, G., Humphreys, E., Kukul, S.S. and Walia, U.S. (2010). Effect of water management on dry seeded and puddle transplanted rice. Part 1. Crop performance. *Field Crops Res.* 120. 112-122.

Walia U S, Gill G, Walia S S and Sidhu A S (2011)Production of direct seeded rice (*Oryza sativa*) under differential plant densities and herbicides in central plains of Punjab *J Crop and Weed* 7(2): 1-5.

Yadav S, Gurjeet G, Kukal SS, Humphreys E, Rangarajan R, Walia US (2010) Water balance in dry seeded and puddled transplanted rice in Punjab, India Soil solution for a changing world.19th World Congress of Soil Science Brisbane, Australia, 43.

Transformation of Indian Agriculture through Innovative Technologies *Pages 57–70*
Editor: **Dr. Shahid Ahamad & Dr. Jag Paul Sharma**
Published by: **ASTRAL INTERNATIONAL PVT. LTD., NEW DELHI**

5 Mushroom Diseases: A Potential Threat to Mushroom Cultivation

Sachin Gupta, Baby Summuna, Ranbir Singh, Moni Gupta and Anil Gupta

Nature has bestowed mankind with thousands of mushroom species which grow in wild. Out of them, roughly 35 mushroom species have been cultivated commercially. Of these, 20 are cultivated on an industrial scale. A majority of cultivated species are edible as well as possess medicinal properties. Like all other crops, mushrooms are also affected adversely by a large number of biotic and abiotic agents/factors. Among the biotic agents, fungi, bacteria, viruses, nematodes, insects and mites cause damage to mushrooms directly or indirectly.

Among the cultivated mushrooms, white button mushroom [(*Agaricus bisporus*) is the most widely cultivated worldwide, accounting for 35-45 per cent of the total mushroom production and has a good market in western countries as well as India.

Fungal diseases commonly occurring in white button mushrooms include dry bubble (*Verticillium* sp.), wet bubble (*Mycogone* spp.), cobweb (*Cladobotryum* spp.), green mould in compost (*Trichoderma harzianum*) and greenmould on casing (*Trichoderma viride*). Among bacterial diseases, bacterial blotch (*Pseudomonas tolaasii*) is most common disease of white button mushrooms. Dieback is the most commonly occurring viral disease which is caused by various virus strains.

Fungal Diseases

There are four important fungal diseases *viz.*, dry bubble, wet bubble, cobweb and green mould, particularly referring to *Agaricus bisporus* (Sharma, 1995).

Dry Bubble

The causal organism of the disease is *Verticillium fungicola*. It is the most widespread disease of commercial mushroom production world wide. If the disease gets out of control it can cause crop losses as high as 20 per cent or more, but 1-5 per cent losses are common (Grogan *et al.*, 2009). A poor understanding of how the disease spreads within and between mushroom crops has been a contributing factor in the persistence of dry bubble problems. Dry bubble disease is prevalent in all mushroom growing areas and has 25-50 per cent incidence (Sharma, 1995).

Symptoms

Dry bubble disease of white button mushroom causing brown spots was reported for the first time by Malthouse in 1901. He found a species of *Verticillium* associated with this disease. Two types of symptoms were observed. Initially fungal growth appeared on the casing soil which later spread and turned greyish yellow. After that, light brown superficial spots appeared on the caps which finally coalesced to become large brown blotches. This disease is transmitted by contaminated compost, casing soil (Kumar *et al.*, 2014), human beings and splash of water (Cross and Jacobs, 1969). Mushrooms infected by *V. fungicola* show typical thickening of stem, resulting in onion shaped fruiting bodies. However, the symptoms vary with the age and the stage of development at which infection takes place. When mushrooms are infected by this fungus at an early stage, symptoms appear as small undifferentiated masses of tissue upto 2 cm diameter. Fruiting bodies are not properly formed and caps are partially differentiated. When infected at a later stage, the stipes are distorted and have tilted caps. Infected mushroom show the presence of grey white mycelia growth and become discoloured and dry but do not rot. They show small pimple like outgrowth or brown grey spots (1-2 cm diameter) on the surface. Such spots often have a yellow or bluish grey halo around them.

Management

Management and control of dry bubble disease relies mainly on hygiene and prevention of introducing inoculum on mushroom farms (Berendsen *et al.*, 2010). Strict hygiene practices are the best mechanism of control once dry bubble has been identified. All wastes need to be immediately removed from the farm to maintain hygiene. Prevention of mites and flies can help eliminate the spread of the pathogen throughout the crop, because whiteflies and springtails are capable of transporting spores from infected to healthy mushrooms (Tsarev, 2014). After spawning, dichlorvas @ 30 ml/100 lit. water/100 m^3 area may be sprayed to check the flies. Casing mixture should be prepared and stored in a clean room away from mushroom wastes, outside soil, insects and rodents that help prevent introduction of the pathogen to the crop (Tsarev, 2014).

The use of select fungicides to control *V. fungicola* can be effective by reducing the amount of present inoculum, but there are many issues with this approach. First, few chemicals can be used because all mushroom hosts are naturally sensitive to fungicides. Second, the pathogen has developed resistance to many fungicides over time and they are found to be increasingly less effective (Berendsen *et al.*,

2010). Third, the use of fungicides to combat *L. fungicola* may be banned in the near future and not be available to control the pathogen (Marlowe and Romaine, 1982). Legislative involvement is restricting the type of chemical controls that can be used to fight this particular pathogen (Berendsen *et al.*, 2010). Carbendazim gives highest per cent growth inhibition of pathogen followed by thiophanate-methyl, Dithane Z-78 and Dithane M-45. Kumar *et al.* (2014) tested the fungitoxicants in bed condition and observed that carbendazim was most effective in reducing the disease incidence. Bhatt and Singh (2002) reported Sporogon (0.075 per cent) to be effective against *V. fungicola*. Efficacy of Bavistin against *V. fungicola* has been reported by Navarro *et al.* (2011), Sharma and Satish (2012).

Wet Bubble

Mycogone perniciosa is the fungal pathogen causing wet bubble disease in crops of commercial *A. bisporus* mushrooms worldwide. The disease is very contagious and results in severe crop loss (Umar *et al.*, 2000). Mushrooms with wet bubble disease are very misshapen and not fit for sale (Dielemann-van Zaayen, 1976).

Symptoms

The pathogen may infect *A. bisporus* at various stages of its development. Infection at pinhead-stage results in undifferentiated, large, irregular forms of mushroom tissue (called sclerodermoid mushrooms) covered with the pathogen's white and fluffy mycelium; later infections lead to deformation of the carpophores and cap spotting (Fletcher and Ganney 1968; Hsu and Han 1981). The characteristic symptom of the disease is also the presence of amber liquid droplets on the surface of distorted mushrooms. It is recognized that the early development of wet bubble is usually associated with the use of infected casing soil or spawn (Fletcher *et al.*, 1989; Sharma and Kumar 2005). *M. perniciosa* produces two types of conidia: small, one-celled, thin-walled phialoconidia and much larger bicellular conidia (aleuriospores), which consist of a dark, spherical thick-walled, verrucose apical cell situated on a thin-walled basal cell. Phialoconidia are formed at the tip of phialides, like in *Verticillium*, whereas aleuriospores develop on short, lateral hyphae. Conidiophores are branched and cylindrical like those of *Verticillium* spp. (Glamoclija *et al.*, 2008; Potocnik 2006).

The development of infection caused by *M. perniciosa* depends on various conditions like temperature and relative air humidity in growing rooms as well as the pathogenicity of isolates. Fletcher *et al.* (1995) showed that pathogenic isolates of *M. perniciosa* are slow-growing on agar and their mycelium is pigmented, because it produces numerous aleuriospores whereas, weakly pathogenic isolates are characterized by fast growth on medium and little pigmented mycelium.

Management

Strict hygienic conditions should be maintained and sterilized casing soil should be used. Plastic pots/cups should be used to cover mushroom showing wet bubble symptoms during the cropping season to prevent spread of disease. Spray of benomyl @ 0.1 per cent immediately after casing has been reported to be very effective for protecting the crop. Application of carbendazim, chlorothalonil, prochloraz manganese complex (Sportak 50 WP) @ 0.1 per cent into casing mixture have also been recommended for the management of wet bubble. A spray of 0.8 per cent formalin on to casing surface, immediately after its application on the beds is also effective.

Cobweb

The disease is caused by *Cladobotryum dendroides*. The genus *Cladobotryum* syn. *Dactylium*, introduced into mycological literature by Nees (1817) was identified to be the imperfect stage of *Hypomyces rosellus* (Fletcher and Atkinson, 1977). In India, it was first recorded in Himachal Pradesh at Chail and Shimla (Seth, 1977) and later in Solan and Kasauli with a natural incidence ranging from 8.17-25.63 per cent in 1986 (Seth and Dar, 1989).

Symptoms

The disease first appears as small circular patches of grayish white mycelium on the casing surface. As the disease progress, a fluffy white mycelium grows over the mushrooms which look like cotton balls. Eventually they turn brown, begin to rot and die-off. High relative humidity and temperature favours the disease. It is normally introduced into the crop room by contaminated casing soil or spores through air. Secondary spread of the disease occurs by air movement, pickers, water splashes, *etc.*

Management

Thorough disinfection of casing soil with live steam or sterilization of casing mixture at 50°C for 4 hours effectively eliminates the pathogen. Regular cleaning, removal of cut mushroom stems and young half dead mushrooms after each break and controlling temperature and humidity helps in controlling the disease (Sharma, 1994). Single application of prochloraz manganese complex (sporogon) at 1.5g a.i./m^2 of bed 9 days after casing gives satisfactory control of the disease. Seth and Dar (1989) obtained best control of disease by applying bavistin + TMTD at 0.9 and 0.6g/m^2 followed by TBZ and benlate (0.9g/m^2). Effective control of C. *verticillatum* was obtained by spraying with 0.05 per cent carbendazim at spawning followed by 0.25 per cent mancozeb at casing and carbendazim again 15 days later (Sharma *et al.*, 1992).

Green Mould

One of the major diseases of the button mushroom (*Agaricus bisporus*) worldwide is caused by *T. harzianum* (Bayer *et al.*, 2000). Green mould epidemics have been reported in the U.S.A., Canada, South America, Asia, Australia and European Countries (Samuels *et al.*, 2002).

Symptoms

Green mould is characterized by large areas of dense green sporulation on the compost and casing surface resulting in a dramatic reduction in mushroom yield (Anderson *et al.*, 2000). Small blue green cushions are seen on spawned and cased trays/bags. It also grows on dead buds of mushrooms and cut stumps. Mushroom caps may turn brown on the top side. Green moulds generally appear in compost, rich in carbohydrates and deficient in nitrogen. High humidity with low pH of casing promotes its development. Pathogenic green moulds may colonize the substrate or grow on the surface of the emerging mushrooms. No symptoms appear in the bags until 10-35 days following the apparently normal spawn run of *A. bisporus*. *Trichoderma* spp. produce whitish mycelia indistinguishable from those of the mushrooms during spawn run, therefore it is difficult to recognize the infection at this stage (Largeteau *et al.*, 2002). Subsequently, large patches of compost turn green rapidly as spore production begins on the *Trichoderma* mycelia which had run through the compost with *Agaricus* (Seaby, 1996 and Rinker, 1996). Morris *et al.* (1995) described the symptoms of green mould disease as the presence of green fungal sporulation in the mushroom compost or casing layer between 2-5 weeks of the production cycle in the mushroom growing unit. The crop loss is proportional to the area infected, where generally no mushrooms are produced in contaminated bags in the case of serious outbreaks. Furthermore, even if mushrooms do appear, they are unsalable as they often become severely spotted, distorted (Seaby, 1989).

Management

Prevention has to play a central role in green mould management, however, if the infection already occurred at a mushroom producing facility, it has to be controlled. Chemical treatments are often the most effective means of managing green mould. Several fungicides have been evaluated for controlling green mould disease and among them Environ (Abosriwil and Clancy, 2002), Prochloraz, prochloraz + carbendazim (Abosriwil and Clancy, 2003) and Thiabendazol (Rinker and Alm, 2008) are reported to be the most effective in reducing compost colonization by *Trichoderma* strains. Romaine *et al.* (2008) recommended imazalil sulfate (an imidazole) against benzimidazole resistant strains. Abosriwil and Clancy (2003) concluded that the placement of fungicide on the spawn gave better reduction in *Trichoderma* colonization than when the fungicide was dispersed throughout the mass of compost. An alternative to chemical control of *Trichoderma* green mould is the application of microorganisms as biocontrol agents. Certain bacteria, including *Bacillus* species that naturally exist in casing are efficient antagonists of aggressive *Trichoderma* strains, therefore they have the potential to be used for the management of green mould disease.

Plaster Moulds

White plaster mould caused by *Scopulariopsis fimicola* and brown plaster moulds caused by *Papulaspora byssina* and *Scopulariopsis brevicaulis* develop in mushroom compost when nitrogen sources from phase I are not completely utilized by the microbes during phase II of composting and in particular, when the nitrogen is not converted into microbial protein.

Brown plaster mould is caused by *Papulospora byssina* and characteristically produces large, dense, roughly circular patches of mycelium on the surface of the casing, initially whitish, but turning brown and powdery with age; it can also colonise the compost (Fletcher *et al.*, 1989). Whitish patches on the compost or casing ultimately turning to rusty brown in colour are observed on the exposed surface of compost and casing as well as on the side in bags due to moisture condensation. Brown mould is silver grey initially, but as the spores mature the colour changes to a dark tan, beige or light brown (Howard *et al.*, 1994).

White plaster mould is caused by *Scopulariopsis fumicola* and produces dense white patches of mycelium and spores on the casing surface and in the compost. In contrast to other weed moulds, the mycelium of *S. fumicola* remains white (Fletcher *et al.*, 1989). Dense white patches of mycelium on compost and casing soil give flour like appearance. If the compost retains smell of ammonia and has pH more than 8.0, white plaster moulds become common. Both plaster moulds grow well in compost with a pH of 8.0 or higher and reduce yields by competing with mushroom mycelium (Fletcher *et al.*, 1989; Howard *et al.*, 1994). It is not possible to control plaster moulds during growing, so great attention must be paid to prevention (Van Griensven, 1988). Modification of composting practices to improve compost quality reduces the occurrence of plaster moulds.

Yellow moulds are caused by *Myceliophthora lutea, Chryosporium luteum* and *Sepedonium* spp. and the symptoms include brownish yellow corky layer of mycelium (stroma) with a white fluffy edge generally observed at the junction of compost and casing. Yellow moulds are generally observed where the compost has more than 70 per cent moisture and more than 20°C temperature in the crop room.

False truffle is caused by *Diehiliomyces microspores* and the symptoms include initially white fluffy mycelium later turning creamy yellow, prominent between compost and casing layers. The mycelium becomes thicker, solid, wrinkled mass resembling peeled walnut or brain like structure. False truffle is very common where mushroom is grown under natural condition at more than 20°C.

Management

For management of false truffle, proper hygienic conditions should be observed. Properly pasteurized compost and casing soil must be used. Compost should be ammonia (NH_3) free (not more than 8-10 ppm) and pH should be in between 7.2-7.5 at spawning. During spawn run the beds covered with papers, should be moistened twice a week with 0.5 per cent formalin. Proper temperature and relative humidity in the crop room should be maintained as per requirement of the crop. Stumps and dead mushrooms must be removed regularly from the beds and on observing the incidence of false truffle, spray with Dithane M-45 (0.15 per cent), bavistin (0.05 per cent) at 8-10 days intervals may be used.

Lipstick Mould

The causal organism of this mould is *Sporendonema purpurascens* which appears on the compost during spawn run or on the casing during production (Howard *et al.*, 1994). The fungus produces a fine, white mycelial growth rather indiscernible from mushroom mycelium (Fletcher *et al.*, 1989). The crystalline colonies may resemble frost on a windshield or small white cotton balls on straws or casing. Its white colour changes gradually to bright pink and finally buff, with a powdery appearance as the spores are produced (Fletcher *et al.*, 1989). Dissemination of this fungus is through spores and any means that transmit them can spread the organism

(Howard *et al.*, 1994). The presence of this mould in a mushroom crop causes great damage (Van Griensven, 1988).

Management

Raising the temperature in peak-heat to 65°C for four hours can eliminate lipstick mould, but this will increase the chances of other weed moulds developing (Fletcher *et al.*, 1989). Formalin can be used to spot-treat small areas of lipstick mould (Howard *et al.*, 1994) but strict attention to hygiene remains the most effective and safest approach to control (Fletcher *et al.*, 1989).

Penicillium Mould

As the name suggests the mould is caused by *Penicillium* spp. *Penicillium* species are opportunistic fungi, which prefer simple carbohydrates but will also grow on cellulose, fats and lignin (Howard *et al.*, 1994). This characteristic green mould is commonly seen growing on trays or the sideboards of shelves, or pieces of mushroom tissue left on the casing surface, and in grains of spawn. Apart from the appearance and some inconvenience, this mould has no measurable effect on crop yield or quality (Fletcher *et al.*, 1989).

Black Whisker Mould

This type of mould is generally caused by *Doratomyces microspores.*

This mould, is so named because of the dark-grey to black whisker-like bristles produced on the surface of the casing (Fletcher *et al.*, 1989). Black whisker mould in compost indicates an unbalanced nutritional base – specifically, the presence of certain carbohydrates – in the compost at spawning time. When *Chaetomium* green mould is present, whisker mould will often also be present, since both are cellulolytic fungi. Heavily infested compost will appear grey to black because of the high density of spores. When disturbed, the spores are released, resembling smoke. Human allergic responses to the spores have been reported (Howard *et al.*, 1994). The fungus is thought to be antagonistic to mushroom mycelium and may affect yield, the full significance of this fungus however, is not understood (Fletcher *et al.*, 1989).

Cinnamon Brown Mould

The causal organism of this mould is *Chromelosporium fulvum* (perfect stage: *Peziza ostracoderma*). Cinnamon brown mould is found mainly on casing soil just after casing (van Griensven, 1988). It has been observed on the surface of the compost during spawn run (Howard *et al.*, 1994). High air humidity in excess of 90 to 95 per cent and a high temperature usually found in the first one and a half weeks after casing, are ideal for the development of this mould (van Griensven, 1988). The fungus first appears as fine, white, aerial mycelium on the compost or casing (Howard *et al.*, 1994). Within a few days spores form and the colour changes from white to light yellow or light golden brown. The spores can be carried along in the air and infect new crops or cause secondary infections (van Griensven, 1988). Serious outbreaks on mushroom farms are usually indicative of poor hygiene or associated with over pasteurized casing material (Howard *et al.*, 1994). Control is obtained by ensuring the relative humidity in the growing room does not exceed 95 per cent (van

Griensven, 1988). The casing material can also be treated with a 2 per cent or less solution of commercial formalin (37 per cent formaldehyde) (Howard *et al.*, 1994).

Bacterial Diseases

Several bacterial diseases of cultivated mushrooms of the genus *Agaricus* and *Pleurotus* are caused by fluorescent pseudomonads (Paine, 1919, Gill, 1995). Studies on bacterial diseases of cultivated mushrooms in southern Italy showed that brown blotch of *Agaricus bisporus* and the yellowing of *Pleurotus ostreatus*, caused by *Pseudomonas tolaasii*, are actually complex diseases since, besides *P. tolaasii*, which may be considered the main causal agent of the above diseases, also *P. reactans* and not yet characterised fluorescent pseudomonads appear to participate in the expression of the diseases symptoms. Furthermore, *P. reactans* appears to be the causal agent of *P. eryngii* yellowing. Strains of *P. tolaasii* produce tolaasin I and II (Tol I and Tol II) and five minor analogs called tolaasins (A-E). Tolaasins are involved in the pathogen virulence causing membranes lysis through biosurfactant activity and transmembrane ion channel formation for which a barrel-stave mechanism has been proposed. Furthermore, strains of *P. tolaasii* produce *in vitro* an array of volatile substances which are apparently involved in the virulence of the producers and the pathogen/mushrooms interaction. Strains of *P. reactans* produce the White Line Inducing Principle (WLIP) which causes the brown discoloration of *A. bisporus* tissues though at lesser extent than tolaasin I. The loss of WLIP production by avirulent morphological variants of *P. reactans* supports its role in the pathogen virulence. The formation of avirulent variants in the cultures of *P. reactans* strains appears to be responsible for the attenuation/loss of virulence and this may tentatively explain why the pathogenicity of *P. reactans* was neglected and/or not well understood in the past. A comparative evaluation of Tol I and WLIP on blood red cells and artificial lipid vesicles demonstrated a detergent-like mechanism for WLIP. REP-PCR analysis showed that *P. reactans*, on the contrary of *P. tolaasii*, is not a genetically uniform group (Iacobellis, 2011). Brown blotch symptoms on *A. bisporus* are also caused by *P. reactans* (Iacobellis and Cantore, 2003; Munsch, 2002) *Pseudomonas* sp. strain NZ17 apparently related to *P. syringae* (Godfrey, 2001) and *P. costantinii* (Munsch, 2002). The ginger blotch disease of *A. bisporus* is caused by *P. gingeri* (Wong and Preece, 1979). The mummy disease of *A. bisporus* is caused by *Pseudomonas* spp. (Fletcher *et al.*, 1989) and the drippy gill of *A. bisporus* (Young, 1970) is caused by *P. agarici*. The latter pathogen was also reported as the causative agent of the yellowing of *P. ostreatus* observed for the first time in California (Bessette, 1985) and of the brown discolouration of mushrooms (Lo Cantore and Iacobellis, 2004) observed in The Netherlands and Italy. Bacterial disease of cultivated mushrooms caused by bacteria other than fluorescent pseudomonads are the soft rot of *Agaricus* spp. caused by *Burkolderia gladioli* and *Janthinobacterium agaricidannosum* (Lincon, 1991). *Ewingella americana* is responsible for the internal stipe necrosis of *A. bisporus* (Inglis, 1996).

Most common bacterial diseases of mushrooms are discussed as under:

Bacterial Blotch

The brown blotch disease of the button mushroom *Agaricus bisporus*, caused by the bacterium *Pseudomonas tolaasii* (Gill, 1995) was observed for the first time in some mushroom farms in the United States of America (Tolaas, 1915) and its etiology was defined a few years later (Paine, 1919). In Italy, the disease has been reported for the first time in 1970 in a mushroom farm in Apuglia (Ercolani, 1970). Circular, yellowish spots develop on the cap or near the margin and coalesce to form chocolate brown spots which penetrate the fleshy tissues. In severe cases brown lesions develop even on the stems. The disease is characterized by brown, irregular, sunken lesions on the pileus and/or stipe. Under favorable environmental conditions the lesions, initially small and separated, coalesce affecting large areas of the pileus which may gradually decay with the formation of a strong and disagreeable smell. Inadequately sterilized soil and contaminated implements are the main sources of infections. The pathogen spreads through splashing of water drops from infected to healthy sporophores, pickers implements, flies and mites also help in spread of the disease to healthy trays. This may occur also after harvest (Wells, 1996). Casing and air-borne dust are the primary means of introducing the blotch pathogen into a mushroom house. The bacterial pathogen is probably present in most casing material, even after pasteurization. Occurrence of disease is associated with the size of the bacterial population on the mushroom cap (pileus), rather than on the population in the casing, which explains why a prolonged wet period on the cap precedes disease occurrence. Once the disease occurs, blotch-causing bacteria are spread by splash-dispersal during watering, upon tools used by pickers and trashes and by mushroom flies and nematodes. Recent observations suggest compost with a moisture content of less than 62 percent at spawning preconditions mushrooms to blotch infection. Bacterial blotch can develop on the outer surface of a mushroom - on cap or stem or both - at any stage of mushroom growth or development. Bacteria splashed onto a mushroom surface will reproduce in moist conditions, such as occur when water condenses or remains on the mushroom surface for a number of hours. Condensation forms when saturated (with water vapor) air is present and warmer than the cap surface. The cap surface is cooler than the surrounding air when water transpires from the mushroom due to active mushroom growth; transpiration produces a cool surface. Slight fluctuations - a few degrees - in air temperature during cropping can cause the air to vacillate between the saturation point and not being saturated, even though the absolute water vapor content remains constant. Warm air holds more water vapor than cool air, so as the air temperature increases the air becomes less saturated with water vapor; the inverse is also true. With Bacterial blotch disease being so strongly influenced by environmental and surface-moisture conditions, disease control requires inhibiting the pathogens' reproduction on the mushroom surface. Air will dry the mushroom surfaces if it can hold the additional moisture coming from transpiration. If air is cooled to a temperature lower than desired and then reheated a few degrees, it can hold more moisture and mushroom surfaces will dry. This heating operation is accomplished by circulating hot water (110 to 120°F) through perimeter heating pipes found in many growing rooms. In a forced-air ventilation system, the air must pass through a heating coil after it has been cooled, so the cooling coils may need a bit more capacity than might be expected. This system of drying is energy intensive, but essential when drying is needed. High humidity (above 85 per cent) and inadequate ventilation during cropping permit the pileus to remain wet for longer periods, which helps in the disease initiation and spread. Lowering of humidity to 80 per cent and running fans immediately after watering to dry the caps prevent bacteria to spread on the growing sporophores. Spray the beds with 100 ppm bleaching powder.

Yellowing of Oyster Mushrooms

It is caused by *Pseudomonas tolaasii*. The yellowing of the oyster mushrooms *Pleurotus ostreatus* may infect all the stages of development of the mushroom sporophores (Gill, 1995).When the disease occurs at the early differentiation stage, the young sporophores turn to a yellowish-reddish colour, show a slow development followed by a rapid wilting. The alteration can affect the whole bunch or parts of it. On developed sporophores, depressed, yellowish-reddish lesions, sometimes surrounded by yellow-reddish halos may interest pilei and/or stipes. Under high temperature and humidity, sporophores rapidly rot with the production of an unpleasant smell. Sometimes, yellow superficial discoloured lesions may cover the whole sporophore or part of it. The disease has an unpredictable course. In the same cultivation, all the substrate bags in production or only part of them may be showing the above symptoms. In some bags, the disease appears severe with a significant loss of production while in others only a change in color of sporophores is observed. The disease may affect the first flush and then disappear or vice versa; in other occasions the disease affects the cultivation throughout the period of production. The yellowing of the king oyster mushrooms (cardoncello) *Peurotus eryngii* is a disease whose etiology was not well defined for many years. Initially it was attributed to fluorescent pseudomonads related to *P. tolaasii* (Ferri, 1985), but other studies indicated that fluorescent pseudomonads different from *P. tolaasii* were associated to the altered mushrooms (Iacobellis and Lavermicocca, 1990). The first symptoms of the disease are light brown discoloration of the pilei which then turn into reddish-brown. Stem symptoms are represented by hydropic areas, often elongated, which may coalesce, interesting the whole stems. Symptoms of the disease may interest primordia still in the casing soil layer or on fully developed sporophores. Mushroom sporophores turn stunted and wilted and then, under high temperature and humidity condition, bacterial exudates form on the sporophores which rapidly rot with the production of an unpleasant smell. The disease shows very unique features since it may occur in a disruptive way in the first flush and then it disappears resulting in normal production and vice versa; in other cases the disease may lead to the entire production loss. The disease initially can be localized to a substrate bags in production possibly interesting only a part of a mushroom bunch and then shortly the whole cultivation.

Virus Diseases

Mushrooms are also subjected to attack by a number of viruses which cause disease commonly known as La France, watery stripe, die back, X-disease or brown disease and which may result into slight or total failure of the crop. On the basis of the size and shape of the particles five different viruses have been reported attacking mushrooms and are known as virus 1,2,3,4 and 5. These viruses may occur alone or in any combination. In India there was no report about the occurrence of this disease, but recently it has been reported from Bangalore. The most common symptoms are the elongation of the stalk with a small, tilted cap (drumstick). Deterioration of the mycelium (die back) is common which increases with the time resulting into bare patches of the crop. Sometimes small brown mushrooms develop which often open prematurely. Affected fruits bodies have a water soaked appearance (watery stripe) which are found to be totally water logged when squeezed. The viruses cannot live in soil or outside the host tissues as these are obligate parasites. Hence affected crop debris, mushroom mycelium and spores survive readily on wooden boxes and in the spent compost which can be easily carried by air to distant places. Transmission of viruses through mushroom spores has been demonstrated. Some strains of mushroom spawn are also known to contain virus particles.

Management

Strict hygiene inside the farm should be maintained and filtered air should be used inside the peak heating, spawn running and cropping rooms. Mushrooms should be picked before they open. All wooden parts of growing units should be thoroughly cleaned and sterilized to kill any mushroom mycelium from the earlier crop. Tolerant or resistant strains should be used.

REFERENCES

Abosriwil, S. O. and Clancy, K. J. (2002). A protocol for evaluation of the role of disinfectants in limiting pathogens and weed moulds in commercial mushroom production. *Pest Management Science* **58:** 282-289.

Abosriwil, S.O. and Clancy, K. J.(2003). A mini-bag technique for evaluation of fungicide effects on *Trichoderma* spp. in mushroom compost. *Pest Management Science* **60:** 350-358.

Anderson, M. G., Beyer, D. M. and Wuest, P. J. (2000). Using spawn strain resistance to manage *Trichoderma* green mold. In *Science and Cultivation of Edible fungi: Mushroom Science* **15**(2): 641-644.

Bayer, D. M., Wuest, P. J. and Kremser, J. J. (2000). Evaluation of epidermilolgical factors and mushroom substrate characteristics influencing the occurrence and development of *Trichoderma* green mold. In *Science and Cultivation of Edible fungi: Mushroom Science* **15**(2): 633-640.

Bessette, A. E. (1985). Yellow blotch of *Pleurotus ostreatus. Appl. Environ. Microbiol.* 50: 1535-1537.

Berendsen, Roeland, L., Baars, Johan J. P., Kalkhove, Stefanie I. C., Lugones, Luis G., Wösten, Han A. B., Bakker, Peter A. H. M.(2010)."Lecanicillium Fungicola: Causal Agent of Dry Bubble Disease in White-button Mushroom". *Molecular Plant Pathology* 17–33.

Bhatt, N. and Singh, R. P.(2002). Chemical control of mycoparasites of button mushroom. *Indian J. Mycol. Plant Pathol.* **32:** 38-45.

Cross, M. J. and Jacobs, L. (1969). Some observations in the biology of the spores of the *Verticillium malthousei. Mushroom Science* **17**: 239-244

Dielemann-Van Zaayen, A.(1976). Diseases and pests of mushrooms. III. Fungus diseases. 1. Literature and research. *Bedrijfsontwikkeling* 7(12): 933-943.

Ercolani, G. L.(1970). Primi risultati di osservazioni sulla maculatura batterica dei funghi coltivati [*Agaricus bisporus* (Lange) Imbach.] in Italia: identificazione di *Pseudomonas tolaasii* Paine. *Phytopathol. Medit.* **9:** 59-61.

Ferri, F.(1985). In: *I funghi. Micologia, isolamento, coltivazione.* Edagricole, Bologna, p. 398.

Fletcher, J. T. and Atkinson, K.(1977). Mushrooms. A guide to the recognition and control of diseases, weed moulds, competitors and pests. *Agri. Dev. Advis. Ser.* 58.

Fletcher, J.T., White, P. F. and Gaze, R. H. (1989). Mushrooms: Pest and Disease Control. Intercept, Andover, UK, p. 174.

Gill, W. M.(1995). Bacterial diseases of Agaricus mushrooms. *Report Tottori Mycological Institute*, 33: 34-55.

Godfrey, S. A. C.(2001). Characterization by 16 S rRNA sequence analysis of pseudomonads causing blotch disease of cultivated Agaricus bisporus. *Appl Environ Microbiol* **67:** 4316–4323.

Grogan, H., Piasecka, J., Zijlstra, C., Baars, J. J. P. and Kavanagh, K. (2009). 'Detection of *Verticillium fungicola* in samples from mushroom farms using molecular and microbiological methods' Agricultural Research Forum, 2009, p131 (http: / /www.agresearchforum.com/publicationsarf/2009/proceedings2009.pdf)

Howard, R. J., Garland, J. A. and Seaman, W. L.(1994). Mushroom Chapter 26.Diseases and Pests of Vegetable Crops in Canada. The Canadian Phytopathological Society, Canada.

Iacobellis, N. S. and Lavermicocca, P.(1990). Batteriosi del cardoncello: aspetti eziologici eprospettive di lotta. *Professione Agricoltore* 2 (2): 32-33.

Icobellis, N. S. and Lo Cantore, P.(2003). *Pseudomonas "reactans"* a new pathogen of cultivated mushrooms. In: *Pseudomonas syringae pathovars and related pathogens* Iacobellis *et al.*, Eds. ISBN-1-4020-1227-6, Kluwer Academic Publishers, Dordrecht, The Netherlands, pp. 595-605.

Icobellis, N. S.(2011). Recent advances of bacterial diseases of cultivated mushrooms. Proceedings of the 7th International Conference on Mushroom Biology and *Mushroom Products (ICMBMP7)*. pp. 452-460.

Inglis, P.W. (1996). Evidence for the association of the enteric bacterium *Ewingella americana* with internal stipe necrosis of *Agaricus bisporus*. *Microbiology* **142**: 3253-3260.

Kumar, N., Mishra, A. B. and Bharadwaj, M. C.(2014). Effect of *Verticillium fungicola* (PREUSS) HASSEBR inoculation in casing soil and conidial spray on white button mushroom *Agaricus bisporus.Afr. J.Agric. Res.***9**: 1141-1143.

Largeteau-Mamoun, M. L., Mata, G. and Savoie, J. M.(2002). Green mold disease: Adaptation of *Trichoderma harzianum* Th2 to mushroom compost. In: Sanchez *et al.* eds. *Mushroom Biology and Mushroom Products. Proceedings of The 4th* International Conference on Mushroom Biology and Mushroom Products, *Cuernavaca, Mexico, 20-22. February 2002.* 179-187.

Lincon, S. P. (1991). Bacterial soft rot of *Agaricus bitorquis. Plant Pathol.* **40**: 136-144.

Lo Cantore, P. and Iacobellis, N. S.(2004). First report of brown discolouration of *Agaricus bisporus* caused by *Pseudomonas agarici* in southern Italy. *Phytopathol. Medit.* **43**: 35–38.

Marlowe, A. and Romaine, C. P. (1982). "Dry Bubble of Oyster Mushroom caused by *Verticillium Fungicola". Plant Disease.* **66** *(9): 859–860.*

Morris, E., Doyle, O and Clancy, K.J. 1995. A profile of *Trichoderma* species. I-Mushroom compost production. *Mushroom Science.* **14**: 611-618.

Morris, E., Doyle, O. and Clancy, K. J. (1995). A profile of *Trichoderma* species. II–

Mushroom growing units. *Mushroom Science* **14**: 619-625.

Munsch, P.(2002). *Pseudomonas costantinii* sp. nov., another causal agent of brown blotch disease, isolated from cultivated mushroom sporophores in Finland. *Int. J. Syst. Evol. Microbiol.* **52**: 1973–1983.

Navarro, M. J., Santos, M., Dianez, F., Tello, J. C. and Gea, F. J.(2011). Toxicity of compost tea from spent mushroom substrate and several fungicide *Agaricus bisporus*. Proceedings of the 7th international conference on mushroom biology and mushroom products (ICMBMP7). 2: 196-201.

Nees, C. G.(1817). *Das system der Pilze and Schwamme, ein Versuch.* Wurzburg, 329.

Paine, S.G. (1919). A brown blotch disease of cultivated mushrooms. *Ann. Appl. Biol.* **5**: 206-219.

Rinker, D. L. and Alm, G.(2008). Management of casing *Trichoderma* using fungicides. *Mushroom Science* 496-509.

Rinker,D.L.(1996). *Trichoderma* disease: progress toward solutions. *Mushroom World* **7**: 46-53.

Romaine, C. P., Royse, D. J. and Schlagnhaufer, C. (2008). Emergence of benzimidazole-resistant green mould, *Trichoderma aggressivum*, on cultivated *Agaricus bisporus* in North America. *Mushroom Science* **17**: 510-523.

Samuels, G. J., Dodd, S. L., Gams, W., Castlebury, L. A. and Petrini. O.(2002). *Trichoderma* species associated with the green mold epidemic of commercially grown *Agaricus bisporus. Mycologia* **94**: 146-170.

Seaby, D. A. (1996). Differentiation of *Trichoderma* taxa associated with mushroom production. *Plant Pathology* **45:** 905-912.

Seaby, D. A.(1989). Further observations on *Trichoderma. Mushroom* **197:** 147-151.

Seth, P. K. (1977). Pathogens and competitors of *Agaricus bisporus* and their control. *Indian Journal of Mushrooms* **3**(1): 31-40.

Seth, P. K. and Dar, G. M. (1989). Studies on *Cladobotryum dendroides* (Bukk. Merat) W.Gams et Hozzem, causing cobweb disease of *Agaricus bisporus* (Lange) Singer and its control. *Mush. Sci.* **12**(2): 711-23.

Sharma, V. P. and Satish, K.(2012). Comparative efficacy and prochloraz-mn against dry bubble, wet bubble and cobweb disease of button mushroom. *Mushroom Res.* **21**(2): 145-149.

Sharma, S. R.(1995). Management of mushroom diseases. In: Advance in Horticulture, Mushroom Eds. K. L. Chadha and S. R. Sharma, Malhotra Publish. House, New Delhi. **13**: 195-238.

Sharma, S. R.(1994). Viruses in mushrooms. In: Mushroom Biotechnology (Nair MC, Gokulapalan C, Das L eds). pp. 658-685. Indus Publishing Company, New Delhi.

Sharma, V. P., Suman, B. C. and Guleria, D. S.(1992). *Cladobotryum verticillium* a new pathogen of *A. bitorquis*. *Indian J. Mycol. Pl.Path.* **22**(1): 62-65

Tolaas, A. G. (1915). A bacterial disease of cultivated mushrooms. *Phytopathology* **5:** 51-54.

Tsarev, A. (2014). "Dry bubble (fungus spot) or Verticillium disease." Dry bubble (fungus spot) or Verticillium disease. N.p., n.d. Web. 14 Oct. 2014. <http: // en.agaricus.ru/cultivation/diseases/

Umar, M. H., Geels, F.P. and Van Griensven, L. J.L. D.(2000). Pathology and pathogenesis of *Mycogone perniciosa* infection in *Agaricus bisporus*. *Mushroom Science* **15:** 561-568.

Van Griensven, L. J. L. D.(1988) Wet bubble (*Mycogone perniciosa*). The Cultivation of Mushrooms, pp. 401-402.

Wells, J. M. (1996) Postharvest discoloration of the cultivated mushrooms *Agaricus bisporus* caused by *Pseudomonas tolaasii, P. 'reactans'* and *P. 'gingeri'*. *Phytopathology* **86:** 1098-1104.

Wong, W. C. and Preece, T. F.(1979). Identification of *Pseudomonas tolaasii*: the white line in agar and mushroom tissue block rapid pitting tests. *J. Appl. Bacteriol.* **47:** 401-407.

Young, J.M. (1970). Drippy gill: a bacterial disease of cultivated mushrooms caused by *Pseudomonas agarici* n. sp. *New Zealand Journal of Agricultural Research* **13:** 977-990.

Transformation of Indian Agriculture through Innovative Technologies *Pages 71–82*
Editor: **Dr. Shahid Ahamad & Dr. Jag Paul Sharma**
Published by: **ASTRAL INTERNATIONAL PVT. LTD., NEW DELHI**

6 Diversification of Agriculture through Mushroom Cultivation in Kashmir

Mohammad Najeeb Mughal and Ali Anwar

Introduction

Mushroom cultivation is one of the biggest money spinning agri-business in the world besides being an important cash crop. Mushrooms often called as 'Queen of vegetables', are in fact a group of higher fungi, belonging to the class Basidiomycetes or Ascomycetes. More than 2000 edible mushrooms have been reported throughout the world while in India 283 have been recorded and eight are being cultivated commonly (Maria *et al.*, 2000). Global mushroom production in 2009 was 2.4 million tonnes which is growing at the rate of 7 per cent per year. China is the leading mushroom producer (1.7 million tonnes) accounting for 70 per cent of world production while India with 1.5 per cent contribution ranks eighth in global mushroom production. The production of mushrooms in India during 2009-10 was 0.040 million tonnes, maximum contribution being from white button mushroom (31.81 per cent), followed by Shiitake (25.2 per cent) and Pleurotus (*Pleurotus* spp.) (14.2 per cent) (Shukla and Jaitly, 2011). Domestic demand of mushroom is growing at the rate of 25 per cent per year and per capita consumption per year is 20-25gms in India, where as in Europe and U.S.A, per capita consumption per year is 2-3 kg. The present share of India in exporting of mushroom is less than 1 per cent (FAO, 2008).

Mushrooms are known to have a broad range of uses both as food and medicine. As a food item, the nutritive value of mushrooms lies between that of meat and vegetables (Maria *et al.*, 2000). The protein content of different mushrooms ranges from 19 to 35 per cent, carbohydrates from 50 to 65 per cent (Ahmed *et al.*, 2010), fat contents 1.1 to 8.3 per cent and fibre content from 7.4 and 2.76 per cent on dry weight basis (Maria *et al.*, 2000). Mushrooms also contain all the essential amino acids and amides. Lysine, which is low in most of the cereals, is the most important amino acid in mushrooms (Babu and Subhasree, 2010). Moreover, mushrooms produce a series of metabolites of pharmacological and medicinal interest such as anti-oxidants, immuno-stimulants and antimicrobials (Moradali *et al.*, 2007 and Israilides *et al.*, 2008).

Jammu and Kashmir State is primarily an agricultural state with vast potential for mushroom cultivation. It has a varied climate across different zones which are ideally suited for the cultivation of different mushrooms. The raw materials in the form of agricultural waste such as paddy straw, wheat straw, seed hulls, corn cobs, tree litter, agro-industrial waste and house waste is available in plenty. Moreover, due to longer winter period, the marginal farmers and field labourers have little work to do for considerable part of the year. Successful mushroom cultivation may provide a helping hand in increasing household income to such farmers/labourers.

Although the climate of Kashmir is suitable for cultivation of many mushrooms *viz.*, button mushroom (*Agaricus bisporus*), pleurotus (*Pleurotus* spp.), winter mushroom (*Flammulina velutipes*), *Ganoderma* spp. *etc.* only button mushroom cultivation has been taken up by farmers for commercial cultivation owing to its wide spread popularity among locals Kashmiris and tourists, high demand and easily marketing. Cultivation of *Pleurotus* spp. (*Pleurotus sajor caju* and *P. ostreatus*) is also coming up fast as the production technology of this mushroom is very simple and local people are also showing their liking towards this mushroom. Keeping this in view, the production technologies button (*Agaricus bisporus*) and dhingri (*Pleurotus sajorcaju* and *P. ostreatus*) mushroom is discussed here.

Production Technology of Button (*Agaricus bisporus*) Mushroom

The whole process of button mushroom production can be divided into the following steps:

1. Spawn production
2. Compost preparation
3. Spawning
4. Spawn run
5. Casing
6. Fruiting and harvesting
7. Packing and storage

1. Spawn Production

Spawn is the seed of mushroom crop. Mycelial culture of mushroom fungus is prepared from selected strains of button mushroom under sterile conditions. Sterilized wheat grains mixed with 2 per cent calcium carbonate and calcium sulphate (1: 4) in bottles/polypropylene bags is then inoculated with mycelial

Figure 6.1. Button Mushrooms in Local Mushroom Farms in Kashmir.

culture under aseptic conditions and incubated at 25±2° C. The inoculated bottles are shacked after every 3-4 days till the spawn run in the bottles is complete. The full run spawn bottles are kept at low temperature till they are used. In Kashmir valley, spawn is available at Mushroom Research and Training Centre, Division of

Figure 6.2

Plant Pathology, SKUAST-K, Shalimar and Department of Agriculture, Lal Mandi, Srinagar. The spawn should be of good quality having potential for high yield and free from any contamination. Fresh spawn preferably spawn prepared in the current season (1-3 months old) should be used for high yield. Strain S-176 is commonly available for commercial cultivation of button mushroom in Kashmir.

2. Compost Preparation

The substrate on which button mushroom is cultivated is known as compost and the process of preparation of compost is known composting. Compost is mainly prepared from a mixture of plant wastes (cereal straw/sugarcane bagasse *etc.*), salts (urea, superphosphate/gypsum *etc.*), supplements (rice bran/wheat bran) and water. The ratio of C: N in a good substrate should be 25-30: 1 at the time of staking and 16-17: 1 in final compost. Keeping all the requirements in view, SKUAST-K have evolved composting formulae SK-3A and SK-3B based on locally available substrates and supplements for optimal production of quality mushroom. The quantity of ingredients required for preparing compost for 65 polythene bags of 10kg capacity is as under:

SK 3-A		*SK 3-B*	
Wheat straw	300 kg	Paddy straw	400 kg
Poultry manure	200 kg	Poultry manure	300 kg
Rice bran	50 kg	Rice bran	50 kg
*Corn liquor	5 lit from 5 kg maize grain	*Corn liquor	5 lit from 5 kg maize grain

SK 3-A		*SK 3-B*	
Linseed meal	7 kg	Linseed meal	7kg
Urea	5 kg	Urea	5 kg
Potash	2 kg	Potash	2 kg
Gypsum	10 kg	Gypsum	15 kg

*15-20 kg of molasses may be used if corn liquor is not available.

For the preparation of compost, the substrate is thoroughly washed with water to remove all the soil/dirt and is chopped (20-30cm) for faster composting and ease of handling during turning. The substrate is then sprinkled with water or soaked overnight in large drums/tubs so that it become soft and can under go faster decomposition during the process of composting. The composting can be achieved by following methods:

a. Long Method of Composting

This method of composting is commonly practised in Kashmir as it does not require pasteurization facilities. Long term composting is achieved by following steps:

i. Pre-wetted and chopped substrate is stacked on a hard clean surface/ surface covered with tarpelene/polythene or cemented floor.

ii. All the ingredients are mixed separately and kept ready for mixing with substrate (straw)

iii. Laying/raising of pile (the heap formed on floor/ground as a result of laying of substrate mixed with ingredients, is called pile).

iv. When the pile is one to one and half feet in height, it is pressed lightly and a portion of mixed ingredients is spread over it and in a similar manner another layers of straw and ingredients are laid over it and so on till the pile is 1.5 to 2 m in height and 1m in width.

v. The pile is given regular turns so that every portion of the pile is uniformly composted.

vi. First four turns are given at an interval of seven days and on 28 day i.e fourth turn, 10 to 15 kg of gypsum is added to the pile.

vii. On the 5th turn *i.e.* on 32 day, pile is sprayed with nematicide

viii. On 6th and final turn on 35th day, pile is sprayed with insecticide and pile is opened and spread on the floor.

ix. At every turn water is sprinkled over the pile to make the water loss caused by evaporation.

In the next few days, pile is regularly checked for smell of ammonia and when found it is free from ammonia, it is ready to spawning.

b. Short Method of Composting

Short term method of composting involve following steps:

i. Chopping: Straw is chopped into 20-30 cm pieces as in long term method

ii. Pre-wetting: Chopped straw is spread over floor and thoroughly wetted by sprinkling water or kept overnight in drums/tubs filled with water as in long term method.

Phase-I

i. Day-4: All ingredients except urea and gypsum are mixed with wetted straw. Heap of mixed straw is raised in such a manner that after every 30 cm thick layer of mixed straw heap is pressed tightly to favour anaerobic fermentation

ii. Day-2: 1st turning is given urea is mixed and compressed heap is again raised

iii. Day-0: 2nd turning is given, but pile is made without any pressure to favour aerobic decomposition

iv. Day-2: 3rd turning is given

v. Day-4: 4th turning is given

vi. Day-6: 5th turning is given

Phase-II

i. Day-8: 6th turning is given but heap is covered with double layer of black plastic terpelene/polythene completely. Perforated pipes (4 inches diameter) are kept horizontal and vertical in the pile at a distance of three feet (32) and the pipes should reach to the interior of the heap to favour continuous fresh air circulation for aerobic decomposition. It has two main purposes.

   a. Conversion of ammonia into microbial protein.

   b. Pasteurization, killing of pests and pathogens and to make the substrate suitable only for button mushroom.

ii. Day-20: Removal of terpelene/polythene covering, 7th turning and mixing of gypsum.

iii. Day-22: Turning and opening of pile.

iv. Day-23: Filling of compost in containers and spawning should be done immediately with a rapid decline of compost temperature to 25°C.

3. Spawning

The process of mixing spawn with compost is called spawning. Different methods of spawning are:

(i) Spot Spawning: Lumps of spawn are planted in 5 cm deep holes made in the compost at a distance of 20-25 cm. The holes are later covered with compost.

(ii) **Surface Spawning:** The spawn is evenly spread in the top layer of the compost and then mixed to a depth of 3-5 cm. The top portion is covered with a thin layer of compost.

(iii) **Through spawning:** The spawn is thoroughly mixed throughout the compost.

(iv) **Layer Spawning:** About 3-4 layers of spawn mixed with compost are prepared which is again covered with a thin layer of compost like in surface spawning. This method of spawning is generally practised all over Kashmir valley as the spawn run by this method results in fast and uniform growth of button mushroom fungus in the compost.

The spawn is mixed through the whole mass of compost at the rate of 500 to 750 g/100 kg compost (0.5 to 0.75 per cent).

4. Spawn Run

After the spawning process is over and compost is filled in polythene bags/trays (mostly wooden trays)/shelves covered either with a newspapers or polythene sheets, a temperature of 24±2°C and relative humidity of around 90 per cent is maintained by spraying water on the papers/polythene sheets which are used to cover the compost in the production room. The fungal mycelium grow out from the spawn and spread to entire compost. This growth of button mushroom fungus from spawn is known as spawn run. It takes about two weeks (12-14 days) to colonise the whole compost.

5. Casing

A thin layer of soil or soil amended with different materials is used to cover the full spawn run compost. This layer is known as casing layer and the process is known as casing. The casing layer should be of 3-5 cm thickness. The casing material should be having high water holding capacity and the pH should range between 7.0-7.5. Generally soil and peat (2: 1) is used as casing material. However, when peat is not available, different mixtures like garden loam soil and sand (4: 1); decomposed cowdung and loam soil (1: 1) and spent compost (2-3 years old); sand and lime may be used.

The casing soil before application should be either pasteurized (at 66-70° C for 7-8 hours), treated with formaldehyde (2 per cent) and carbendazim 50 WP (75 ppm.) or steam sterilized. The treatment needs to be done at least 15 days before the material is used for casing. After casing is done the temperature of the room is again maintained at 23±2°C and relative humidity of 85-90 per cent for another 8-10 days. Low CO_2 concentration is favourable for reproductive growth at this stage.

6. Fruiting and Harvesting

Normally small white pin like structures start emerging 21 days after casing, provided above mentioned congenial conditions for fruiting are maintained. This stage is known as 'pin heads stage'. It generally take 3-4 days for a pin head to reach 'button stage' which is considered the right stage for harvesting the fruitbodies of button mushroom. Generally cropping season continues for 6-8 weeks (six week under controlled conditions and eight weeks under seasonal growing conditions) and a yield of 15 to 20 kg per kg of dry weight of compost is attainable under Kashmir conditions. The fruitbodies, what is generally known as mushrooms, are harvested by twisting the mushrooms from the base first anti clock wise and then clock wise and finally by pulling upwards. If the mushrooms are in clusters than a sharp edged cutter/blade is used to harvest then and care is taken to cut the mushroom as close to casing soil as possible. The areas from where mushrooms have been harvested are again covered by sterilized casing soil so that the production continues.

7. Packing and Storage

a. Short-Term Storage

Button mushrooms are highly perishable in nature and need to be immediately transported to the retailer or consumer as the whole sale market for mushrooms does not exist in Kashmir. Harvested mushrooms are washed in a plain water to remove the soil attached to the mushrooms. Sometimes, mushroom after harvesting are washed with a solution of potassium meta bisulphite (KMS) (5g in 10 litre water) for removing the soil particles and to induce whiteness. For removing the excess water after washing mushrooms are spread on news papers to remove excess moisture. Mushrooms are then packed in perforated polythene or polypropylene bags, each containing 250 -500g and transported to retailer or consumer. Mushrooms can be stored in polythene bags at 4-5° C for a short period of 3-4 days.

b. Long-Term Storage

White button mushrooms are canned for long term storage. Canning is the most popular method of preserving the white button mushrooms and sizeable quantity of canned produce are exported to international markets. Besides pickling of button mushroom is also practiced by some units.

Economics of a Small Scale Model

The demand for button mushroom is fast increasing and a big gap exists between supply and demand. There is great opportunity for marginal farmers, agricultural labourers, unemployed youth of Kashmir valley to take advantage of this situation and adopt mushroom cultivation as an enterprise as it is a highly economically viable venture. Economics of cultivation of white button mushroom on small scale in polythene bags, calculated by MRTC, Division of Plant Pathology, SKUAST-K, Shalimar, shows benefit cost ratio of 1: 1.11.

General recommendations for button mushroom cultivation

i. **Mushroom Cultivation Planning**: Mushroom cultivation should be planned. Climatic conditions, substrate availability, market facility, transportation to market should be taken into consideration before starting the cultivation. Although mushroom being a fungus can be cultivated any time during the year if it is given the required temperature and humidity. But maintenance of specific temperature and humidity increases the cost of production. Therefore cultivation of mushroom should be taken up when climatic conditions are suitable. Best season for cultivation of button mushroom in Kashmir valley prevails twice in a year. First cultivation should be started i.e pile should be laid in the first fortnight of January so that the crop can be harvested well before rise of temperature in April and May. Second Pile should be laid in Ist fortnight of August so that crop is harvested well before sharp decline of temperature in December.

ii. Piles should laid on raised, clean and hard surface or surface covered by tarplene or polythene sheets or preferably cemented floor so that nutrients are not leached out to soil and also prevent the entry of harmful microorganisms from soil.

iii. If piles are raised in open, they should be covered during snow or rains.

iv. All the mentioned ingredients should be added to pile so that a good crop can be achieved.

v. Turning of pile should be such that each part of the pile gets place in the centre of pile turn by turn because temperature is highest in the centre of the pile as a result of microbial activity. This will result in uniform composting of whole pile.

vi. Sterilized casing material should be used.

vii. Temperature during spawn run and few days after casing should be around 25°C and 16-18°C during cropping period. Higher temperature may result in incidences of diseases and pests while lower temperature may delay spawn run and fruiting

viii. Production rooms should be sterilized after every crop.

ix. Crop should be harvested at button stage. If bigger sized mushroom are harvested, they will open during transportation and result in less selling price.

x. If there is any incidence of diseases and pests, it should dealt with immediately otherwise whole crop loss may be lost.

xi. It is always better to get practical training before starting any enterprise. Training can be had from MRTC, Division of Plant Pathology, Shalimar/ KVKs of SKUAST-K. or from Department of Agriculture.

Production Technology of Dhingri (*Pleurotus isajor caju* and *P. ostreatus*) Mushroom

Cultivation technology of dhingri mushroom is very simple as it does not involve preparation of compost. Its cultivation process can be divided into following steps:

i. Washing of substrate to remove soil/dirt attached to it.

ii. Chopping of substrate into 3-5 cm pieces and its wetting either by sprinkling of water or soaking over night in large drums.

iii. Dipping of substrate in hot water (65 to 70°C for 20 minutes) for sterilization.

Figure 6.3. ***Pleurotus sajor-caju.***

Figure 6.4. ***Pleurotus ostratus.***

iv. Light squeezing of substrate to adjust moisture level around 80per cent. This is indicated when after squeezing water droplets do not trickle down the substrate freely.

v. Filling of substrate in polythene bag/baskets or any container and spawning in layers as is done incase of button mushrooms. Spawn is used @ 30g spawn/kg of substrate.

vi. Researcher at SKUAST-K recommend that addition of black gram powder @ 1kg/q dry substrate at the time of spawning increases the yield of dhingri mushroom.

vii. If polythene bags are used, they should be cut at lower corners to make 2-3cm diameter holes so that aeration is not hindered and excess of water comes out of the bags.

viii. The bags/container should be closed from top to maintain high humidity.

ix. Temperature around 25°C should be maintained in the production room.

x. In around 15 days whole of the substrate will turn white because of spawn run.

xi. After full spawn run, the polythene bags are removed, as the dhingri mushroom at this stage require more fresh air and less humidity for fruit formation.

xii. 'Pin heads' formation starts after 15 days and within next 3-5 days first flush of crop can be harvest.

xiii. Total harvest comprises of 3-5 flushes and total yield q^{-1} dry substrate is 50-55 kg.

Economics of cultivation of dhingri mushroom on small scale in polythene bags, calculated by MRTC, Division of Plant Pathology, SKUAST-K, Shalimar, shows benefit cost ratio of 1: 1.

General Recommendations for Dhingri Mushroom Growers

i. Best period dhingri mushroom cultivation in Kashmir is from April to September

ii. Spawn should be booked in advance as availability of dhingri spawn is limited in Kashmir.

iii. Careful survey of market for selling of mushroom should be made in advance.

iv. Incase of any incidence of diseases and pests, please contact scientists at MRTC, Division of Plant Pathology, SKUAST-K, Shalimar.

REFERENCES

Ahmad, B.W., Bodha, R.H. and Wani, A.H. 2010. Nutritional and medicinal importance of mushrooms. *Journal of Medicinal Plants Research* **4**(24): 2598-2604.

Babu, D. and Subhasree, R.S. 2010. Valuing the Suitable Agro Industrial Wastes for Cultivation of *Pleurotus platypus* and *P. eous*. *Advances in Biological Research* **4**(4): 207-210.

FAOSTAT, 2008. *Statistics*. Food and Agriculture Organization of the United Nations. Rome, Italy.

Israilides, C., Kletsas, D., Arapoglou, A., Philippoussis, H., Pratsinis, H., Ebringerova, A., Hribalova, V. and Harding, S.E. 2008. *In vitro* cytostatic and immunomodulatory properties of the medicinal mushroom *Lentinula edodes*. *Phytomedicine* **15:** 512-519.

Maria, E.J., Florence, M. and Balasundaran. 2000. Mushroom cultivation using forest litter and waste wood. *KFRI Research Report* 195.

Moradali, F., Mostafavi, H., Ghods, S. and Hedjaroude, A. 2007. Immunomodulating and anticancer agents in the realm of macromycetes fungi (macrofungi). *International Immunopharmacology* **7**: 701-724.

Shukla, S. and Jaitly, A.K. 2011. Morphological and Biochemical Characterization of Different Oyster Mushroom (*Pleurotus* spp.). *Journal of Phytology* **3**(8): 18-20.

Transformation of Indian Agriculture through Innovative Technologies *Pages 83–99*
Editor: **Dr. Shahid Ahamad & Dr. Jag Paul Sharma**
Published by: **ASTRAL INTERNATIONAL PVT. LTD., NEW DELHI**

7 Role of Endophytes in Plant Disease Management

Mudasir Bhat, Ali Anwar, Arif Hussain, Mudasir Hassan and Saleem Dar

Introduction

Plant pathogens include fungi are the most visible threats to sustainable food production. The decreasingefficacy of the fungicides as well as risks associated with fungicide residues on the leaves and fruit, have highlighted the need for a more effective and safer alternative control measures. In recent years, endophytes have received increasing attention as a promising supplement or alternative to chemical control. The strategic use of naturally occurring organisms to control pest populations and increase production of major crops represents a viable option to host-plant resistance and pesticide-based pest and pathogen control. Endophytic microorganisms that grow in the intercellular spaces of higher plants are recognized as one of the most chemically promising groups of microorganisms in terms of diversity and pharmaceutical potential (Wagenaar and Clardy, 2001). Beneficial endophytic microorganisms comprise especially fungi and bacteria that colonize internal plant tissues without causing visible damage to their hosts (Petrini, 1991). Furthermore, the endophytic microorganisms are not considered as saprophytes since they are associatedwith living tissues, and may in some way contribute to thewell being of the plant. Endophytes exist in a range of tissue types within a broad range of plants, colonizing the plant systemically with bacterial colonies and biofilms, residing latently in intercellular spaces, inside the vascular tissue or within cells (Ulrich *et al., 2008*). Endophytic microorganisms that reside in the tissues of living plantsare relatively unstudied and potential sources of novel natural products for exploitation in agriculture. That is, theplant is thought to provide nutrients to the microbe, while the microbe may produce factors that protect the host plantfrom attack by animals, insects or microbes (Yang *et al.*, 1994). Studies on microorganisms from plant species arerecently becoming more frequent, since these fungi and bacteria have been studied for biological control andproduction of compounds with pharmacological properties. They are different from phyto-pathogenic microorganisms because they are not detrimental, do not cause diseases to plants and are distinct from epiphytic microorganismswhich live on the surface of plant organs and tissues (Hallmann *et al.*, 1997). Endophytic bacteria are able topenetrate and become systemically disseminated in the host plant, actively colonizing the apoplast (Quadt-Hallmann*etal.*, 1997b), conducting vessels (Hallmann *et al., 1997*) and occasionally the intracellular spaces (Quadt-Hallmann *et al.*, 1997a). This colonization presents an ecological niche, similar to that occupied by plant pathogens and these endophytic bacteria can, therefore, act as biological control agents against pathogens (Hallmann *et al.*, 1997). In thissense, the suppression of plant diseases due to the action of endophytic microorganisms has been demonstrated inseveral pathosystems (Narisawa *et al.*, 1998). Severalmechanisms may control this suppression, either directlyon the pathogen inside the plant by antibiosis (Sturz *et al.*, 1996) and competition for nutrients (Puentea, *et al.*, 2009), or indirectly by induction of plant resistance response (M'Piga *et al., 1997*) and more recentlytheir potential for enhanced degradation of several pollutants has also been investigated (Doty, 2008). Thereare many reports demonstrating that many bioactivecompounds could be produced by endophytic microorganisms (Huang *et al.*, 2001). At the same time, molecular markers provide gigantic sources of data that can assist scientists in developing tools to monitor thegenetic and environmental fate of these agents. In the present review we will focus on examples of associations between endophytic microorganisms and plants, especially those that result in diseases control. The intent of this review is to provide insights into the presence of endophytes in nature, the products that they make and how some of these organisms are beginning to show some potential for control of plant pests and diseases.

What are Endophytes?

The term endophyte refers to interior colonization of plants by bacterial or fungal microorganisms. Endophytic microorganisms, microorganisms that grow in the intercellular spaces of higher plants, are recognized as one of the most chemically promising groups of microorganisms in terms of diversity and pharmaceutical potential (Wagenaar and Clardy, 2001). Furthermore, the endophytic microorganisms are not considered as saprophytes since they are associated with living tissues, and may in some way contribute to the well being of the plant. It seems that other microbial forms, *e.g.*, mycoplasmas and archaebacteria, most certainly exist in plants as endophytes, but no evidence for them has yet been presented. The most frequently isolated endophytes are the fungi. Endophytic bacteria colonize an ecological niche similar to that colonized by plant pathogens but do not cause damage to their hosts. It turns out that the vast majority of plants have not been studied for their endophytes. Thus, enormous opportunities exist for the recovery of novel fungal forms, taxa and biotypes. Hawksworth and Rossman estimated there may be as many as 1 million different fungal species, yet only about 100,000 have been described (Hawksworth, 1991). As more evidence accumulates, estimates keep rising as to the actual number of fungal species. It seems obvious that endophytes are a rich and reliable source of genetic diversity and novel, undescribed species. The endophytes that we are most concerned with are the ones growing inside a turfgrass plant. Finally, in our experience, novel microbes usually have associated with them novel natural products. This fact alone helps eliminate the problems of dereplication in compound discovery.

Fungal Endophytes

Endophytic fungi are taxonomically and biologically diverse but all share the character of colonizing internal plant tissues without causing apparent harm to their host (Wilson, 1995). The best understood of these are members of the Clavicipitaceae (Ascomycota), which are endophytes of some temperate grasses. In these systems, there is usually only one endophytic fungal species per host and these fungi appear to be highly coevolved with their host. Generally, these fungi are transmitted vertically (from mother to offspring through seeds, as reviewed by Clay and Schardl (2002); see also Saikkonen *et al.* (2004)). This transmission pattern is thought to promote beneficial relationships with the host plant (Herre *et al., 1999*). Nonetheless, in grasses, the net effect of endophyte associations can range from parasitic (*e.g.*, choke disease) to strongly mutualistic (Clay and Schardl, 2002). Beneficial effects for hosts include increased drought tolerance (Arechavaleta *et al.*, 1989), deterrence of insect herbivores (Breen, 1994; Rowan and Latch, 1994), protection against nematodes (Pedersen *et al.*, 1988; West *et al.*, 1988; Kimmons *et al.*, 1990), and resistance against fungal pathogens (Gwinn and Gavin, 1992; Bonos *et al.*, 2005; Clarke *et al.*, 2006). The last is also true for endophytes found in some tropical grasses (Kelemu *et al.*, 2001). Anti-pathogen protection mediated by endophytes has been observed also in non-gramineous hosts. For example, endophytic fungi have been found to protect tomatoes (Hallman and Sikora, 1995) and bananas (Pocasangre *et al.*, 2001; Sikora *et al.*, 2008) from nematodes, and beans and barley (Boyle *et al., 2001*) from fungal pathogens. However, even with the accumulating evidence that endophytic fungi can reduce pathogen damage in grasses and other host plants, little is known about the generality of this role in natural systems and whether it can be exploited as a biocontrol strategy in crop protection.

How Fungal Endophytes can Affect Plant Disease

Evidence showing that endophytes have a role in the outcome of plant – pathogen interactions leading to disease has been increasing in recent years. Diverse mechanisms by which they may counteract pathogen development have been observed. For example, some endophytic species may induce plant defense mechanisms which counteract pathogen attack, others produceantibiotic substances which inhibit pathogen growth,competition for plant space and resources may also occur between resident endophytes and incoming pathogens; finally, some parasites of plant pathogens are known to behave as endophytes.

Interactions with Plant Pathogenic Fungi

Many endophytic species produce antibiotic substances (Strobel, 2002; Schulz and Boyle, 2005; Wang *et al.*, 2007). Liquid extracts from endophyte cultureshave been found to inhibit the growth of several speciesof plant pathogenic fungi (Liu *et al.*, 2001; Park *et al.*, 2005; Inácio *et al.*, 2006; Kim *et al.*, 2007). If suchcompounds where produced by endophytes*in planta*, this could constitute a defense mechanism against fungalpathogens. Experiments where plant protection against pathogenic fungi is observed after the inoculation of plants with endophytes, as well as after the application of endophytic culture filtrates, suggest that the endophyte may produce an antifungal compound or a substance that induces plant defense mechanisms in the plant. This is the case with *Chaetomium* and *Phoma* endophytes of wheat, when these fungi were previously inoculated in plants, reduced severity of foliar disease caused by *Puccinia* and *Pyrenophora* spp. was observed and, the same protective effect was observed when only endophytic culture filtrates were applied to the plants(Dingle and McGee, 2003; Istifadah and McGee, 2006). In these experiments the effects of culture filtratesupon plant pathogens were not tested.When a mixture of six species of endophytes frequently isolated from cacao (*Theobroma cacao* L.) trees was used to inoculate leaves of endophyte-free seedlings of this plant species, the severity of a leaf disease caused by a *Phytophthora* sp. was significantly reduced in endophyte-inoculated leaves. A mechanism of induced plant resistance did not seem to be involved,because differences in disease severity were observed between endophyte-inoculated and non-inoculated leaves of the same plant. In this case, the protection against a pathogen could be the result of direct competition among endophytes already present in leaves and the pathogen (Arnold *et al.*, 2003). For instance, most tissue available for infection may be already occupied, or endophytesmay produce zones of inhibition restricting the entry of other fungi. Endophyte infection may alter plant biochemistry in a way that defense mechanisms against pathogens are induced. *Piriformo sporaindica* Sav. Verma, Aj. Varma, Rexer, G. Kost and P. Franken is a root endophyte with a wide host range, including several species of cereals and *Arabidopsis*. Barley plants inoculated with this endophyte have shown resistance to a vascular [*Fusarium culmorum* (W.G.Sm.) Sacc.] and a leaf pathogen [*Blumeri agraminis* (DC.) Speer], in addition to an increase in yield and salt stress tolerance (Waller *et al.*, 2005). The protection against the leaf pathogen appears to be mediated by a mechanism of induced resistance, because in the pathogen-inoculated plants there is a defense response involving the death of host cells. Some endophytes may be mycoparasites. *Acremonium strictum* W. Gams is an endophyte which has been frequently isolated from *Dactylis glomerata* L. and other grasses (Sánchez Márquez *et al.*, 2007); recently it has been shown that this fungus is a mycoparasite of *Helminthosporiumt solani* Durieu and Mont., a potato pathogen (Rivera Varas *et al.*, 2007). A significant increase in resistance to dollar spot disease, caused by *Sclerotinia homoeocarpa* F.T. Benn., has been observed in *Festu carubra* L. cultivars infected by *Epichloë festucae* Leuchtm., Schardl and M.R. Siegel. (Clarke *et al.*, 2006). Cultivars of several turfgrass species infected by *Epichloe* and *Neotyphodium* endophytes are commercially available at the present time. The efficient vertical transmission of these endophytes has allowed the production of infected seed at a commercial scale. Since *Neotyphodium* and *Epichloë* infected cultivars have shown increased resistance to herbivores, plant pathogens, and some conditions ofabiotic stress, the use of such symbiotic cultivars canresult in a reduction in the use of insecticides and fungicides in lawns (Brilman, 2005). Similar applications of other species of endophytes may be seen in the future. The above studies suggest that the outcome of some pathogen attacks may be dependent on the endophytic mycobiota associated to a host plant. Therefore, the endophytic assemblage of a given species may represent a source of organisms with potential applications for disease control in the same plant species. Out of the multiple species that can penetrate and infect a plant, only a relatively small subset, that of thepathogens, produces disease. This shows that part of the plant disease cycle is shared by pathogens of endophytes. Once a fungus enters a plant it can behave as an endophyte or as a pathogen, and it seems that a majority of plant associated fungi act as endophytes. What is the difference between infection processes caused by endophytes and pathogens, is a good question for plant pathologists. Some studies directed to answer such question point out to fungal as well as to host characteristics. For instance, a mutation in a single locus can convert a pathogen such as *Colletotrichum magna* Jenkins and Winstead in a mutualistic endophyte (Freeman and Rodríguez, 1993). However, some isolates of the above species may behave as a pathogen in cucurbits or as an endophyte in some species of other plant families (Redman *et al.*, 2001).

Interactions with Nematodes

Inhibitory effects against some species of migratory and sedentary endoparasites occur in grasses infected by *Neotyphodium* endophytes (West *et al.*, 1988; Kimmons *et al.*, 1990). *Neotyphodium* species infect aerial tissues, not roots. Therefore, the inhibitory effects observed in infected plants were interpreted as the result of fungal alkaloids being translocated to roots. This was supported by the fact that some naturally occurring *Neotyphodium* strains deficient in the production of ergot alkaloids do not show protective effects as good against *Pratylenchus* sp. as those observed in ergot alkaloid producing strains (Timper *et al.*, 2005). In contrast, other experiments showed that the amount of ergot alkaloids translocated to roots is very small, and experiments with *Neotyphodium* knockout mutants having their pathway of ergot alkaloid synthesis disrupted suggested that these alkaloids are not responsible forthe inhibition of nematode populations in endophyteinfected plants (Panaccione *et al.*, 2006). Nevertheless, other types of alkaloids with antiherbivore activity are produced by *Neo typhodium* species, and chemical changes such as the production of phenolic compounds do occur in *Neotyphodium*-infected roots (Malinowski and Belesky, 2000). In conclusion, *Neotyphodium* endophytes provide host plants with protection against several nematode species, but the mechanism of action underlying this process is still unknown. Non pathogenic strains of *Fusarium oxysporum* E.F. Sm. and Swingle isolated from plant roots are other group of endophytes known to be implicated in anti-nematode activity. Culture filtrates of *F. oxysporum* have an inhibitory effect on *Meloidogyne incognita* Kofoid and White, suggesting that fungal toxins could be the mechanism of interaction (Hallmann and Sikora, 1996). However, the mechanism of *Fusarium* inhibition of nematodes appears to be more complex than a toxin operated system. In an experimental setup where banana plants were grown in a split root system, the plants were resistant to *Radopholussimilis* Cobb, Thorne in the root half which was not inoculated with a *Fusarium* endophyte. In this case, a phenomenon of systemic plant resistance induced by the endophyte appeared to be the mechanism of resistance to the nematode pathogen (Vu *et al.*, 2006). Another type of plant protection mediated by endophytic fungi may come from nematophagous fungi which can inhabit plant roots as endophytes (Bordallo *et al.*, 2002). In a similar fashion, some species of entomophagous fungi [*e.g. Beauveria bassiana* (Bals.-Criv.) Vuill., *Torrubiella confragosa* Mains, *Metarhizium anisopliae* (Metschn.) Sorokin] have been isolated from several host plants, and appears that part of their life cycle can be endophytic (Bills, 1996). In conclusion, it is very likely that fungal endophytes affect the outcome of nematode attacks in plants, and certain endophytes could be used for nematode protectionin agriculture.

Interactions with Bacteria and Viruses

Tests of the influence of endophytes upon bacterial and viral pathogens are not as numerous as those made with other plant pathogens. Bactericidal effects of endophyte culture extracts have been demonstrated and do not seem to differ from those observed for fungi or nematodes (Wang *et al.*, 2007).In the case of viruses, the incidence of *Barley yellowdwarf virus* (BYDV) was lower in *Loliumpratense* infected by *Neotyphodium* than in endophyte free plants. Since BYDV is transmitted by means of aphid vectors,toxic fungal alkaloids may be the reason for this effect,in fact, aphid reproduction was lower in endophyte infected plants than in those free of endophyte (Lehtonen *et al.*, 2006).A very interesting connection of a different kind exists among endophytes and viruses. A *Curvularia* endophyte of the plant *Dichantelium lanuginosum* (Elliott) Gould was found to confer tolerance to high soil temperatures to the plant. Further observation of this system led to the discovery that a virus infecting the endophyte was an important factor contributing to the heat toleranceobserved in the plants. Furthermore, the virus-infected endophyte could be used to confer heat tolerance to tomato plants (Márquez *et al.*, 2007). *Epichloë festucae* virus 1 (EfV1) is another virus which asymptomatically infects the grass endophyte *Epichloë festucae*, in this case it is not known if the presence of the virus in the endophyte affects the plant host (Romo *et al.*, 2007).

Bacteria Endophytes

Endophytic bacteria are able to penetrate and become systemically disseminated in the host plant, actively colonizing the apoplast (Quadt-Hallmann *et al.*, 1997b), conducting vessels (Hallmann *et al.*, 1997), and occasionally the intracellular spaces (Quadt- Hallmann *et al.*, 1997a). Endophytic bacteria colonize an ecological niche similar to that of plant pathogens, especially vascular wilt pathogens, which might favor them as candidates for biocontrol agents. In addition. intensive work on rhizosphere biocontrol agents has recently shown that presence of six rhizobacteria, induced systemic resistance in cucumber and exhibited both external and internal root colonization (Kloepper *et al.*, 1992b). Exploiting an additional microbial habitat for biocontrol purposes might enhance overall disease control and increase control consistency, since the control agent could avoid unfavorable conditions in one habitat by escaping into the other habitat.

Fungal Disease Suppression by Endophytes

Wilt Diseases

Studies using endophytic bacteria as biocontrol agents are conducted mostly on wilt diseasesof different commercial and food crops. Endophytic bacteria isolated from potato tubers showed *in vitro* antibiosis against *F. avenaciarum, F. sambucinum*and *F. oxysporum*. The antibiosis of the isolates decreased progressively with depth of their site of isolation from the tuber surface. Possibly, the plants might have adopted bacteria as part of a disease suppressive response to pathogenic attack (Sturz *et al.*, 1999.). Endophytic bacteria isolated from live oak stems, showed *in vitro* antagonism against oak wilt pathogen, *C.fagacearum* (Brooks *et al.*, 1994). A pre-inoculation with endophytic isolates *P. denitrificans*and *P. putida* significantly reduced crown loss. *Bacillus* spp. injected into Spanish oak stems, enabled more efficient colonization of the plants by *Pseudomonas spp.* (Brooks *et al.*, 1994). Endophytic bacteria isolated from rape and tomato plants, inhibited mycelial growth and reduced the incidence and severity of diseases caused by *V. dahliae*and *F. oxysporum*f sp. *lycopersici*in these crops by more than 75 per cent and increased the plant height and shoot dry weight (Nejad and Johnson, 2000). These strains produce volatile metabolites other than hydrogen cyanide. *Bacillus* spp., present in xylem vessels, reduced the percentage of silvermaple stem colonization by *V. dahlia* (Hall *et al.*, 1986.).

Rots and Damping-off Diseases

Endophytic bacteria isolated from rice seeds, when applied as seed treatment, colonized thestellar region of the root and exhibited strong anti-fungal activity against *R. solani, Pythiummyrotylum, Guamanomyces graminis* and *Heterobasidiu mannosum* (Mukhopadhyay *et al.*, 1996). Two endophytic bacterial strains isolated from cotton, which were antagonistic to *R. solani in vitro*, reduced infection of cotton plants from *R. solani*infection by 60 per cent (Qui *et al.*, 1990.), Of the 170 endophytic bacterial strains isolated from cotton, 40 strains protected cotton plants from *R. solani* infection (Chen *et al.*, 1995.). Endophytic bacteria from potato tubers were inhibitory to *Phytophthora infestans* (Sturz *etal.*, 1999). Endophytic *Bacillus* and *Pseudomonas* spp. isolated from seeds and plants of different crops were tested for their antagonistic potential against various pathogenic fungi including *P. ultimum, R. solani* and *S. rolfsii* (Pleban *et al.*, 1995). *P. fluorescens*, isolated from cotton, *B. cereus* strain 65 from Sinapis. *B. cerells* from cauliflower. *B. pumulis*from sunflower, *B. subtilis*from onion were found antagonistic to one or more of the fungal pathogens tested. Radioactive labelling showed that *B. cereus* strain 65, when introduced into cotton, was present at the root stem junction for 16 days. The bacteria were found up to 72 days after their introduction into the root and stem atlevels of 2.8X 105 and 5X104 CFU g-1 fresh weight of root and stem tissue, respectively. Various endophytic bacterial strains tested, including *B. cereus* strain 65, when introduced into the plant during the seed germination period, protected cotton and bean seedlings from infection with *R. solani* and *S. rolfsii* by more than 50 per cent. Strain 65 produced a 36 kDachitinolytic enzyme characterized as chitobiosidase and the crude preparation of extracellular proteins significantly decreased spore germination of *F. oxysporum* f.sp. *meloni* (Pleban *et al.*, 1997). *P. fluorescens*, isolated from the internal tissues of micro propagated apple plantlets, effectively colonized the root tips of beans. This strain was cloned to express the chiA gene coding for the major chitinase of *Serratia marcescens*, and the recombinant strain when introduced into bean seedlings provided effective protection from *R. solani*. Addition of this bacterium to soil did not confer protection indicating that internal colonization by *P. fluorescens* was essential for disease control (Downing and Thomson, 2000).

Galls and Abnormal Growth

Lima *et al.* (1994) isolated 160 bacterial strains from the woody tissues of lemon and sourorange, 55 of which were antagonistic to *Phoma tracheiphila* the casual agent of citrus malsecco disease. Nine of the most effective antagonistic strains were tested for disease controlby inoculating them into the stem of sour orange seedlings 15 days before pathogeninoculation. Three isolates of *B. subtilis* and one isolate of *P. fluorescens* significantly lowered the disease symptoms and maintained higher populations in the internal tissues of thehost plant (Lima *et al.*, 1994.)

Leaf Spots and Leaf Blights

Limited attempts have been made to use endophytic bacteria as biocontrol agents of leaf spotand leaf blight diseases. *Bacillus subtilis*, isolated from xylem fluid of chestnuts, suppressedthe growth of chestnut blight pathogen, *Cryphonectria parasitica* and reduced the lesion areason stems, when applied 3 days prior to fungal challenge, in an *in vitro* bioassay (Wilhelm *etal.*, 1998). Also. *B. subtilis* induced the synthesis of acidic chitinase and β- 1,3-glucanase inchestnut (Wilhelm *et al.*, 1998). Krishna Murthy and Gnanamanickam (1997) reported that *Psuedomonas* spp. induced systemic resistance in rice against sheath blight diseases causedby *R. solani*. Inspite of the absence of this bacterium on plant surfaces, its presence in theinternal stem led to suppression of disease. In a survey of antagonistic microorganisms, 8bacteria and 24 fungi were found antagonistic to *P. infestans*in the phyllosphere, rhizosphere and endoshpere of tomato (Garita *et al.*, 1988) indicated the endophytic nature of antagonistic organisms. Crown gall caused by *Agrobacterium tumefaciens*is one of the most extensively worked outsystem for biological control using rhizosphere isolates (Kerr, 1980). Endophytic bacteria isolated from xylem sap of grapevine plants were screened for their antagonistic activityagainst a range of tumerogenic *A. tumefaciens* biovar 3 strains (Bell *et al.*, 1995). Despite the variabilities in *in vitro* antibiosis, 24 of the 851 strains were strongly inhibitory to *A. vitis*, which produces galls in grapevine. These bacteria were identified as *E. agglomerans*, *Rahnella aquatilis* and *Pseudomonas* spp. An isolate of *P. corrugate* effectively reduced the populations of *Agrobacterium* strains *in situ*, In further trials with grape cultivar, chardonnay,where the incidence of galled vines was moderate, three endophytic bacterial strains gavesignificant protection against the disease. But these strains were ineffective in protecting gewurztraminer vines. In which the incidence of galls was severe (Bell *et al.*, 1995). Thesestudies also suggest that the performance of endophytic bacteria as biocontrol agents islargely dependent on the host genotype.

Nematode Diseases

Endophytic bacteria have an additional advantage in control of nematodes since the wounds produced by nematodes favour bacterial colonization of the root surface and their introductioninto the root tissue (Bookbinder *et al.*, 1982; Khan, 1993). Inoculation of cotton and cucumber plants with *Meloidogyne incognita* resulted in a higher density and diversity of totalendophytic bacteria in these plants. To determine, the interaction between Meloidogyne andendophytic bacteria *E. asburiae* Strain JM22 was used as a model system (Hallmann *et al.*, 1998). A large number of JM22 were observed on the surface of nematode galls, especiallywhere the root epidermis had disrupted due to gall expansion. Accumulation of bacterial cellsaround necrotic plant cells in the vicinity of galls was observed by electron microphotographs (Hallmann *et al.*, 1998). Soil amendment with chitin (1 per cent w/w), protected cotton from the attack of plant-parasiticnematodes (Hallmann *et al.*, 1999). When the soil bacteria and endophytic microbial communities in chitin-amended and non-amended soils were compared. *B. cepacia* populations were found to be same in both the soils, but it effectively colonized theinternal tissues of cotton only in chitin-amended soil (Hallmann *et al.*, 1999). Sevenendophytic bacteria, *Aerococcus viridans, B. megaterium, B. subtilis,P. chlororaphis, P. vesicularis, S. marcescens* and *Sphingo monaspaucimobilis* isolated from cotton and cucumber plants, on seed bacterization, significantly protected cucumber seedlings from *M. incognata* infection (Hallmann *et al.*, 1995).

Effects of Endophytic Microorganisms towards Pathogens

Indeed, intensive work has shown that endophytic microorganisms can have the capacity to control pathogens (Duijff, *et al.*, 1997 and Sturz, and Matheson. 1996) and nematodes (Hallmann *et al.*, 1998). The first record of an endophyte affecting a plant disease was that by Shimanuki (1987) who showed that timothy (*Phleumpratense*) plant sinfected with the choke fungus, *Epi chloetyphina*, were resistant to the fungus *Cladosporium phlei*. In some cases, they can also accelerate seedling emergence and promote plant establishment under adverse conditions and enhance plant growth and development (Lazarovits, and Nowak. 1997, and Pillay, and Nowak. 1997). Furthermore, several antagonistic entophytes bacterial species have been isolated from the xylem of lemon roots (*Citrus jambhiri*), including *Achromo bacter* spp., *Acinetobacter baumannii, A. lwoffii, Alcaligenes-Moraxella* spp., *Alcaligenes* sp., *Arthrobacter* spp., *Bacillus* spp., *Burkholderia cepacia, Citrobacter freundii, Corynebacterium* spp., *Curtobacterium flaccumfaciens, Enterobacter cloacae, E. aerogenes, Methylobacterium extorquens, Pantoea agglomerans, Pseudomonas aeruginosa,* and *Pseudomonas* spp. against root pathogens (Araújo *et al.*, 2001, and Lima *et al.*, 1994). Several bacterial endophytes have been reported to support growth and improve the health of plants (Hallmann *et al.*, 1997, Stoltzfus *et al.*, 1998) and therefore may be important sources of biocontrol agents. *Erwiniacarotovora*, for example, is inhibited by numerous endophytic bacteria, including several strains of *Pseudomonas* sp.,*Curtobacteriumluteum*, and *Pantoeaag glomerans* (Sturz *et al.*, 1999). Furthermore, Wilhelm *et al.* (1997)demonstrated that *Bacillus subtilis*strains isolated from thexylem sap of healthy chestnut trees exhibit antifungaleffects against *Cryphonectriaparasitica*causing chestnutblight. Endophytic bacteria have the ability to promote growth and inhibit plant disease, and as they are in intimatecontact with the plant they are an attractive choice asbiological control agents. For example, Sturz *et al.* (1999)found that 61 of 192 endophytic bacterial isolates frompotato stem tissues were effective biocontrol agents against*Clavibactermichiganensis*subsp. *sepedonicus*. In oak,endophytic bacteria biologically active against the oak wiltpathogen *Ceratocystisfagacearum*have been isolated(Brooks *et al.*, 1994). A number of the biologically active endophytes and root-colonizing microorganisms that have been isolated or detected belong to the actinobacterialphylum, specifically the genus *Streptomyces* (Coombs, and

Franco. 2003, Sessitsch *et al.*, 2001, Xaio *et al.*, 2002). The first actinobacterial endophyte isolated, belonging to the genus *Frankia*, is a nitrogen-fixing actinobacterium that forms actinorhizae with eight families of angiosperms (Provorov *et al.*, 2002). A number of endophyticactinobacteria were previously isolated by culturedependentmethods, with the major genera being *Streptomyces*, *Microbispora*, *Micromonospora*, and *Nocardioides*(Coombs and. Franco. 2003). A number of these isolates were capable of suppressing fungal pathogens of wheat in vitro and in planta, including *Rhizoctonia solani*, *Pythium*spp., and *Gaeumannomycesgraminis*var*tritici*, indicating their potential use as biocontrol agents (Coombs *et al.*, 2003).

Mechanisms of Diseases Control Displayed by Endophytes

In this sense, the suppression of plant diseases due to the action of endophytic microorganisms has been demonstrated in several pathosystems (Narisawa *et al.*, 1998). Several mechanisms may control this suppression,either directly on the pathogen inside the plant by antibiosis and competition for nutrients, or indirectly by induction of plant resistance response (M'Piga *et al.*, 1997). Endophytes usually occur in above-ground plant tissues, but also occasionally in roots (for example, dark septateendophytic fungi have been isolated from various plants), and are different from mycorrhizae by lacking external hyphae (LehoTedersoo *et al.*, 2009). Although some root endophyticfungus requires host cell death for proliferation during forming mutualistic symbiosis with plant (Deshmukh *et al.*, 2006), it is universally hypothe-sized that endophyte-hostinteractions involve a balance of antagonism and exhibit great phenotypic plasticity compared to plant pathogens (Schulz and Boyle, 2005). Only few documents refer to the plant secondary metabolism mediated by the fungal endophytes. Currently, endophytes are viewed as an outstanding source of bioactive natural products becausethere are so many of them occupying literally millions of unique biological niches (higher plants) growing in so many unusual environments. Thus, it appears that these biotypical factors can be important in plant selection, since they may govern the novelty and biological activity of the products associated with endophytic microbes. Peppermint growth and terpene production of *in vitro* generated plants (*Mentha piperita*) in response to inoculation with a leaf fungal endophyte indicate variation of the essential oil profile by fungal infection. The other study showed that the weight of roots, seedlings and terpenoid production of *Euphorbia pekinensis* increased after they were inoculated with an extensive host range endophytic *Phomopsis* sp. Meanwhile, microbial elicitor derived from some fungal endophytes also promotes biomass and induces the terpenoids (artemisinin) biosynthesis and production in plant suspension cells (Wang *et al.*, 2006). It seems likely that both mycorrhizal fungi and fungal endophytesinfection might result in specific-enhancement of the MEP pathway metabolic flux in plants.

The red resin of *Dracaena cochinchinensis*is commonly used in traditional Chinese medicine for the treatment of traumatic and visceral hemorrhages. Chemical studies have revealed that the resin contains various flavonoids (Zheng *et al.*, 2004).In addition, endophyticactinomycetes may also affect plant growth either by nutrient assimilation or enhancedsecondary metabolites (anthocyanin) synthesis. Furthermore, the production of antimicrobialsubstances, such as antibiotics or HCN, is an important mechanism to fight phytopathogens (Blumer, and Haas. 2000). Koshino *et al.* (1989) have described compounds, toxic to some fungi, which include sesquiterpenes, chokols, hydroxyl-unsaturated fats, phenolic glycerides and an aromatic sterol which are produced in the mycelial-choked heads of timothy. Endophytes effectively inhibit and killcertain other fungi and bacteria by producing a mixture ofvolatile compounds (Strobel *et al.*, 2001). The majority of these compounds have been identified by gas chromatography-mass spectrometry, synthesized or acquired, and then ultimately made into an artificial mixture. This mixture mimicked the antibiotic effects of the volatile compounds produced by the fungus. The newly described *Muscodor roseus* was twice obtained from tree species growing in the Northern Territory of Australia. This fungus is just as effective in causing inhibition and death of test microbes in the laboratory as *Muscodor albus* (Worapong *et al.*, 2002). Another endophytics *treptomycete* (NRRL 30566), from a fern-leaved *Grevillea*tree (*Grevillea pteridifolia*) growing in the Northern Territory of Australia, produces, in culture, novel antibiotics called kakadumycins (Castillo *et al.*, 2003). Each of these antibiotics contains, by virtue of their amino acid compositions, alanine, serine, and an unknown amino acid. Colletotric acid, a metabolite of *Colletotrichum gloeosporioides*, an endophytic fungus in *Artemisiamongolica*, displays antimicrobial activity against bacteria as well as against the fungus *Helminthsporium sativum* (Zou *et al.*, 2000). Another *Colletotrichum* sp., isolated from *Artemisia annua*, produces bioactive metabolites that showed varied antimicrobial activity as well. Yue *et al.*(2000) have identified a number of compounds produced by cultures of *Epichloe* and *Neotyphodium* species that have antifungal activity against the chestnut blight fungus *Cryphonectria parasitica* and suggest that they may play a similar role against other pathogens, the compounds in this study which showed the greatest antifungal activity were the indole derivatives indole-3-acetic acid and indole-3-ethanol, a sesquiterpene and a diacetamide. Indirect disease control is achieved by mechanisms modulating the plant immune response, including the induction of systemic acquired resistance (van Wees *et al.* 1999).

Genetic and Environmental Modifications Influencing Diseases Control by Endophytes

Identification of endophytes has relied mainly upon cultivation-based methods (Bell *et al.*, 1995). Molecular techniques based on the rRNA gene as a phylogeneticmarker (Amann *et al.*, 1995) provide a powerful approachto circumvent drawbacks related to cultivation. Molecular markers provide the means to assess genetic variation in endophytes and host plants, providing an insight into the relationship between variation in endophyte and host plants and the variability of agronomic traits (Gamper *et al.*, 2008). Researchers have endeavored to elucidate the molecular mechanisms during the establishment of plantendophyticassociation

(Bailey *et al.*, 2006). Techniques such as terminal restriction fragment length polymorphism (T-RFLP) analysis or denaturing gradient gel electrophoresis (Smalla, *et al.*, 2001) in combination with sequence analysis of rRNA genes allow rapid characterization of microbial communities. Comparison with data from amplified fragment length polymorphism (AFLP) data demonstrated that the SSR markers are informative for assessing genetic variation within and between endophyte species. Following the development of these markers for the sensitive detection of endophytes *inplanta*, the assessment of endophyte diversity in a globally distributed pool of perennial ryegrass germplasm are reported. Recently, Garbeva *et al.* (2001) monitored endophytic populations of potato by PCR-denaturing gradient gel electrophoresis, which revealed the occurrence of a range of organisms falling into several distinct phylogenetic groups. Their results also suggested the presence of nonculturable endophytes in potato.

REFERENCES

Amann, R. I., W. Ludwig, and K. H. Schleifer. 1995. Phylogenetic identification and in situ detection of individual microbial cells without cultivation. *Microbiol. Rev.***59**: 143-169.

Araújo, W. L., H. O. Saridakis, P. A. V. Barroso, C. I. Aguilar-Vildoso, and J. L. Azevedo. 2001. Variability and interactions between endophytic bacteria and fungi isolated from leaf tissues of citrus rootstocks. *Can. J. Microbiol.***47**: 229-236.

Arechavaleta, M., Bacon, C.W., Hoveland, C.S., Radcliffe, D., 1989. Effect of the tall fescue endophyte on plant response to environmental stress. *Agronomy Journal* **81**: 83–90.

Bailey, B.A., Bae, H., Strem, M.D., Robert, D.P., Thomas, S.E., Crozier, J., Samuels, G.J., Choi, I.k.- Young, Holmes, K.A. 2006. Fungal and plant gene expression during the colonization of cacao seedlings by endophytic isolates of four *Trichoderma*species. *Planta.***224**: 1449–1464.

Bell, C. R., G. A. Dickie, W. L. G. Harvey, and J. W. Y. F. Chan. 1995. Endophytic bacteria in grapevine. *Can. J. Microbiol.***41**: 46-53.

Blumer, C., and D. Haas. 2000. Mechanism, regulation, and ecological role of bacterial cyanide biosynthesis. *Arch. Microbiol.***173**: 170-177.

Bonos, S.A., Wilson, M.M., Meyer, W.A., Reed Funk, C., 2005. Suppression of red thread in fine fescues through endophyte-mediated resistance. Applied Turfgrass Science. doi: 10.1094/ATS-2005-0725-01-RS.

Boyle, C., Gotz, M., Dammann-Tugend, U., Schultz, B., 2001. Endophyte-host interaction III. Local vs. Systemic colonization. *Symbiosis* **31**: 259–281.

Breen, J.P., 1994. Acremoniumendophyte interactions with enhanced plant resistance to insects. *Annual Review of Entomology* **39**: 401–423.

Brilman, L.A., 2005. Endophytes in turfgrass cultivars. In: *Neotyphodium*in cool season grasses (Roberts C.A., West C.P., Spiers D.E., eds). Blackwell Publishing, Iowa, USA. pp. 341-349.

Brooks, D. S., C. F. Gonzalez, D. N. Appel, and T. J. Filer. 1994. Evaluation of endophytic bacteria as potential biological control agents for oak wilt. *Biol. Control* **4**: 373-381.

Castillo, U., J. K. Harper, G. A. Strobel, J. Sears, K. Alesi, E. Ford, J. Lin, M. Hunter, M. Maranta, H. Ge, D. Yaver, J. B. Jensen, H. Porter, R. Robison, D. Millar, W. M. Hess, M. Condron, and D. Teplow. 2003. Kakadumycins, novel antibiotics from Streptomyces sp. NRRL 30566, an endophyte of *Grevillea pteridifolia. FEMS Lett.***224**: 183-190.

Clarke B.B., White J.F., Hurley H., Torres M.S., Sun S., Huff D.R., 2006. Endophyte-mediated suppression of dollar spot disease in fine fescues. *Plant Dis***90**: 994-998.

Clarke, B.B., White Jr., J.F., Hurley, R.H., Torres, M.S., Sun, S., Huff, D.R., 2006. Endophyte-mediated suppression of dollar spot disease in fine fescues. Plant Disease 90, 994–998.

Clay, K., Schardl, C., 2002. Evolutionary origins and ecological consequences of endophyte symbiosis with grasses. *The American Naturalist***160**: 99–127.

Clay, K., Schardl, C., 2002. Evolutionary origins and ecological consequences of endophyte symbiosis with grasses. *The American Naturalist***160**: 99–127.

Coombs, J. T., and C. M. M. Franco. 2003. Isolation and identification of actinobacteria isolated from surface-sterilized wheat roots. *Appl. Environ. Microbiol.***69**: 5303-5308

Coombs, J. T., P. P. Michelsen, and C. M. M. Franco. 2003. Evaluation of endophyticactinobacteria as antagonists of *Gaeumannomycesgraminis var. tritici*in wheat. *Biol. Control***29**: 359-366.

Deshmukh, S., Hückelhoven, R., Schäfer, P., Imani, J., Sharma, M., l. Weiss, M., Waller, F., Kogel, K.H. 2006. The root endophytic fungus *Piriformos-poraindica*requires host cell death for proliferation during mutualistic symbiosis with barley. *Proc. Natl. Acad. Sci.***103**: 18450–18457.

Doty S.L. 2008. Tansley review: enhancing phytoremediation through the use of transgenics and endophytes. *New Phytol***179**: 318–333.

Duijff, B. J., V. Gianinazzi-Pearsonand, and P. Lemanceau. 1997. Involvement of the outer membrane lipopolysaccharides in the endophytic colonization of tomato roots by biocontrol *Pseudomonas fluorescens* strain WCS417r. *New Phytol.***135**: 325-334

Freeman S., Rodríguez R.J., 1993. Genetic conversion of a fungal plant pathogen to a non-pathogenic, endophyticmutualist. *Science***260**: 75-78.

Gamper H.A., Young J.P.W., Jones D.L., Hodge A. 2008. Real-time PCR and microscopy: are the two methods measuring the same unit of arbuscularmycorrhizal fungal abundance?*Fungal Genetics andBiology* **45**: 581–596.

Garbeva, P., L. S. van Overbeek, J. W. L. van Vuurde, and J. D. van Elsas. 2001. Analysis of endophytic bacterial communities of potato by plating and denaturing gradient gel electrophoresis (DGGE) of 16S rDNA based PCR fragments. *Microb. Ecol.***41**: 369-383.

Gwinn, K.D., Gavin, A.M., 1992. Relationship between endophyte infestation level of tall fescue seed lots and Rhizoctoniazeaeseedling disease. *Plant Disease***76**: 911–914.

Hallman, J., Sikora, R., 1995. Influence of Fusarium oxysporum, a mutualistic fungal endophyte, on Meloidogyne incognita infection of tomato. *Journal of Plant Disease and Protection***101:** 475–481.

Hallmann, J., A. Quadt-Hallmann, R. Rodríguez- Kábana, and J. W. Kloepper. 1998. Interactions between *Meloidogyne incognita* and endophytic bacteria in cotton and cucumber. *Soil Biol. Biochem.***30**: 925-937.

Hallmann, J., A. Quadt-Hallmann, W. F. Mahaffee, and J. W. Kloepper. 1997. Bacterial endophytes in agricultural crops. *Can. J. Microbiol.* 43: 895-914.

Hawksworth, D.L. 1991. The fungal dimension of biodiversity: magnitude, significance, and conservation. *Mycol. Res.*,**95**: 641-655.

Herre, E.A., Knowlton, N., Muller, U., Rehner, S., 1999. The evolution of mutualisms: exploring the paths between conflict and cooperation. *Trends in Ecology and Evolution***14**: 49–53.

Huang Y., Wang J., L.I. G., Zheng Z., Su W. 2001. Antitumor and antifungal activities in endophytic fungi isolated from pharmaceutical plants Taxusmairei, Cephalotaxusfortunei and Torreyagrandis. *FEMS Immunol. Med. Microbiol.*,**34**: 163-167.

Kelemu, S., White, J.F., Mun˜oz, F., Takayama, Y., 2001. An endophyte of tropical forage grass Brachiariabrizantha. Isolating, identifying and characterizing the fungus and determining its antimycotic properties. *Canadian Journal of Microbiology* **47**: 55–62.

Kim, H.Y., Choi G.J., Lee H.B., Lee S.W., Kim H.K., Jang K.S., Son S.W., Lee S.O., Cho K.Y., Sung N.D., Kim J.C., 2007. Some fungal endophytes from vegetable crops and their anti-oomycete activities against tomato late blight. *Lett Appl Microbiol* **44**: 332-337.

Kimmons, C.A., Gwinn, K.D., Bernard, E.C., 1990. Nematode reproduction on endophyte-infected and endophyte-free tall fescue. *Plant Disease***74**: 757–761.

Koshino, H., Yoshihara, T.,Sakamura, Y., Shimanuki, S., Sato, T. and Tajimi. A. 1989. A ring B aromatic sterol from stromata of *Epichloetyphina. Phytochemistry* **28**: 771-772

Lazarovits, G., and J. Nowak. 1997. Rhizobacteria for improvement of plant growth and establishment. *Hortscience* **32**: 188-192.

LehoTedersoo, Pärtel, K., Jairus, T., Gates, G., Põldmaa, K. and Tamm, H..2209. Ascomycetes associated witectomycorrhizas: molecular diversity and ecology with particular reference to the *Helotiales*. *Environmental Microbiology*. 11(12), 3166–3178.

Lima, G., A. Ippolito, F. Nigro, and M. Salerno. 1994. Attempts in the biological control of citrus mal secco (*Phomatracheiphila*) using endophytic bacteria. *Dif. Piante* 17: 43-49.

LIU C.H., ZOU W.X., LU H., TAN R.X., 2001. Antifungal activity of *Artemisia annua*endophyte cultures against phytopathogenic fungi. *J Biotech* 88, 277-282.

M'piga, P., Bélanger, R.R., Paulitz, T.C. and Benhamou, N. 1997. Increased resistance to *Fusarium oxysporum* f. sp. *radicis-lycopersici* in tomato plants treated with the endophytic bacterium *Pseudomonas fluorescens* strain 63-28. *Physiological and Molecular Plant Pathology*, V.50, pp.301-320.

Narisawa, K., Tokumasu, S.and Hashiba, T. 1998. Suppression of clubroot formation in chinese cabbage by the root endophytic fungus, *Heteroconium chaetospira*. *Plant Pathology*, v.47, pp. 206-210.

PARK J.H., CHOI G.J., LEE H.B., KIM K.M., JUNG H.S., LEE S.W., JANG K.S., CHO K.Y., 2005. Griseofulvin pathogenic fungi. *J Microbiol Biotech* 15, 112-117.

Pedersen, J.F., Rodriguez-Kabana, R., Shelby, R.A., 1988. Ryegrass cultivars and endophyte in tall fescue affect nematodes in grass and succeeding soybean. *Agronomy Journal* 80, 811–814.

PFENNING L.H., ARAUJO A.R., 2006. Antifungal metabolites from *Colletotrichum gloeosporioides*, an endophytic fungus in *Cryptocaryam andioccana* Nees (Lauraceae). *Bioch. Syst. Ecol.* 34, 822-824.

Pillay, V. J., and J. Nowak. 1997. Inoculum density, temperature and genotype effects on in vitro growth promotion and epiphytic and endophytic colonization of tomato (*Lycopersicum esculentum* L.) seedlings inoculated with a pseudomonad bacterium. *Can. J. Microbiol.* **43**: 354-361.

Pocasangre, L., Sikora, R.A., Vilich, V., Schuster, R.P., 2001. Survey of banana endophytic fungi from Central America and screening for biological control of burrowing nematode (*Rhadopholus similis*). *ActaHorticulturae***531**: 283–290.

Provorov, N. A., A. Y. Borisov, and I. A. Tikhonovich. 2002. Developmental genetics and evolution of symbiotic structures in nitrogen-fixing nodules and arbuscular mycorrhiza. *J. Theor. Biol.* 214: 215-232.

Puentea, M., Ching Y. Li b,1, YoavBashana,C. 2009. Endophytic bacteria in cacti seeds can improve the development of cactus Seedlings. *Environmental and Experimental Botany* **66:** 402–408.

RiveraVaras V.V, Freeman T.A., Gusmestad N.C., SecorG.A., 2007. Mycoparasitism of *Helminthosporium solani* by *Acremonium strictum*. *Phytopathology* **97**: 1331-1337.

Rowan, D.D., Latch, G.C.M., 1994. Utilization of endophyte-infected perennial ryegrasses for increased insect resistance. In: Bacon, C.W., White, J.F., Jr. (Eds.), Biotechnology of Endophytic Fungi of Grasses. CRS Press, Boca Raton, pp. 169–183.

Saikkonen, K., Wali, P., Helander, M., Faeth, S.H., 2004. Evolution of endophyte–plant symbioses. *Trends in Plant Science* **9**: 275–280.

Sánchez Márquez S., Bills G.F., Zabalgogeazcoa, I., 2007. The endophytic mycobiota of the grass *Dactylis glomerata*. *Fung Divers* **27**: 171-195.

schulz B., Boyle C., 2005. The endophytic continuum. *Mycol Res* **109**: 661-686.

Sessitsch, A., B. Reiter, U. Pfeifer, and E. Wilhelm. 2001. Cultivation-independent population analysis of bacterial endophytes in three potato varieties based on eubacterial and Actinomycetes-specific PCR of 16S rRNA genes. *FEMS Microbiol. Ecol.* **1305**: 1-10.

Shimanuki, T. 1987. Studies on the mechanisms of the infection of timothy with purple spot disease caused by *Cladosporium*(Gregory) de Vries. *In* Res. Bull. 148 Hokkaido Natl. Agric. Exp. Sta. pp. 1-56.

Smalla, K., G. Wieland, A. Buchner, A. Zock, J. Parzy, S. Kaiser, N. Roskot, H. Heuer, and G. Berg. 2001. Bulk and rhizosphere soil bacterial communities studied by denaturing gradient gel electrophoresis: plant-dependent enrichment and seasonal shifts revealed. *Appl. Environ. Microbiol.***67**: 4742-4751.

Stoltzfus, J. R., R. So, P. P. Malarvithi, J. K. Ladha, and F. J. de Bruijn. 1998. Isolation of endophytic bacteria from rice and assessment of their potential for supplying rice with biologically fixed nitrogen. *Plant Soil* **194**: 25-36.

Strobel G.A., 2002. Rainforest endophytes and bioactive products. *Crit Rev Biotech* **22**: 315-333.

Strobel, G. A., Dirksie, E., Sears, J. and. Markworth, C. 2001. Volatile antimicrobials from a novel endophytic fungus. *Microbiology***147**: 2943-2950.

Sturz, A. V., and Matheson. B. G. 1996. Populations of endophytic bacteria which influence host-resistance to *Erwinia*-induced bacterial soft rot in potato tubers. *Plant Soil***184**: 265-271.

Sturz, A. V., B. R. Christie, B. G. Matheson, W. J. Arsenault, and N. A. Buchanan. 1999. Endophytic bacterial communities in the periderm of potato tubers and their potential to improve resistance to soil-borne plant pathogens. *Plant Pathol.* **48**: 360-369.

Ulrich, K., Ulrich A. and Ewald, D. 2008. Diversity of endophytic bacterial communities in poplar grown under field conditions. *FEMS Microbiol Ecol* **63**: 169– 180.

Wagenaar, M.M.andClardy, J. 2001. Dicerandrols, new antibiotics and cytotoxic dimmers produced by the fungus *Phomopsislongicolla* isolated from an endangered mint. *J. Nat. Prod.,* **64**: 1006-1009.

Waller F., Achatz B., Baltruschat H., Fodor j., Becker K., Fischer M., Heier T., Hückelhoven Wang F.W, Jiao R.H., Cheng A.B., Tan S.H., Song Y.C. 2007. Antimicrobial potentials of endophytic fungi residing in *Quercus variabilis* and brefeldina obtained from *cladosporium* sp. *World J. Microbiol Biotech* **23:** 79-83.

West, C.P., Izekor, P., Robbins, R.T., Oosterhuis, D.M., Robbins, R.T., 1988. The effect of *Acremonium coenophialum* on the growth and nematode infestation of tall fescue. *Plant and Soil* **112:** 3–6.

Wilson, D., 1995. Endophyte–The evolution of a term and clarification of its use and definition. *Oikos***73**: 274–276.

Xaio, K., L. L. Kinkel, and D. A. Samac. (2002). Biological control of *Phytophthora*root rots on alfalfa and soybean with *Streptomyces*. *Biol. Control* **23**: 285- 295.

Yang, X., Strobel, G., Stierle, A., Hess, W.M., Lee, J. and Clardy, J. 1994. A fungal endophyte-tree relationship: Phoma sp. in Taxus wallachiana. *Plant Sci.,***102**: 1-9.

Yue, Q., Miller, C. J.,White, J. F. and Richardson. M. D. 2000. Isolation and characterization of fungal inhibitors from *Epichloe festucae*. *J. Agric. Food Chem.* **48**: 4687-4692.

Zheng, Q.A., Li, H.Z., Zhang, Y.J., Yang, C.R. (2004). Flavonoids from the resin of *Dracaenacochinchinensis*. *Helvetica Chim. Acta.***87**: 1167–1171.

Zou, W. X., J. C. Meng, H. Lu, G. X. Chen, G. X. Shi, T. Y. Zhang, and R. X. Tan. 2000. Metabolites of *Colletotrichum gloeosporioides*, an endophytic fungus in *Artemisia mongolica*. *J. Nat. Prod.***63**: 1529-1530.

Transformation of Indian Agriculture through Innovative Technologies *Pages* **101–105**
Editor: **Dr. Shahid Ahamad & Dr. Jag Paul Sharma**
Published by: **ASTRAL INTERNATIONAL PVT. LTD., NEW DELHI**

8 Empowerment of Rural Youths for Rural Transformation

Shahid Ahamad, Banarsi Lal and Akas Sharma

According to the UN youth is defined as a person in 15-24 years of age group whereas the census of India treats people in the age group of 15-29 years as youths. According to Baizerman (1991), in the third world countries youth have not been able to play a vital role in the developmental process and has remained a marginalized group. The participation of the rural Indian youth can be more impressive if more attention is paid on them. Mass media does not pay more attention on rural youths. Due to their proximity to institutions like universities, colleges, institutions of mass communication, political centres and by virtue of being an integral part of modernization process like developing a rational outlook, adopting different sub-cultures and promoting cosmopolitanism, the urban youth continuously remains in the limelight. Rural areas are slow in their socio-economic development, slow pace of modernization, industrialization, overpowering presence of powerful but retrogressive social institutions, inequitable distribution of land and most significantly and underutilization of human resources. The stereotype of rural youth is gullible, conservative and he himself tries to fastly change but cannot escape from the adverse impact of the aforementioned problems.

The globalization process has brought certain changes whose implications for the rural society and the rural youths need to be thoroughly understood. The onset of market economy has witnessed a massive growth in industrial activity which requires large workforce. Migration of rural youths in urban areas in search of employment has risen fastly. This has resulted a decline in workforce in the agricultural sector. The advent of free market has been able to generate employment opportunities in the private sector which requires highly skilled workers. A majority of rural youth may not meet the criteria of private sector for a variety of reasons, like lack of advance technical expertise, inadequate trainings *etc.*

Present era is an era of information. Communication network has combined the world into a cyber-frame. The transaction in all the sectors is now being carried through computers. The rural youths in this field are left behind their urban brethren. The urban youths have access to computer education while the exposure of youth to computers in rural areas is still limited. The 1990s witness a rapid expansion of television networks. The cable and satellite television made a mark in India and television emerged as the most effective medium of entertainment. It plays a key role in dissemination of information and entertainment. In some of the rural areas still the rural youths miss the opportunity to view the informative programmes. Even in some of the downtrodden rural areas still the youths do not get information through print media. The latest information should be reached to the rural youths of these remote areas. There has been a sustained campaign by the market forces to increase their rural marketing operations as three-fourths of the consumers live in rural areas and more than half of the national income is generated in rural areas. It has been observed that mostly television is the forte of the market forces and continuously promote consumerism which stimulates unrealistic desires in the rural youths. The political processes at the village level are intertwined with the operational aspects of the existing social hierarchy. It has been observed that rural youths are encouraged during elections campaign but the number of elected representatives from this segment is very low. Thus, despite their sincere efforts in the democratic processes they have a long way to go in holding positions and decentralization of political power has not resulted in major changes in the social structure of villages. It has been observed that increasing population, overexploitation of biological resources, construction activities and changing consumption has led to the loss of bio-diversity. For all this rural population cannot be held responsible for excessive consumption as it is the youth of urban areas whose consumption levels reached new heights leading to enormous pressure on the natural resources. The industries release untreated effluents contaminating the water reserves which will expose the nearby rural population to waterborne diseases. The toxic wastes of the industries are dumped in the waste lands on the outskirts of the urban areas or nearby villages which poses a serious impediment to the wasteland development projects. Watershed development is a major agricultural activity. It is the youth of village who take the responsibility in the success of watershed projects like construction of check dams and water tanks. The educated rural youths can prevail on the rest of population to take steps to preserve ground water resources. Joint forest management is another dimension of environment protection. The youth in the villages can be mobilized to take care of illegal felling of trees and make social forestry programme successful.

In rural areas gender discrimination is very high. Less literacy rate and traditional thinking of the people are the major reasons for the inferior status to the rural women. It has been observed that a large number of women representatives are chosen. Despite their success the rightful ascendancy of women is questioned by the male dominated society and the position of rural female remains secondary in all aspects of life. Rural female youths in many cases have to marry below the stipulated age because of family pressure and obsolete norms of the community. The higher death rate of the rural female youths indicates the lack of availability of proper medical facilities during pregnancy and delivery, poor diet and lack of care of their family members towards their health. Early motherhood combined with lack of proper education and inadequate physical and mental maturity will compound the problems of women. There is a dire need to motivate the rural female youth to take part in her decision making. With the implementation of employment generating schemes, efforts should also be made to sensitize the rural youths on various social issues. The mass media, whose influence on society is quite significant, should be prevailed upon to increase the focus on developmental issues and social concerns.

Rural India represents around 70 per cent of the Indian population. It is very important to engage the rural youth in a productive way in tandem by providing them credible opportunities for the growth and development. Agriculture is losing the attraction among the rural youths due to less profit. If the youths will generate extra income in agriculture then the interests of the youths in agriculture can be retained. There is need to make agri-based industries in the rural areas so that employment opportunities can be generated for the rural youths. India is having the youngest workforce in the world and our nation can become the human resource capital in the world by creating skill among the youths and convert the trained manpower for the growth of the Indian economy. In India around 51 per cent of the households survive on income from manual casual labour, 30 per cent from cultivation and 19 per cent from other sources. Around 74 per cent of the rural households earn less than Rs.5000/- per month. Around 35. 73 per cent of the rural people are illiterate and about 67 per cent have education below or till primary. Women constitute about 48.5 per cent of the population and play important role in the Indian economy. According to NSSO Employment-Unemployment Survey (EUS) 2011-12, due to low level of education and skill, around 49 per cent of the workforce is engaged in agriculture, 12.60 per cent in manufacturing, 10.60 per cent in construction and 27 per cent in the services sector. Presently around 10 per cent of the workforce is trained which include about 3 per cent formally trained and 7 per cent informally trained. Large number of the workforce does not get opportunity for trainings.

It has been observed that around two million people are shifting from rural to urban areas annually. From 2001 migration from rural to urban areas has increased from 27.8 to 31 per cent. Rural migrations to the urban areas have direct impact on agricultural productivity. Rural India remains in the focus of the policy makers as 10. 1 per cent of its labour force is unemployed as compared to 7. 3 per cent in the urban areas. Agriculture is the dominant employer in the rural areas followed by construction, manufacturing and community services. Rural areas are also the source of workforce in the adjoining towns and cities. In rural areas income from the agriculture is not sufficient especially for the small and marginal farmers who constitute 85 per cent of the farm holdings and for the dry land farmers who constitute more than half of the cultivated land. There is dire need to increase the employment opportunities for the rural youths to stop the rural migration. Rural youths migrate from rural areas to urban areas in search of employment as the agriculture in the rural areas does not suffice their basic needs. In urban areas they face the problem of housing, language and skill especially in the beginning. Due to these factors they are under paid and their growth is hampered. There is need to address the issue of rural migration. There is need to establish youth hostels in the urban areas which can give shelter to the rural youths especially in the beginning. There is also the need to change the set up of some departments to help the rural youths. The employment office established at each district should be changed for the career guidelines department. Some states in the country have taken initiative in this direction. From thousands of years farming has been sustaining life on the earth. We observe various success stories of the progressive farmers across the nation. Some success stories from Reasi district of J&K where organic vegetables growing, poultry farming, floriculture and diversified farming have transformed the lives of the farmers. These farmers have achieved the great success with the adoption of innovative technologies. With the adoption of innovative technologies they have increased their crops production and productivity. It occurred because of their consistent and arduous efforts. The scientific approach adopted by these farmers has resulted maximum output with the minimum inputs. There is need to make our agriculture more entrepreneurial and profitable. Value addition in the agricultural crops can create more income and employment for the rural youths. Farmers need to adopt the modern agricultural technologies to make farming more profitable. Protected cultivation is high-tech cultivation which produces 5-12 times higher output than cultivation in open fields. The demand of mushroom is increasing in the market and mushroom cultivation has become the profitable business. Important species of mushrooms are button, oyster, wood ear, shitake and paddy straw. India produces only 0.12 million tonnes of mushrooms out of which button mushroom contributes about 85 per cent of the total mushroom in the country. Out of the total agricultural residue, if one per cent is utilized for mushroom cultivation, the country can produce over three million tonnes of mushrooms and 10 million tonnes of organic manure annually. Production and supply of inputs required in the agriculture is the commercial ventures with lot of scope especially for the rural youths. Farm machinery can also create employment for the rural youths. In addition repair of farm machinery can also provide commercial ventures to the farmers. Farm machinery on the one hand will benefit the individuals engaged in the different ventures and on the other hand will help to increase our farm incomes. There is need to combine the farm and non-farm income at the household level which will provide resilience against adverse situations. If the present trend of migration continues then it is estimated that over 50 per cent of our population will be living in the urban areas. People will prefer to live in rural areas, semi-rural area or a small town if there is good telecom connectivity, a good road network and proper education and health care facilities. Rural areas need to be equipped with technical and educated manpower by expanding the network of industrial training institutes with better facilities. Industries like software, textiles, leather, electronics, pharmaceutical and many others can create such infrastructure. Such efforts can also restrict the migration of rural youths from the rural areas.

Tourism contributes around 11 per cent of the world work force and 10.2 per cent of the global gross domestic product. Our nation has a vast and varied agriculture landscape with natural beauty of blooming mustard fields of the northern India to blooming horticultural trees in the hilly areas to the tulip garden in Kashmir. Our rural areas have immense scope in agri-tourism which can help to create new employment for the rural youths. Tourism industry in India is growing at a rate of 10.1 per cent. Some states are making some strenuous efforts for rural tourism. In Maharashtra, people residing in the rural areas have formed Maharashtra State Agri and Rural Tourism. There are 150 agri-tourism centres in the state and all these centres are running without the aid of government. In Kerala rural tourism is attracting tourists across the globe. Similarly Rajasthan has also some rural tourism destinations. Our adjoining state Himachal Pradesh is also promoting rural tourism and natural tourism in the hilly areas of the state is attracting the tourists. Gujarat rural areas are also attracting the tourists. Promotion of rural tourism needs conceptual convergence with rural tourism, health tourism, eco-tourism, adventure tourism *etc.* Rural tourism can flourish only when the rural infrastructure such as road connectivity, communication and sector is created in the rural areas. There is need to search the potential areas in the rural areas for the rural tourism and infrastructure required in these areas should be made. There is need to do every effort to make the lives of rural people comfortable and lucrative. There is need to launch some more rural centric schemes to enhance the rural infrastructure.

Transformation of Indian Agriculture through Innovative Technologies *Pages* **107–121**
Editor: **Dr. Shahid Ahamad & Dr. Jag Paul Sharma**
Published by: **ASTRAL INTERNATIONAL PVT. LTD., NEW DELHI**

9 Biological Control of Temperate Fruit Crop Diseases

Mudasir Bhat, Ali Anwar, V.K. Ambardar and Arif Hussain

Introduction

Agro-climatic conditions in hilly states offer immense natural potential for increasing area and production under temperature fruits, especially apple. The production of these fruits is influenced to a great deal due to various diseases caused by fungi, bacteria and viruses. These diseases result in heavy losses to the farmers. To minimize the losses due to the diseases, farmers employ chemical control measures in fruit production system. Chemical fungicides which are used on fruit crops are posing threat in the form of environment and health hazards. In modern era of organic farming the use of chemicals is being discouraged. Also the development of resistant pathogen populations against the chemicals has stimulated interest in the formulation of integrated disease control system. Biological control would appear to have significant potential in terms of both environmental and economic issues for incorporation into organic and conventional temperate fruit production systems. Perennial cropping systems such as temperate fruits production provide unique opportunities and impediments to the use of biological control measures especially during the process of plant establishment. Several opportunities will exist in the nursery environment for the application of microbial inoculants such as biocontrol agents or mycorrhizal fungi. An additional opportunity for introduction of biocontrol agents exists at the time of orchard establishment.

Soil-borne Diseases

Replant Problem

Cultivation of fruit trees is rapidly changing, with growers planting high density orchards of trees on dwarfing rootstocks instead of low density orchards with vigorous trees. Because of the limited land available to plant new orchards, growers are compelled to lay out new plantation on the old orchard site. Due to this replant problem has become major concern of growers. The problem of establishing fruit trees in old nurseries or orchard sites is generally known as replant problem or replant disease. Replant problem includes biotic and abiotic factors that cause poor growth and delayed fruit production. Replant disease is caused by biotic factors and is one of the components of the replant problem. Replant problem is a complex malady of temperate fruit crops. It is caused by abiotic and biotic factors. Abiotic factors including phyto-toxins, imbalance nutrition, poor soil structure, poor drainage, cold or draught stress and excess or lack of soil moisture contribute to the occurrence of replant problem. Among biotic factors various fungi, bacteria, nematodes and actinomycetes have been reported to be associated with replant disease. Various species of fungi like *Fusariumequiseti, F. oxysporum, F. solani, Rhizoctonia* spp., *Cylindro cladium* spp., *Rosellinia necatrix, Penicillium claviforme, P. janthinellum, Phytophthora* spp., *Puthium* spp., *Cylindrocarpon* spp., of bacteria like *Pseudomonas* spp. and *Bacillus* spp., and of nematodes like *Pratylenchuspenetrans*and *Xiphinema* spp. have been found associated with replant disease by various workers (Utkhede, 1996). The cause of the replant problem varies from region to region. Depending upon the causal factor involved in the replant problem different control measures including cultural practices, chemicals, bioagents and host resistance have been suggested to control this. Chemical control with soil fumigants is the most adopted method of controlling replant disease all over the world. But it is not an attractive approach as the effectiveness of volatile fumigants is influenced by temperature and moisture. Also the application of fumigants is difficult, expensive and hazardous. Moreover, soil fumigation is believed to destroy the natural equilibrium between pathogens and antagonistic microorganisms in soil. In this regards attempts have been made to develop a biological control of replant problem. A number of studies have demonstrated benefits resulting from application of plant growth promoting and disease suppressive rhizobacteria to subsequent growth of apple in replant soil (Caesar and Burr, 1987; Utkhede and Li, 1989b; Janisiewicz and Covey, 1983). A diversity of bacterial species has been identified that suppress individual causal elements and enhance growth of plants in replant soil. Biological control of *Phytophthoracactorum* which contributes to replant disease (Mazzola, 1998) has been reported in response to application of *Enterobacteraerogenes* (Utkhede and Smith, 1991). Application of strain B8 of *E. aerogenes*alone was found increasing growth of apple seedlings in replant soil (Utkhede and Li, 1989a). Similarly, BACT-1, EBW and B10 strains of *Bacillus subtilis* and B8 strain of *E. aerogenes* applied as soil drench increased plant growth over and above that of formalin fumigation (Utkhede and Li, 1989b). The inoculation of roots of young apple plants with *Agrobacterium radiobacter* has eliminated replant problems in greenhouse and nursery experiments (Catska and Hudska, 1990). This biocontrol agent reduced the number of colonies of phytotoxic micromycetes which contribute towards replant disease. *Pseudomonas putida*strain 2 CB isolated from apple roots was found to inhibit growth of each element of the fungal complex reported to incite replant disease and enhanced growth of M-26 root stock in multiple apple replant soil (Mazzola *et al.*, 2002). Casear and Burr (1987) identified two fluorescent pseudomonads and an enteric bacterium possessing the ability to promote growth of apple in replant soils. Enhanced growth by these rhizobacteria was associated with a reduction in root infection of *Cylindrocarpon destructans*, a fungal pathogen known to contribute apple replant disease (Jaffe *et al.*, 1982; Mazzola, 1998). Application of *Bacillus subtilis* and *Enterobacter aerogenes* has also been found effective in promoting growth of apple plants even under filed condition in British Columbia, Canada (Utkhede and Smith, 1994). These results indicate that these species of rhizobacteria have a potential for biological control of replant disease. Various studies have indicated that phosphorus is an essential nutrient for early growth of young plants in replant soil. Mycorrhizae symbiosis can improve nutrient uptake, particularly for immobile ions such as phosphate (Mosse, 1973). As a result of increased uptake of mineral nutrients from soil, mycorrihizal plants grow more vigorously and appear healthier than non mycorrhizal plants. However, these beneficial fungi are eliminated when replant disease is controlled by soil fumigation (Nemec, 1980). Inoculation of apple seedlings with arbuscularmycorrhizal fungi (AMF) was increased their growth is replant disease soil (Catska and Taube-Baab, 1994). Inoculation of apple seedlings with AMF *Glomus fasciculatum* and *G. Macro carpus* suppressed the population of phytotoxic micromycetes, responsible for replant disease and increased plant biomass (Catska, 1994). Two AMF, *Glomus intraradices* and *G. Mosseae* significantly increased total shoot length and number of shoots per rootstock in replant soil. The seedlings inoculated with *G. Mosseae* showed increased growth in replant soil which was neither pasteurized nor fertilized (Utkhede *et al.*, 1992). Preplant sterilization of soil and subsequent inoculation with AMF, *Glomus epigaeum* significantly controlled the replant problem of apple and peach. It was observed that the growth promotion by inoculation with AMF was more in autoclaved replant soil (Bingye and Shengrui, 1998).

Root Rot

Root rot is a very serious soil-borne disease infecting temperate fruits especially apple. It is caused by *Rosellinianecatrix* Berl. exPrill. (Anam. *Dematophora necatrix* Hartig).The perfect stage of the fungus is not known to occur in India. The fungus has a wide host range of about 158 plant species belonging to over 45 families (Ito and Nakamura, 1984) comprising of fruit plants, forest trees and vegetable and field crops. In India the disease was first observed by Singh in 1939. Agarwala (1961) observed the disease on apple trees in Himachal Pradesh. The annual losses estimated due to this disease are about Rs. 1.3 million (Agarwala and Sharma, 1966) which are expected to be much more as the disease is reported to occur in all apple growing regions of the country. The pathogen affects the underground parts of the trees. The lateral roots turn dark brown and are covered with greenish gray or white mycelial mat and with the progress of disease all the roots are attacked and fibrous root system disappears. Whitish mycelial mat like fungal growth is visible during monsoon on the affected parts. The affected plants show bronzing of the leaves and progressive decline and ultimately die within 2-3 years of infection. The pathogen survives in the form of mycelium or sclerotia in the infected roots. The infection of new roots takes place by the fungal mycelium present in the soil on debris or by the contact of new plant roots with the old dead roots. The disease is more serious in water logged acidic soils. Management of root rot is very difficult because of deep seated infection. It is very difficult to make the reach ofremedial measures up to the point of infection. It can be managed by practicing preventive as well as curative measures consisting of cultural, biological and chemical methods and resistant root stocks. Biocontrol of soil borne pathogens with antagonistic fungi and bacteria has been under intensive investigations for the last many years. This method has gained considerable attention and appears to be promising as a viable supplement or alternative to chemical control. Ieki *et al.* (1969) have reported the inhibition of *R. Necatrix* by different isolates of *Trichoderma* spp. Species of *Trichoderma viz., T. harzianum* and *T. hamatum* isolated from naturally roots rot infected root inhibited the growth of the fungus *R. Necatrix* (Freeman *et al.*, 1986.) The use of antagonistic *Trichoderma* spp. against the disease has also been explored along with soil solarisation (Sztejnberg *et al.*, 1987). Reduction of root rot by using the antagonists like *T. harzianum, T. koningii* and *T. viride* has also been noticed under pot culture studies (Sharma, 1993). Fungal antagonists *T. harzianum* and *T. viride* and bacterial antagonists *Pseudomonas fluorescens* and *Bacillus* spp. have proved effective in controlling root rot under pot culture studies (Sharma, 2000). Other bacterial antagonists *Enterobacter aerogenes* have been found to protect the plants from *D. Necatrix* (Gupta and Jindal, 1989). Repeated application of these antagonists enhanced their efficacy against the pathogen and the effect persisted for longer period. Besides antagonistic fungi and bacteria the arbuscularm ycorrhizal fungi (AMF) have also been used against soil borne diseases. Bharat and Bhardwaj (2001) while studying the interactions between AMF and root rot pathogens *D. Necatrix*on apple seedlings under pot culture found that the apple seedlings previously inoculated with local AMF isolate of *Glomus* spp. showed less root rot severity as compared to the seedlings which were not inoculated with *Glomus* spp. The mycorrhizal seedlings also exhibited increased growth.

Crown Rot

Crown rot, also known as collar rot is prevalent in all apple growing regions of the world. It is caused by a fungus *Phytophthora cactorum* (Lebert- Cohn) Schroeter. The disease causes extensive losses sometimes resulting in death of trees. The infection starts from the collar region and spread mostly to underground parts and the above ground stem. Bark at the soil level becomes slimy and rots resulting in cankered area. The attacked trees show chlorotic foliage with red colouration of veins and margins. The causal fungus in known to survive in the orchard soils as chlamydo spores in plant debris or soil. The fungus produces oospores which serve as the source of primary inoculum. Moderate temperature and high soil moisture favour the disease. For the management of this disease attempts have been made by using chemicals, cultural practices and through host resistance. But the use of biological agents for controlling *P. cactroum* has also been tested by various workers.

Roiger and Jeffer (1991) evaluated *Trichoderma* spp. against *Phytophthora* crown rot of apple and found that *T. virens*, *T. koningii* and *T. harzianum* were effective in controlling the disease. Other species of *Trichoderma*, *T. longibrachiatum*, *T. viride* also proved effective in controlling *P. cactorum* in apple (Kumar, 2002). Some bacterial antagonists have also been found effective in checking the disease. Utkhede (1983) showed that bacterium *Enterobacter aerogenes* was antagonistic to *P. cactorum*. When this antagonistic bacterial cells were applied as soil drench, reduced the infection of *P. cactorum*in McIntosh apple seedlings. Other bacterial antagonists such as *Pseudomonas* spp., *Bacillus subtilis* were also found affective against *Phytophthora* crown rot (Janisiewicz and Covey, 1983; Utkhede, 1984). Antifungal activity of *E. aerogenes* and *Bacillus* spp. against *P. cactorum* was also observed by Gupta and Utkhede (1986) and with the treatments of apple plants with these antagonists even managed the crown rot infection under field conditions (Utkhede, 1987). Effect of soil drenching of apple trees with *E. aerogenes* was evaluated for three years by Utkhede and Smith (1991) in British Columbia with god control of the disease. Apple trees treated with strain B8 of *E. aerogenes* for two consecutive years not only reduced *P. cactorum* infection but also increased the plant height (Levesque *et al.*, 1993). It has also been demonstrated that antibiotics produced by *E. aerogenes* provide protection to apple trees from infection by other soil borne pathogens (Marchi and Utkhede, 1994). Another species of antagonistic bacterium *Enterobacter*, *E. agglomerans*have also been reported to reduce disease severity and increase the trunk girth and fruit yield in McSpur apple on MM 106 rootstock which is otherwise susceptible rootstock for crown rot (Utkhede and Smith, 1997). Lyophilized dry formulation of strain B8 of this antagonist proved effective. Another product Binab-1 formulated from antagonist *Trichoderma viride* have proved successful in controlling *P. cactorum* infection in apple seedlings (Orlikowski and Schmidle, 1985). Application of bacterial antagonistic *E. Agglomerans* and *B. subtilis* along with arbuscular mycorrhizal fungus *Glomus intraradices* in a six year experiment significantly reduced the infection of apple plants with *P. cactorum* and increased fruit yield and tree trunk growth (Utkhede and Smith, 2000).The application of this antagonist has also been tried in combination with fungicide metalaxyl as soil trunk drench twice in a year for three years and once in spring for seven years. The combined application reduced disease ratings of *P. cactorum* (Utkhede and Smith, 1993). Similarly the same combination (each 1g/tree) when applied once in a year resulted higher annual increase in trunk diameter of effected apple plants (Levesque *et al.*, 1993). Kumar (2002) has observed that pre inoculation of apple seedlings with *Trichoderma longibrachiatum* along with simultaneous inoculation with *Bacillus subtilis* effectively controlled *P. cactorum* infection in apple seedlings under pot cultures.

Foliar and Fruit Diseases

Apple Scab

Apple scab occurs worldwide and is the most important disease of apples. Its primary effect is reduction of the quality of fruits. It also reduces size and results in premature fruits drop, defoliation and poor fruit bud development for the next year, and it reduces the length of time infected fruit can be kept in storage. In

India the disease was first noticed in Kashmir in 1930 (Nath, 1935). This disease appeared in epidemic form in 1973 and caused large scale damage to apple crop in Kashmir. In Himachal Pradesh the disease was noticed in 1977 (Gupta, 1978) and spread to epidemic form in 1983 (Gupta, 1989). The disease destroyed most of the apple crop in the state. The disease was declared as one of the five main plant diseases of national importance of Government of India. The disease symptoms are observed on leaves, petioles, blossoms, fruits and pedicels. On emerging leaves velvety brown to olive green lesions appear first on lower surface and later on the both the surfaces. As the infected leaf ages, the tissue adjacent to the lesion thicken resulting in deformed leaves. The number of lesions may range from one to many and sometimes the entire surfaces become covered with scab and the term sheet scab is used to refer this type of symptoms. Infection of petioles and pedicels result in premature defoliation of leaves and fruits. On fruit similar lesions appear as the infected fruit enlarge the lesions become brown and corky. The disease is caused by a fungus *Venturia inaequalis* (Cke.) Wint. Both the imperfect and perfect stages of the fungus are encountered in nature. The perfect stage is produced in overwintered leaves or fruits on the orchard floor. The pseudothecia formed of diseased fallen leaves in the presence of moisture continue to mature with development of asci and ascospores. The mature ascospores are discharged over a period of 5 to 9 weeks which coincides with time between pink and full bloom stages of apple bud development. Once the primary infection has taken place, the secondary infection is through the conidia produced on the primary lesions. Adequate moisture is essential for the discharge of ripe ascospores from the pseudothecia. The germination of ascospores takes place under suitable temperature and moisture (leaf wetness) condition. The conditions of temperature and leaf wetness duration for scab infection are well defined on Mills period or Tables. On the basis of weather data and amount of primary inoculum scab prediction and warning has become possible and an advantage in the monitored apple scab control programme especially in Himachal Pradesh. For the management of the disease in India a protective fungicide spray programme is being adopted. The first spray of protective fungicide must start at silver tip stage. And this way, 6-7 sprays of various fungicides are needed during the growing season which causes pollution and health hazards. Hence the work on resistance breeding and biological control of apple scab was initiated. The work on biological control is primarily focused on the control of overwintering stage of the pathogen on leaf litter. Carisse *et al.* (2000) have studied the effect of antagonistic fungi *viz., Microsphearopsis* spp., *Diplodia* spp. and *Trichoderma* spp. on ascospore production under natural conditions either as foliar postharvest sprays or as a ground application at 90 per cent leaf fall. They found significant reduction in ascospore production. Antagonistic fungus *Microsphaeropsis ochraceae* has been considered as potent bio-sanitation agent against apple scab as this fungus kills the resting structures of *V. inaequalis* and thereby reduces disease in the subsequent crop reason by lowering the initial inoculums (Carisse *et al.*, 2007). In Canada use of biocontrol agents for the management of apple scab has become popular in organic apple production (Carisse and Dewdney, 2002). Use of antagonistic fungi to control overwintering stage of scab pathogen in fallen leaves is seen as an alternative to chemical control measures (Bengtson *et al.*, 2001). Yeasts isolated from apple leaves have also been evaluated against *V. inaequalis* for their antagonistic activity (Fiss *et al.* 2003). Three strains H 10, H 15 and H 25 were found to suppress scab severity on apple seedlings under greenhouse trials. Application of a concentration of 1.5 x 107 yeast cells per ml on 9 year old apple tree cv. Golden Delicious led to significant reduction in scab. Altinok and coworkers (2002) studied antagonistic effect of 30 different fungal isolates against apple scab on three cultivars of apple *viz.* granny Smith, Stark spur Golden and Starkrimson. Some isolates such as *Cryptococcus* spp. (white yeast), *Sporobolomyces* (pink yeast), *Alternaria* spp., *Epicoccum* spp. and *Popularia* spp. were found to produce volatile antibiotics and resulted 100 per cent inhibition of colony development of scab pathogen. Some biofungicides have also been evaluated against apple scab. Pleskatsevich and Berlinchick (2004) evaluated Fruitine, a biofungicide formulation of *Bacillus sublilis* strain BIMV 262 under integrated disease control system against scab and observed reduction in the disease. Other biofungicide formulations *viz.*, Dizofungin, Biostat and Narciss of fungal and bacterial species were found to induce resistance in apple plants against scab in Russia (Nadykta, 2004).

Apple Powdery Mildew

Powdery mildew of apple is a major foliar disease prevalent in all apple growing countries of the world. The disease was first noticed on apple seedlings in Iova, USA in 1871 (Bessey, 1877). It is caused by the biotrophic fungus *Podosphaera leucotricha* (Ell. and Ev.) Salm. It can be a significant commercial problem in every stage of apple development, from reducing the growth of nursery stock to causing fruit russetting (Jones and Aldwinckle, 1990). In apple the pathogen overwinters in dormant buds, and infected buds may fail to develop new shoots during the subsequent growing season. Conidia developing from overwintering mycelium serve as the primary source of inoculum, and secondary spread of the pathogen is incited by inoculum that develops from infection of young leaves, blossoms and young fruits. Perfect stage of the pathogen is also reported in nature (Bharat and Bhardwaj, 2000) but is not believed to have a major role in the disease cycle. For biological control of powdery mildew all agents that have been reported to provide biocontrol have been fungal in nature. As powdery mildews are biotrophs and typically do not require exogenous nutrients for germination and initial penetration, control through competition for nutrients is not a viable strategy. Likewise as exposure of the pathogen on the leaf surface after spore germination is limited, control through antibiosis is not likely to be a suitable mechanism for disease control. As such the greatest attention has been focused on the use of mycoparasites for the control of powdery mildews. These include *Ampelomyces quisqualis* (Novitskaya and Puzanova, 1992). The mechanism of biocontrol by the fungus has been established as hyper-parasitism as this fungus possess the ability to colonize the mycelium of powdery mildew and produce reproductive structures. This fungus is naturally occurring hyper-parasite of both sexual and asexual structures of powdery mildew pathogen. It parasitizes and form pycnidia with in powdery mildews hyphae, conidiophores and cleistothecia. The parasitized colonies of powdery mildew are dull in appearance, flattened off-white to gray in colour with reduced spore production (Falk *et al.*, 1995). Vaidya and Thakur (2005) have also isolated *A. quisqualis* from affected apple plants and other plant species belonging to rosaceae family. It showed the natural occurrence of this hyper-parasite in western Himalayan region for managing powdery mildew disease. Use of plant extracts along with hyper-parasite especially walnut extract has also been found to reduce the primary infection of apple foliar diseases including powdery mildew and (Meszka and Bielenin, 2006). The mycoparasite *A. quisqualis* isolate A-10 has been released as a commercial product AQ 10 TM for biological control of powdery mildew (Grove and Boal, 1997). Another formulation Ampelomitsin form *Ampelomyces* spp. has been observed achieve 70-80 percent control of powdery mildew of temperate fruits (Smol-Yokova *et al.*, 2004).

Postharvest Diseases

Amongst temperate fruits stone fruits are comparatively more perishable and prone to postharvest diseases. Postharvest diseases result in 10-50 per cent losses to temperate fruits. (Steppe, 1976). Many fungi and bacteria can cause postharvest spoilage and the major pathogens are species of *Alternaria, Aspergillus, Botyosphaeria, Botrytis, Colletotrichum, Monilinia, Mucor, Penicillium, Rhizopus and Trichothecium.* In most instances the postharvest pathogens gain entry to susceptible fruit tissues and cause infection via entry through fruits surface wounds that are generated through fruits surface wounds that are generated through the process of harvesting and postharvest handling. Some pathogens can also enter through natural opening like lenticels or decay may be initiated in the sinus between the calyx and core cavity (Spotts*et al*, 1988). Attention to fruits handling practices in the field and during storage to reduce mechanical and physical injuries and management of controlled atmospheric condition can be effective techniques in reducing postharvest diseases. However, these methods do not ensure effective protection of stored fruits. Hence postharvest disease control has been reliant on the use of fungicides (Eckert and Ogawa, 1988). But the use of fungicides is being avoided in modern era of organic farming due to perceived or actual hazards to human health as well as environment. As the consumption of fruits can be sources of direct ingestion of fungicides, the use of biological methodologies for control of postharvest diseases has enjoyed significant attention and even a modicum of success over the past two decades. The primary sources of biological agents for thecontrol of postharvest pathogens have been the microbial community resident to the phyllosphere and fruit surface (Janisiewicz, 1987). Bacteria and yeasts or yeast like organisms have been the most commonly employed agents for the biological control of postharvest diseases of temperate fruits (Sharma and Kaul, 1999). Several studies have identified bacterial agents for control of postharvest diseases of apple pear and peaches. A saprophytic strain of *Pseudomonas syringe* provided biological control of gray mould, blue mould and mucor rot on pear and also on apple (Janisiewicz and Marchi, 1992; Jeffers and Wright, 1994). The strain *P. syringe* ESC-11 is currently registered for postharvest application to apples and marketed as the product Bio-save 110. An isolate of *Burkholderi acepacia* provided biological control of blue mould and gray mould of Golden delicious apples (Jamisiewicz and Roitman 1988). *Bacillus subtilis* applied to wounded apple reduced fruit rot caused by *Botrytis cineria, Penicillium expansum* and *P. malicorticis* (Leibinger *et al.*, 1997). Biological control of postharvest brown rot of stone fruits has been reported by the application of *Bacillus subtilis,*

Epicoccum nigrum and *Pseudomonas* spp. (Pusey and Wilson, 1984; Smilarick *et al.*, 1993; Madrigal *et al.*, 1994; Foschi *et al.*, 1995). *B. subtilis* can also be incorporated into wax which is normally used for the control of brown rot of peaches (Pusey *et al.*, 1986). Successful biological control of postharvest diseases of fruits has been obtained through application of yeasts. In apple diversity of yeasts have been studies for biological controls of postharvest decays. Control of *Penicillium expansum, Botrytis cineria and Mucor spp.* has been reported using yeasts *Condida guilliermondii* (McLaughlin *et al.*, 1992), *C. Oleophila* (Mercier and Wilson, 1994), *Cryptococcus laurentii* (Roberts, 1990), *Kloeckeraapiculata* (McLaughlin *et al.*, 1992) and *Sporobolomycesroseus* (Janisiewicz *et al.*, 1994). The yeast *Candida oleophila* strain 182 has been commercialized as postharvest fungicide Aspire for conduct of *Botrytis* spp. and *Penicillium* spp. However, Biological control of postharvest diseases has been obtained through the use of individual isolates, the application of strain mixtures has been promoted the application of multiple agents, which in concert expand the range of biocontrol activity. Janisiewicz (1988) found that combination of *Pseudomonas* spp. and *Acremonium breve* gave complete control of *B. cineria* and *P. expansum* on apple. A co-application involving the bacterial antagonist *P. syringe* and the yeast *S. roseus* applied in equal biomass provided control of blue mould that was superior to that obtained by treatment with the individual agents applied separately using a biomass equivalent to that of the mixture (Janisiewicz and Bors, 1995). Mixture of two *Aureobasidium pullulans* strains and an isolate of *Rhodotorula glutinis* were superior to any of the strains applied individually in controlling decay caused by *B. cineria, P. expansum and P. malicorticis* (Leibinger *et al.*, 1997). Mechanisms reported to contribute to biological control of postharvest diseases include antibiosis competitive exclusion, induced resistance in the host and production of hydrolytic enzymes suppression of *B. cinerea* and *P. Expansum* by *Burkholderia cepacia* is reported to involve impart the production of antibiotic pyrrolnitrin (Janisiewicz *et al.*, 1991). *Bacillus subtilis* produced antifungal substance *i.e.* iturin peptides which possess a wide spectrum of antifungal activity. This substance was determined to be primary mode of action of this antagonistic bacterium to *Monilinia fructicola* causing brown rot of stone fruits (McKeen *et al.*, 1986). Utilization of carbon or nitrogen sources, nutrient competition and physical exclusion of pathogen are considered the possible mechanism in control of postharvest diseases of fruits by yeasts (Roberts, 1990). Droby and Chalutz (1994) suggested that application of yeasts to wounds on different fruit resulted in enhanced ethylene production. As ethylene is reported to have a role in induction of the resistance process, it was proposed that induction of the host resistance may operate in the disease suppression resulting from yeast application.

There has been increasing interest in determining the practicality of applying bio-control agents in the fields prior to harvest as a mean to control postharvest pathogens. This interest emanates from the fact that the infection of fruit by postharvest pathogens may occur in the field prior to harvest or as result of wound created during harvesting and handling but preceding the movement of fruit into dump tanks where bio-control agents have been applied. Biological control of postharvest diseases of fruit in a preharvest setting does appear in option on the basis of limited studies. Application of *Trichoderma harzianum* to apple in field significantly reduced symptoms of bulls eye rot in storage. Leibinger *et al.* (1997) applied antagonists to Golden Delicious apple in field to control postharvest disease and application of a combination containing two antagonists *Aurobasidium pullans* and *Rhodotorula glutinis* was as effective as chemical fungicides used against these diseases. These findings suggest that application of certain biocontrol agents that are a characteristic element of the resident micro flora of fruits may be useful in reducing infection in the field and subsequent infections during fruit storage.

REFERENCES

Agarwala, R. K. (1961). Problems of root rot of apple in Himachal Pradesh and Prospectus of its control with antibiotics. *Himachal Hortic.* 2, 171-178.

Agarwala, R. K. and Sharma, V. C. (1966). White root rot disease of apple in Himachal Pradesh. *Indian Phytopath.* 19, 82-86.

Ali A, Amina, F., Moncef, H. and Ali, B. (2004). Efficacy of non-pathogenic *Agrobacterium* strain K84 and K1026 against crown gall in Tunisia. *Phytopathol. Mediter.*43, 167-176.

Al-Momani, F., Saadoun, I. and Malkawi, H. I. (1999). *Streptomyces* spp. from Jordon soils with *in vitro* inhibitory activity against *Agrobacterium tumefaciens*Ab. 136 pti 854. *African Plant Production* 5, 129-130.

Altinok, H. H., Erkilik, A. and Canihos, Y. (2002). Antagonistic effect of volatile and non-volatile antibiotics produced by fungi isolated from apple phyllosphere on *Venturiaina equalis* (Cke). Wint. Bulletin-0ILB/SROP 25(10), 249-252.

Bengtson, M., Green, H., Hockenhull, J. and Pedersen, H. L. (2001). Microbiological control of cherry leaf spot and apple scab. DJE Rapport Markbrug 49, 33-38.

Bessy, C. W. (1877). On injurious fungi: The blight (*Erysiphe*): Iowa State College of Agriculture Bienn. Rep. pp 185-204.

Bharat, N. K. and Bhardwaj, L. N. (2001). Interactions between VA-mycorrhizal fungi and *Dematophoranecatrix*and their effect of health of apple seedlings. Indian J. Plant Pathol. 19, 47-51.

Bharat, N. K. and Bhardwaj, L. N.(2000). Occurrence of perfect stage of powdery mildew of apple in North Western Himalaya. *Plant Dis. Res.* 15, 81-82.

Bingye, X. and Shengrui, Y. (1998). Studies on replant problems of apple and peach. *Acta Hortic.* 477, 83-88.

Caesar, A. J and Burr, T. J. (1987). Growth promotion of apple seedling rootstocks by specific strains of rhizobacteria. *Phytopathology* 77, 1583-1588.

Carisse, O. and Dewdbney, M. (2002). A review of non-fungicidal approaches for the control of apple scab. *Phytoprotection* 83, 1-29.

Carisse, O., Holloway, G. and Leggett, M. (2007). Potential and limitations of *Microsphaeropsis ochraceae*, an agent for bio-sanitation of apple scab. Biological Control-a Global Perspective 2007, 234-240.

Carisse, O., Philion, V., Rolland, D. and Bernier, J. (2000). Effect of fall application of fungal antagonists on spring ascospore production of apple scab pathogen, *Venturia inaequalis. Phytopathology* 90, 31-37.

Catska, V. (1994). Interrelationship between vesicular-arbuscularmycorrhiza and rhizospheremicroflora in apple replant disease. *Biologia Plantarum* 36, 99-104.

Catska, V. and Hudska, G. (1990). The possibility of using biological preparations of *Agrobacterium radiobacter*for biological control of apple replant problem. *Replant Newsletter* 3, 3.

Catska, V. and Taube-Baab, H. (1994). Biological control of replant problems. *Acta Hortic.* 363, 115-119.

Droby, S. and Chalutz, E. (1994). Mode of action of biocontrol agents of post harvestdiseases.In: Biocontrol of postharvest diseases Theory and Practice, Wilson, C.L.Q. and Wisniewski, M. E. (eds.). Boca Raton, F L: CRC Press, pp. 365-389.

Eckert, J. W. and Ogawa, J. M. (1988). The chemical control of postharvest diseases: deciduous fruit, berries, vegetables and root/tuber crops. *Annu. Rev. Phytopathol.* 26, 433-469.

Falk, S. P., Gadoury, D. M. and Pearson, R. C. (1995). Partial control of grape powdery mildew by the mycoparasite *Ampelomyces quisqualis. Plant Dis.* 79, 483-490.

Fiss, M., Barckhausen, O., Gherbawy, Y., Kollar, A., Hamamoto, M. and Auling, G. (2003). Characterization of apiphytotic yeasts of apple as potential biocontrol agents against apple scabs.(*Venturiainaequalis*).Zeitsch.Pflanz.Krankh. Pflanzensch.110(6), 513-523.

Foschi, S., Roberti, R., Bremelli, A. and Flori, P. (1995). Application of antagonistic fungi against *Monilinialexa* agent of fruit rot of peach. Bull OILB/SROP. 18, 79-82.

Freeman, S., Sztejnberg, A. and Chet, I. (1986). Evaluation of *Trichoderma* as a biocontrol agent for *Rosellinia necatrix. Plant Soil* 94, 163-170.

Grove, G. G. and Boal, R.J. (1997). Apple powdery control trials using the mycroparasite. *Ampelomyces quisqualis* (AQ 10).F and N Tests 52, 7.

Gupta, A. K. and Khosla, K. (2007). Integration of soil solarization and potential native antagonist for the management of crown gall or cherry root stock colt. *Scientia Hortic.* 112, 51-57.

Gupta, G. K. (1978). Present status of apple scab (*Venturia inaequalis*). in Himachal Pradesh and strategy for its control. *Pesticides* 12, 13-14.

Gupta, G. K. (1989). Apple scab-economics and Technology. *Pesticides* 23, 33-39.

Gupta, V. K. and Jindal, K. K. (1989). Management of white root rot of apple (*Rosellinianecatrix*). by biological and cultural methods. *Proc. Int. Hortic.Cong.* Italy 2, 3229.

Gupta, V. K. and Utkhede, R. S. (1986). Factors affecting the production of antifungal compounds by *Enterobacter aerogenes* and *Bacillus subtilis*, antagonists of *Phytophthora cactorum*. *J. Phytopathol.* 117, 9-16.

Ieki, H., Kubomura, Y. and Koi, S. (1969). Detection and vertical distribution of white root rot fungus in forest soils. *Ann. Phytopathol. Soc.* Japan 35, 76-81.

Ito, S. I. and Nakamura, N. (1984). An outbreak of white root rot and its environmental conditions in the experimental arboretum. *J. Japan For. Sci.* 66, 262-267.

Jaffee, B. A., Abawi, G. S. and Mai, W. F. (1982). Role of soil microflora and *Pratylenchuspenetrans*in the apple replant disease. *Phytopathology* 72, 247-251.

Jamisiewicz, W. J. and Covey, R. P. (1983). Biological control of collar rot caused by *Phytophthoracactorum*. *Phytopathology* 73, 822.

Janisiewicz, W. J.(1987).Postharvest biological control of blue mold on apples. *Phytopathology* 77, 481-485.

Janisiewicz, W. J. (1988). Biocontrol of postharvest disease of apples with antagonist mixtures. *Phytopathology* 78, 194-198.

Janisiewicz, W. J. and Bors, B. (1995). Development of a microbial community of bacterial and yeast antagonists to control wound invading postharvest pathogens of fruit. *Appl. Environ. Microbiol.*61, 3261-3267.

Janisiewicz, W. J. and Marchi, A. (1992). Control of storage rots on various pear cultivars with a saprophytic strain of *Pseudomonas syringe*. *Plant Dis.* 76, 55-560.

Janisiewicz, W. J. and Roitman, J. (1988). Biological control of blue mold and gray mold on apple and pear with *Pseudomonas cepacia*. *Phytopathology* 78, 1697-1700.

Janisiewicz, W. J., Peterson, D. L. and Bors, R. (1994). Control of storage decay of apples with *Sporobolomycesroseus*. *Plant Dis.* 78, 466-470.

Janisiewicz, W. J., Yourman, L., Roitman, J. and Mahoney, N. (1991). Postharvest control of blue mold and gray mold of apple and pears by dip-treatment with pyrrolnitrin, a metabolite of *Pseudomonas cepacia*. *Plant Dis.* 75, 490-494.

Janisiewicz, W. Z. and Covey, R. B. (1983b). Biological control of *Phytophthora cactorum* *Z. Pflanz.Pflanzenschutz*. 90, 140-145.

Jeffers, S. N. and Wright, T. S. (1994). Comparison of four promising biological control agents for managing postharvest diseases of apples and pears. *Phytopathology* 84, 1082.

Johnson, K. B. and Dileone, J. A. (1999). Effect of antibiosis on antagonist dose plant disease response relationships for biological control of tomato and cherry. *Phytopathology* **89,** 974-980.

Jones, A. L. and Aldwinckle, H. S. (1990). Compendium of apple and pear diseases. APS Press. St. Paul, Minnesota 101pp.

Jones, D. A. and Kerr, A. (1989). *Agrobacterium radiobacter*strain K1026, a genetically engineered derivative of strain K-84, for biological control of crown gall. *Plant Dis* 73,15-18.

Kerr, A. (1980). Biological control of crown gall through production of agrocin 84. *Plant Dis.* 64, 25-30.

Kumar, B. (2002). Studies on collar rot of apple. Ph.D. Thesis, Dr. Y.S. Parmar UHF, Nauni, Solan H.P., India.

Leibinger, W., Breuker, B., Hahn, M. and Mendgen, K. (1997). Control of postharvest pathogens and colonization of the apple surface by the antagonistic microorganisms in the field. *Phytopathology* 87, 1103-1110.

Levesque, C. A., Holley, J. D. and Utkhede, R. S. (1993). Individual and combined effect of *Eaterobacteraerogenes*andmetalaxyl on apple tree growth and Phytophthora crown and root rot development. *Soil Biol. Biochem.* 25, 975-979.

Madrigal, C., Pascual, S. and Melgarejo, P. (1994). Biological control of peach twig blight (*Monilinialaxa*). with *Epicoccumnigrum*. *Plant Pathol.*43, 554-561.

Marchi, A. and Utkhede, R. S. (1994). Effect of *Enterobacter aerogenes*or rhizosphere microflora of apple trees. *J. Phytopathol.* 141, 127-132.

Mazzola, M. (1998).Elucidation of the microbial complex having a causal role in the development of apple replant disease in Washington. *Phytopathology.* **88**, 930-938.

Mazzola, M., Granastein, D. M. Elfving, D. C., Mullinix, K. and Gu, Y. H. (2002). Cultural management of microbial community structure to enhance growth of apple in replant soils. *Phytopathology* 92, 1363-1366.

McKeen, C. D., Reilly, C. L. and Pusey, P. L. (1986). Production and partial characterization of antifungal substances antagonistic to *Monilinia fructicola* from *Bacillus subtilis*. *Phytopathology* 76, 136-139.

McLaughlin, R. J., Wilson, C. L., Droby, S., Ben-Arie, R. and Chalutz, E. (1992). Biological control of postharvest diseases of grape, peach and apple with the yeasts *Kloeckerae piculata* and *Candida guilli4er mondii*. *Plant Dis.* 76, 470-473.

Mercier, J. and Wilson, C. L. (1994). Colonization of apple wounds by naturally occurring and introduced *Candida oleophila* and their effect on infection by *Botrytis cineria* during storage. *Biol. Control* 4, 138-144.

Meszka, B. and Biclenin, A. (2006). Non-chemical possibilities for control of apple fungal diseases. *Phytopathologia Polonica* 39, 63-70.

Moore, L. W., and Warren, G. (1979). *Agrobacterium radiobacter*strain 84 and biological control of crown gall.*Annu. Rev. Phytopathol* 17, 163-179.

Mosse, B. (1973). Advances in the study of vesicular- arbuscularmycorrhiza. *Annu. Rev. Phytopathol.* 11, 171-192.

Nadykta, V. D. (2004). Prospects of biological plant protection from phytopathogenic organisms. *Zashchita-i-Karantin Rastenii* **6**, 26-28.

Nath, P. (1935). Studies in the disease of apple in Northern India. II. A short note on apple scab due to *Fusicladium dendriticum* Fuckel. *J. Indian Bot. Soc.* 14, 121-124.

Nemec, S. (1980). Effect of fungicides on endomycorrhizal development in sour orange. *Can. J. Bot.* 58, 522-526.

Novitskaya, L. N. and Puzanova, L. A. (1992). Biological protection of fruit nursery from powdery mildew. Zashch Rust. (Mosc.).6, 25.

Orlikowski L B and Schmidle A.(1985).On the biological control of *Phytophthora cactorum* with *Trichoderma viride. Deutsch. Pflanzensch.* 37, 78-79.

Pleskatsevich, R. I. and Berlinchik, E. E. (2004). Evaluation of the efficiency of local paste like biofungicidefruitine against apple scab. *Zashchita Rastenii* 28, 132-138.

Pusey, P. L., Wilson, C. L., Hotchkiss, M. W. and Franklin, J. D. (1986). Compatibility of *Bacillus subtilis*for postharvest control of peach brown rot with commercial fruit waxes, dichloran and cold storage conditions. *Plant Dis.* 70, 587-590.

Pussey, P. L. and Wilson, C. L. (1984). Postharvest biological control of stone fruit brown rot by *Bacillus subtilis. Plant Disease* 68, 753-756.

Roberts, R. G. (1990). Postharvest biological control of gray mold of apple by *Cryptococcus laurentii. Phytopathology.* 80, 526-530.

Roiger, D. J. and Jeffer, S. N. (1991). Evaluation of *Trichoderma* spp. for biological control of *Phytophthora*crown rot of apple seedlings. *Phytopathology* 81, 910-917.

Sharma, M. (2000). Non chemical methods for the management of white root rot of apple, Ph.D. Thesis, Dr. Y.S. Parmar UHF, Nauni, Solan (HP), India.

Sharma, R. C. and Kaul, J. L. (1999). Postharvest diseases of temperature fruits and their management.In: Diseases of Horticultural Crops Fruits, Verma, L. R. and Sharma, R. C. (eds.). Indus Pub.Co.New Delhi, 582-623pp.

Sharma, S. K. (1993). Studies on management of white root rot of apple. Ph.D. Thesis, Dr. Y.S. Parmar UHF, Nauni, Solan (HP), India, 146p.

Smilarick, J. L., Deris-Arrue, R., Bosch, J. R., Gonzalez, A. R., Herson, D. and Janisiewicz, W. J. (1993.). Control of postharvest brown rot of nectarines and peaches by *Pseudomonas* Species. *Crop Prot.* 12, 313-320.

Smol-Yakova, V. M., Podogornaya, M. E., Puzanova, L. A., Cherkezova, S. R. and Yakuba, G. V. (2004). Ecologization of production of stone fruits from diseases. Sadovodstovo-i-Vinogradorstvo 5, 5-8.

Spotts, R. A., Holmes, J. R. and Washington, W. S. (1988). Factors effecting wet core rot of apples. *Aus. Plant Pathol.* 14, 53-57.

Steppe, H. M. (1976). Postharvest losses of agricultural products.Rep W.P./225/76 Serial No. 240 UNDP, Tehran, Iran.

Sun, Y., Wang, H., Wang, J. and Song, X. (2000). Pathogenic bacteria of apple crown gall and their biological control. *ActaPhytopathol.Sinica* 30, 332-336.

Sztejnberg, A., Freeman, S., Chet, I. and Katan, J. (1987). Control of *Roselliniannecatrix*in soil and in apple orchard by solarization and *Trichoderma harzianum. Plant Dis.* 71, 365-369.

Utkhede R S and Smith E M. (1991a). Biological and chemical treatment for control of *Phytophthora*crown and root rot caused by *Phytophthora cactorum* in high density apple orchard. *Can. J. Plant Pathol.*13: 267-270.

Utkhede, R S. and Smith, E. M. (1994). Development of biological control of apple replant disease. *Acta Hortic.* 363, 129-133.

Utkhede, R. S. (1984). Effect of bacterial antagonists on *Phytophthora cactorum* and apple crown rot. *Phytopathol. Z.* 109, 169-176.

Utkhede, R. S. (1987). Chemical and biological control of crown and root rot of apple caused by *Phytophthora cactorum. Can. J. Bot.* 61, 3343-3348.

Utkhede, R. S. (1996). Replant disease and soil sickness. In: Management of soil-borne diseases (Eds. Utkhede RS and Gupta VK), Kalyani Publishers Ludhiana pp. 21-39.

Utkhede, R. S. and Li, T. S. C. (1989b). Evaluation of *Bacillus subtilis* for potential control of apple replant disease. *J. Phytopathol.* 126, 305-312.

Utkhede, R. S. and Li, T.S.C. (1989). Chemical and biological treatments for control of apple replant disease in British Columbia. *Can. J. Plant Pathol.*11: 143-147.

Utkhede, R. S. and Smith, E. M. (1991b). *Phytophthora* and *Pythium* spp. associated with root rot of young apple trees and their control. *Soil Biol. Biochem.* 23, 1059-1063.

Utkhede, R. S. and Smith, E. M. (1993). Long term effect of chemicals and biological treatment on crown rot and root rot of apple trees caused by *Phytophthora cactorum. Soil Biol. Biochem.* 25, 383-386.

Utkhede, R. S. and Smith, E. M. (1997).Effectiveness of dry formulation of *Enterobacteraerogenes*for control of crown and root rot of apple trees. *Can. J. Plant Pathol.* 19, 397-401.

Utkhede, R. S. and Smith, E. M. (2000). Impact of chemical, biological and cultural treatments on growth and yield of apple in replant disease soil. *Aus. Plant Path.* 29, 129-136.

Utkhede,R.S., Li,T.S.C. and Smith, E.M. (1992). The effect of *Glomus mosseae* and *Enterobacter aerogenes* on apple seedlings grown in apple replant disease soil. *J. Phytopathol.* 135, 281-288.

Vaidya, S. and Thakur, V. S. (2005). *Ampelomyces quisqualis* Ces.-a mycoparasite of apple powdery mildew in western Himalayas. *Indian Phytopath.* 58, 250-251.

Wazir, F. K., Meladul, K., Qureshi, J. A., Barech, A. R. and Kakar, K. M. (2000). Effect of physical and biological control measures soil. *Sarhad Journal Agriculture* 16: 49-51.

Zambryski, P. C. (1992). Chronicles from the *Agrobacterium* plant cell DNA transfer strong. *Annu. Rev. Plant Mol. Biol.* 43, 465-490.

Transformation of Indian Agriculture through Innovative Technologies *Pages* **123–150**
Editor: **Dr. Shahid Ahamad & Dr. Jag Paul Sharma**
Published by: **ASTRAL INTERNATIONAL PVT. LTD., NEW DELHI**

10 Role of Nanotechnology in Plant Disease Management

Mudasir Bhat, Ali Anwar, V.K. Ambardar, Arif Hussain and Ghulam Hassan

Introduction

Nanotechnology, the brain child of Richard Feynman is the frontier science of materials having unique properties than their macroscopic or bulk counterparts. The ability of nano-materials to work at the molecular level, atom by atom, to create large structures with fundamentally new molecular organization is the key to its applications. It has a significant effect in the food industry - development of new functional materials, product development and design of methods and instrumentation for food safety and bio-security (Moraru *et al.*, 2003). According to National Science Foundation (NSF) and National Nanotechnology Initiative (NNI), nanotechnology is defined as the ability to understand, control and manipulate matter at the level of individual atoms and molecules, as well as at the supramolecular level involving clusters of molecules (in the range of 0.1 to 100 nm), in order to create materials, devices and systems with fundamentally new properties and functions because of their small structure. Nanotechnology is the creation and utilization of materials, device system, through the control of the properties and structuresthe matter at the nanometric scale (Ajayan *et al.*, 2003; Astruc *et al.*, 2010). Nanotechnology is an interdisciplinary science encompassing areas like physics, chemistry, biology, material science and engineering (Roco, 2007). Nanotechnology, focusing on special properties of materials emerging from nanometric size has the potential to revolutionize the agricultural and food sectors, biomedicine, environmental engineering, water resources, energy conversion and numerous other areas. The application of nanotechnology to the agriculture and food industry was first addressed by a United States Department of Agriculture in 2003. There is application of nanotechnology in disease control, slow release of pesticides and developing diagnostic tools and development of functional food systems, to produce interactive, edible nano wrappers to keep the pathogens away, targeted release of chemicals, packaging, extensive nano surveillance, interactive agrochemicals as herbicides and pesticides. Hence, nanotechnology can be used for combating the plant diseases either by controlled delivery of functional molecules or as diagnostic tool for disease detection (Tarafdar and Raliya, 2012). The new nanotechnology with materials having unique properties than their macroscopic or bulk counter parts, has promised applications in various fields. The essence of nanotechnology is the ability to work at the molecular level, atom by atom, to create large structures with fundamentally new molecular organization. The aim is to exploit these properties by gaining control of structures and devices at atomic, molecular and supra-molecular levels and to learn to efficiently manufacture and use these devices. Nanotechnology has provided new solutions to problems in plants and food science (post-harvest products) and offers new approaches to the rational selection of raw materials, or the processing of such materials to enhance the quality of plant products. The heart of nanotechnology lies in the ability to compress the tools and devices to the nanometre range and to accumulate atoms and molecules into bulkier structures while the size remains very small.

Properties of Nanoparticles

Nanoparticles show properties sharp contrast to their bulk in many respects which is utilised for its use in nanotechnology *e.g.*

- Small size (1-100nm)
- Large surface to volume ratio
- Chemically alterable physical properties
- Change in the chemical and physical properties with respect to size and shape
- Structural sturdiness in spite of atomic granularity
- Enhanced or delayed particle aggregation depending on the type of the surface modification, enhanced photoemission, high electrical and heat conductivity and improved surface catalytic activity (Roco, 2007).

These nanostructures may be zero dimensional (nanoparticles), one dimensional (nanowires), two dimensional (thin films) or three dimensional (arrays, hierarchical structures). A comparative dimension of different nano and micro structures in biology are given in Figure 10.1

Techniques for Preparation of Nanoparticles

Nanomaterials are abundant in nature as living organisms operate basically at a nanoscale level. Nanotechnologists seek to produce and utilize both novel nanomaterials and some natural nanomaterials in larger quantities and within a more consistent size range. Numerous techniques are used to fabricate different nanomaterials. Nanostructures can be obtained through two different approaches

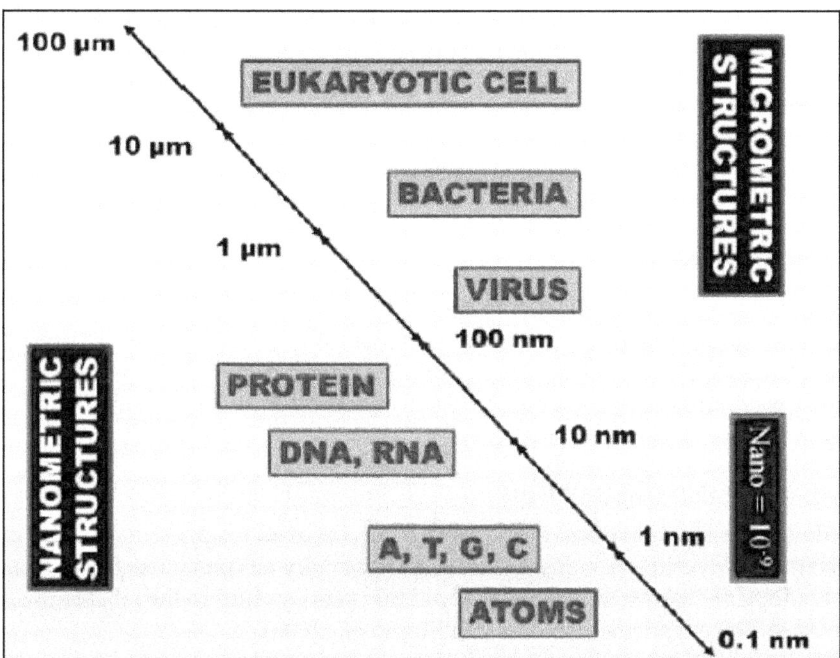

Figure 10.1. Scale Showing the Dimensions of different Nanometric and Micrometric Objects in Biological Materials (A, T, G, C are nucleotides molecules).

which have been termed as top down process and bottom up process. The top down approach usually involves breaking down of big chunks of material (physically or chemically) into smaller objects of desired shapes and sizes through mechanical milling, ion implantation, *etc.* The bottom up approach uses self-assembly to build up the nanostructures by bringing in individual atoms and molecules together.

Various methods involved in preparation of nanoparticles are:

a) Solvent Extraction/Evaporation

Nanoparticles of some organic polymers can be fabricated by solution in a

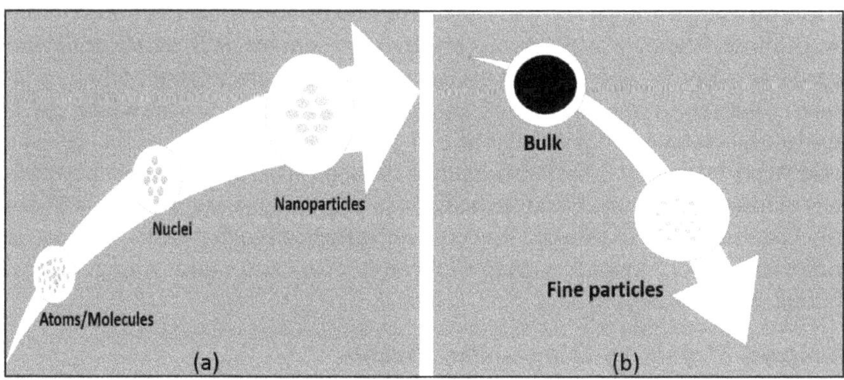

Figure 10.2

solvent such as dichloromethane followed by sonication, evaporation, filtration and freeze-drying (Zhang and Feng, 2006).

b) Crystallization

Hydroxyapatite-aspartic acid (or glutamic acid) crystals were synthesized in the presence of solutions containing different amounts of the amino acids (Boanini *et al.*, 2006).

c) Self-assembly

Manipulation of physical and chemical conditions such as pH, temperature and solute concentrations can induce self-assembly of molecules to form fibrous nanostructures (Boanini *et al.*, 2006). Vesicles, called polymerosomes, that may be useful for encapsulation, can also be self-assembled by slow evaporation of an organic solvent (Lorenceau *et al.*, 2005).

d) Layer-by-layer Deposition

Platforms for bilayer membranes that can be used for protein analysis can be fabricated by layering of sodium silicate and poly ally amine hydrochloride on gold followedby calcination in a furnace. Lipid bilayers can fuse to the silicate layer and be used to detect specific proteins (Phillips *et al.*, 2006).

e) Microbial Synthesis

Living cells have been harnessed to produce nanoparticles, for example, silver nanoparticles produced extra cellularly by the fungus *Aspergillus fumigatus* (Bhainsa and D'Souza, 2006). Gold and silver nanoparticles can also be produced by other fungi and a number of bacterial species (Bhattacharya and Gupta, 2005).

Relevance of Nanotechnology with Plant Pathology

Detection and Diagnosis of Plant Pathogens

Early detection of plant diseases has prompted nanotechnologists to look for a nano solution for protecting the food and agriculture from bacteria, fungi and viral agents through autonomous nanosensors linked into a GPS system for real-time monitoring throughout the field to monitor soil conditions and crop. The union of biotechnology and nanotechnology in sensors will create equipment of increased sensitivity, allowing an earlier response to environmental changes and diseases. There is an urgent need of ultrasensitive diagnostic tool which can detect the molecular defects, be it at genomic or biochemical level, rapidly. Bio-systems are endowed with functional nanometric devices such as enzymes, proteins and nucleic acids, which detect vital processes in plants. Disease diagnosis is difficult mainly because of the extremely low concentration of biochemicals and also due to the presence of very low amount of detectable virus and many fungal or bacterial infections (Misra *et al.*, 2013).

Nanosized Metals as Diagnostic Probes

As the techniques have not been perfected yet and things are in budding stage in terms of plant pathogen detection, researchers are looking forward to harness the

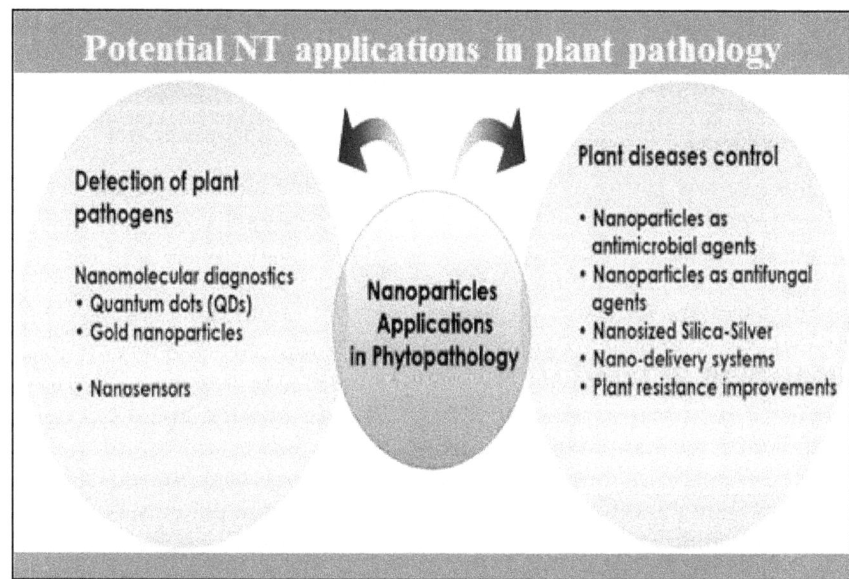

Figure 10.3

benefits of nanomaterials by addressing all possible drawbacks as found in currently available diagnostic tools. Nanoparticles are different from their bulk counterparts, which, when reduced to nanosize (1-100 nm) achieve certain properties which make them suitable for development as diagnosticprobes (Sharon *et al.*, 2010). Fluorescent silica nanoprobes have potential for rapiddiagnosis of plant diseases. Fluorescent silica nanoprobesconjugated with the secondary antibody of goat anti-rabbitIgG (Yao *et al.*, 2009) was used for detection of a bacterialplant pathogen *Xanthomonasa xonopodis* pv. *Vesicatoria* (bacterial spot on solanaceous plants). An organic dye tris-2, 2′-bipyridyldichlororuthenium (II) hexahydrate (Rubpy)was incorporated into the core of circular silicananoparticles with average diameter of 50 ± 4.2 nm. Thussilica nanoparticles became fluorescent which wasphotostable.

Nanoscale Biosensor/Nanosensors

The nanosensors which would be small and portable would provide rapid response and real-time processing with accurate, quantitative, reliable and stable results. Detection of infection in nonsymptomatic plant followed by targeted delivery of treatment would be an essential component for precision farming. Use of micromechanical cantilever arrays for detection of fungal spore (*Aspergillus niger* and *Saccharomyces cerevisiae*) was demonstrated by Nugaeva *et al.* (2005). Proteins like concanavalin A, fibronectin or immunoglobulin G were surface grafted on micro-fabricated uncoated as well as gold-coated silicon cantilevers. These proteins were found to have different affinities to bind to the molecular structures present on fungal cell surface. Spore immobilization and germination of the test fungi led to shift in resonance frequency which was measured by dynamically operated cantilever arrays. This took a few hours in contrast to several days in conventional techniques. The finding that shift was proportional to the mass of single fungal spore can be used for quantitative estimation. The biosensors detected the target fungi in the range of 10^3-10^6cfu ml^{-1} in the investigation made by Nugaeva *et al.* (2005).

Quantum Dots

"QDs are few nm in diameter, roughly spherical, fluorescent, crystalline particlesof semiconductors whose excitons are confined in all thethree spatial dimensions". QDs have emerged as importanttool for detection of a specific biological marker in medicalfield with extreme accuracy. They have been used in celllabelling, cell tracking, in vivo imaging and DNA detection(Sharon *et al.*, 2010).

Management of Plant Diseases

Some of the nano particles that have entered into the arena of controlling plant diseases are nanoforms of carbon, silver, silica and alumino-silicates. Recently scientists have reported that when they planted tomato seeds in a soil that contained carbon nanotubes; these CNTs could not only penetrate into the hard coat of germinating tomato seeds but also exerted growth enhancing effect. They envisaged that the enhanced growth was due to increased water uptake caused by penetration of CNT. This could be a boon for using CNT as vehicle to deliver desired molecules into the seeds during germination that can protect them from the diseases. Since it is growth promoting, it will not have any toxic or inhibiting or adverse effect on the plant. A wide variety of bacterial and fungal pathogens spoil vegetables. Among them the most common bacterial agents are *Erwinia carotovora, Pseudomonas spp., Corynebacterium* and *Xanthomonascampestris* which attack most vegetables. Fungal pathogens causing spoilage of vegetables are species belonging to genera *Alternaria, Aspergillus, Cladosporium, Colletotrichum, Phomopsis, Fusarium, Penicillium, Phoma, Phytophthora, Pythium, Rhizopusspp., Botrytis cinerea, Ceratocystisfimbriata, Rhizoctonia solani, Sclerotiniasclerotiorum* and some mildews. Some of these organisms are host specific whereas others affect a wide variety of vegetables causing huge economic losses. Some pathogens produce toxic metabolites and adversely affect human health. Many of these agents enter the plant tissue over mechanical or chilling injuries and cause overwhelming losses (Tournas, 2005). With the estimated doubling in global food demand in next 50 years huge challenges have been posed in food production. In the year 2000, the pesticide production was about three million tons of active ingredients worldwide (Tilman *et al.*, 2002). It is reported that very small amount (less than 0.1 per cent) of pesticide reaches the site of action, due to loss of pesticide in air during application and as run-off, spray drift, off-target deposition and photo degradation affecting both the environment and application costs (Pimentel, 1995; Castro *et al.*, 2013). With the growing demand of pesticides worldwide to control the pathogens and pests, there is an urgent need to tackle the excessive usage of pesticides and fertilizers by finding alternatives. Table 1 shows some of the nanomaterials used for control of plant pathogens. Potential applications of nanotechnology in crop protection include controlled release of encapsulated pesticide, fertilizer and other agrochemicals in protection against pests and pathogens, early detection of plant disease and pollutants including pesticide residues by using nanosensors (Ghormade *et al.*, 2011). The potential applications of nano-materials in crop protection help in the development of efficient and potential approaches for the management of plant pathogens.

Nanoparticles in Disease Management

Various nanoparticles employed in plant disease management are:

a) Biopolymer Nanoparticles

Development of nano-formulation for field application of agrochemicals requires the use of readily biodegradable, nontoxic, environment friendly, safe and low-cost materials. So, use of biopolymers produced by natural sources with good physical and chemical properties is a fascinating approach to prevent the use of petrochemical and toxic chemical substances in production of nanomaterial.

b) Chitosan

Chitosan nanoparticles have got various applications in biology due to its biodegradable and nontoxic properties. In acidic condition the free amino group of chitosan protonates and contributes to its positive charge (Phaechamud and Ritthidej, 2008). The inhibition mode of chitosan against fungi is defined by the following three mechanisms;

i) The positive charge of chitosan interacts with negatively charged phospholipid components of fungi membrane, which in turn alter cell permeability of plasma membrane and causes the leakage of cellular contents, which consequently leads to death of the cell (García-Rincón *et al.*, 2010).

ii) Chitosan chelates with metal ions, which has been implicated as a possible mode of antimicrobial action (Rabea *et al.*, 2003). On binding to trace elements, it interrupts normal growth of fungi by making the essential nutrients unavailable for its development (Roller and Covill, 1999).

iii) It is suggested that chitosan could penetrate fungal cell wall and bind to its DNA and inhibit the synthesis of mRNA and in turn, affect the production of essential proteins and enzymes (Sudarshan *et al.*, 1992; Kong *et al.*, 2010). In search of natural antimicrobials to avoid harmful synthetic chemicals, chitosan and chitosan nanoparticles are found to be more effective against plant pathogens like *Fusarium solani*. Inhibitory effect was also influenced by particle size and zeta potential of chitosan nanoparticles. The chitosan therefore could be formulated and applied as a natural antifungal agent in nanoparticle form to enhance its antifungal activity (Ing *et al.*, 2012). Chen *et al.* (2010) studied antibacterial activities of low molecular weight chitosan products with average nanoparticle sizes of 117 to 965 nm. They emphasised that antimicrobial activity of chitosan nanoparticles depends on its zeta potential, which plays a significant role in binding with negatively charged microbial membrane. Antimicrobial activity of chitosan, chitosan derivatives, bound metal ions and nanoparticles are well studied (Sanpui *et al.*, 2008; Jagadish *et al.*, 2012; Kaur *et al.*, 2012). In our recent study, we have used chitosan-silvernanoparticle (chitosan

-Ag NP) composite for inhibition of conidial germination in *Colletotrichum gloeosporioides*. We observed complete inhibition of spores at 100µg/ml concentration of the composite, but chitosan alone did not show significant inhibition at the same concentration. We also found the shrinking of spores and othermorphological changes in the spores treated with CS-Agnanocomposite (Figure 10.4) (Chowdappa *et al.*, unpublished data).

c) Metallic Nanoparticles

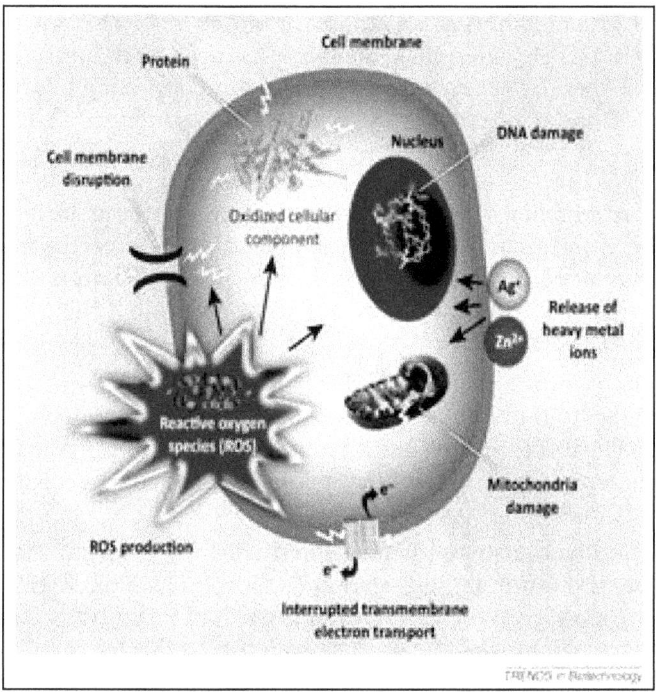

Mode of Action of Nanoparticles.

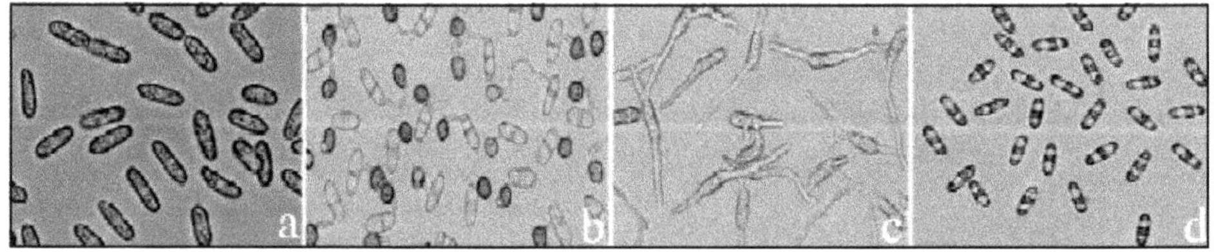

Figure 10.4. Effect of Chitosan-AgNP Composite on Conidial Germination of *Colletotrichum gloeosporioides*.

a) Normal conidia; b) Water control with 0.1 per cent (v/v) acetic acid (appressoria formation); c) Conidial germination in chitosan (100µg/ml) with 0.1 per cent (v/v) acetic acid; d) Complete inhibition of conidial germination by chitosan-AgNP composite at concentration of 100 µg/ml.

Metallic nanoparticles possess unique chemical and physical properties, small size, huge surface to volume ratio, structural stability and strong affinity to their targets (Kumar *et al.*, 2010). Metal nanoparticles can be used as new antimicrobial agents and an alternative to synthetic fungicides to delay or inhibit the growth of many pathogens because of its multiple mode of inhibition.

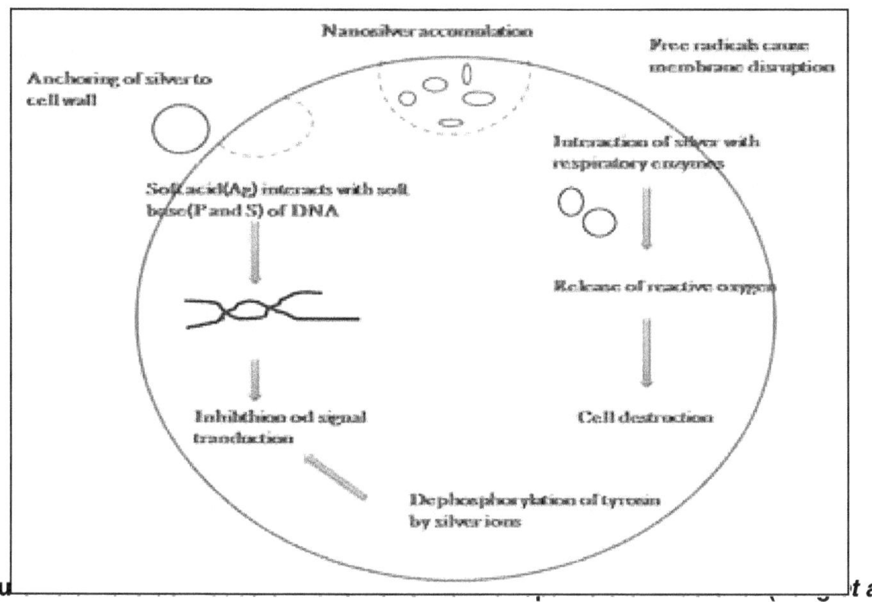

Figu ... **t al., 2008).**

d) Carbon Nanoparticles

Recently scientists have reported that when they planted tomato seeds in a soil that contained carbon nanotubes; these CNTs could not only penetrate into the hard coat of germinating tomato seeds but also exerted growth enhancing effect. They envisaged that the enhanced growth was due to increased water uptake caused by penetration of CNT. This could be a boon for using CNTs to deliver desired molecules into the seeds during germination that can protect them from the diseases. Since it is growth promoting, it will not have any toxic or adverse effect on the plant. It is possible that at high CNM concentrations, water uptake as well as spore development could be impeded due to the increased blockage of water channels imposed by surface-adsorbed CNMs. Therefore, the water channel blockage decreased the water content of spores during incubation with CNMs, which could be one factor leading to plasmolysis. Another explanation could be that CNMs were able to regulate the gating of existent water channels (aquaporins) of spores and modify related biological pathways before impacting spore development. It is confirmed that the abnormal expression of several water channel genes, including the important water-channel LeAqp2 gene in tomato plants, was induced by MWCNTs However, the spore-specific water channel aquaporins (Aqy1) were shown to be produced during the later stages of sporulation rather than the subsequent maintenance or germination, suggesting that the CNMs cannot regulate the expression of aquaporins. Therefore, water channel blockage imposed by surfaceadsorbed CNMs of spores can be speculated to just inhibit the water uptake inside spores, which could be one main factor for plasmolysis and the inhibition of sporegermination. Figure 10.6 showed that the effect of CNMs on the spore germination of *Fusarium poae* was dose-dependent. 93.55 per cent of the spores inthe control conditions possessed germ tubes after germination, while the germination of spores was inhibited by >90.8 per cent with the highest dose of SWCNTs tested (500 lgmL_1) (Figure 10.6A). Meanwhile, the germination of spores was reduced by 84.4 per cent, 82.1 per cent, and 32 per cent with the highest dose of MWCNTs, GO and rGO, respectively (Figure 10.6B–D). However, at the highest concentration, C60 and AC showed no significant difference from the control in their effects on the spore germination of *F. poae* (Figures 10.6E and F).

e) Nano Alumino-Silicate

Leading chemical companies are now formulating efficient pesticides at nano scale. One of such effort is use of Alumino-Silicate nanotubes with active ingredients. The advantage is that Alumino-Silicate nanotubes sprayed on plant surfaces are easily picked up in insect hairs. Insects actively groom and consume pesticide-filled nanotubes. They are biologically more active and relatively more environmentally-safe pesticides. Mesoporous Silica Nanoparticles (Wang *et al.*, 2002) have shown that these particles can deliver DNA and chemicals into plants thus, creating a powerful new tool for targeted delivery into plant cells. Lin's research group has developed porous, silica nanoparticle systems that are spherical in shape and the particles have arrays of independent porous channels. The channels form a honeycomb-like structure that can be filled with chemicals or molecules. These nanoparticles have a unique "capping" strategy that seals the chemical inside. They have also demonstrated that the caps can be chemically activated to pop open and release the cargo inside the cells where it is delivered. This unique feature provides total control for timing the delivery. Plant cells have rigid cell wall. Hence to penetrate it they had to modify the surface of the particle with a chemical coating. It has been successfully used to introduce DNA and chemicals into arabidopsis, tobacco and corn plants. The other advantage is that with the mesoporous nanoparticles, one can deliver two biogenic species at the same time.

f) Silver Nanoparticles

Due to emerging plant diseases, agricultural production is reduced worldwide. Every year millions of dollars have been spent to control plant diseases. Various natural and artificial control measures have been used for plant protection. Use of pesticides is the most prevalent method for disease control at present. Scientists are searching for alternative measures against pesticide application, due to its environmental hazards and residual problem. As an alternative to chemical pesticides, use of silver nanoparticles as antimicrobial agents has become more common (Jo *et al.*, 2009; Kim *et al.*, 2012). Silver has been used as an antimicrobial agent since ancient civilizations; it has been used extensively due to its broad

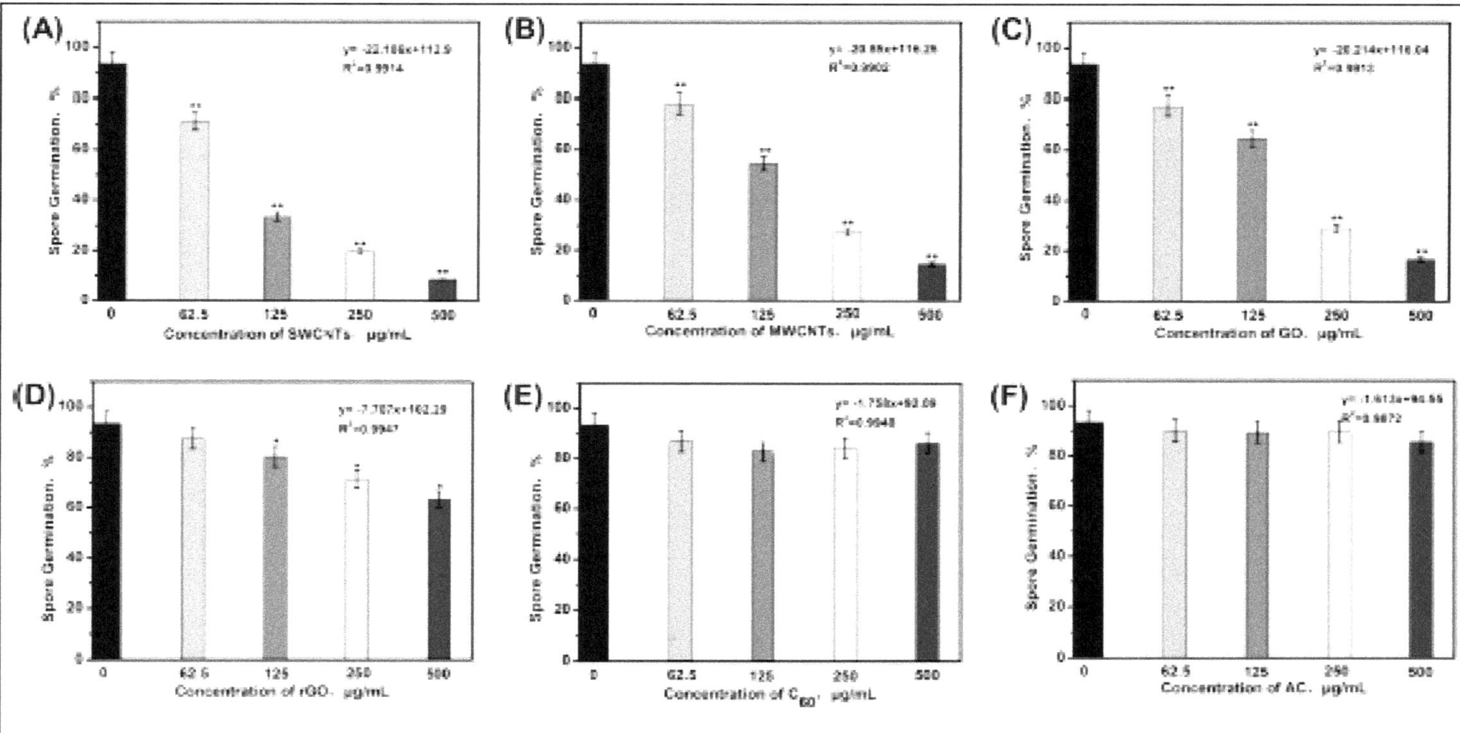

Figure 10.6. Effect of CNMs on Spore Germination of *Fusarium poae*.

Spores were germinated on distilled water at 28 _C in darkness at different concentrations of (A) SWCNTs, (B) MWCNTs, (C) GO, (D) rGO, (E) C60, and (F) AC dispersions. Germination was evaluated after 3 h incubation. Error bars represent the standard deviation (N = 4). Where appropriate, statistical significance is indicated: *p < 0.05; **p < 0.01.

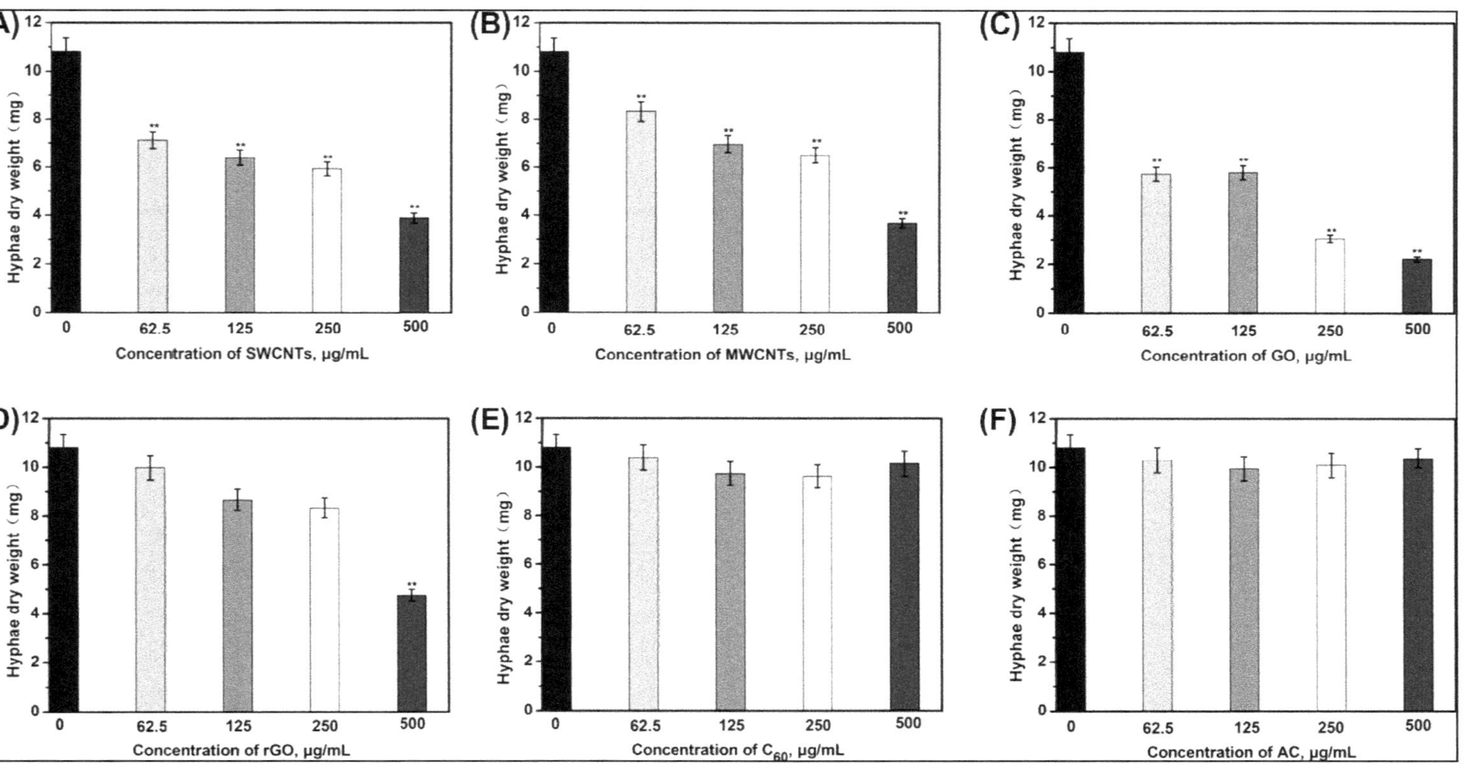

Figure 10.6. Effect of CNMs on the Mycelial Biomass of *F. poae*.

Mycelial biomass was tested at different concentrations of (A) SWCNTs, (B) MWCNTs, (C) GO, (D) rGO, (E) C60, and (F) AC dispersions for 120 h at 24 ± 2 _C. Error bars represent the standard deviation (N = 4). Where appropriate, statistical significance is indicated: $*p < 0.05$; $**p < 0.01$.

spectrum and multiple modes of antimicrobial activity (Wei *et al.*, 2009). Silver exhibits higher toxicity to microorganisms and lower toxicity to mammalian cells. The application of silver nanoparticles as antimicrobial agents is because of its economical production and multiple modes of inhibitory action to microorganisms (Clement and Jarrett, 1994). Silver nanoparticles are the most studied and utilized nano particles in bio-system because of its strong inhibitory and antimicrobial activities. Silver nanoparticles, which have highsurface area and high fraction of surface atoms, have high antimicrobial effect as compared to the bulk silver. Scientists studied theantifungal effectiveness of colloidal nano silver solution against rose powdery mildew caused by *Sphaerotheca pannosa* Var*rosae* (Figure 10.7). It is a very wide spread and common disease of both green house and outdoor grown roses.It causes leaf distortion, leaf curling, early defoliation and reduced flowering. Double capsulized nanosilver was prepared by chemical reaction of silver ion with aid of physical method, reducing agent and stabilizers. They were highly stable and verywell dispersive in aqueous solution. The nano silver colloidal solution of 5000 ppmconcentration was diluted in 10 ppm of500 kg and sprayed at large area of 3306 m^2 polluted by rose powdery mildew. Two days after the spray more than 95 per cent ofrose powdery mildew faded out and did not recur for a week. Nano silver colloid is a well dispersed and stabilized silver nanoparticle solution and is more adhesive on bacteria and fungus, hence are better fungicides. It eliminates unwanted microorganisms in planter soils and hydroponic systems. It is being used as foliarspray to stop moulds, rot and several other plant diseases. Moreover, silver is an excellent plant-growth stimulator.There are literally thousands of other essential uses for this odorless, nearly tasteless and colorless, totally benign,powerful, non-toxic disinfectant and healing agent. Nano Silica-Silver composite Silicon (Si) is known to be absorbed intoplants to increase disease resistance and stress resistance. Aqueous silicate solution, used to treat plants, is reported toexhibit excellent preventive effects on pathogenic microorganisms causing powdery mildew or downy mildew in plants. Moreover, it promotes the physiological activity and growth of plants and induces disease and stress resistance in plants. But, since silica

Figure 10.7. Leaves with Powdery Mildew.

has no direct disinfection effects on pathogenic microorganisms in plants, it does not exhibit anyeffect on established diseases. Further, the effects of silica significantly vary with the physiological environment and thus, they are not registered as agricultural chemicals. As mentioned above, silver is known as a powerful disinfecting agent. Itkills unicellular microorganisms by inactivating enzymes having metabolic functions in the microorganisms by oligodynamication and is known to exhibit superb inhibitory effects on algal growth also. Silver in an ionic state exhibits highantimicrobial activity. However, ionic silver is unstable due to its high reactivity and thus gets easily oxidized orreduced into a metal depending on the surrounding media and it does not continuously exert antimicrobial activity. Silver inthe form of a metal or oxide, is stable in the environment, but because of its low antimicrobial activity it is used in relativelyincreased amounts, which is not very desirable. A new composition of nano-sized Silica Silver for control of various plantdiseases has been developed, which consisted of nano-silver combined with silica molecules and water solublepolymer, prepared by exposing a solution including silver salt, silicate and water soluble polymer to radioactive rays. Itshowed antifungal activity and controlled powdery mildews of pumpkin at 0.3 ppm in both field and greenhouse tests. The pathogens disappeared from the infected leaves 3days after spray and the plants remained healthy thereafter,also studied the 'effective concentration' of nano sized silica silver on suppression of growth of many fungi and found that *Pythium ultimum, Magnaporthe grisea, Colletotrichum gloeosporioides, Botrytis cinerea*and *Rhizoctonia solani,* showed 100 per cent growth inhibition at 10 ppm of the nanosized silica-silver. Whereas, *Bacillus subtilis, Azotobacter chrococum, Rhizobium tropici, Pseudomonas syringae* and *Xanthomonas compestris* pv. *Vesicatoria* showed 100 per cent growth inhibition at 100 ppm. They have also reported chemical injuries caused by a higher concentration of nanosized silica-silver on cucumber and pansy plant, when they were sprayed with a high concentration of 3200 ppm. Inthis study, we analyzed the inhibition effect of three different AgNPs (WA-CV-WA13B, WA-AT-WB13R, and WA-PR-WB13R) against various plant pathogenic fungi *in vitro*. The results suggest that AgNPs are capable of inhibiting these pathogens; however, results vary according to the concentration and type of AgNPs applied to pathogens. Most fungi showed a high inhibition effect at 100 ppm concentration of silver nanoparticles. In addition,results indicate that a higher inhibition rate was observedon PDA media, compared with others. Among AgNPs, WA-CV-WA13B showed the highest inhibition effect. In most cases, inhibition increased as the concentration of AgNPs increased. This could be due to the high density at which the solution was able to saturate and cohere tofungal hyphae and to deactivate plant pathogenic fungi. Reports on the mechanism of inhibitory action of silver ions on microorganisms have shown that upon treatment with Ag+, DNA loses its ability to replicate, resulting in inactivated expression of ribosomal subunit proteins as well as certain other cellular proteins and enzymes essentialto ATP production. It has also been hypothesized that Ag^+ primarily affects the function of membrane-bound enzymes, such as those in the respiratory chain. In summary, AgNPs exerted potent antifungal effects on fungi tested *in vitro,* probably through destruction of membrane integrity; therefore, it was concluded that AgNPs have considerable antifungal activity. Kim *et al.* (2008) evaluated the antifungal efficacy of colloidal nano silver solution, against rose powdery mildew. Nano silver colloid is more adhesive on bacterial and fungal cell surface; hence acts as better fungicide because of its well dispersed and stabilized silver nanoparticle solution. Nano silver is classified as pesticide (Baier, 2009). Since silver acts as an excellent antimicrobial agent it is now an accepted agrochemical replacement. It acts as plant-growth stimulator and reduces unwanted micro-organisms in soils and hydroponic systems (Sharma *et al.*, 2012). Relatively few studies were reported on the applicability of silver in controlling various plant pathogens in a relatively safer way compared to synthetic fungicides (Park *et al.*, 2006). Since nanoparticles efficiently penetrate into microbial cells, lower concentrations of silver nanoparticles are sufficient for microbial control. This would be effective, especially for those organisms that are less sensitive to antibiotics because of poor penetration of some antibiotics into microbial cells (Samuel and Guggenbichler, 2004). Lamsal *et al.* (2011a) showed the effective usage of silver nanoparticles instead of commercial fungicides. They evaluated the effect of silver nanoparticles against six *Colletotrichum* species associated with pepper anthracnose under different culture conditions and found that, application of 100 ppm concentration of silver nanoparticles inhibited the growth of fungal hyphae as well as conidial germination *in vitro* when compared to the control. Silver nanoparticles showed significantly high inhibition of fungi in field conditions when applied on the plants before disease outbreak. Recently, Aguilar-Méndez *et al.* (2011) studied the dose-dependent fungistatic activity of the silver nanoparticles on *Colletotrichum gloesporioides*. Jo *et al.* (2009) tested various forms of silver ions and nanoparticles to examine their antifungal activity on two plant-pathogenic fungi, *Bipolaris sorokiniana* and *Magnaporthe grisea*. The *in vitro* and *in planta* evaluations of silver showed that both silver ions and nanoparticles effect colony formation of spores and disease progress of fungi. Kim *et al.* (2012) reported the inhibitory effect of three different silver nanoparticles (WA-CV-WA13B, WA-AT-WB13R and WA-PR-WB13R) against eighteen different commercially important plant pathogenic fungi on potato dextrose agar (PDA), malt extract agar and corn meal agar. They found that inhibition of fungal pathogens with silver nanoparticles is concentration dependent and also on type of silver nanoparticles used. Most fungi showed a good inhibitory effect at 100 ppm concentration of silver nanoparticles on PDA, compared with others. WA-CV-WA13B showed the highest inhibition effect compared to other silver nanoparticles. Effect of silver nanoparticles on the growth of sclerotium-forming species *Rhizoctonia solani, Sclerotinia sclerotiorum* and *S. minor*, revealed that silver nanoparticles effectively inhibit the hyphal growth in a dose-dependent manner. Further, the microscopic observation of hyphae exposed to silver nanoparticles showed severe damage and resulted in the separation of layers of hyphal wall and collapse of fungal hyphae (Min *et al.*, 2009). A recent study on *in vitro* and *in vivo* efficacy of silver nanoparticles against powdery mildew before and after disease outbreak in plants under different cultivation conditions, showed maximum inhibition of fungal hyphae and conidial germination with less concentration of nanoparticle on cucumbers and pumpkins (Lamsal *et al.*, 2011b). The preventative and post-inoculation application of the silver nanoparticles effectively reduced disease severity on plants at all concentrations. A mechanism of this antifungal activity is suggested by the direct effect on germination and infection process in the fungi. *Magnaporthe grisea* can cause foliar disease and reproduces as asexual conidia. Disease infection is initiated by the attachment of spores to the plant surface and formation of germ tubes (Tucker and Talbot, 2001). Under favourable conditions of high humidity and warm temperature (25°C), conidia germinate and the resulting germ tubes penetrate plant surfaces within 24 hrs(Howard and Ferrari, 1989). Antifungal efficiency of silver nanoparticles was observed at 24 h after inoculation, suggesting that direct contact of silver with spores or germ tubes is critical in inhibiting disease development (Young *et al.*, 2009). Moreover, antifungal efficiency of silver was also observed at 5 days after inoculation, suggesting that silver nanoparticles could have penetrated the plant cell wall and inhibited the disease development. It could be concluded that, silver nanoparticles can be used effectively in the control of rice blast disease and the prevention of deleterious infections, even though there are no phytotoxicity appeared on rice.

Table 10.1. Prevention of Plant Diseases with 'Nano-5'

Plant pathogens/plant disease	Mode of application	Killing time
Gray mold, blast, Fusarium wilt, early blight	Spray 'Nano-5' onto the surface of leaves once every 3 days	1-2 hrs
Late blight, Phytophthora diseases, southern blight, white root rot, blister blight of tea, rust	Apply to the roots twice	-do-
Sclerotinia rot, ergot, powdery mildew, Fusarium root rot, Downy mildew,	Spray onto the surface of leaves once every 5-7days	Stops infection within 1-2hr
Bakanae disease, white rust, leaf blight, soft rot,	Apply to the roots twice	-do-
Bacterial wilt, leaf spot, rot, brown leaf spot, black rot, canker	-do-	-do-
Mosaic, ringspot, transitory yellowing, tristeza virus, exocortis viroid	Spray 'Nano-5' onto the surface of leaves and apply to the roots once every 3 days	7 days perfect control
Stem and bulb nematode, cyst nematode, spiral nematode	Apply to the roots until very wet	1-2 hr

Source: http: //www.unofortune.com.tw/index.htm.

In this study, we evaluated the effect of silver nanoparticles biosynthesised by *Aspergillus terreus* (KC462061) on growth and aflatoxin production by five isolates of *A.flavus* which were isolated from nuts. TEM micrographs observed that the particles of AgNPs are spherical in shape and without significant agglomeration and the particle size ranges from 5 to 30 nm. Our results showed that all five *A. flavus* isolates were inhibited to various extents by different concentrations of silver nanoparticles but the best inhibition by 150 ppm differ significantly. In general, inhibition per cent of aflatoxin production at 50ppm ranged from 48.2 to 61.8 per cent, at 100 ppm ranged from 46.1 to 82.2 per cent whereas at 150ppm inhibition per cent reached to 100 per cent. Scanning electron microscopy (SEM) was used to study antifungal activities of silver nanoparticles and to characterize the changes in morphology. SEM images indicate two different antifungal activities of AgNPs against *A. flavus*. AgNPs inhibited the growth of *A. flavus* by affecting cellular functions which caused deformation in fungal hyphae. In comparison, AgNPs cause reduction in spores number, malformation and hypertrophy, these effects lead to destruction and damaging of spores.

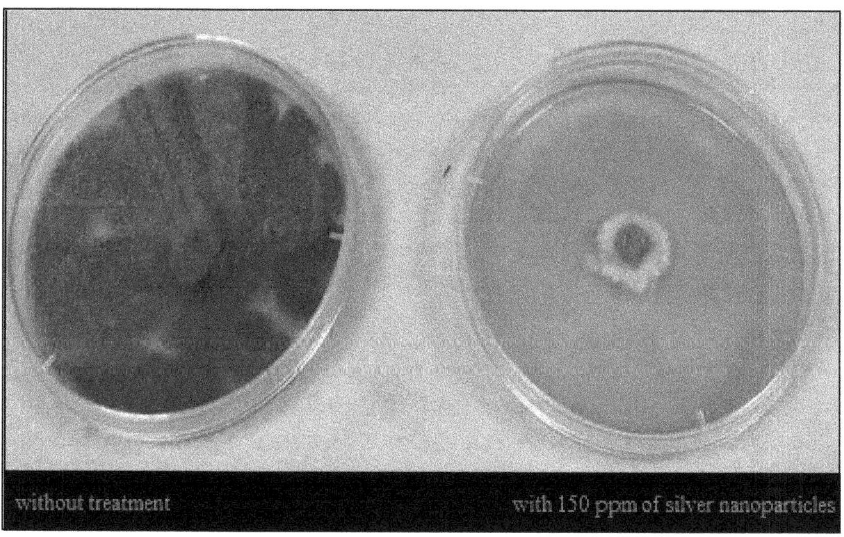

Figure 10.8. ***Aspergillus flavus.***

g) Silica Nanoparticle

Silicon (Si) increases disease resistance and stress resistance in plants (Brecht *et al.*, 2004). It also stimulates the physiological activity and growth of plants (Carver *et al.*, 1998). Torney *et al.* (2007) used honeycomb mesoporous silica nanoparticle (MSN) system with 3nm pores to deliver DNA and chemicals into plant cells and intact leaves. They loaded the gene and its chemical inducer into the MSN system

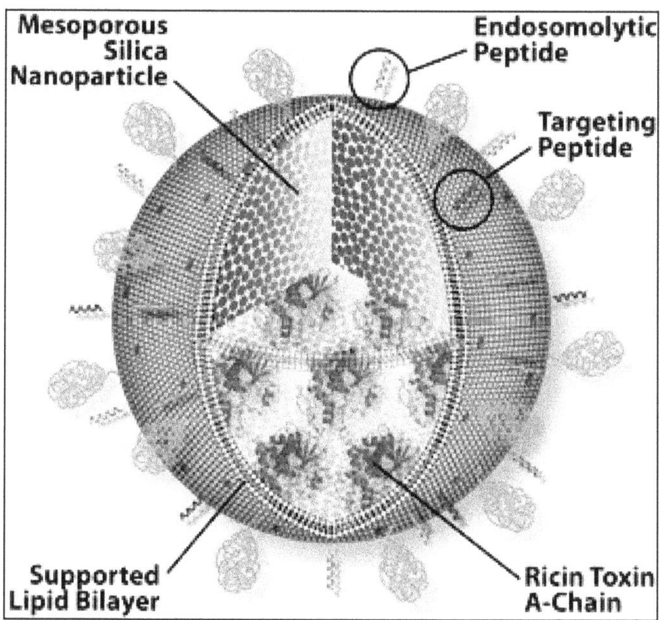

Figure 10.9

and capped the ends with gold nanoparticles and studied the release pattern of chemicals and induction of gene expression in the plants under controlled-release conditions. Their study showed an application of silica nanoparticles in target-specific delivery of proteins, nucleotides and chemicals in plant biotechnology. Silicon (Si) is known to be absorbed into plants to increase disease resistance and stress resistance by promoting the physiological activity and growth of plants. Aqueous silicate solution is reported to exhibit exceptional preventive effects on pathogenic microorganisms causing powdery mildew or downy mildew in plants. Additionally, it promotes the physiological activity and growth of plants and induces disease and stress resistance in plants.

h) Copper Nanoparticle

Copper-based fungicides produce highly reactive hydroxyl radicals which can damage lipids, proteins, DNA and other biomolecules. It plays an important role in disease prevention and treatment of large variety of plants (Borkow and Gabbay, 2005). Complexation of copper with chitosan nanogels was shown to have strong synergistic effect between chitosan and copper in inhibiting the growth of phytopathogenic fungus *Fusarium graminearum*. Because of its bio-compatibility, these nanohydrogels are included as a new generation of copper-based bio-pesticides and it could also be developed into an efficient delivery system for copper based fungicides for plant protection (Brunel *et al.*, 2013). Low melting point soda-lime glass powder containing copper nanoparticles showed efficient antimicrobial activity against gram-positive, gram-negative bacteria, yeast and fungi; key reason for the increased antimicrobial activity is because of inhibitory synergistic effect of the Ca^{2+} lixiviated from the glass (Esteban-Tejeda *et al.*, 2009).

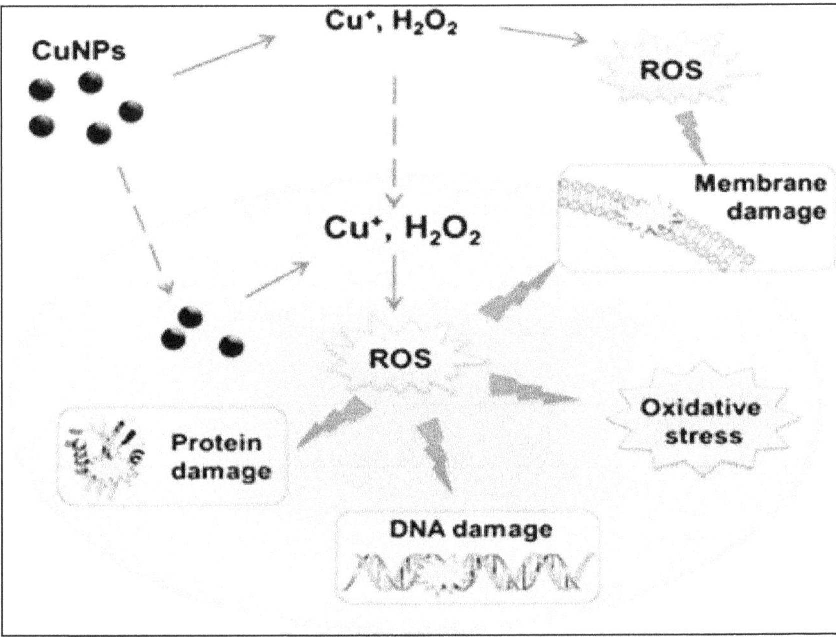

Figure 10.10

i) Zinc Nanoparticle

Mechanism of action of zinc nitrate derived nano-ZnO on important fungal pathogen *Aspergillus fumigatus* showed hydroxyl and superoxide radicals mediated fungal cell wall deformity and death due to high energy transfer (PrasunPatra and Goswami, 2012). Zinc oxide nanoparticles (ZnO NPs) could be used as an effective fungicide in agricultural and food safety applications. Recent study by He *et al.* (2011) showed significant inhibition of two postharvest pathogenic fungi *Botrytis cinerea*and *Penicillium expansum* with ZnONPs with sizes of approximately 70 nm at less concentration, their mode of action was confirmed by SEM and Raman spectroscopy. ZnO nano particles cause deformation of fungal hyphae and prevent the conidiophores and conidial development which ultimately leads to the death of fungal hyphae.

j) Nano Composites

Silver has been studied as an antibacterial agent and less as an antifungal agent. Pinto *et al.* (2013) described preparation and antifungal activity of composite films of pullulan and Ag nanoparticles (NP) against *Aspergillus niger* as a model system. They found that these composite films show strong inhibitory action on fungal sporulation, which was confirmed by disruption of the spore cells when observed under SEM. Silver in an ionic state exhibits high antimicrobial activity (Thomas and McCubbin, 2003). Park *et al.* (2006) developed a new nano-sized Silica-Silver composite for control of various plant diseases. Composite showed good antifungal activity where pathogens disappeared from the infected leaves within three days of spraying and the plants remained healthy thereafter. They also attempted to determine the effective concentration of composites and also used it effectively for suppression of growth of many pathogens. Nano composites showed 100 per cent growth inhibition of *Pythium ultimum, Magnaporthe grisea, Colletotrichum gloeosporioides, Botrytis cinerea* and *Rhyzoctonia solani* at 10 ppm concentration, whereas *Bacillus subtilis, Azotobacter chrococuum, Rhizobium tropici, Pseudomonas syringae* and *Xanthomonas compestris pv. vesicatoria* showed 100 per cent growth inhibition at 100 ppm concentration. Nanosized silica-silver (Si-Ag) particles were producedand tested by Park *et al.* (2006) against a number of fungal and bacterial pathogens. In vitro test showed higher effectiveness of silica-silver nanoparticles towards fungi at the dose of 10 ppm causing 100 per cent inhibition of vegetative growth (Table 10.2). It was found that smaller size of silver nanoparticles was more effective against fungi. Most of the bacteria tested were inhibited completely with only 100 ppm of silica-silver nanoparticles. When nanosized silica-silver particles were applied in field condition to control powdery mildew diseases of cucurbits, 100 per cent control was achieved after 3 weeks (Park *et al.*, 2006). These nanoparticles were found to be phytotoxic only at a very high dose of 3200 ppm when tested in cucumber and pansy plants. Nanosized silica silver inhibited the growth and development of both Gram-positive and Gram-negative bacteria.

Nanoparticles in Post-harvest Disease Management

Ever increase in human population, depleting naturalresources and emergence of new resistant pathogens has made the supply of sufficient and healthy food a

Table 10.2

Microorganisms	Percent growth inhibition in presence of different concentration of nanosized silica-silver			
	0.3 ppm	3.0 ppm	10 ppm	100 ppm
Pythium ultimum	15.5	66.7	100	100
Magnaporthe grisea	1.4	27.0	100	100
Colletotrichum gloeosporioides	11.8	21.6	100	100
Botrytis cinerea	2.6	82.7	100	100
Rhizoctonia solani	54.8	94.8	100	100
Bacillus subtilis	0	0	50	100
Azotobacter chroococcum	0	0	0	100
Rhizobium tropici	0	0	0	100
Pseudomonas syringae	0	0	0	100
Xanthomonas campestris pv. *vesicatoria*	0	0	0	100

Source: Park *et al.*, 2006.

daunting task. This problem might be magnified several folds in near future. Now, there is a need to increase production efficiency and decrease post-harvest wastage with application of emerging technologies like biotechnology and nanotechnology in post-harvest products. Nanotechnology has been effectively applied in agricultural and horticultural products by increasing shelf life, controlling growth of microorganisms by nanofilms and coatings, controlling influence of gases and the harmful rays (UV), using Nano biosensors for detection of quality and spoilage (Yadollahi *et al.*, 2009). Nanotechnology can be applied in postharvest operations such as drying,storage and preservation of agricultural products. Chitosan, a de acetylated derivative of chitin, is found to be very effective in reducing postharvest decay of fruits and vegetables (Liu *et al.*, 2007). Chitosan at a concentration of 1g/L has found to be very effective in reducing the growth of several phytopathogenic fungi causing post-harvest spoilage of fruits and vegetables (Hirano, 1997). Yu *et al.* (2012) investigated the effect of 1 per cent chitosan film with 0.04 per cent nano-silicon dioxide on the qualitative properties of harvested jujube after 32 d of storage under ambient temperature; they studied the related defence enzymes in fruits and found that coated sample showed lower red indices, decay incidence, respiration rate and weight loss. Shi *et al.* (2013) studied the novel chitosan/nanosilica hybrid film and its effect on preservation quality of longan fruits under ambient temperature. Coating extended shelf life, reduced browning index, retarded weight loss and inhibited the increase of malondialdehyde amount and polyphenoloxidase activity in fresh longan fruit. Liu *et al.* (2009) investigated the effects of nanosilveron post-harvest shelf life of cut gerbera (*Gerbera jamesonii*) cv. Ruikou flowers. They observed that, pulsing for 24 h with 5 mg/L nanosolution extended vase life and inhibited the bacteria growth in vase solution for initial 2 days when observed *in vitro* under microscope. A postharvest treatment of nano silver was extensively studied. Nano silver wasfound to be very significant in prolonging the vase life by inhibiting the growth of bacteria on several cut flowers with different cultivar variety, including Rose, Gerbera and *Acacia holosericea,* (Lu *et al.*, 2010; Li *et al.*, 2012; Liu *et al.*, 2012; Mohsen Kazemi., 2012; Nazemi Rafi and Ramezanian, 2013). In our recent study, we have evaluated the applicability of the CS-Ag Np composite as a fruit coating material to inhibit the growth of *Colletotrichum gloeosporioides* associated with mango anthracnose. We found that nano-composites showed a significant effect in reducing the percentage of rotting fruit tissue (71.28 per cent at 1 per cent concentration). The addition of 0.1 per cent non-ionic surfactant tween 80 enhanced the wettability and adhesion property of coating solution and exhibited significant disease reduction compared to control (84.55 per cent at 1 per cent concentration). Thus, these nano-composites can be utilized as coating material in preventing quiescent infections of *C.gloeosporioides* on mango to prevent post-harvest losses (Chowdappa *et al.*, unpublished data).Nanomaterial has important implication in management of postharvest diseases. Research findings showed the better applicability and advantages of nanopacking materials over conventional normal packing material on physicochemical and physiological quality of stored fruits, vegetables and other horticultural crops.

Nanostructures in Association-Colloidal Forms for Delivery of Functional Ingredients

Surfactant micelles, vesicles, bilayers, reverse micelles and liquid crystals have been found to be idealnanomaterials for nano-dispersions and nano-capsulation for delivery of functional ingredients. C*olloid* is a stable system of asubstance containing small particles dispersed throughout in a liquid. Association colloids have been used for many years todeliver polar, non-polar and amphiphilic functional ingredients (21-24). Size of nanoparticles in colloids, range from 5 to 100nm. The major disadvantage of colloids is that they can spontaneously dissociate if diluted.

Nano-emulsions

It is a mixture of two or more liquids (such as oil and water) that do not easily combine. In nanoemulsion, thediameters of the dispersed droplets are 500 nm or less. Nano-emulsions can encapsulate functional ingredients within theirdroplets, which can facilitate a reduction in chemical degradation (25).

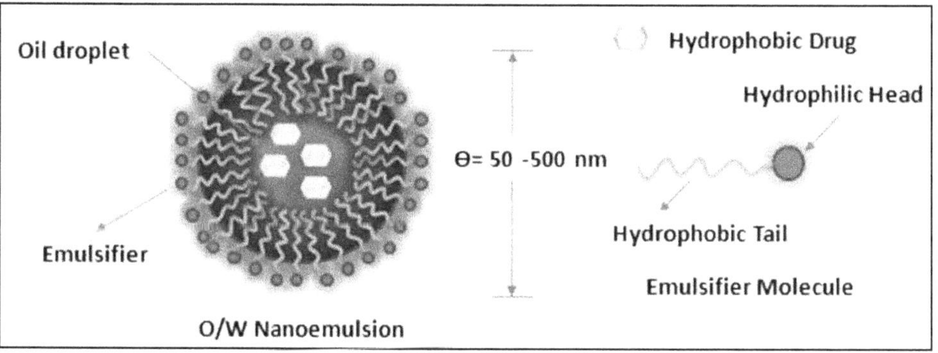

Figure 10.11

Nanoparticles Acting as Smart Delivery Systems

Syngenta is using nanoemulsions in its growth regulator Primo MAXX®, which if applied prior to the onset of stress such as heat, drought, disease or traffic can strengthen the physical structure of turfgrass and allow it to withstand ongoing stresses throughout the growing season. Another encapsulated product Karate® ZEON from Syngenta delivers a broad control spectrum on pesticide which breaks open on contact with leaves. However, the encapsulated product "gutbuster" only breaks open to release its contents when it comes into contact with alkaline environments. The ultimate aim is to tailor these products is a controlled release in response to different signals *e.g.* magnetic fields, heat, ultrasound, moisture, *etc.* New research also aims to make plants use water, pesticides and fertilizers more efficiently, to reduce pollution and to make agriculture more eco-friendly.

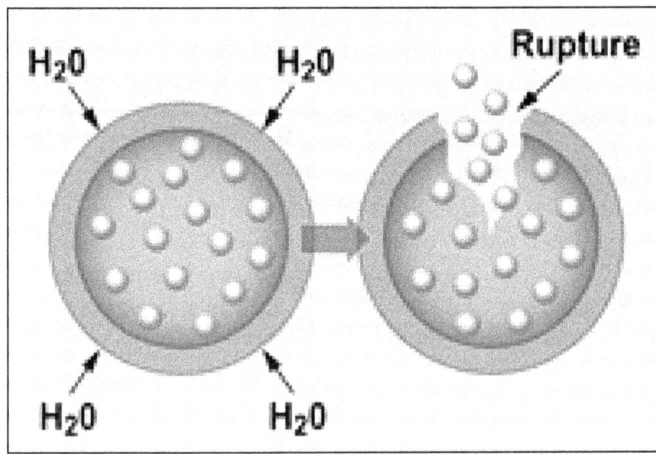

Figure 10.12

Plant Pathogens in Biosynthesis of Nanoparticles

The research on nanoscience and nanotechnologyessentially involves preparation and use of nanoparticlesof various elements and compounds. Among various uses,nanoparticles are also being used as antimicrobial agentsfor plant disease management. Formation of nanoparticlescan be achieved via several processes which may be eitherphysical or chemical.The safe method of nanoparticle production is thebiological systems especially microorganisms (Mansoori, 2005). Microorganisms offer several advantages like i)manoeuvrability for desired result using biotechnology, ii)ease of handling especially fungi (Vigneshwaran *et al.*, 2006) iii) cheapness of production iv) easy scaling up of the process iv) high efficiency (Goodsell, 2004) v) simplicity and vi) nature of green chemistry or eco-friendliness. Microorganisms have been regarded as 'biofactories' for production of metallic nanoparticles.

Fungi

Fungi are relatively recent in their use in synthesis ofnanoparticles. There has been a shift from bacteria to fungito be used as natural 'nanofactories' owing to easydownstream processing, easy handling (Mandal *et al.*, 2006) and their ability to secrete a large amount of enzymes.However, fungi being eukaryotes are less amenable togenetic manipulation compared to prokaryotes. Therefore,any alteration of fungi at genetic level for synthesis of morenanoparticles would not be so easy. A good number of fungihave been tried to synthesise metallic nanoparticles till date (Table 10.3). It is important to know the mechanism of synthesisof nanoparticles in microbial systems to get better controlover shape, size and other desired properties of thesynthesized nanomaterials.

Bacteria

Among microbes, prokaryotes have received the mostattention for biosynthesis of nanoparticles (Mandal *et al., 2006*) some of which are presented (Table 3). Bacteria havebeen used to biosynthesize mostly silver, gold, FeS and magnetite nano particles and quantum dots of cadmiumsulphide (CdS), zinc sulphide (ZnS) and lead sulphide(PbS).

Table 10.3

Sl No.	Nanoparticle	Fungus	Reference
1	Silver nanoparticles	*Verticillium* sp	Sastry *et al.*, 2003
2	Silver nanoparticles	*Phoma* sp.	Chen *et al.*, 2003
3	Silver nanoparticle	*Fusarium oxysporum*	Durán *et al.*, 2005
4	Silver nanoparticles	*Phaenerochaete chrysosporium*	Vigneshwaran *et al.*, 2006
5	Silver nanoparticle	*Aspergillus flavus*	Vigneshwaran *et al.*, 2007
6	Nanocrystalline silver	*Trichoderma asperellum*	Mukherjee *et al.*, 2008
7	Silver nanoparticles	*Fusarium semitectum*	Basavaraja *et al.*, 2008
8	Silver nanoparticles (3-30 nm)	*Aspergillus niger*	Gade *et al.*, 2008
9	Silver nanoparticle	*Fusarium solani*	Gade *et al.*, 2009
10	Silver nanoparticle	*Fusarium oxysporum*	Khosravi and Shojaosadati, 2009
11	Silver nanoparticle (10-100 nm), extracellular	*Cladosporium cladosporioides*	Balaji *et al.*, 2009
12	Silver nanoparticle (5-50 nm)	*Pleurotus sajor caju*	Nithya and Ragunathan, 2009
13	Silver nanoparticles	*Alternaria alternata*	Gajbhiye *et al.*, 2009
14	Silver nanoparticle	*Penicillium brevicompactum*	Shaligram *et al.*, 2009
15	Silver nanoparticle	*Bipolaris nodulosa*	Saha *et al.*, 2010
16	Silver nanoparticle (5-40 nm)	*Trichoderma viride*	Fayaz *et al.*, 2010
17	Silver nanoparticle (10-25 nm)	*Aspergillus clavatus*	Verma *et al.*, 2010
18	Silver nanoparticles (3-30 nm)	*Aspergillus niger*	Jaidev and Narasimha, 2010
19	Gold nanoparticles	*Colletotrichum* sp.	Shankar *et al.*, 2003
20	Gold nanoparticles	*Verticillium* sp.	Mukherjee *et al.*, 2001
21	Gold and gold-silver alloy nanoparticles	*Fusarium semitectum*	Sawle *et al.*, 2008
22	Bimetallic gold-silver alloy nanoparticle	*F. oxysporum*	Senapati *et al.*, 2005
23	Gold, silver (5-50 nm) and gold-silver alloy nanoparticle (8-14 nm)	*F. oxysporum*	Mandal *et al.*, 2006
24	Cadmium sulphide	*Coriolus versicolor*	Sanghi and Verma, 2009
25	Cadmium sulphide nanoparticles	*Fusarium* sp.	Ahmad *et al.*, 2002; Reyes *et al.*, 2009
26	Zirconia nanoparticles	*Fusarium oxysporum*	Bansal *et al.*, 2004
27	Nanoparticulate magnetite	*Fusarium oxysporum* and *Verticillium* sp	Bharde *et al.*, 2006
Bacteria			
28	Silver nanoparticle	*Clostridium versicolor*	Sanghi and Preetiverma, 2009
29	Silver nanoparticle (5-60 nm)	*Bacillus subtilis*	Saifuddin *et al.*, 2009
30	Silver nanoparticle (50 nm)	*Brevibacterium casei*	Kalishwaralal *et al.*, 2010
31	Silver nanoparticle (1-100 nm)	*Escherichia coli*	Gurunathan *et al.*, 2009
32	Silver nanoparticle (1-100 nm)	*Staphylococcus aureus*	Nanda and Saravanan, 2009
33	Silver, silver sulphide	*Pseudomonas stutzeri*	Slawson *et al.*, 1992
34	Silver, gold, and alloy of silver and gold	*Lactobacillus*	Nair and Pradeep, 2002
35	Triangular gold nanoprisms	*Actinomycete*	Shankar *et al.*, 2004
36	Gold nanoparticle	*Rhodococcus* sp.	Ahmad *et al.*, 2003
37	Gold nanoparticle (5-25 nm)	*Bacillus subtilis* 168	Fortin and Beveridge, 2000

Plant Virus

Plant virus especially spherical/icosahedral virusesrepresent the examples of naturally occurring nanomaterials or nanoparticles. The smallest plant viruses known till date is satellite Tobacco necrosis virus measuring only 18 nm in diameter (Hoglund, 1968). By nature's design, plant viruses are ought to be used for the advancement of nanoscience and nanotechnology. Plant viruses are made up of single or double stranded RNA/DNA as genome which is encapsidated by a protein coat. The protein coat/shell structurally and functionally appears like a container carrying the nucleic acid molecule as cargo from one host to another. Their ability to infect, deliver nucleic acid genome to a specific site in host cell, replicate, package nucleic acid and come out of host cell precisely in an orderly mannerhave necessitated them to be used in nanotechnology. Plant viruses have been used as template for synthesis of various types of nanomaterials (Table 10.4). A complete review on use of plant viruses as biotemplates for nanomaterials and their application has been done by Young *et al.* (2008).

Table 10.4

Sl No.	Plant virus	Application	Reference
1	*Cowpea mosaic virus* (CMV), an engineered CMV	Iron-platinum nanoparticle (30 nm diameter) synthesis	Shah *et al.*, 2009
2	*Cowpea chlorotic mottle virus* (CCMV)	Gold nanoparticle synthesis	Slocik *et al.*, 2005
3	*Cowpea chlorotic mottle virus* (CCMV)	As reaction vessel for nanomaterial synthesis	Douglas and Young, 1998
4	*Tobacco mosaic virus* (TMV)	Ag and Ni nanoparticle synthesis	Dujardin *et al.*, 2003
5	*Tobacco mosaic virus* (TMV)	Synthesis of nanowire of nickel and cobalt	Young *et al.*, 2008
6	*Tobacco mosaic virus* (TMV)	Synthesis of bimetallic alloys of CoPt, $CoPt_3$ and $FePt_3$ nanowires	Tsukamoto *et al.*, 2007
7	*Brome mosaic virus*	Gold nanoparticle synthesis	Chen *et al.*, 2005; Dragnea *et al.*, 2003; Sun *et al.*, 2007
8	-do-	Iron oxide synthesis	Huang *et al.*, 2007
9	*Red clover necrotic mosaic virus*	Au, $CoFe_2O_4$, and CdSe nanoparticles synthesis	Loo *et al.*, 2007

Conclusion

New tools with nano-devices capable of replacing many cellular types of machinery efficiently are underway. Use of nanotechnology could permit rapid advances in agricultural research. Still, the full potential of nanotechnology in the agricultural and food industry is yet to be realised. It is gradually moving from theoretical knowledge towards the application regime. Smart sensors and smart delivery systems will help the agricultural industry combat viruses and other crop pathogens. Nanostructured catalysts will be available which will increase the efficiency of pesticides, allowing on demand doses to be used. Nanotechnology holds the promise of controlled delivery of agrochemicals to improve disease resistance. Nanotechnology in conjunction with biotechnology has significantly extended the applicability of nanomaterials in crop protection and production. There is big future waiting for the tiny technology. It has potential to change the course of time. As the size decreases computing speed and computing power will increase, materials will be stronger and small doses of fungicides will cure plant diseases rapidly and more efficiently than ever. The technology that works at the nanometre scale of molecules and atoms will be a large part of this in the future.

REFERENCES

Aguilar-Mendez, M., San Martín-Martínez, E., Ortega- Arroyo, L., Cobián-Portillo, G. and Sánchez-Espíndola, E. 2011. Synthesis and characterization of silver nanoparticles: effect on phytopathogen *Colletotrichum gloesporioides*. *Journal of Nanoparticle Research* **13**: 2525-2532.

Ajayan, P.M., Schadler, L.S. and Braun, P.V. 2003.Nanocomposite science and technology, Wiley, p. 2.

Astruc, D., Boisselier, E. and Ornelas, C. 2013. Dendrimers designed for functions: from physical, photophysical, and supramolecular properties to applications in sensing, catalysis, molecular electronics, and nanomedicine. *Chemical Review* **110**: 1857-1959.

Baier, A. C. 2009. Regulating nanosilver as a pesticide., Environmental Defense Fund, February 12.

Bhainsa, K.C. and D'Souza, S. F. 2006. Extracellular biosynthesis of silver nanoparticles using the fungus *Aspergillus fumigatus*. Colloids Surfaces B. *Biointerfaces* **47**: 160–164.

Bhattacharya, D. and Gupta, R.K. 2005. Nanotechnology and potential of microorganisms. *Crit Rev Biotechnol* **25**: 199–204.

Boanini, E., Torricelli, P., Gazzano, M., Giardino, R. and Bigi, A. 2006. Nanocomposites of hydroxyapatite with aspartic acid and glutamic acid and their interaction with osteoblast-like cells. *Biomaterials* **27**: 4428– 4433.

Borkow, G. and Gabbay, J. 2005. Copper as a biocidaltool.*Current medicinal chemistry* **12**: 2163-2175.

Brunel, F., El Gueddari, N.E. and Moerschbacher, B.M. 2013.Complexation of copper(II) with chitosan nanogels: Toward control of microbial growth. *Carbohydrate Polymers* **92**: 1348-1356.

Carver, T. L. W., Thomas, B. J., Robbins, M. P. and Zeyen, R. J. 1998. Phenylalanine ammonia-lyase inhibition, autofluorescence and localized accumulation of silicon, calcium and manganese in oat epidermis attacked by the powdery mildew fungus *Blumeria graminis* (DC) Speer. *Physiological and Molecular Plant Pathology* **52**: 23-243.

Chen, L.C., Kung, S.K., Chen, H.H. and Lin, S.B. 2010. Evaluation of zeta potential difference as an indicator for ant ibacterialstrength of low molecular weight chitosan. *Carbohydrate Polymers* **82**: 913-919.

Clement, J.L. and Jarrett, P.S. 1994. Antibacterial silver.*Metal Based Drugs* **1**: 467-482.

Esteban-Tejeda, L., Malpartida, F., Esteban-Cubillo, A., Pecharromán, C. and Moya, J. S. 2009.Antibacterial and antifungalactivity of a soda-lime glass containing copper nanoparticles. *Nanotechnology* **20**: 505701.

García-Rincón, J., Vega-Pérez, J., Guerra-Sánchez, M. G., Hernández-Lauzardo, A. N., Peña-Díaz, A. and Velázquez- Del Valle, M. G. 2010. Effect of chitosan on growth and plasma membrane properties of *Rhizopusstolonifer* (Ehrenb.: Fr.) Vuill. *Pesticide Biochemistry and Physiology* **97**: 275-278.

He, L., Liu, Y., Mustapha, A. and Lin, M. 2011.Antifungal activity of zinc oxide nanoparticles against *Botrytis cinerea* and *Penicillium expansum*.*Microbiological Research* **166**: 207-215.

Hu, C., Lan, Y.Q., Qu, J.H., Hu, X.X. and Wang, AM. 2006. Ag/AgBr/TiO2 visible light photocatalyst for destruction of azodyes and bacteria. *J Physical Chem B* **110**: 4066–4072.

Ing, L. Y., Zin, N. M., Sarwar, A. and Katas, H. 2012. Antifungal activity of chitosan nanoparticles and correlation with their physical properties. *International Journal of Biomaterials* **55**: 9.

Jo, Y.K., Kim, B. H. and Jung, G. 2009.Antifungal activity of silver ions and nanoparticles on phytopathogenic fungi. *Plant Disease* **93**: 1037-1043.

Kaur, P., Thakur, R. and Choudhary, A. 2012.An *In vitro* Study of The Antifungal activity of silver/chitosan nanoformulat ions against important seed borne pathogens. *International Journal of Scientific and Technology Research* **1**: 83-86.

Kim, H. S., Kang, H. S., Chu, G. J. and Byun, H. S. 2008. Antifungal effectiveness of nanosilver colloid against rose powdery mildew in greenhouses. *Solid State Phenomena* **135**: 15-18.

Kim, S. W., Jung, J. H., Lamsal, K., Kim, Y. S.,Min, J. S. and Lee, Y.S. 2012. Antifungal Effects of silver nanoparticles (AgNPs) against various plant pathogenic fungi. *Mycobiology***40**: 53-58.

Kong, M., Chen, X. G., Xing, K. and Park, H. J. 2010. Antimicrobial properties of chitosan and modeof action: a state of the art review. *International Journal of Food Microbiology* **144**: 51-63.

Kumar, R., Sharon, M. and Choudhary, A. K. 2010.Nanotechnology in agricultural diseases and food safety. *Journal of Phytology* **2**: 83-92.

Lamsal, K., Kim, S.W., Jung, J. H., Kim, Y. S., Kim, K. S. and Lee, Y. S. 2011a. Application of silver nanoparticles for the control of *Colletotrichum* species *In vitro* and pepper anthracnose disease in field. *Mycobiology* **39**: 194-199.

Lamsal, K., Kim, S.W., Jung, J. H., Kim, Y. S., Kim, K. S. and Lee, Y.S. 2011b. Inhibition Effects of Silver nanoparticles against powdery mildews on cucumber and pumpkin. *Mycobiology***39**: 26-32.

Lorenceau, E., Utada, A.S., Link, D.R., Cristobal, G., Joanicot, M. and Weitz, D.A. 2005.Generation of polymerosomes from double emulsions. *Langmuir* **21**: 9183–9186.

Min, J. S., Kim, K. S., Kim, S. W., Jung, J. H., Lamsal, K., Kim, S. B., Jung, M. and Lee, Y. S. 2009. Effects of colloidal silver nanoparticles on sclerotium-forming phytopathogenic fungi. *Journal of Plant Pathology* **25**: 376-380.

Misra, A.N., Misra, M. and Ranjeet Singh. 2013. Nanotechnology in agriculture and food industry. *International Journal of Pure and Applied Sciences and Technology* **16**(2): 1-9.

Moraru, C.I., Lee, T.C., Karwe, M.V. and Kokini, J.L. 2003. Plasticizing and antiplasticizing effects of water and polyols on a meat-starch extruded matrix, *J. Food Sci.* **44**: 3396-3401.

Navrotsky, A. 2000.Technology and applications Nanomaterials in the environment, agriculture, and technology (NEAT). *J Nanopart Res* **2**: 321–323.

Park, H. J., Kim, S. H., Kim, H. J. and Choi, S.H. 2006.A New Composition of nanosized silica-silver for control of various plant diseases. *Journal of Plant Pathology* **22**: 295-302.

Phaechamud, T. and Ritthidej, G. C. 2008. Formulation variables influencing drug release from layered matrix system comprising chitosan and xanthan gum. *AAPS Pharm SciTech*, **9**: 870-877.

Phillips, K.S., Han, J.H., Martinez, M., Wang, Z.Z., Carter, D. and Cheng, Q. 2006. Nanoscaleglassification of gold substrates for surface plasmon resonance analysis of protein toxins with supported lipid membranes. *Analytical Chemistry* **78**: 596–603.

Pinto, R. J., Almeida, A., Fernandes, S. C., Freire, C. S., Silvestre, A. J., Neto, C. P. and Trindade, T. 2013. Antifungal activity of transparent nanocomposite thin films of pullulan and silver against *Aspergillus niger*. *Colloids and Surfaces B: Biointerfaces* **103**: 143-148.

Raliya, R., Tarafdar, J.C., Gulecha, K., Choudhary, K., Ram, R., Mal, P. and Saran, R.P. 2013. Review Article: Scope of Nanoscience and Nanotechnology in Agriculture. *Journal of Applied Biology and Biotechnology* **1** (3): 41-44.

Roco, M.C. 2007. Handbook on Nanoscience, Engineering and Technology, Taylor and Francis.

Roller, S. and Covill, N. 1999.The antifungal properties of chitosan in laboratory media and apple juice. *International Journal of Food Microbiology* **47**: 67-77.

Sah, S.K., Kaur, A. and Wani, S.H. 2014. Nanotechnology: changing horizons of science. *Biolife* **2**(3): 905-916.

Samuel, U. and Guggenbichler, J. P. 2004. Prevention of catheter-related infections: the potential of a new nanosilver impregnated catheter. *International Journal of Antimicrobial Agents*, 23, Supplement **1**: 75-78.

Sanpui, P., Murugadoss, A., Prasad, P. V. D., Ghosh, S. S. and Chattopadhyay,A. 2008.The antibacterialproperties of a novel chitosan.Ag-nanoparticle compos ite. *International Journal of Food Microbiology* **124**: 142-146.

Sharma, K., Sharma, R., Shit, S. and Gupta, S. 2012.Nanotechnological application on diagnosis of a plant disease.International Conference on Advances in Biological and Medical Sciences (ICABMS.2012) July 15-16, 2012 Singapore.

Sharonl M., Kr. Choudhary A. and Rohit K. 2010. Nanotechnology in agricultural diseases and food safety. *Journal of Phytology* **2**(4): 83–92.

Sudarshan, N. R., Hoover, D. G. and Knorr, D. 1992. Antibacterial action of chitosan. *Food Biotechnology* **6**: 257-272.

Tarafdar, J.C. and Raliya, R. 2012. The Nanotechnology. Scientific Publisher, Jodhpur, India. ISBN: 9788172337582. 214 pp.

Thomas, S. and McCubbin, P. 2003. A comparison of the antimicrobial effects of four silver-containing dressings on three organisms. *Journal of Wound Care* **12**: 101-107.

Torney, F., Trewyn, B. G., Lin, V. S.Y. and Wang, K. 2007.Mesoporous silica nanoparticles deliver DNA and chemicals into plants. *Nature Nanotechnology* **2**: 295-300.

Wang, Y.A., Li, J.J., Chen, H.Y. and Peng, X.G. 2002.Stabilization of inorganic nanocrystals by organic dendrons. *J Am Chem Soc.***124**: 2293-2298.

Wei, D., Sun, W., Qian, W., Ye, Y. and Ma, X. 2009. The synthesis of chitosan-based silver nanoparticles and their antibacterial activity. *Carbohydrate Research* **344**: 2375-2382.

Zhang, Z.P. and Feng, S.S. 2006. The drug encapsulation efficiency, in vitro drug release, cellular uptake and cytotoxicity of paclitaxel-loaded poly(lactide)-tocopheryl polyethylene glycol succinate nanoparticles. *Biomaterials* **27**: 4025–4033.

Transformation of Indian Agriculture through Innovative Technologies *Pages* **151–189**
Editor: **Dr. Shahid Ahamad & Dr. Jag Paul Sharma**
Published by: **ASTRAL INTERNATIONAL PVT. LTD., NEW DELHI**

11 Role of PR Proteins in Host Defence Mechanism

Mudasir Bhat, Mudasir Hassan, Ali Anwar, Arif Hussain and Saleem Dar

Introduction

Since their discovery in tobacco leaves hypersensitively reacting to TMV by two independently working groups (Van Loon and Van Kammen, 1970; Gianinazzi *et al.*, 1970), pathogenesis-related proteins (initially named "b" proteins) have focused an increasing research interest in view of their possible involvement in plant resistance to pathogens. This assumption flowed from initial findings that these proteins are commonly induced in resistant plants, expressing a hypersensitive necrotic response (HR) to pathogens of viral, fungal and bacterial origin. Later, however, it turned out that b-proteins are induced not only in resistant, but also in susceptible plant pathogen interactions, as well as in plants, subjected to abiotic stress factors (Van Loon, 1985). Thus, still in 1980 Antoniw *et al.* coined the term "pathogenesis-related proteins" (PRs), which have been defined as "proteins encoded by the host plant but induced only in pathological or related situations", the latter implying situations of non-pathogenic origin. To be included among the PRs, a protein has to be newly expressed upon infection but not necessarily in all pathological conditions. Pathological situations refer to all types of infected states, not just to resistant, hypersensitive responses in which PRs are most common; they also include parasitic attack by nematodes, insects and herbivores(Van Loon *et al.*, 1994).The term "PR-like proteins" was proposed to accommodate proteins homologous to PRs as deduced from their amino acid sequences or predicted from the nucleotide sequence of their corresponding cDNA or gene, but which are induced in developmentally controlled, tissue-specific manner. However, the designation "PR-likeproteins" neverbecame popular; some proteins that were originally considered to be Pathogenesis-related proteins: research progress in the last 15 years 107 "PR-like" on the basis of sequence homology, have been shown to be strongly induced by infections, and hence are best regarded as genuine PRs.Higher plants have a broad range of mechanisms to protect themselves against various threats including physical, chemical and biological stresses, such as wounding, exposures to salinity, drought, cold, heavy metals, air pollutants and ultraviolet rays and pathogen attacks, like fungi, bacteria and viruses(Van Loon, 1999). Plant reactions to these factors are very complex, and involve the activation of set of genes, encoding different proteins. These stresses can induce biochemical and physiological changes in plants, such as physical strengthening of the cell wall through lignification, suberization and callose deposition; by producing phenolic compounds, phytoalexins and pathogenesis-related (PR) proteins which subsequently prevent various pathogen invasions. Among these,production and accumulation of pathogenesis related proteins in plants in response to invading pathogen. PR protein in the plants was first discovered and reported in tobacco plants infected by tobacco mosaic virus. Most PR proteins in the plant species are acid-soluble, low molecular weight, and protease-resistant proteins. PR proteins depending on their isoelectric points may be acidic or basic protein. Most acidic PR proteins are located in the intercellular spaces, whereas, basic PR proteins are predominantly located in the vacuole. The PR proteins have been classically divided initially into 5 families based on molecular mass, isoelectric point, and localization and biological activity. Currently PR-proteins were categorized into 17 families according to their properties and functions including β-1,3-glucanases,chitinases,thaumatin-likeproteins,peroxidases,ribosome inactivating proteins, defenses, thionins, nonspecific lipid transfer proteins, oxalate oxidase and oxalate-oxidase-like proteins. Among these PR proteins chitinases and β-1,3-glucanases are two important hydrolytic enzymes that are abundant in many plant species. The amount of them significantly increase and play main role of defense reaction against fungal pathogen by degrading cell wall, because chitin and β-1,3-glucan is also a major structural component of the cell walls of many pathogenicfungi. β-1,3-glucanases appear to be coordinately expressed along with chitinases after fungal infection. This co-induction of the two hydrolytic enzymes has been described in many plant species, including pea,bean, tomato, tobacco, maize, soybean, potato and wheat (Glazebrook, 1999).The defense strategy of plants against stress factors involves a multitude of toolsincluding various types of stress proteins with putative protective functions. A groupof plant-coded proteins induced by different stress stimuli, named "pathogenesis related proteins" (PRs) is assigned an important role in plant defense against pathogenic constraints environment. Plants are exploited as a source of food and shelter by a wide range of parasites, including viruses, bacteria, fungi, nematodes, insects and even other plants. Plants lack a circulating adaptive immune system to protect themselves against pathogens. They have evolved other mechanisms of antimicrobial defense which are either constitutive or inducible. Plants are resistant to most pathogens in their environment, as they are not host plants for particular pathogen or are host plants, but harbor resistance genes, allowing them to recognize specifically distinct pathogen races (Scheel, 1998). Two types of plant resistance responses can be distinguished: nonhost and host or race/cultivar specific resistance response. In both cases, the biochemical processes involved in pathogen resistance are very similar (Somssich and Hahlbrock, 1998). Resistance in plants is manifested by the inability of the pathogen to grow or multiply and spread and often takes the form of a hypersensitive reaction (Agrois, 1988). The hypersensitive response is characterized by localized cell and tissue death at the site of infection (Van Loon, 1997). As a result the pathogen remains confined to necrotic lesions near the site of infection. A ring of cells surrounding necrotic lesions become fully refractory to subsequent infection, known as localized acquired resistance (Hammon-Koasack and Jones, 1996; Baker *et al.*, 1997; Fritig *et al.*, 1998). These local responses often trigger nonspecific resistance throughout the plant, known as systemic acquired resistance, providing durable protection against challenge infection by a broad range of pathogens (Ryals *et al.*, 1996; Sticher *et al.*, 1997; van Loon, 1997; Fritig *et al.*, 1998). The metabolic alterations in localized acquired resistance include: cell wall reinforcement by deposition and crosslinking of polysaccharides, proteins, glycoproteins and insoluble phenolics; stimulation of secondary metabolic pathways, some of which yield small compounds with antibiotic activity (phytoalexins) but also defense regulators such as salicylic acid, ethylene and lipid-derived metabolites; accumulation of broad range of defense-related proteins and peptides (Hahn, 1996; Fritig *et al.*, 1998). Understanding of the plant response to the pathogen attack has advanced rapidly in recent years. Bacterial and fungal pathogenicity factors have been isolated, and mechanisms utilized by the plant to recognize the pathogen and initiate a plethora of defense mechanisms have been identified. The present review is focused on recent advances in the study of molecular mechanisms and components involved in pathogen defense in plants.

Role in Plant Disease Management

Pathogen Recognition

Activation of inducible defenses is triggered by a specific recognition of pathogen invasion by plants. Perception in host specific resistance involves receptors with high degrees of specificity for pathogen strains, which are encoded by constitutively expressed defense resistance (R) genes, located either on the plasma membrane or in the cytosol (Edreva, 1991; Martin, 1999; McDowell and Dangl, 2000). Large repertoires of distantly related individual R genes with diverse recognition specificities are found within a single plant species (Ellis *et al.*, 2000). Individual R genes have narrow recognition capabilities and they trigger resistance when the invading pathogen expresses a corresponding avirulence (Avr) gene. Avr genes from different pathogen classes are structurally very diverse and have different primary functions in the biology of these organisms. Specific recognition of the aggressor by the plant requires the presence of matching Avr and R genes in the two species and is thought to be mediated by ligand receptor binding (Glazebrook, 1999). Over 20 R genes with recognition-specificity for defined Avr genes have been isolated from seven plant species, including both monocots and dicots (Milligan *et al.*, 1998; Rossi *et al.*, 1998; Martin, 1999). These genes are effective against bacterial, viral, and fungal pathogens and against both nematodes and aphides species (Mi gene from tomato) (Martin, 1999). In spite of the great diversity in lifestyles and pathogenic mechanisms of dise- ase-causing organisms, R genes were found to encode proteins with certain common motifs. Five classes R proteins are now recognized: intracellular protein kinases; receptor-like protein kinases with an extracellular leucine-rich repeat (LRR) domain; intracellular LRR proteins with a nucleotide binding site (NBS) and a leucine zipper (LZ) motif; intracellular NBS-LRR proteins with a region with similarity to the Toll and interleukin-1 receptor (TIR) proteins from Drosophila and mammals; and LRR proteins that encode membrane bound extracellular proteins. Interestingly, the NBS- LRR class of R genes represents as much as 1 per cent of the Arabidopsis genome (Ellis *et al.*, 2000). Plant R genes encode proteins that both determine recognition of specific Avr proteins and initiate signal transduction pathways leading to complex defense responses (Zhou *et al.*, 1998; del Pozo and Estelle, 1999; Martin, 1999). Despite these significant insights into R gene structure, much remains to be elucidated about the molecular mechanisms by which R proteins recognize and transduce this information in the plant cell. In addition to gene for gene recognition mediated by R and Avr genes, nonhost resistance is achieved through the recognition of specific pathogen or plant cell wall derived signal molecules, termed exogenous or endogenous elicitors, respectively. These elicitors are often low-molecular-weight compounds that are either synthesized as such or are liberated from polymeric precursors during infection. The chemical structure of different elicitors is of great variety, such as glycoproteins, peptides and oligosaccharides. Some proteinaceous elicitors are directly produced by bacterial or fungal pathogens, whereas biologically active oligo-saccharides are released from pathogen and plant cell walls by hydrolases secreted by the two organisms. Complex and largely unresolved perception systems exist for these elicitors on the plant cell surface that activate multiple intracellular defense signaling pathways. In conclusion, the multicomponent response of plants to pathogens in host and nonhost resistance appears to be activated by ligand/receptor interactions, in which Avr gene and pathogen or plant surface-derived elicitors serve as ligands for plasma membrane located or cytosolic receptors(Somssich and Hahlbrock, 1998).

Signal Transduction

Receptor-mediated recognition at the site of infection initiates cellular and systemic signaling processes that activate multicomponent defense responses at local and systemic levels, resulting in rapid establishment of local resistance and delayed development of systemic acquired resistance (Scheel, 1998). The earliest reactions of plant cells include changes in plasma membrane permeability leading to calcium and proton influx and potassium and chloride efflux (McDowell and Dangl, 2000). Ion fluxes subsequently induce extracellular production of reactive oxygen intermediates, such as superoxide ($O2^-$), hydrogen peroxide and hydroxyl free radical (OH•), catalyzed by a plasma membrane-located NADPH oxidase and/or apoplastic-localized peroxidases (Somssich and Hahlbrock, 1998). The initial transient reactions are, at least in part, prerequisites for further signal transduction events resulting in a complex, highly integrated signalling network that triggers the overall defensive response (Figure 11.1). The role of calcium is shown in experiments with calcium channel inhibitors, which prevent increase of cytosolic calcium concentrations, delay the development of the hypersensitive response. Heterotrimeric GTP-binding proteins and protein phosphorylation/

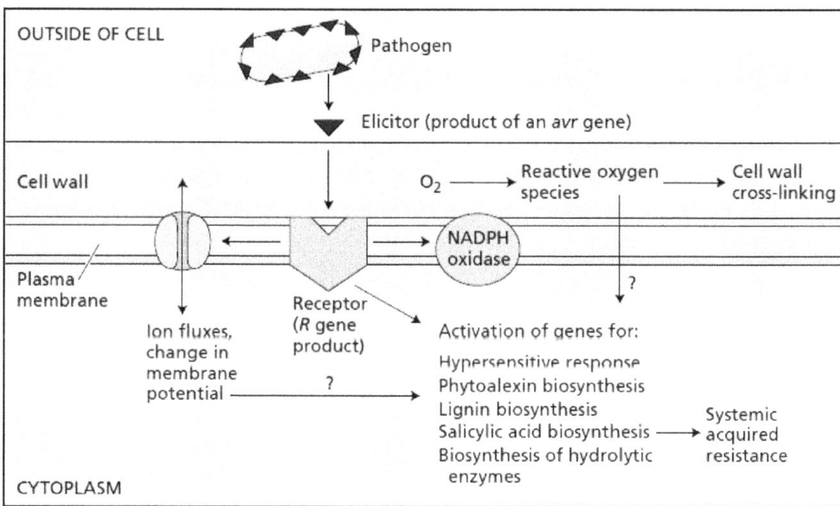

Figure 11.1. Major Components of the Signal-Transduction Chain from Elicitor Perception to Gene Activation.

dephosphorylation are probably involved in transferring signals from the receptor to calcium channels that activate downstream reactions (Legendre *et al.*, 1992). The changes in ion fluxes trigger localized production of reactive oxygen intermediates and nitric oxide, which act as second messengers for hypersensitive response induction and defense gene expression (Piffanelli *et al.*, 1999). Synergistic interactions between reactive oxygen intermediates, nitric oxide and salicylic acid have been postulated (McDowell and Dangl, 2000). Other components of the signal network are specifically induced phospholipases, which act on lipid bound unsaturated fatty acids within the membrane, resulting in the release of linolenic acid, which serves as a substrate for the production of jasmonate, methyl jasmonate and related molecules via a series of enzymatic steps. Oxidative burst is a central component of plants defense machinery (Lamb and Dixon, 1997; Alvarez *et al.*, 1998). Reactive oxygen intermediates have been associated with apoptosis of mammalian cells, indicating a role in cell death during the hypersensitive response in plants (Heath, 1998; Richberg *et al.*, 1998). This analogy is sup- ported by the identification of a plant equivalent to the mammalian NADPH oxidase complex that produces respiratory burst in neutrophils (Keller *et al.*, 1998). The burst of hydrogen peroxide production at the plant cell surface drives rapid peroxidase-mediated oxidative cross-linking of structural proteins in the cell wall, thereby reinforcing this physical barrier against pathogen ingress (Scheel, 1998). Additionally, low doses of reactive oxygen metabolites act as signals for the induction of detoxification mechanisms involving superoxide dismutase and glutathione-S-transferase and activation of other defense reactions in neighbouring cells. As in mammalian phagocytizing cells, superoxide radicals are first generated and then rapidly converted to hydrogen peroxide and oxygen, probably by extracellular superoxide dismutase (Scheel, 1998). Most of the inducible, defense-related genes are regulated by signal pathways involving one or more of the three regulators jasmonate, ethylene and salicylic acid (Sticher *et al.*, 1997; Reymond and Farmer, 1998;Ananieva and Ananiev, 1999). The exact role of ethylene as a defense regulator is not clear but it has been shown that this hormone preferentially induces basic pathogenesis-related proteins. Jasmonate is essential for the defense of tomato against tobacco hornworn larvae and for defense of Arabidopsis against fungal pathogens as *Phythium mastophorum* and the fly Bradysia (Reymond and Farmer, 1998). Jasmonate and ethylene co-operate to regulate the expression of many genes and at least some jasmonate-inducible genes are not inducible in plants.

Recognition of the elicitor by its plasma membrane receptor stimulates transient influxes of H^+ and Ca^{++} and effluxes of K^+ and Cl^-. These ion fluxes are prerequisite for the activation of specific MAP (mitogen-activated protein) kinases and for the generation of reactive oxygen intermediates (the oxidative burst). Phosphorylation and dephosphorylation of some proteins is also observed. Binding of the elicitor stimulates the generation of jasmonic acid via a membrane associated phospholipase. However, activation of jasmonate pathway is not essential for initiating the various defense responses (Somssich and Hahlbrock, 1998)to produce or sense ethylene (Reymond and Farmer, 1998). During systemic acquired response salicylic acid levels rise throughout the plant. Defense genes such as pathogenesis related genes are expressed and the plant becomes more resistant to pathogen attack. Plants that cannot accumulate salicylic acid due to the presence of transgene that encodes salicylic acid-degradating enzyme develop hypersensitive response after challenge to avirulent pathogens, but do not exhibit systemic expression of defense genes and do not develop resistance to subsequent pathogen attack (Glazebrook, 1999). Arabidopsis mutants, compromised in their ability to respond to jasmonate or to produce salicylic acid, have been used to demonstrate that the two pathways are utilized differentially against contrasting modes of attack (McDowell and Dangl, 2000). The ethylene-jasmonate-dependent pathway is activated by pathogens that kill plant cells to obtain nutrients. In contrast, salicylic acid-dependent response is triggered by a pathogen that obtains nutrients from living plant tissue. It has been also suggested that ethylene-jasmonate and salicylic acid pathways are mutually inhibitory. Such cross-talk probably implies a capacity for a selective defense against specific types of parasites.

Changes in Gene Activity

Activation of signal transduction network after pathogen recognition results in reprog- ramming of cellular metabolism, involving large changes in gene activity. Plants contain many defense related proteins. In addition to resistance R genes and genes encoding signal transduction proteins, they possess downstream defense genes, such as pathogenesis-related proteins (PRs), enzymes involved in the generation of phytoalexins, the enzymes of oxidative stress protection, tissue repair, lignification, and others. It should be stressed that many of these genes are involved in secondary metabolism, such as shikimate and phenylpropranoid pathways (Somssich and Hahlbrock, 1998). Induction of defense gene transcripts is observed sometimes in the attacked cells, but mostly in surrounding plant tissues (Scheel, 1998). Accumulation of the pathogenesis-related proteins represents the major quantitative change in protein composition that occurs in non-inoculated plant parts that, upon challenge, exhibit acquired resistance (Van Loon, 1997). Eleven pathogenesis-related protein families from different plant species have been characterized and classified according to sequence similarities (Fritig *et al.*, 1998), although additional pathogen-induced proteins with potential anti-pathogenic action keep being described. Within one family several members may share similar biological activities but differ substantially in other properties such as substrate specificity, physicochemical properties or subcellular localization. The inducible pathogenesis-related proteins are mostly acidic proteins that are secreted into the intercellular space (Van Loon, 1997). In addition, basic pathogen-related proteins occur at relatively low levels in the vacuole.Most pathogen-related proteins have a damaging action on the structures of the parasite: PR-1 and PR-5 interact with the plasma membrane, whereas β-1,3- glucanases (PR-2) and chitinase (PR-3, PR-4, PR-8 and PR-11) attack β-1,3-glucans and chitin, which are components of the cell walls in most higher fungi. PR-5 proteins are thought to create transmembrane pores and have therefore been named permatins. Chitinases can also display lysosyme activity and hydrolyze bacterial peptidoglycan. Microbial proteinases involved in pathogenesis are completely inhibited by a tobacco proteinase inhibitor. Plants also synthesize inhibitors of fungal polygalacturonases considered as pathogenicity factors. The PR-10 family has sequence similarity to ribonucleases and is the only family consisting of cytoplasmic proteins. In addition to pathogenesis-related proteins, small peptides with antimicrobial activity, such as thionins, defensins and lipid transfer proteins, also accumulate in infected plants and are probably components of the induced defense system (Bergey *et al.*, 1996; Broekaert *et al.*, 1997; Fritig *et al.*, 1998). The reprogramming of cellular metabolism comprises not only positive, but also negative regulatory mechanisms. For example, in potato the mRNA and protein levels of Rubisco are drastically reduced by pathogen infection or elicitor treatment. In pars- ley, the expression of several genes in cell proliferation and cell-cycle regulation and as well flavanoid biosynthesis, are repressed to a large extent during defense response (Somssich and Hahlbrock, 1998).

Biochemical and Structural Characteristics of PR Proteins

Cellular and Tissue Localisation

PRs are distinguished by specific biochemical properties. They are low-molecular proteins (6-43 kDa), extractable and stable at low pH (< 3), thermostable, and highly resistant to proteases (Van Loon, 1999). The structure of a PR-1 family member (tomato PR1-b) was solved by nuclear magnetic resonance and found to represent a unique molecular architecture. The protein contains four α-helices and four β-strands arranged antiparallel between helices. The tight packing of the α-helices on both sides of the central β-sheet (α–β–α sandwich structure) results in a compact, bipar- tite molecular core, which is stabilized by hydrophobic interactions and multiple hy- drogen bonds (Fernández *et al.*, 1997). This compact structure probably determines the high stability of PRs and their insensitivity to proteases. PRs have dual cellular localisation – vacuolar and apoplastic, the apoplast being the main site of their accumulation (Van Loon, 1999). Apart from being present in the primary and secondary cell walls of infected plants, PRs are also found in cell wall appositions (papillae) deposited at the inner side of cell wall in response to fungal attack (Benhamou *et al.*, 1991; Jeun, 2000). Interestingly, they are detected in the cell walls of invading fungal pathogens and in the space formed between cell walls and invaginated plasma membrane of fungi (Jeun, 2000; Jeun and Buchenauer, 2001). Acidic and basic PRs are identified, each of these counterparts having both apoplastic and vacuolar localisation (Buchel and Linthorst, 1999). Earlier data show that acidic tobacco PR-1 are localized in the apoplast, whereas basic tobacco PR-1 accumulate in the vacuole (Bol *et al.*, 1990). This may be valid for one PR family (PR-1) in a host plant, such as tobacco, but cannot be generalized as a differential localization feature of acidic and basic proteins in plants. Presently, PRs are established in all plant organs – leaves, stems, roots, flowers (Van Loon, 1999), being particularly abundant in the leaves, where they can amount to 5-10 per cent of total leaf proteins. Thus, the original claiming that PRs occurrence is limited to photosynthetizing tissues (Asselin *et al.*, 1985) has been abolished. The application of sensitive immunological techniques allowed the detection of PRs in roots of tobacco and tomato plants inoculated with the fungal pathogens Chalara elegans and Fusarium oxysporum, respectively (Tahiri-Alaoui *et al.*, 1990; Benhamou *et al.*, 1991), as well as in lupine and birch roots exposed to abiotic stress (Utriainen *et al.*, 1998; Przymusinski *et al.*, 2004). In the leaves PRs are present in mesophyll and epidermal tissues. They are also localized in the abscission zone of leaves and inflorescence, abscission zone at the stem-petiole junction, and vascular tissue of stems and petioles (Del Campillo and Lewis, 1992; Eyal *et al.*, 1993). In inflorescences PRs are detected in sepals, pedicels, anthers, pistils, stigmata and ovaries (Van Loon, 1999; Buchel and Linthorst, 1999). In seeds of maize, sorghum, oat, barley, and wheat a group of PRs is established,commonly named permatins, characterized as PR-5 thaumatin-like proteins (Vigers *et al.*, 1991). Linusitin from flax seeds is referred to the same group (Anlovar *et al.*, 1998). Noteworthy, specific cell types, such as cultured plant cells, are highly active in PRs expression (Singh *et al.*, 1987).

Induction

Besides the known PRs inducers of biotic origin (pathogens, insects, nematodes, her- bivores) (Fidantsef *et al.*, 1999; Van Loon, 1999; Robert *et al.*, 2001; Reiss and Horstmann, 2001; Schultheiss *et al.*, 2004), a new type of biotic inducers, Orobanche weeds have been reported in tobacco (Joel and Portnoy, 1998). Pathogen-derived elicitors are potent PRs inducers. Well-characterized are glucan and chitin fragments derived from fungal cell walls, fungus-secreted glycoproteins, peptides, and proteins of elicitin family (Honée *et al.*, 1998; Zhou, 1999; Kombrink *et al.*, 2001; Edreva *et al.*, 2002). Protein products of avirulence genes in fungi and bacteria are capable of PRs inducing (Staskawicz *et al.*, 1995; Hennin *et al.*, 2001). Earlier data pointed that cell wall splitting enzymes, such as polygalacturonases, induced PRs accumulation (Pierpoint *et al.*, 1981). Later it has been shown that polygalacturonases release biologically active pectic fragments from plant cell walls, named endogenous elicitors (Mc Neil *et al.*, 1984), capable of inducing a set of defense responses in plants, including PRs accumulation (Boudart *et al.*, 1998). Chemicals, such as salicylic, polyacrylic and fatty acids, inorganic salts, as well as physical stimuli (wounding, UV radiation, osmotic shock, low temperature, water deficit and excess), are involved in PRs induction. A special class of PRs inducers are hormones (ethylene, jasmonates, abscisic acid, kinetin, auxins) (Edreva, 1991; Tamás *et al.*, 1997; Van Loon, 1999; Buchel and Linthorst, 1999; Fujibe *et al.*, 2000). Recently, the dissipation of the proton gradient across the plasma membrane, provoked by the fungal toxin fusicoccin, activator of the plasma membrane H+-ATPase, was reported to induce PRs (Schaller *et al.*, 2000). Reactive oxygen species (ROS)-mediated PRs-formation has largely been recognized (Anderson Grant and Loake, 2000; Schulteiss *et al.*, 2004). Besides being induced by a wide array of environmental/external cues, PRs synthesis can be triggered by internal plant developmental stimuli. Fraser (1981) was the first to report the formation of a set of PRs in leaves of healthy tobacco plants as they reached the flowering and senescing stage. Similar data are reported by Hanfrey *et al.* (1996) for senescing Brassica napus leaves. The presence of PRs in different flower parts, their appearance in abscission zones (Buchel and Linthorst, 1999), as well as their relation to seed germination and somatic embryogenesis (Kragh *et al.*, 1996) point that they are developmentally con-trolled. It is noteworthy that developmentally-induced PRs are accumulated in an organ and tissue-specific manner (Ekramoddulah *et al.*, 2000; Kombrink *et al.*, 2001). PRs in plants are coded by a small multigene family. Since their discovery, regulation of PRs has been a highly active research area. Putative plasma membrane- localized receptors of PRs inducers are suggested and secondary signals of PRs induction, such as salicylic acid (SA), jasmonic acid and ethylene are established. Many of these secondary signals are well-known inducers of PRs expression (Zhou, 1999; Cameron *et al.*, 2000; Poupard *et al.*, 2003). Cross-talks are common between signaling pathways mediated by these secondary messengers. Thus, SA-independent/jasmonate depen- dent and vice-versa pathways of PRs induction have been demonstrated (Mitsuhara *et al.*, 1998; Fidantsef *et al.*, 1999). It has been proven that PRs synthesis is regulated at transcriptional level; the exact mechanisms of transcriptional regulation have been ones of the most active fields of PR gene studies. Several cis regulatory elements in PR-promotors mediating PR gene expression have been identified. These include W-box, GCC box, G box, MRE-like sequence, SA-responsive element (SARE) (Zhou, 1999). New mutants are developed providing clues into the better understanding of the regulation of PRs (Delaney, 2000). PRs are synthesized following a long lag period (Matsuoka and Ohashi, 1986); the synthesis proceeds *in situ, i.e.* PRs are not translocated from the site of their induction to other plant parts, as proven by elegant grafting experiments (Gianinazzi *et al.*, 1982).

Occurrence and Properties of PR Proteins

PR-1 Proteins

The PR-1 family contains the first discovered PRs and is also the most predominant. Despite extensive studies, no biochemical function is known for any of the PR-1 proteins (van Loon *et al.*, 2006). Tobacco plants transformed to constitutively express PR-1 showed enhanced resistance specifically against two oomycete fungi, *Peronospora tabacina* and *Phytophthora parasitica var nicotinae.* However, the transformants were as susceptible as the non-transformants when challenged with other types of fungi, bacteria or viruses (Linthorst *et al.*, 1989;

Alexander *et al.*, 1993). When aliquots of tobacco PR-1 were applied to leaf discs of tobacco, it significantly reduced the development of *Phytophthora infestans* and an additional in vitro study showed that especially a basic tobacco PR-1 exhibited negative effects on this oomycete (Niderman *et al.*, 1995).Reports on the localization of different PR-1 proteins in infected tissues show multiple localizations. PR-1 has been localized to vacuoles in tomato infected by citrus exocortis viroids and in tobacco after being induced by darkness (Vera *et al.*, 1989; Sessa *et al.*, 1995). In potato leaves infected by *P. infestans,* PR-1 was localized in epidermis including stomata guard cells and glandular trichomes, the intercellular space, crystal idioblasts and in both phloem and xylem tissues of vascular bundles (Hoegen *et al.*, 2002). PR-1 proteins have also been found in cell walls of *Phytophthora capsici* and *Chalara elegans* (Tahiri-Alaoui *et al.*, 1993; Hong and Hwang, 2002). A PR-1 protein in maize, PRm, has been localized specifically to the plasmodesmata of the phloem in both maize and transgenic tobacco plants (Murillo *et al.*, 1997; Bortolotti *et al.*, 2005). PR-1 gene transcriptshave been localized to special phloem cells in the vascular bundle of pepper stems infected with P. capsici (Lee *et al.*, 2000a). Additionally, PR-1 has been detected in the xylem sap from non-infected Brassica napus and in the guttation fluid from non- infected barley seedlings, indicating long-distance transport of PR-1 in the transpiration stream (Grunwald *et al.*, 2003; Kehr *et al.*, 2005).

PR-2 Proteins

The PR-2 proteins are -1,3-glucanases, endoglucanases that can catalyze hydrolytic cleavage of -1,3-D-glucosidic linkages in -1,3-glucans (Leubner-Metzger and Meins, 1999). PR-2 is believed to act primarily on glucans present in the cell wall of most fungal pathogens to release oligosaccharides (Mauch and Staehelin, 1989). The plant may then perceive these fragments as elicitors that serve to trigger further defense responses. PR-2 is also active in plant reproductive processes and ripening of fruits (Leubner-Metzger and Meins, 1999). The PR-2 family is divided into three structurally distinct classes of β-1,3- glucanases, with acidic and basic counterparts that significantly differ in their specific enzymatic and antifungal activity (Kauffmann *et al.*, 1987; Sela-Buurlage *et al.*, 1993). Several in vitro experiments have demonstrated antifungal effects mainly by basic class I β -1,3-glucanases against a wide range of fungi, either alone or in combination with PR-3 (Mauch *et al.*, 1988; Ludwig and Boller, 1990; Sela-Buurlage *et al.*, 1993). The synergistic effect between PR-2 and PR-3 has also been shown in transgenic plants (Zhu *et al.*, 1994). Morphological studies on the influence of PR-2 on hyphal tips of Trichoderma longibrachiatum showed that PR-2 and PR-3 together are particularly effective at the hyphal tip causing balloon-like swelling and lysis of the tip (Mauch *et al.*, 1988; Arlorio *et al.*, 1992). Both PR-2 and PR-3 are likely to play a dual role in plant defense both directly by hydrolyzing structural components from fungal cell walls and indirectly by releasing elicitors that may amplify the defense response in the plant (Stintzi *et al.*, 1993). PR-2 induced by ethylene has been restricted to the vacuoles of lower epidermal cells and parenchyma cells adjacent to vascular bundles and over the middle lamella in the intercellular space in bean leaves (Mauch and Staehelin, 1989; Mauch *et al.*, 1992). In an incompatible reaction in wheat leaves against Puccinia recondita, PR-2 was mainly recovered in the domain of the host cell wall closest to plasmalemma, cell wall appositions, intercellular space, guard cells and secondary thickening of xylem vessels as well as in the hyphal cytoplasm and cell wall (Hu and Rijkenberg, 1998). Tomato roots infected with Fusarium oxysporum showed PR-2 predominantly localized in the cell walls and vacuoles of the host, and in the cell wall and septa of the fungus. Thickened secondary cell walls of the xylem vessels were also heavily labeled (Benhamou *et al.*, 1989). PR-2 has been studied in floral organs of barley where it is developmentally regulated. It was localized in the anther and pistil tissues, including the stigmatic hairs. Besides cell walls, PR-2 was also recovered in plastids in cells of the style and ovary cell wall (Liljeroth *et al.*, 2005).

PR-3 Proteins

Proteins of the PR-3 family are endochitinases, which hydrolyze -1,4-linkages between N-acetylglucosamines of chitin, releasing oligosaccharides from the cell walls of many fungi (Boller, 1993). Chitin is not a natural component of plant cells but is present in most fungal cell walls and the insect cuticula (Stintzi *et al.*, 1993). The division of chitinases into different classes (I-VII) is mainly based on the presence or absence of a cysteine-rich domain and of a C-terminal extension providing a signal for vacuolar targeting. The cysteine-rich domain is believed to be the chitin-binding part targeting PR-3 to chitin-containing pathogens (Neuhaus, 1999). Class I chitinases contain a cysteine-rich domain and has a 10-15 fold higher chitinase activity than class II chitinases that lack this domain (Sela-Buurlage *et al.*, 1993). The high chitinase activity of PR-3 class I is also reflected in the antifungal activity demonstrated in in vitro studies of Fusarium solani and Rhizoctonia solani (Broglie *et al.*, 1991; Sela-Buurlage *et al.*, 1993). In fact, PR-3 has been shown to inhibit growth of most fungi but not of the oomycetes Phytophthora and Pythium, which lack chitin in their cell wall (Mauch *et al.*, 1988). Several studies have also shown the synergistic effect of PR-3 and PR-2 as mentioned above. It has been proposed that the thinning of the fungal cell wall by PR-2 exposes the chitin present in the inner parts of the wall, making it accessible to chitinases to hydrolyze the fungal cell wall as well as to release elicitors (Kombrink and Somssich, 1997). PR-3 has been localized to the vacuole of ethylene-treated bean leaves and tomato leaves infected with Cladosporium fulvum (Mauch *et al.*, 1992; Wubben *et al.*, 1992). Wheat spikes infected with *Fusarium culmorum* showed PR-3 mainly on the cell walls, over cell wall appositions, intercellularly and on cell walls of the hyphae (Kang and Buchenauer, 2002). PR-3 has also been localized specifically to the cell wall of the style and stigmal branches of flower organs in barley early in the development. The labeling intensity of PR-3 did however decrease significantly during the later stages of flower development (Liljeroth *et al.*, 2005). Chitinase labeling has been reported on host cell walls and in the intercellular space of pepper stems infected with P. capsici, including the presence of chitinase mRNA in phloem-related cells (Lee *et al.*, 2000b). The presence of chitinase mRNA in phloem cells has also been described in potato leaves challenged with P. infestans and in pepper leaves infected with Colletotrichum coccodes (Büchter *et al.*, 1997; Hong and Hwang, 2002). Xylem sap from tomato infected with F. oxysporum revealed that the presence of PR-3 was very low in the fluid compared to PR-1, PR-2 and PR-5 (Rep *et al.*, 2002). In contrast, chitinase-antifreeze proteins (AFP) induced by cold acclimation of rye were found in cell walls of all leaf tissues particularly abundant in epidermal cell walls and xylem vessels. The corresponding mRNAs were found in the same cell types as the chitinase-AFPs and in vascular parenchymal cells surrounding xylem vessels (Pihakaski-Maunsbach *et al.*, 2001).

PR-5 Proteins

Proteins that belong to the PR-5 family are also known as thaumatin-like (TL) proteins as they show sequence similarities to the sweet-tasting plant protein thaumatin (Linthorst, 1991). Osmotins, proteins induced by salt stress, also belong to the PR-5 family (Velazhahan *et al.*, 1999). Like other PRs the PR-5 proteins constitute of acidic-neutral and basic isoforms. In some dicotyledonous plants the extracellular PR-5 proteins tend to be acidic while the vacuolar ones tend to be basic. The various isoforms of PR-5 are associated with diverse functions such as antifungal activity, protection against osmotic stress (Kononowicz *et al.*, 1992) and freezing tolerance (Hon *et al.*, 1995). Several PR-5 proteins display significant activity *in vitro* in inhibiting hyphal growth, spore germination or development of germ tubes, probably by a fungal plasma membrane permeabilizing mechanism (Velazhahan *et al.*, 1999). A basic barley PR-5 possesses inhibitory activity *in vitro* against germ tube development of *Blumeria graminis* (Tandrup Poulsen, 2001). A basic PR-5, osmotin from tobacco, has been shown to inhibit growth of *P. infestans, Neurospora crassa, Trichoderma reesei* and *Candida albicans* in vitro (Woloshuk *et al.*, 1991, Vigers *et al.*, 1992). Interestingly, two basic barley PR-5 proteins inhibited growth of *Trichoderma viride* and *C. albicans* (Hejgaard *et al.*, 1991). However, one of their homologous acidic tobacco PR-5 did not show any activity against these fungi. Instead it was most potent against *Cercospora beticola* (Vigers *et al.*, 1992). Tobacco osmotin induced spore lysis, inhibited spore germination or reduced spore viability in different species of *Bipolaris, Fusarium* and *Phytophthora*. However, the hyphal growth of *Aspergillus, Rhizoctonia* and *Macrophomina* was not affected by osmotin (Abad *et al.*, 1996). The first leaf of transgenic barley plants, with a pathogen-inducible epidermis- specific promoter fused to a basic PR-5, showed enhanced resistance against *B. graminis, Rynchosporium secalis* and *Drechslera teres*, while no disease reduction was observed on infection with Puccinia hordei (Tandrup Poulsen, 2001). Overexpression of PR-5 in potato delayed development of disease symptoms of *P. infestans* (Liu *et al.*, 1994), whereas transgenic potato plants expressing antisense PR-5 did not exhibit any higher susceptibility (Zhu *et al.*, 1996). Tobacco plants constitutively overexpressing a rice PR-5 showed enhanced resistance to *Alternaria alternata* (Velazhahan and Muthukrishnan, 2003). Overexpression of a specific fungal cell wall protein in Saccharomyces cerevisiae, otherwise susceptible to tobacco osmotin, increased the resistance, whereas deletion of the genes in a tolerant strain, resulted in sensitivity towards tobacco osmotin (Yun *et al.*, 1997). The resistance or susceptibility of different fungi towards the different PR-5 isoforms indicates specificity in recognition between potentially antifungal proteins and certain binding features of different fungal cell wall proteins (Vigers *et al.*, 1992; Yun *et al.*, 1997). Few localization studies have been reported with PR-5. A basic PR-5 was found on the cell wall of *P. infestans* and in starch granules of chloroplasts and in papilla of tomato leaves expressing SAR (Jeun and Buchenauer, 2001). In barley, PR-5 mRNA was specifically expressed in the mesophyll early after infection with the necrotrophic fungus *Rhynchosporium secalis* (Steiner-Lange *et al.*, 2003). It has been shown that PR-5 proteins are among the most abundant proteins in the xylem sap of healthy *Brassica napus* (Kehr *et al.*, 2005).

PR Protein-10

Ribonuclease like

Ribosome inactivating proteins Ribosome-inactivating proteins are a class of translational inhibitors that share a site- specific RNA N-glycosidase activity (Barbieri and Stirpe 1982; Stirpe *et al.*, 1992). Specifically, RIPs depurinate a universally conserved adenine residue of the large ribosomal RNA (Endo *et al.*, 1987; Endo and Tsurugi 1987) and make susceptible ribosomes impaired in translational elongation processes (Wool *et al.*, 1992). Plant RIPs are stable, basic (pI >8) proteins found in a large number of species (Barbieri and Stirpe 1982; Stirpe and Barbieri 1986) with greater sfpecificity and varying activities towards non host ribosomes of distantly related species including fungi (Roberts and Selitrennikof 1986; Stirpe and Hughes 1989). RIPs exist has single-chain monomeric enzymes with apparent molecular masses of 30 kDa (Type I) or as two polypeptides in which one polypeptide with RIP activity (30 kDa A-chain) is linked by a disulphide bridge to a carbohydrate-binding lectin (30 kDa B-chain) (Type II). The B chain facilitates membrane recognition and transport across cell membranes (Stirpe *et al.*, 1992). Type II RIPs are potent toxins the best known being ricin. Plant RIPs may function as antimicrobial or antiviral defense proteins, consistent with their usefulness in engineered plant disease resistance (Chen *et al.*, 2002; Hong *et al.*, 1996; Jach *et al.*, 1995; Kim *et al.*, 2003; Lodge *et al.*, 1993; Logeman *et al.*, 1992; Veronese *et al.*, 2003; Yuan *et al.*, 2002; Zoubenko *et al.*, 2000). Plant RIPs have also been used to develop therapeutic reagents with anti-HIV, anticancer, or immunomodulatory properties (Di Massimo *et al.*, 1997; French *et al.*, 1995; Stirpe *et al.*, 1992). Large-scale surveys for plant RIPs revealed that a wide variety of plants express RIPs (Gasperi-Campani *et al.*, 1985; Stirpe and Barbieri 1986) and that some plants, including maize, express multiple RIPs (Van Damme *et al.*, 2001). Because of their wide distribution and conserved enzymatic activity, RIPs have generally been presumed to make some important, contribution to the plant defense (Hartley and Lord 1993; Hartley *et al.*, 1996; Nielsen and Boston 2001; Veronese *et al.*, 2003).

PR Protein-12

Defensin

The plant defensin family is quite diverse with the sequence conservation restricted only to eight structurally important cysteines (Lay and Anderson 2005; Thomma *et al.*, 2002). Two dimensional 1H NMR studies revealed that defensins have similar tertiary structures though the primary sequences differ greatly (Bloch *et al.*, 1998; Bruix *et al.*, 1993; Teeter *et al.*, 1990, Vermeulen *et al.*, 1987). Based on the morphology of hyphal branching of *Fusarium culmorum*, defensins could be grouped as morphogenic and non- morphogenic (Osborne *et al.*, 1995). As this grouping depends on the test fungus and medium used. Lay *et al.* (2003) divided defensins into two major classes according to the structure of precursor proteins predicted from their cDNA clones. Precursor proteins with an endoplasmic reticulum signal sequence and defensin domain form the first and largest class, where as members of the second class have larger precursor with both an ER signal sequence and a C-terminal domain (Lay *et al.*, 2003). The prodomain may function as a target sequence for sub-cellular targeting, mostly as a vacuolar targeting signal in monocotyledons as well as dicotyledonous plants (Raikhel 1992). The prodomain may also help in the maturation of defensins by acting as an intra-molecular chaperone and prevents deleterious interaction with other cellular proteins or lipid membranes (Florack *et al.*, 1994). Molecular bases for the antifungal inhibitory activity of some plant defensins, such as DmAMP1 from Dahlia (*Dahlia merckii*) (Osborne *et al.*, 1995), RsAFP2 from radish (*Raphanus sativus*) (Terras *et al.*, 1992), and HsAFP1 from coral bells (*Heuchera sanguinea*) (Osborne *et al.*, 1995), PsDef1 from pea (*Pisum sativum*) (Almeida *et al.*, 2000) and MsDef1 from alfalfa (*Medicago sativa*) (Gao *et al.*, 2000) have been analysed. Concentration of cations and salt in the medium have an antagonizing effect on the antimicrobial activity of plant defensisns (Osborne *et al.*, 1995; Terras *et al.*, 1992; Terras *et al.*, 1993) which is a common feature among insect (Cociancich *et al.*, 2003) and mammalian defensins (Bals *et al.*, 1998; Garcia-Olmedo *et al.*, 2001; Vylkova *et al.*, 2007). Plant defensins fail to induce ion permeable pores or change the electrical properties of artificial phospholipid membranes (Caaveiro *et al.*, 1997; Thevissen *et al.*, 1996); these defensins have a relatively long lag phase before they induce membrane permeabilization indicating that the permeabilization is only secondary effect rather than the cause of their antifugnal activity (Theis and Stahl 2004). This predicted the existence of specific binding sites formed by sphingolipids on the fungal envelop (Thevissen *et al.*, 1999; Thevissen *et al.*, 2000a; Thevissen *et al.*, 2000b). Sphingolipids, having a structural role, are important signaling molecules in cell regulation, cell growth and cell stress response (Bieberich 2004; Dickson *et al.*, 2006). These specific sphingolipids are identified to be mannosyl di-inositol phosphoryl-ceramide in DMAmp1 (Thevissen *et al.*, 2005; Thevissen *et al.*, 2004; Thevissen *et al.*, 2000b; Thevissen *et al.*, 2003; Aerts *et al.*, 2007). RS AFP2 interacts with Glucosyl ceramides (Glc Cer) of fungi (Saito *et al.*, 2006; Sakaki *et al.*, 2001; Terras *et al.*, 1992a; Thevissen *et al.*, 2004; Thevissen *et al.*, 2007) resulting in cell lysis. Where as it is non-toxic to plants or humans as it cannot bind to Glucosyl ceramides of plant or human (Thevissen *et al.*, 2004; Terras *et al.*, 1992a; Terras *et al.*, 1995). RS AFP2 is also known to induce intracellular signaling cascade, leading to ROS induction, membrane permeabilization, fungal cell growth arrest, cell death (Aerts *et al.*, 2007) and apoptosis (Madeo *et al.*, 1997). In *Candida albicans* it also triggered the activation of caspases or caspase like proteases (Aerts *et al.*, 2009). Pea defensin Psd1 (Almeida *et al.*, 2000) in *Neurospora crassa* suppress normal progression of the cell cycle by interfering with cyclin F and blocks conidial formation (Berrocol-Lobo *et al.*, 2007). MsDef1 is a broad-spectrum antifungal defensin from alfalfa (*M. sativa*) seed, previously referred to as Alf AFP (Gao *et al.*, 2000). It is a morphogenic defensin on *Fusareium graminearum*. MS Def 1 targets Ca^{2+}channels in susceptible fungi (Tsien and Tsien 1990) and mammalian cells (Spelbrink *et al.*, 2004). Ca^{2+} is a ubiquitous signaling molecule in fungi, regulates cAMP levels (Iida *et al.*, 1990), bud formation

(Davis1995) and hyphal elongation (Jackson and Health 1993). Hyphal elongation is a process that is controlled by a gradient in cytosolic Ca^{2+} generated by hyphal tip-localized Ca^{2+}channels (Tsien and Tsien 1990). MsDef1 binds to the extracellular side of the Ca^{2+} channel and causes hyper branching. MsDef1, like RsAFP2, targets GlcCer in membranes of susceptible yeast (Ramamoorthy *et al.*, 2007a) and regulate mitogen activated protein kinase (MAPK) signaling pathways (Ramamoorthy *et al.*, 2007b) which control multiple developmental processes related to cell wall integrity, sexual reproduction and pathogenicity (Lengeler *et al.*, 2000; Xu 2000). Different plants have been transformed with plant defensin genes. Constitutive expression of the radish defensin enhanced resistance of tobacco plants to the fungal leaf pathogen *Alternaria longipes* (Terras *et al.*, 1995) and to *A. solani* in tomato (Ohtani *et al.*, 1977). Transgenic Canola (*Brassica napus*) constitutively expressing a pea defensin had slightly increased resistance against blackleg (*Leptosphaeria maculans*) disease (Wang *et al.*, 1999). Constitutive expression of MsDef1 in potatoes (Gao *et al.*, 2000), showed resistance against the fungus *Verticillium dahliae.* Infection levels of fungus in the transformed plants was reduced six-fold compared to the non-transformed plants and the resistance was stable and equal to, or greater than, the level of resistance obtained with non-transgenic plants grown in fumigated, non-infested soil, both in the glass house and field conditions (Gao *et al.*, 2000). Transformed plants of *Solanum melongena* expressing the defensin Dm-AMP1 showed resistance against the pathogenic fungus Botrytis cinerea and also to the root pathogen, *Verticellium albo-atrum.* However, symbiotic relation with the mycorrhizal fungus *Glomus mosseae* was not affected even though the defensin was expressed in root exudates of transformants (Turrini *et al.*, 2004).

PR-13 Protein

(Thionin)

Thionins During fermentation of wheat flour, in 1895, Jago and Jago observed decreased CO^2 release and suggested the existence of substances with antimicrobial activity in wheat (Jago and Jago 1911). A low molecular weight protein with high sulfur content was crystallized from wheat endosperm, and named as purothionin (from Greek puro– wheat and thio- sulfur) (Balls *et al.*, 1942). Proteins similar to purothionins were isolated from the endosperm of the *Triticum* and *Aegilops species* (Carbonero and Garcia-Olmedo 1961), α and β-hordothionins from barley (Ponz *et al.*, 1983; Ponz *et al.*, 1986; Redman and Fisher, 1969; Rodriguez- Palenzuela *et al.*, 1988), α and β-avenohtionins from oat (Bekes and Lasztity, 1981), secale thionins (Hermandez-Lucas *et al.*, 1978) from rye, viscotoxins from mistletoe (Samuelsson and Pettersson, 1970; Samuelson, 1973), phoratoxins A and B from *Phoradendron tomentosum* (Mellstrand and Samuelson 1973; Mellstrand and Samuelson 1974; Thunberg 1983), denclatoxin B from *Dendrophtora clavata* (Samuelsson and Pettersson, 1977) and ligatoxins A and B from Phoradendron liga (Li *et al.*, 2002; Thunberg and Samuelsson 1982a; Thunberg and Samuelsson 1982b), crambin from *Crambe abyssinica* (Hendrickson and Teeter 1981; Teeter and Hendrickson 1979; Teeter *et al.*, 1981; Van etten *et al.*, 1965; Teeter *et al.*, 1981f), pyrularia thionins from *Pyrularia pubera* (Vernon *et al.*, 1985) and leaf thionins from barley (Bohlman and Apel 1987;

Gausing 1987). Except crambin, thionins inhibit growth of phyto-pathogenic fungi (Bohlman *et al.*, 1988), bacteria (Fernandez-deCaleya *et al.*, 1972), Yeast (Hermandez-Lucas *et al.*, 1974), and insect larvae (Kramer *et al.*, 1979). Mammalian cells subjected to different thionins showed increased cell membrane permeability (Carrasco *et al.*, 1981). Thionins act by showing strong immunomodulatory effects as in the case of viscotoxins (Stein *et al.*, 1999a; Stein *et al.*, 1999b; Tabiasco *et al.*, 2002), inhibiting protein synthesis in cell free systems through direct interaction with mRNA or at initial translational level (Brummer *et al.*, 1994; Garcia-Olmedo *et al.*, 1983), interfering with DNA synthesis by inhibiting ribonucleotide reductase (Johnson *et al.*, 1987) and irreversible inhibition of β- glucuronidase (Diaz *et al.*, 1992). Thionins might act as regulatory proteins in the redox regulation of enzymes as thioredoxins (Johnson *et al.*, 1987; Wada and Buchanan 1981), as storage proteins especially as sources of sulfur as evidenced by viscotoxins (Schrader-Fischer and Apel 1993), and as defense proteins with generalized effects whose transcription levels increase after inoculation with pathogen (Bohlmann *et al.*, 1988). Thionins have been deployed to obtain transgenic plants resistant to phytopathogenic fungi. The α-hordothionin gene expression in tobacco increased the resistance against Pseudomonas syringae (Carmona *et al.*, 1993), viscotoxin expression in Arabidopsis thaliana conferred resistance to *Plasmodiophora brassicae* (Holtorf *et al.*, 1998), and expression of an oat thionin in transgenic rice conferred protection against phytopathogenic bacteria, *Burkholderia plantarii* and *B. glumae* (Iwai *et al.*, 2002).

PR Protein-14

Lipid Transfer Proteins

Lipid Transfer Proteins (LTPs) LTPs have an important role in plant defense against pathogens and possess inhibitory activities with increased expression levels in plant tissues soon after infection. Plant LTPs are a homogeneous class of small (9-10 kDa), ubiquitous, abundant (≥ 4 per cent of total soluble proteins), basic proteins containing eight cystine residues with four conserved disulfide bridges, capable of transferring lipids between membranes. Structural studies of LTPs in rice (Lee *et al.*, 1998), maize (Shin *et al.*, 1995) and wheat (Simorre *et al.*, 1991) have revealed a conserved four α helix bundle stabilized by four disulfide bonds, which forms a hydrophobic internal cavity accommodating interaction between the aliphatic chain of lipids and the hydrophobic residues exposed in the cavity (Lee *et al.*, 1998). The cavity is also essential for cell wall loosening and extension (Nieuwland *et al.*, 2005). LTPs also carry a signal peptide indicating their secretory nature (Mundy and Rogers 1986; Skriver *et al.*, 1992; Tchang *et al.*, 1988; Vignols *et al.*, 1994). Due to their non-specific nature, they are involved in transferring various types of polar lipids, like phosphatidyl choline, phosphatidyl ethanol amine and phosphatidyl inositol, besides galactolipids (Douliez *et al.*, 2001; Kader 1997; Kader 1996). They function as a carrier of the acyl monomer, which is an essential building block in cutin biosynthesis and cell wall strengthening through the secretion or deposition of cutin (Kader 1996). Immunocytolocalization studies revealed the presence of LTPs in the cell wall of several species (Pyee *et al.*, 1994; Sossountzov *et al.*, 1991; Thoma *et al.*, 1993; Thoma *et al.*, 1994). LTPs structurally resemble elicitins of Phytophtora and Pythium (Ponchet *et al.*, 1999) which bind the plant sterols that are necessary for successful reproduction by the oomycete (Hendrix 1970). LTPs are essential to develop SAR and express PR genes in leaves by acting as a translocator for the release of mobile signal in Arabidopsis (Maldonado *et al.*, 2002). Various LTPs were purified from different plant tissues (Bouillon *et al.*, 1987; Cammue *et al.*, 1995; Castro *et al.*, 2003;Conti *et al.*, 2001; Desormeaux *et al.*, 1992; Douliez *et al.*, 2001; Jones and Marinac 2000; Kader *et al.*, 1984; Liu *et al.*, 2002; Molina *et al.*, 1993; Nielsen *et al.*, 1996; Ostergaard *et al.*, 1993; Ramirez-Medeles *et al.*, 2003; Takishima *et al.*, 1988; Terras *et al.*, 1992b). Transcripts of three LTPs accumulated in leaf, stem and fruit tissues of pepper plants following infection with Xanthomonas campestris pv. vesicatoria, Phytophthora capsici and Colletotricum gloeosporioides (Jung *et al.*, 2003). Increased expression of LTPs following inoculation with tobacco mosaic virus (TMV) was observed in pepper plants (Park *et al.*, 2002b). Suppression of endogenous pepper lipid transfer proteins CALTPI and CALTPII by virus induced gene silencing (VIGS), resulted in enhanced susceptibility to *Xanthomonas campestris* pv. vescatoria and pepper mosaic mottle virus (Sarowar *et al.*, 2009). LTP110 from rice was able to inhibit the germination of *P. oryzae* spores, but only slightly inhibited the growth of Xanthomonas in vitro (Ge *et al.*, 2003). Overexpression of barley LTP2 in transgenic tobacco and Arabidopsis enhanced resistance to *P. syringae* pv. *tabaci* and *P. syringae* pv. tomato DC3000 (Molina and Garcia-Olmedo 1997). Overexpression of the pepper LTP1 gene has been shown to confer resistance to *P. syringae* pv. tomato DC3000 and *Botrytis cinerea* and tolerance to high salinity and drought stress in Arabidopsis (Jung *et al.*, 2005). The constitutive expression of CALTPI and CALTPII genes in tobacco plants showed enhanced resistance to oomycete pathogen, *Phytophthora nicotianae* and bacterial pathogen, *Pseudomonas syringae* pv. *Tabaci* (Sarowar *et al.*, 2009).

Table 11.1. Lipid Transfer Proteins Reported against Phytopathogens

Sl.No.	Source of LTP	Target Pathogen	Reference
1.	Onion	*Rhizoctonia solani*	Patkar and Chatto, 2006
		X. Oryzae pv. oryzae	
2.	Rice	*Pyricularia oryzae*	Ge *et al.*, 2002
		X. oryzae	
3.	Mung bean seeds	*Fusarium solani*	Weng *et al.*, 2004
4.		*F. oxysporium*	
		Pythium aphandermatum	

Functions

The assumption that PRs are devoided of enzymatic functions was challenged by Legrand *et al.* (1987) detecting chitinase activity in four members of group 3 tobacco PRs. The same research team established β-1,3-glucanase activity in four members of group 2 tobacco PRs (Kauffmann *et al.*, 1987). Later on chitinase activity was detected in PR-4, PR-8 and PR-11, PR-4 being referred to as chitin-binding proteins. Proteinase, peroxidase, ribonuclease and lysozyme activities were established in PR-

7, PR-9, PR-10 and PR-8, respectively. PR-6 was assigned proteinase-inhibitory properties. Membrane-permeabilizing functions are characteristic of defensins, thiols and lipid-transfer proteins (LTPs), referred to as PR-12, PR-13 and PR-14, respectively, and of osmotins and thaumatin-like proteins (PR-5). Multiple enzymatic, structural and receptor functions are detected in "do-all" germins and germin-like proteins re- ferred to as PR-15 and PR-16, respectively (Van Loon and Van Strien, 1999, and references therein; Van Loon, 2001; Selitrennikoff, 2001; Bernier and Berna, 2001; Park *et al.*, 2004 a, b). An important common feature of most PRs is their antifungal effect; some PRs exhibited also antibacterial, insecticidal, nematicidal, and – as recently shown – antiviral action. Toxicity of PRs can be generally accounted for by their hydrolytic, proteinase-inhibitory and membrane-permeabilizing ability. Thus, hydrolytic enzymes (β-1,3-glucanases, chitinases and proteinases) can be a tool in weakening and de- composing of fungal cell walls, containing glucans, chitin and proteins, while PR-8 can disrupt gram-positive bacteria due to lysozyme activity (Van Loon and Van Strien, 1999; Van Loon, 2001; Selitrennikoff, 2001). Last year PR-10 (named Ca PR-10), induced in hot pepper (*Capsicum annuum*) by incompatible interactions with TMV- Po and *Xanthomonas campestris* pv. *vesicatoria*, was shown to function as a ribonu- clease. Data are presented that subsequent phosphorylation of Ca PR-10 increases its ribonucleolytic activity to cleave invading viral RNAs, and this activity is important to its antiviral pathway in vivo. Antibiotic activity of Ca PR-10 is also exerted against oomycete fungi (Park *et al.*, 2004 a). Proteinase-inhibitory properties of PR-6 may confer anti-insect and anti-nematode effects, inactivating the proteins secreted by these parasites in the invaded plant tissues. Plasma membrane-permeabilizing ability proper to PR-5, PR-12, PR-13 and PR-14 contributes to plasmolysis and damage of fungal and bacterial pathogens, inhibiting their growth and development (Vigers *et al.*, 1992; Abad *et al.*, 1996; Van Loon and Van Strien, 1999, and references therein; Van Loon, 2001; Selitrennikoff, 2001). This effect may be due to electrostatic inter- actions of PRs with membrane components, leading to conformatial changes, dissi- pation of membrane gradient, and formation of pores in membranes (Abad *et al.*, 1996; Cheong *et al.*, 1997; Anlovar *et al.*, 1998). Multifaceted functionality of PR- 15 and PR-16, including cell wall remodeling ability, can be directed against patho- gens and may have protective role (Park *et al.*, 2004 b). Interestingly, it was recently found that among the identified plant allergens 23 per cent belong to the group of PRs. So far, plant-derived allergens have been identified with similarities to PR families 2, 3, 4, 5, 10 and 14 (Hoffmann-Sommergruber, 2001). This property may confer defensive functions displayed by the plant in hostile environment. Apart from the above-described mechanisms of direct break-down or damage of pathogens, PRs can operate in a distinct pathway involving the hydrolytic release of chitin and glucan fragments from fungal cell walls. These oligosaccharides are endowed with elicitor activity and can induce a chain of defense reactions in the host plant (Ham *et al.*, 1991; Lawrence *et al.*, 2000; Kombrink *et al.*, 2001). The peroxidase activity of PR-9 can contribute to the rigidification and strengthening of plant cell wall in response to pathogen attack (Lagrimini *et al.*, 1987). The defensive functions of PRs against pathogens are presumably corroborated by data about their constitutive expression in seeds and plant organs; high fungitox- icity of seed osmotins and thaumatin-like proteins has been established (Vigers *et al.*, 1992; Abad *et al.*, 1996). Protective role of PRs can also be inferred from their accumulation in plant cell wall appositions formed against pathogen ingress, as well as from their release into fungal structures penetrating plant tissues (Benhamou, 1991; Jeun, 2000; Jeun and Buchenauer, 2001). The cell-, tissue-, organ- and development-specific expression pattern of PRs suggests important functions beyond defense against pathogens. Thus, basic tobacco glucanase PR-2d functions developmentally in seed germination by weakening the endosperm, thus allowing the radicle to protrude (Vögeli-Lange *et al.*, 1994). Chitinases homologous to PR-3 and PR-4 act as morphogenetic factors in carrot embryogenesis (Kragh *et al.*, 1996), and several PRs accumulate upon the transition of plants to flowering and senescence (Fraser, 1981; Hanfrey *et al.*, 1996), also suggestive of a developmental role. Basic PR-5 (osmotins) are abundantly induced in tobacco and tomato cells in response to osmotic stress, thus contributing to osmotic adaptation (Singh *et al.*, 1987). These findings raise the question as to whether PR genes evolved primarily to limit damage by invading pathogens, or were adapted from other functions to serve an accessory protective role (Van Loon and Van Strien, 1999).Until now the functions of the most abundant PRs family, PR-1, remain obscure. A direct inhibitory effect of basic tomato and broad bean PR-1 family members against fungal pathogens (*Phytophthora infestans* and *Uromyces fabae*, respectively) has been demonstrated by *in vitro* and *in vivo* experiments (Niderman *et al.*, 1995; Rauscher *et al.*, 1999), but the mode of action as well as the cellular and molecular targets of PR-1 proteins are still unknown. Comparative "screening" of PR-1-type proteins from various plant taxa indicates that PR-1 family is highly conserved in plants. Moreover, as already mentioned, it is related to sequences present in yeast, insects and verte-brates. The corresponding proteins in insects are major venom allergens presumably directed to other organisms. It may be speculated that related protein family in verte-brates encodes lytic enzymatic and antimicrobial activity. The widespread occur-rence of PR-1 family suggests that these proteins share an evolutionary origin and possess activity essential to the functioning and surviving of living organisms (Van Loon, 2001). Presently, it is still difficult to assign a causative role of PRs in plant resistance to pathogens. The reason for this is that the numerous data on PRs as disease resistance factor are mostly of correlative character. Four lines of supporting evidence can be outlined.

a) Stronger accumulation of PRs in inoculated resistant as compared to suscep- tible plants. Besides previous data, substantiating this statement (Van Loon, 1985, and references therein), differential responses of resistant/susceptible plants were recently reported in tomato plants, inoculated with *Cladosporium fulvum* (Wubben *et al.*, 1996); Phytophthora infestans-infected potato (Tónon *et al.*, 2002); *Venturia inaequalis*-inoculated apple (Poupard *et al.*, 2003); *Pseudomonas syringae*-infected grapevine (Robert *et al.*, 2001); *Xanthomonas campestris* pv. *vesicatoria* and TMV- Po-infected hot pepper (Park *et al.*, 2004 a, b), *etc.* Additional resistance gene(s) against *Cladosporium fulvum* present on the Cf-9 introgression segment have been shown to be associated with strong PR-protein accumulation (Laugé *et al.*, 1998). In some pathosystems mRNAs for certain PRs members accumulate to similar levels in compatible and incompatible interactions, but the maximum level of expression is reached much faster in the latter (Van Kan *et al.*, 1992).

b) Important constitutive expression of PRs in plants with high level of natural disease resistance. This correlation was observed in several pathosystems, such as apple – *Venturia inaequalis* (Gau *et al.*, 2004), tomato – *Alternaria solani* (Lawrence *et al.*, 2000), and potato – *Phytophthora infestans* (Vleeshouwers *et al.*, 2000), the last authors proposing PR mRNAs as molecular marker in potato breeding programs.

c) Significant constitutive expression of PRs in transgenic plants overexpressing PR genes accompanied by increased resistance to pathogens. Thus, increased toler-ance to Peronospora tabacina and *Phytophthora parasitica* var. *nicotianae* was dem-onstrated in tobacco overexpressing PR1a gene (Alexander *et al.*, 1993). Transgenic rice and orange plants overexpressing thaumatin-like PR-5 possessed increased tol- erance to *Rhizoctonia solani* and *Phytophthora citrophthora*, respectively (Datta *et al.*, 1999; Fagoaga *et al.*, 2001), while transgenic potato overexpressing PR-2 and PR-3 had improved resistance to *Phytophthora infestans* (Bachmann *et al.*, 1998). PR-2 and PR-3 genes, coding for β-1,3-glucanase and chitinase, respectively, confer resistance of carrot to several fungal pathogens. The simultaneous expression of to- bacco β-1,3-glucanase and chitinase genes in tomato plants results in increased [re- sistance to fungal pathogens] (Melchers *et al.*, 1998). On the contrary, silencing of PR-1b gene in barley facilitates the penetration of the fungal pathogen *Blumeria graminis* f. sp. *hordei* in the leaves (Schultheiss *et al.*, 2003).

d) Accumulation of PRs in plants in which resistance is locally or systemically induced. Generalizing this broad research area it can be stated that PRs are recog- nized as markers of the systemic acquired resistance (SAR), and PRs genes are in-volved in the list of the so-called SAR-genes (Ward *et al.*, 1991). It has largely been demonstrated that SAR and the accompanying set of PRs are induced by different pathogens, as well as by a range of chemicals predominantly in a salicylic acid- dependent pathway; SAR is active against a broad spectrum of pathogens. Some SAR-inducing chemicals, such as benzothiadiazole (BTH), β-aminobutyric acid (BABA) or 2,6-dichloroisonicotinic acid (DCINA) are harmless commercially sup- plied compounds and have promising practical application as novel tools in plant protection (Van Loon, 1997; Edreva, 2004 and references therein). It is essential to underline, that PRs members induced in resistant or SAR-ex- pressing plants, as well as PRs from transgenic resistant plants exhibit high antimi- crobial activity (Enkerli *et al.*, 1993; Anfoka and Buchenauer, 1997; Rauscher *et al.*, 1999; Tonón *et al.*, 2002; Anand *et al.*, 2004), this suggesting their direct role in disease resistance. In contrast to the above data, a view has been developed that PRs are rather related to the severity of symptom expression than to resistance. This assumption flows from experiments with PVY-infected potato (Naderi and Berger, 1997), PVY-,

Table 11.2.Transgenic Plants Devloped using PR Proteins.

Host/Source	Gene	Transgenic plant	Targeted pathogens	Reference
Alfa alfa and rice	*alAFP* and rice chitinase gene.	Tomato	*Botrytis cineria*	Chen *et al.*, 2009
PR4 class Wheatwin1 and Wheatwin2	*wPR4a* and *wPR4b*	Tobacco	*Phytophthora nicotianae*	Fiocchetti *et al.*, 2008
Tobacco	chitinase and _-1,3-glucanase gene	*Oryza sativa* L.	*Rhizoctonia solani.*	Sridevi *et al.*, 2008
Trichoderma harzianum	chitinase gene.	lemon plants	*Phoma tracheiphila* and *Botrytis cinerea*	Gentile *et al.*, 2007
Balsam pear (*Momordica charantia* L.) chitinase	*Mcchit1*	Overexpression in *N. benthamiana*	*Phytophthora nicotianae*	Xiao *et al.*, 2007
	Mcchit1	Cotton	*Verticillium wilt*	Xiao *et al*, 2007
Tobacco, alfalfa	Tobacco β-1,3-glucanase (*GLU*), alfalfa defensin *alfAFP*, and their bivalent *GLU-AFP*	Tomato	*Ralstonia solanacearum*	Chen *et al.*, 2006
Phaseolus vulgaris	ch5B (Chitinase)	Strawberry cv.pajara	Resistant against *Botrytis cinera* but no enhanced resistance to *Colletotrichum acutatum*	Vellice *et al.*, 2006
Nicotiana tobacum	gln2 (Glucanase) and ap24 (TLP)	Strawberry cv.pajara	*Botrytis cinera (No effect)*	Vellice *et al.*, 2006

PVX-, and CMV-infected tobacco (Röhring, 1998), viroid-infected tomato (Camacho Henriquez and Sänger, 1982), *etc.* The strong induction of PRs in tobacco leaves by PVY and necrosis-inducing abiotic factors (Edreva, 1990) is in line with this idea. It may be speculated that the induction of PRs accompanying the symptom expression could also have a protective role of "last barrier", impeding the full destroyment of stressed plants by both internal and environmental constraints. In the last years it was surprisingly established that PRs are not synthesized dur-ing the expression of a newly-reported "induced systemic resistance" (ISR) in Arabidopsis (Pieterse *et al.*, 1996), as well as in crop plants (Hoffland *et al.*, 1995; Reitz *et al.*, 2001; Siddiqui and Chaukat, 2004). ISR is induced during root coloniza-tion by non-pathogenic rhizobacteria of Pseudomonas spp., and is effective against a broad range of pathogens. A novel signaling pathway, salicylic acid-independent but jasmonate- and ethylene-dependent, is engaged in the control of ISR (Pieterse *et al.*, 1998). Besides by non-pathogenic Pseudomonas bacteria, ISR is also elicited by growth-promoting Bacillus spp. (Kloepper *et al.*, 2004; Silva *et al.*, 2004), with this approach having a beneficial effect in field conditions. The non-involvement of PRs in ISR is indicative of the diversity of plant defense strategies, pointing that PRs are only one of the multiple means employed by plants.

REFERENCES

Abad, L., D´ Urzo, M.P., Liu, D., Narasimhan, M.L., Reuveni, M., Zhu, J.K., Niu, X., Singh, N.K., Hasegawa, P.M. And Bressan, R.A. 1996. Antifungal activity of tobacco osmotin has specificity and involves plasma membrane permeabilization. *Plant Science* **118:** 11- 23.

Agrios, G.N. 1997. Plant Pathology. Fourth edition. Academic Press, London, UK. 635 p.

Agrois, G. N., 1988. Plant Pathology. 3rd Ed. Academic Press Inc., San Diego, USA.

Ahl, P., A. Cornu, S. Gianinazzi. 1982. Soluble proteins as genetic markers in studies of resistance and phylogeny in Nicotiana. *Phytopathology* **72**: 80-85.

Åkesson, H. 1995. Infection of barley by Bipolaris sorokiniana: toxin production and ultrastructure. PhD thesis. Lund University, Sweden.

Alexander, D., Goodman, R.M., Gut-Rella, M., Glascock, C., Weymann, K., Friedrich, L. Maddox, D., Ahl-Goy, P., Luntz, T., Ward, E. And Ryals, J. 1993. Increased tolerance to two oomycete pathogens in transgenic tobacco expressing pathogenesis-related protein 1a. Proceedings of the National Academy of Sciences USA. **90**: 7327-7331.

Allan, A.C. and Fluhr, R. 1997. Two distinct sources of elicited reactive oxygen species in tobacco epidermal cells. *The Plant Cell* **9**: 1559-1572.

Almgren, I., Gustafsson, M., Fält, A.-S., Lindgren, H. And Liljeroth, E. 1999. Interaction between root and leaf disease development in barley cultivars after inoculation with different isolates of *Bipolaris sorokiniana*. *Journal of Phytopathology* **147**: 331-337.

Alvarez, M. E., R. I. Pennel, P. J. Meijer, A. Ishikawa, R. A. Dixon, C. Lamb, 1998. Reactive oxygen intermediates mediate a systemic signal network in the establishment of plant immunity. *Cell* **92**: 773–784.

Anand, A., Z.T. Lei, L.W. Summer, K.S. Mysore, Y. Arakane, W.W. Backus, S. Muthukrishnan, 2004. Apoplastic extracts from a transgenic wheat line exhibiting lesion-mimic phe- notype have multiple pathogenesis-related proteins that are antifungal. *Plant-Microbe Interact* **17**: 1306-1317.

Ananieva, K. I., E. D. Ananiev.1999. Effect of methyl ester of jasmonic acid and benzyl- aminopurine on growth and protein profile of excised cotyledons of Cucurbita pepo (zucchini). *Biol. Plant* **42**: 549–557.

Anderson, M.D., Z. Chen, D.F. Klessig, 1998. Possible involvement of lipid peroxidation in salicylic acid-mediated induction of PR-1 gene expression. *Phytochemistry* **47**: 555- 566.

Anfoka, G., H. And Buchenauer, 1997. Systemic acquired resistance in tomato against *Phytophthora infestans* by pre-inoculation with tobacco necrosis virus. *Physiol. Mol. Plant Pathol.* **50**: 85-101.

Anlovar, S., M.D. Serra, M. Dermastia, G. Menestrina, 1998. Membrane permeabilizing activity of pathogenesis-related protein lunusitin from flax seed. *Mol. Plant-Microbe Intera.* **11**: 610-617.

Antoniw, J.F., C.E. Ritter, W.S. Pierpoint, L.C. Van Loon, 1980. Comparison of three patho- genesis-related proteins from plants of two cultivars of tobacco infected with TMV. J. *Gen. Virol.* **47**: 79-87.

Apoga, D. and Jansson, H.-B. 2000. Visualization and characterization of the extracellular matrix of *Bipolaris sorokiniana. Mycological Research* **104**: 564-575.

Arlorio, M., Ludwig, A., Boller, T. And Bonfante, P. 1992. Inhibition of fungal growth by plant chitinases and -1,3-glucanases. A morphological study. *Protoplasma* **171:** 34-43.

Asselin, A., J. Grenier, F. Cté, 1985. Light-influenced extracellular accumulation of b (patho- genesis-related) proteins in Nicotiana green tissue induced by various chemicals or prolonged floating on water. *Can. J. Bot***63**: 1276-1283.

Bachmann, D., E. Rezzonico, D. Retelska, A. Chételat, S.Schaerer, R. Beffa, 1998. Improve- ment of potato resistance to Phytophthora infestans by overexpressing antifungal hydrolases. 5th International Workshop on pathogenesis-related proteins. Signalling pathways and biological activities. March 29-April 2, 1998, Aussois, France. Abstracts, p.-57.

Baker, B., P. Zambryski, B. Staskawicz, S. P. Dinesh-Kumar, 1997. Signaling in plant-microbe interactions. *Science* **276**: 726–733.

Baker, C.J. and Orlandi, E.W. 1995. Active oxygen in plant pathogenesis. *Annual Review of Phytopathology* 33, 299-321.

Bakonyi, J., Aponyi, I. and Fischl, G. 1998. Diseases caused by Bipolaris sorokiniana and Dreschslera tritici repentis in Hungary. In: Duveiller, E., Dubin, H.J., Reeves, J. and McNab, A. (Eds.). Helminthosporium Blights of Wheat: Spot Blotch and Tan Spot. Mexico, D.F.: CIMMYT. 80-87 pp.

Benhamou, N. 1995. Immunocytochemistry of plant defense mechanisms induced upon microbial attack. *Microscopy Research and Technique* **3**: 63-78.

Benhamou, N. 1996. Elicitor-induced plant defence pathways. Trends in Plant Science 1, 233- 240. Boller, T., Gehri, A., Mauch, F. and Vögeli, U. 1983. Chitinase in bean leaves: induction by ethylene, purification, properties, and possible function. *Planta* **157:** 22-31.

Benhamou, N., J. Grenier, A. Asselin, 1991. Immunogold localization of pathogenesis-related protein P14 in tomato root cells infected by *Fusarium oxysporum* f. sp. radicis- lycopersici. *Physiol. Mol. Plant Pathol* **38:** 237-253.

Benhamou, N., Joosten, M.H. and De Wit, P.J.G.M. 1990. Subcellular localization of chitinase and of its potential substrate in tomato root tissues infected by Fusarium oxysporum f.sp. racicis-lycopersici. *Plant Physiology* **92**: 1108-1120.

Bergey, D. R., G. A. Howe, C. A. Ryan, 1996. Polypeptide signaling for plant defensive genes exhibits analogies to defense signaling in animals. *Proc. Natl. Acad. Sci. USA* **93**: 12053– 12058.

Bernier, F., A. Berna, 2001. Germins and germin-like proteins: Plant do-all proteins. But what do they do exactly? *Plant Physiol. Biochem.***39**: 545-554.

Bol, J.F., H.J.M. Linthorst, B.J.C. Cornelissen, 1990. Plant pathogenesis-related proteins in- duced by virus infection. *Annu. Rev. Phytopathol* **28***:* 113-138.

Boller, T. 1993. Antimicrobial functions of the plant hydrolases chitinase and -1,3-glucanase. In: Fritig, B. and Legrand, M. (Eds.). *Developments in Plant Pathology.* **391** p.

Bortolotti, C., Murillo, I., Fontanet, P., Coca, M. and San Segundo, B. 2005. Long-distance transport of the maize pathogenesis-related PRms protein through the phloem in transgenic tobacco plants. *Plant Science* **168:** 813-821.

Boudart, G., C. Lafitte, J.-P. Barthe, D. Fraser, M.-T. Esquerré-Tugayé, 1998. Differential elicitation of defense responses by pectic fragments in bean seedlings. *Planta* **206**: 86-94.

Briquet, M., Vilret, D., Goblet, P., Mesa, M. and Eloy, M.-C. 1998. Plant cell membranes as biochemical targets of the phytotoxin Helminthosporol. *Journal of Bioenergetics and Biomembranes* 30, 285-295.

Brisson, L.F., Tenhaken, R. and Lamb, C. 1994. Function of oxidative cross-linking of cell wall structural proteins in plant disease resistance. *The Plant Cel* **6:** 1703-1712.

Broekaert, W. F., B. P. A. Cammue, M. F. C. De Bolle, K. Thevissen, G. W. Desamblanx, R. W. Osborn, 1997. Antimicrobial peptides from plants. *Crit. Rev. Plant Sci.,***16**: 297–323.

Broglie, K., Chet, I., Holliday, M., Cressman, R., Biddle, P., Knowlton, S., Mauvais, C.J. and Broglie, R. 1991. Transgenic plants with enhanced resistance to the fungal pathogen *Rhizoctonia solani. Science* **254**: 1194-1197.

Bryngelsson, T. and Collinge, D.B. 1992. Biochemical and molecular analyses of the response of barley to infection by powdery mildew. In: P.R. Shewry (Ed.). *Barley Genetics, Molecular Biology and Biotechnology.* CAB International, Wallingford. 459-480 pp.

Bryngelsson, T. and Greén, B. 1989. Characterization of a pathogenesis-related thaumatin-like protein isolated from barley challenged with an incompatible race of mildew. *Physiological and Molecular Plant Pathology* **35**: 45-52.

Bryngelsson, T., Sommer-Knudsen, J., Gregersen, P.L., Collinge, D., Ek, B. and Thordal- Christensen, H. 1994. Purification, characterization, and molecular cloning of basic PR- 1-type pathogenesis-related proteins from barley. *Molecular Plant Microbe Interactions* **7:** 267-275.

Buchel, A.S., H.J.M. Linthorst, 1999. PR-1: A group of plant proteins induced upon pathogen infection. In: Pathogenesis-related proteins in plants. Eds. S.K. Datta, S. Muthukrishnan, CRC Press LLC, Boca Raton, 21-47.

Büchter, R., Strömberg, A., Schmelzer, E. and Kombrink, E. 1997. Primary structure and expression of acidic (class II) chitinase in potato. *Plant Molecular Biology* **3:** 7459-761.

Buhtz, A., Kolasa, A., Arlt, K., Walz, C. and Kehr, J. 2004. Xylem sap protein composition is conserved among different plant species. *Planta* **219**: 610-618.

Butt, A., Mousley, C., Morris, K., Beynon, J., Can, C., Holub, E., Greenberg, J.T. and Buchanan- Wollaston, V. 1998. Differential expression of a senescence-enhanced metallothionein gene in Arabidopsis in response to isolates of *Peronospora parasitica* and *Pseudomonas syringae. The Plant Journal* **16**: 209-221.

Camacho Henriquez, A., H.L. Sänger, 1982. Analysis of acid-extractable tomato leaf proteins after infection with a viroid, two viruses and a fungus and partial purification of the ' pathogenesis-related' protein p.14. *Arch. Virol* **74**: 181-196.

Cameron, R.K., 2000. Salicylic acid and its role in plant defense responses: what do we really know? *Physiol. Mol. Plant Pathol* **56**: 91-93.

Carlson, H., Nilsson, P., Jansson, H.-B. and Odham, G. 1991b. Characterization and determination of prehelminthosporol, a toxin from the plant pathogenic fungus Bipolaris sorokiniana, using liquid chromathography/mass spectrometry. *Journal of Microbiological Methods* **13**: 259-269.

Carlson, H., Stenram, U., Gustafsson, M. and Jansson, H-B. 1991a. Electron microscopy of barley root infection by the fungal pathogen *Bipolaris sorokiniana. Canadian Journal of Botany* **69**: 2724-2731.

Cheong, N.E., Y.O. Choi, W.Y. Kim, I.S. Bae, M.J. Cho, I. Hwang, J.W. Kim, S.Y. Lee, 1997. Purification and characterization of an antifungal PR-5 protein from pumpkin leaves. *Mol. Cells* **7**: 214-219.

Christensen, A.B., Ho Cho, B., Næsby, M., Gregersen, P.L., Brandt, J., Madriz-Ordeñana, K., Collinge, D. and Thordal-Christensen, H. 2002. The molecular characterization of two barley proteins establishes the novel PR-17 family of pathogenesis-related proteins. *Molecular Plant Pathology* **3**: 135-144.

Clark, T. A., Zeyen, R. J., Smith, A. G., Bushnell, W. R., Szabo, L. J., and Vance, C. P. 1993. Host response gene transcript accumulation in relation to visible cytological events during *Erysiphe graminis* attack in isogenic barley lines differing at the Ml-a locus. *Physiological and Molecular Plant Pathology* **64**: 169- 178.

Clay, R.P., Bergmann, C.W. and Fuller, M.S. 1997. Isolation and characterization of an endopolygalacturonase from *Cochliobolus sativus* and a cytological study of fungal penetration of barley. *Phytopathology* **87**: 1148-1159.

Clay, R.P., Enkerli, J. and Fuller, M.S. 1994. Induction and formation of *Cochliobolus sativus* appressoria. *Protoplasma* **178**: 34-47.

Collinge, D.B., Gregersen, P.L. and Thordal-Christensen, H. 1993. The induction of gene expression in response to pathogenic microbes. In: Basra, A.S (Ed.). Mechanisms of plant growth and improved productivity: Modern approaches and perspectives. Marcel Dekker. 1-35 pp.

Conrath, U., Silva, H. and Klessig, D.F. 1997. Protein dephosphorylation mediates salicylic acid- induced expression of PR-1 genes in tobacco. *The Plant Journal* **11**: 747-757.

Cornelissen, B.J.C., R.A.M. Hooft van Huijuijnen, J.F. Bol, 1986. A tobacco mosaic virus- induced tobacco protein is homologous to the sweet tasting protein thaumatin. *Nature* **321:** 531-532.

Datta, K., R. Velazhahan, N. Oliva, I. Ona, T. Mew, G.S. Khush, S. Muthukrishnan, S.K. Datta, 1999. Over-expression of the cloned rice thaumatin-like protein (PR-5) gene in transgenic rice plants enhances environmental friendly resistance to *Rhizoctonia solani* causing sheath blight disease. *Theor. Appl. Genetics* **98**: 1138-1145.

Deacon, J.W. 1987. Programmed cortical senescence: a basis for understanding root infection. In: Pegg, G.F. and Ayres, P.G. (Eds.). Fungal infections of plants. Cambridge University Press, Cambridge, UK. 285-297 pp.

Dehne, H-W. and Oerke, E-C. 1985. Investigations on the occurrence of *Cochliobolus sativus* in barley and wheat. II. Infection, colonization and damage of stems and leaves. *Journal of Plant Diseases and Protection* **92**: 606-617.

Del Campillo, E., L.N. Lewis, 1992. Identification and kinetics of accumulation of proteins induced by ethylene in bean abscission zone. *Plant Physiol* **98**: 955-961.

Del Pozo, J.C., M. Estelle, 1999. Function of the ubiquitin-proteosome pathway in auxin res- ponse. *Trends Plant Sci* **4**: 107–112.

Delaney, T. P., S. Uknes, B. Vernoij, L. Friedrich, K. Weyman, D. Negrotto, T. Gaffney, M. Gut-Rella, H. Kessmann, E. Ward, J. Ryals, 1994. A central role of salicylic acid in plant disease resistance. *Science* **266**: 1247–1250.

Delaney, T.P., 2000. New mutants provide clues into regulation of systemic acquired resis-tance. *Trends Plant Sci* **5**: 49-51.

Doke, N., Miura, Y., Sanchez, L.M., Park, H.-J., Toritake, T., Yoshioka, H. and Kawakita, K. 1996. The oxidative burst protects plants against pathogen attack: Mechanism and role as an emergency signal for plant bio-defense – a review. *Gene* **179**: 45-51.

Duveiller, E. and Altamirano, G.I. 2000. Pathogenecity of *Bipolaris sorokiniana* isolates from wheat roots, leaves and grains in Mexico. *Plant Pathology* 49, 235- 242. FAOSTAT, 2005. http: / /www.fao.org

Edreva, A. 1991. Stress proteins of plants - PR(b)-proteins. *Sov. Plant Physiol* **38**: 579-588.

Edreva, A. 2004. A novel strategy for plant protection: induced resistance. *J. Cell Mol. Biol* **3**: 61-69.

Edreva, A., D. Blancard, R. Delon, P. Bonnet, P. Ricci, 2002. Biochemical changes in β- cryptogein-elicited tobacco: a possible basis of acquired resistance. *Beitr. Tabakforsch. Internat* **20**: 53-59.

Edreva, A.M., 1990. Induction of "pathogenesis - related" proteins in tobacco leaves by physi- ological (non-pathogenic) disorders. *J. Exp. Bot* **41**: 701-703.

Edreva, M. 1991. Plant stress proteins PRPs. *Fiziologia Rastenii* **38**: 788–800.

Ekramoddoulah, A.K.M., X.S. Yu, R. Sturrock, A. Zamani, D. Taylor, 2000. Detection and seasonal expression pattern of a pathogenesis-related protein (PR-10) in Douglasfir (*Pseudotsuga menziesiti*) tissues. *Physiol. Plant* **110:** 240-247.

Enkerli, J., U. Gisi, E. Mösinger, 1993. Systemic acquired resistance to Phytophthora infestans in tomato and the role of pathogenesis-related proteins. *Physiol. Mol. Plant Pathol* **43**: 161- 171.

Evidence for the occurrence of the "pathogenesis-related proteins" in leaves of healthy tobacco plants during flowering. *Physiol. Plant Pathol* **19**: 69-76.

Eyal, Y., Y. Meller, S. Lev Yadun, R. Fluhr, 1993. A basic-type PR-1 promotor directs ethyl-ene responsiveness, vascular and abscission zones specific expression. *Plant J* **4**: 225-234.

Fagoaga, C., I. Rodrigo, V. Conejero, C. Hinarejos, J.J. Tuset, J. Arnau, J.A. Pina, L. Navarro, L. Peña, 2001. Increased tolerance to *Phytophthora citrophthora* in transgenic or- ange plants constitutively expressing a tomato pathogenesis related protein PR-5. *Mol. Breed* **7**: 175-185.

Felle, H.H. and Zimmermann, M.R. 2007. Systemic signaling in barley through action potentials. *Planta* **226**: 203-214.

Fernández, C., T. Szyperski, T. Bruyère, P. Ramage, E. Mösinger, K. Wütrich, 1997. NMR solution structure of the pathogenesis-related protein p 14a. *J. Mol. Biol* **266**: 576- 593.

Fidantsef, A.L., M.J. Stout, J.S. Thaler, S.S. Duffey, R.M. Bostock, 1999. Signal interactions in pathogen and insect attack: expression of lipoxygenase, proteinase inhibitor II, and pathogenesis-related protein P4 in tomato, *Lycopersicon esculentum*. *Physiol. Mol. Plant Pathol* **54**: 97-114.

Fujibe, T., K. Watanabe, N. Nakajima, Y. Ohashi, I. Mitsuhara, K.T. Yamamoto, Y. Takeuchi, 2000. Accumulation of pathogenesis-related proteins in tobacco leaves irradiated with UV-B. J. *Plant Res* **113**: 387-394.

Fujita, K., Suzuki, T., Kunoh, H., Carver, T. L. W., Thomas, B. J., Gurr, S. and Shiraishi, T. 2004. Induced inaccessibility in barley cells exposed to extracellular material released by non- pathogenic powdery mildew conidia. *Physiological and Molecular Plant Pathology* **64**: 169-178.

Garcia-Brugger, A., Lamotte, O., Vandelle, E., Bourqoue, S., Lecourieux, D., Poinssot, B., Wendehenne, D. and Pugin, A. 2006. Early signaling events induced by elicitors of plant defenses. *Molecular Plant-Microbe Interactions* **19**: 711-724.

Gau, A.E., M. Koutb, M. Piotrowski, K. Kloppstech, 2004. Accumulation of pathogenesis- related proteins in apoplast of a susceptible cultivar of apple (Malus domestica cv. Elstar) after infection by *Venturia inaequalis* and constitutive expression of PR genes in the resistant cultivar *Remo. Eur. J. Plant Pathol* **110**: 703-711.

Gianinazzi, S., 1982. Antiviral agents and inducers of virus resistance: analogies with inter- feron. In: Active defence mechanisms in plants. Ed. R.K.S. Wood, Plenum Press, L.- N.Y., 275-298.

Glazebrook, J., 1999. Genes controlling expression of defense responses in Arabidopsis. *Curr. Opin. Plant Bio* **2**: 280–286.

Gregersen, P. L., Thordal-Christensen, H. Förster, H. and Collinge, D.B. 1997. Differential gene transcript accumulation in barley leaf epidermis and mesophyll in response to attack by *Blumeria graminis* f.sp. *hordei* (syn. *Erysiphe graminis* f.sp. *hordei*). *Physiological and Molecular Plant Pathology* **51:** 85-97.

Grison, R., B. Besset-Grezes, M. Schneider, N. Lucante, L. Olsen, J.-J. Leguay, A. Toppan, 1996. Field tolerance to fungal pathogens of *Brassica napus* constitutively express- ing a chimeric chitinase gene. *Nature Biotechnol* **14**: 643-646.

Grunwald, I., Rupprecht, I., Schuster, G. and Kloppstech, K. 2003. Identification of guttation fluid proteins: the presence of pathogenesis-related proteins in non-infected barley plants. *Physiologia Plantarum* **119**: 192-202.

Hahn, M. G., 1996. Microbial elicitors and their receptors in plants. *Annu. Rev. Phytopathol* **34**: 387–412.

Ham, K.S., S. Kauffmann, P. Albersheim, A.G. Darvill, 1991. A soybean pathogenesis-re- lated protein with β-1,3-glucanase activity releases phytoalexin elicitor-active heat- stable fragments from fungal wall. *Mol. Plant-Microbe Interact* **4**: 545-552.

Hammond-Kosack, K. E., J. D. G. Jones, 1996. Resistance gene-dependent responses. *Plant Cell***8**: 1773–1791.

Hanfrey, C., Fife, M. and Buchanan-Wollaston, V. 1996. Leaf senescence in Brassica napus: expression of genes encoding pathogenesis-related proteins. *Plant Molecular Biology* **30**: 597-609.

He, D.Y., Yazaki, Y., Nishizawa, Y., Takai, R., Yamada, K., Sakano, K., Shibuya, N. and Minami, E. 1998. Gene activation by cytoplasmic acidification in suspension-cultured rice cells in response to the elicitor N-acetylchitoheptaose. *Molecular Plant-Microbe Interactions* **11**: 1167-1174.

Heath, M. C., 1998. Apoptosis, programmed cell death and the hypersensitive response. *Eur. J. Plant Pathol.*, 104, 117–124. Fritig, B., T. Heitz, M. Legrand, 1998. Antimicrobial proteins in induced plant defense. *Curr. Opin. Immunol* **10**: 16–22.

Heath, M.C. 2000. Nonhost resistance and nonspecific plant defenses. *Current Opinion in Plant Biology* **3**: 315-319.

Hodges, C.F. and Campbell, D.A. 1999. Endogenous ethane and ethylene of Poa pratensis leaf blades and leaf chlorosis in response to biologically active products of *Bipolaris sorokiniana*. *European Journal of Plant Pathology* **105**: 825-829.

Hoegen, E., Strömberg, A., Pihlgren, U. and Kombrink, E. 2002. Primary structure and tissue- specific expression of the pathogenesis-related protein PR-1b in potato. *Molecular Plant Pathology* **3**: 329-345.

Hon, W.-C., Griffith, M., Mlynarz, A., Kwok, Y.C. and Yang, D.S.C. 1995. Antifreeze proteins in winter rye are similar to pathogenesis-related proteins. *Plant Physiology* **109**: 879-889.

Hong, J.K. and Hwang, B.K. 2002. Induction by pathogen, salt and drought of a basic class II chitinase mRNA and its in situ localization in pepper (*Capsicum annuum*). *Physiologia Plantarum* **114**: 549-558.

Hu, G. and Rijkenberg, F.H.J. 1998. Subcellular localization of -1,3-glucanase in Puccinia recondita f.sp. tritici-infected wheat leaves. *Planta* **204**: 324-334.

Ibeas, J.I., Lee, H., Damsz, B., Prasad, D.T., Pardo, J.M., Hasegawa, P.M., Bressan, R.A. and Narasimhan, M.L. 2000. Fungal cell wall phosphomannans facilitate the toxic activity of a plant PR-5 protein. *The Plant Journal* **23**: 375-383.

J., P. Dodds, T. Pryor, 2000. Structure, function and evolution of plant disease resistance genes. *Curr. Opin. Plant Biol* **3**: 278–284.

Jacobsen, S., Mikkelsen, J.D. and Hejgaard, J. 1990. Characterization of two antifungal endochitinases from barley grain. *Physiologia Plantarum* **79**: 554-562.

Jansson, H.-B., Johansson, T., Nordbring-Hertz, B., Tunlid, A. and Odham, G. 1988. Chemotrophic growth of germ-tubes of *Cochliobolus sativus* to barley root exudates. *Transactions of the British Mycological Society* **90**: 647-650.

Jeun, Y.Ch. and Buchenauer, H. 2001. Infection structures and localization of pathogenesis-related protein AP24 in leaves of tomato plants exhibiting systemic acquired resistance against *Phytophthora infestans* after pre-treatment with 3- aminobutyric acid or tobacco necrosis virus. *Journal of Phytopathology* **149**: 141- 153.

Jeun, Y.-Ch., 2000. Immunolocalization of PR-protein P14 in leaves of tomato plant exhibit-ing systemic acquired resistance against Phytophthora infestans induced by pretreat- ment with 3-aminobutyric acid and preinoculation with Tobacco necrosis virus. *J. Plant Dis. Prot* **107**: 352-367.

Joel, D.M., V.H. Portnoy, 1998. The angiospermous root parasite Orobanche L. (Orobanchaceae) induces expression of a pathogenesis-related (PR) gene in suscep- tible tobacco roots. *Ann. Bot* **81**: 779-781.

Kang, Z. and Buchenauer, H. 2002. Immunocytochemical localization of -1,3-glucanase and chitinase in *Fusarium culmorum*-infected wheat spikes. *Physiological and Molecular Plant Pathology* **60**: 141-153.

Kauffmann, S., M. Legrand, P. Geoffroy, B. Fritig, 1987. Biological function of "pathogenesis- related" proteins: four PR proteins of tobacco have β-1,3-glucanase activity. *EMBO J* **6:** 3209-3212.

Kauss, H. 1990. The role of plasma membrane in host-pathogen interactions. In: Larsson, C. and Møller, I.M. (Eds.). The plant plasma membrane, structure and function and molecular biology. Springer-Verlag, Berlin. 320-350 pp.

Kehr, J., Buhtz, A. and Giavalisco, P. 2005. Analysis of xylem sap proteins from Brassica napus. *BMC Plant Biology*. 5: 11. Knoche, H.W. and Duvick, J.P. 1987. The role of fungal toxins in plant disease. In: G.F. Pegg and P.G. Ayres (Eds.). Fungal infection of plants. Cambridge University Press, Cambridge. 158-192 pp.

Kehr, J., Buhtz, A. and Giavalisco, P. 2005. Analysis of xylem sap proteins from Brassica napus. *BMC Plant Biology* **5**: 11.

Keller, T., H. G. Hamude, D. Werner, P. Doerner, R. A. Dixon, C. Lamb, 1998. A plant hom- olog of the neutrophil NADPH oxidase gp91phox subunit gene encodes a plasma membrane protein with calcium binding motifs. *Plant Cell* **10**: 255–266.

Kloepper, J.W., C.M. Ryu, S.A. Zhang, 2004. Induced systemic resistance and promotion of plant growth by *Bacillus* spp. *Phytopathol* **94**: 1259-1266.

Knoche, H.W. and Duvick, J.P. 1987. The role of fungal toxins in plant disease. In: G.F. Pegg and P.G. Ayres (Eds.). Fungal infection of plants. Cambridge University Press, Cambridge. 158-192 pp.

Kombrink, E and Somssich, *I.E.* 1997. Pathogenesis-related proteins and plant defense. In: Carroll, Tudzynski (Ed.). The Mycota V Part A. Plant Relationships. Springer-Verlag, Berlin, Heidelberg. 107-128 pp.

Kombrink, E., G. Ancillo, R. Büchter, J. Dietrich, E. Hoegen, Y. Ponath, E. Schmelzer, A. Strömberg, S. Wegener, 2001. The role of chitinases in plant defense and plant de- velopment. 6th International Workshop on PR-proteins. May 20-24, 2001, Spa, Belgium. Book of abstracts, p. 11.

Kononowicz, A.K., Nelson, D.E., Singh, N.K., Hasegawa, P.M. and Bressan, R.A. 1992. Regulation of the osmotin gene promoter. *The Plant Cell* **4**: 513-524.

Kragh, K.M., T. Hendriks, A.J. De Jong, F. Lo Schiavo, N. Bucherna, P. Hojrup, J.D. Mikkelsen, S.C. De Vries, 1996. Characterization of chitinases able to rescue somatic embryos of the temperature-sensitive carrot variant ts11. *Plant Mol. Biol* **31**: 631-645.

Kumar, J., Hückelhoven, R., Beckhove, U., Nagarajan, S. and Kogel, K-H. 2001. A compromised Mlo pathway affects the response of barley to the necrotrophic fungus *Bipolaris sorokiniana* (Telemorph: *Cochliobolus sativus*) and its toxins. *Phytopathology* **91**: 127- 133.

Kumar, J., Schäfer, P., Hückelhoven, R., Langen, G., Baltruschat, H., Stein, E., Nagarajan, S., and Kogel, K-H. 2002. Pathogen profile. Bipolaris sorokiniana, a cereal pathogen of global concern: cytological and molecular approaches towards better control. *Molecular Plant Pathology* **3**: 185-195.

Lamb, C. and Dixon, R.A. 1997. The oxidative burst in plant disease resistance. *Annual Review of Plant Physiology and Plant Molecular Biology* **48**: 251-257.

Laugé, R., A.P. Dmitriev, M.H.A.J. Joosten, P.J.G.M. de Wit, 1998. Additional resistance gene(s) against *Cladosporium fulvum* present on the Cf-9 introgression segment are associated with strong PR protein accumulation. *Mol. Plant-Microbe Inter* **11**: 301- 308.

Lawrence, C.B., N.P. Singh, J. Qiu, R.G. Gardner, S. Tuzun, 2000. Constitutive hydrolytic enzymes are associated with polygenic resistance of tomato to Alternaria solani and may function as an elicitor release mechanism. *Physiol. Mol. Plant Pathol* **57**: 211- 220.

Lee, Y.K., Hippe-Sanwald, S., Jung, H.W., Hong, J.K., Hause, B. and Hwang, B.K. 2000b. In situ localization of chitinase mRNA and protein in compatible and incompatible interactions of pepper stems with Phytophthora capsici. *Physiological and Molecular Plant Pathology* **57**: 111-121.

Lee, Y.K., Hippe-Sanwald, S., Lee, S.C., Hohenberg, H. and Hwang, B.K. 2000a. In situ localization of PR-1 mRNA and PR-1 protein in compatible and incompatible interactions of pepper stems with *Phytophthora capsici*. *Protoplasma* **211**: 64-75.

Legendre, L., P. F. Heinstein, P. S. Low, 1992. Evidence for participation of GTP-binding proteins in elicitation of rapid oxidative burst in cultured soybean cells. *J. Biol. Chem* **267**: 20140–20147.

Legrand, M., S. Kauffmann, P. Geoffroy, B. Fritig, 1987. Biological function of pathogenesis-related proteins: four tobacco pathogenesis-related proteins are chitinases. *Proc. Natl. Acad. Sci. USA*, 84: 6750-6754.

Leubner-Metzger, G. and Meins, F. Jr. 1999. Functions and regulation of plant -1,3-glucanases (PR-2). In: Datta, S.K. and Muthukrishnan, S. (Eds.). Pathogenesis-related proteins in plants. CRC Press. 49-76 pp.

Liljeroth, E. and Bryngelsson, T. 2001. DNA fragmentation in cereal roots indicative of programmed root cortical cell death. *Physiologia Plantarum* **111**: 365-372.

Liljeroth, E. 1995. Comparison of early root cortical senescence among barley cultivars, Triticum species and other cereals. *New Phytopathology* **130**: 495-501.

Liljeroth, E., Marttila, S. and von Bothmer, R. 2005. Immunolocalization of defence-related proteins in the floral organs of barley (*Hordeum vulgare* L.). *Journal of Phytopathology* **153**: 702-709.

Lin, T.S. and Kolattukudy, P.E. 1980. Isolation and characterization of a cuticular polyester (cutin) hydrolyzing enzyme from phytopathogenic fungi. *Physiological Plant Pathology* **17**: 1- 15.

Linthorst, H.J.M. 1991. Pathogenesis-related proteins of plants. *Critical Reviews in Plant Sciences* 10, 123-150.

Linthorst, H.J.M., Meuwissen, R.L.J., Kauffmann, S. and Bol, J.F. 1989. Constitutive expression of pathogenesis-related proteins PR-1, GRP, and PR-S in tobacco has no effect on virus infection. *The Plant Cell* **1**: 285-291.

Liu, D., Raghothama, K.G., Hasegawa, P.M. and Bressan, R.A. 1994. Osmotin over expression in potato delays development of disease symptoms. *Proceedings of the National Academy of Sciences* USA 91: 1888-1892.

Lotan, T., N. Ori, R. Fluhr, 1989. Pathogenesis-related proteins are developmentally regulated in tobacco flowers. *Plant Cell* **1:** 881-887.

Ludwig, A. and Boller, T. 1990. A method for the study of fungal growth inhibition by plant proteins. FEMS Microbiology Letters 69, 61-66. Luttenberger, A.S. 1992. Bipolaris sorokiniana. Faktablad om växtskydd. Jordbruk. Sveriges lantbruksuniversitet. 64 J.

Luttenberger, A.S. 1992. *Bipolaris sorokiniana.* Faktablad om växtskydd. Jordbruk. Sveriges lantbruksuniversitet. 64 J. Mathre, D.E., Johnston, R.H. and Grey, W.E. 2003. Diagnosis of common root rot of wheat and barley. Online. *Plant Health Progress* DOI: 10.1094/PHP- 2003-0819- 01-DG.

Ma, Q.H., Y.R. Son, J.S. Sun, 1996. The role of cytokinin biosynthetic gene in regulation the expression of a class of pathogenesis-related protein genes in tobacco plants. *Acta Bot. Sin* **38**: 870-874.

Martin, G. B., 1999. Functional analysis of plant disease genes and their downstream effectors. *Curr. Opin. Plant Biol* **2**: 273–279.

Matsuoka, M., Y. Ohashi, 1986. Induction of pathogenesis-related proteins in tobacco leaves. *Plant Physiol* **80**: 505-510.

Mauch, F. and Staehelin, A. 1989. Functional implications of the subcellular localization of ethylene-induced chitinase and -1,3-glucanase in bean leaves. *The Plant Cell* **1**: 447-457.

Mauch, F., Mauch-Mani, B. and Boller, T. 1988. Antifungal hydrolases in pea tissue. II. Inhibition of fungal growth by combinations of chitinase and -1,3-glucanase. *Plant Physiology* **88**: 936-942.

Mauch, F., Meehl, J.B. and Staehelin, A. 1992. Ethylene-induced chitinase and β-1,3-glucanase accumulate specifically in the lower epidermis and along vascular strands. *Planta* **186**: 367-375.

Mc Neil, M., A.G. Darvill, S.C. Fry, P. Albershein, 1984. Structure and function of primary cell walls of plants. *Annu. Rev. Biochem.*, 53, 625-663.

McDowell, J. M., J. L. Dangl, 2000. Signal transduction in the plant immune response. *Trends Biochem. Sci* **25**: 79–82.

Melchers, L.S., W. Lageweg, M.H. Stuiver, 1998. The utility of PR genes to develop disease resistance in transgenic crops. 5th International Workshop on pathogenesis-related proteins. Signalling pathways and biological activities. March 29-April 2, 1998, Aussois, France. Abstracts, P-46.

Memelink, J., Linthorst, H.J.M., Schilperoort, R.A. and Hoge, J.H.C. 1990. Tobacco genes encoding acidic and basic isoforms of pathogenesis-related proteins display different expression patterns. *Plant Molecular Biology* **14**: 119-126.

Milligan, S. B., J. Bodeau, J. Yaghoobi, I. Kaloshian, P. Zabel, V. M. Williamson, 1998. The root knot nematode resistance gene Mi from tomato is a member of the leucine zip- per, nucleotide binding, leucine-rich family of plant genes. *Plant Cell* **10**: 1307–1319.

Murillo, I., Cavallarin, L. and San Segundo, B. 1997. The maize pathogenesis-related PRms protein localizes to plasmodesmata in maize radicles. *The Plant Cell* **9**: 145- 156.

Muthukrishnan, S., Liang, G.H., Trick, H.N. and Gill, B.S. 2001. Pathogenesis-related proteins and their genes in cereals. Plant Cell, Tissue and Organ Culture 64, 93- 114.

Naderi, M., P.H. Berger, 1997. Pathogenesis-related protein 1a is induced in potato virus Y- infected plants as well as by coat protein targeted to chloroplasts. *Physiol. Mol. Plant Pathol* **51**: 41-44.

Narasimhan, M.L., Lee, H., Damsz, B., Singh, N.K., Ibeas, J.I., Matsumoto, T.K., Woloshuk, C.P. and Bressan, R.A. 2003. Overexpression of a cell wall glycoprotein in *Fusarium oxysporum* increases virulence and resistance to a plant PR-5 protein. *The Plant Journal* **36**: 390-400.

Narvaéz-Vásquez, J. 1995. Autoradiographic and biochemical evidence for the systemic translocation of systemin in tomato plants. *Planta* **195**: 593-600.

Neale, A.D., J.A. Wahleithner, M. Lund, H.T. Bonnett, A. Kelly, D.R. Meeks-Wagner, W.J. Peacock, E.S. Dennis, 1990. Chitinase, β-1,3-glucanase, osmotin, and extensin are expressed in tobacco explants during flower formation. *Plant Cell* **2**: 673-684.

NIderman, T., I. Genetet, T. Bruyère, R.Gees, A. Stintzi, M. Legrand, B. Fritig, E. Mösinger, 1995. Pathogenesis-related PR-1 proteins are antifungal. Isolation and characterization of three 14-kilodalton proteins of tomato and of a basic PR-1 of tobacco with inhibitory activity against *Phytophthora infestans. Plant Physiol* **108**: 17-27.

Olbe, M., Sommarin, M., Gustafsson, M. and Lundborg, T. 1995. Effect of the fungal pathogen Bipolaris sorokiniana toxin prehelminthosporol on barley root plasma membrane vesicles. *Plant Pathology* **44**: 625-635.

Pääkkönen, E., S. Seppänen, T. Holopainen, H. Kokko, S. Kärenlampi, L. Kärenlampi, J. Kangasjärvi, 1998. Induction of genes for the stress proteins PR-10 and PAL in relation to growth, visible injuries and stomatal conductance in birch (*Betula pendula*) clones exposed to ozone and/or drought. *New Phytol* **138**: 295-305.

Park, C, J., K, J. Kim, R. Shin, J.M. Park, Y.-C. Shin, K.-H. Peak, 2004a. Pathogenesis-related protein 10 isolated from hot pepper functions as a ribonuclease in an antiviral pathway. *Plant J.*, **37**: 186-198.

Park, C.-J., J.-M. An, Y.-C. Shin, K.-J. Kim, B.-J. Lee, K.-H. Paek, 2004b. Molecular charac- terization of pepper germin-like protein as the novel PR-16 family of pathogenesis- related proteins isolated during the resistance response to viral and bacterial infection. *Planta* **219**: 797-806.

Peltonen, S., Karjalainen, R. and Niku-Paavola, M.-L. 1994. Purification and characterization of a xylanase from *Bipolaris sorokiniana*. *Mycological research* **98**: 67-73.

Perfect, S.E., Green, J.R. and O´Connell, R.J. 2001. Surface characteristics of necrotrophic secondary hyphae produced by the bean anthracnose fungus, *Colletotrichum lindemuthianum*. *European Journal of Plant Pathology* **107**: 813- 819.

Piening, L. 1997. Common root rot and seedling blight. In: Mathre, D.E, (Ed.). Compendium of barley diseases, second edition. *The American Phytopathological Society Press*, Minnesota, USA. 10-13 pp.

Pierpoint, W.S., N.P. Robinson, M.B. Leason, 1981. The pathogenesis-related proteins of tobacco: their induction by viruses in intact plants and their induction by chemicals in detached leaves. *Physiol. Plant Pathol* **19**: 85-97.

Pieterse, C.M., S.C.M. Van Wees, J.A. Van Pelt, M. Knoester, R. Laan, H. Gerrits, P.J. Weisbeek, L.C. Van Loon, 1998. A novel signaling pathway controlling induced systemic resistance in Arabidopsis. *Plant Cell* **10**: 1571-1580.

Piffanelli, P., A. Devoto, P. Schulze-Lefert, 1999. Defence signalling pathways in cereals. *Curr. Opin. Plant Biol* **2:** 295–300.

Pihakaski-Maunsbach, K., Moffatt, B., Testillano, P., Risueno, M., Yeh, S., Griffith, M. and Maunsbach, A.B. 2001. Genes encoding chitinase-antifreeze proteins are regulated by cold and expressed by all cell types in winter rye shoot. *Physiologia Plantarum* **112**: 359- 371.

Poupard, P., L. Parisi, C. Campion, S. Ziadi, P. Simoneau, 2003. A wound- and ethephon- inducible PR-10 gene subclass from apple is differentially expressed during infec- tion with a compatible and incompatible race of *Venturia inaequalis*. *Physiol. Mol. Plant Pathol* **62:** 3-12.

Pryce-Jones, E., Carver, T. and Gurr, S.J. 1999. The roles of cellulase enzymes and mechanical force in host penetration by *Erysiphe graminis* f. sp. hordei. *Physiological and Molecular Plant Pathology* **55**: 175-182.

Przymusinski, R., R. Rucinska, E.A. Gwozdz, 2004. Increased accumulation of pathogenesis-related proteins in response of lupine roots to various abiotic stresses. *Env. Exp. Bot* **52**: 53-62.

Rauscher, M., A.L. Ádám, S. Wirtz, R. Guggenheim, K. Mendgen, H.B. Deising, 1999. PR-1 protein inhibits the differentiation of rust infection hyphae in leaves of acquired resistant broad bean. *Plant J* **19**: 625-633.

Rep, M., Dekker, H.L., Vossen, J.H., de Boer, A.D., Houterman, P.M., Speijer, D., Back, J.W., de Koster, C.G. and Cornelissen, B.J.C. 2002. Mass spectrometric identification of isoforms of PR proteins in xylem sap of fungus-infected tomato. *Plant Physiology* **130**: 904-917.

Reymond, P., E. E. Farmer, 1998. Jasmonate and salicylate as global signals for defense gene expression. *Curr. Opin. Plant Biol* **1**: 404–411.

Richberg, M. H., H. A. Daniel, L. D. Jeffrey, 1998. Dead cells do tell tales. *Curr. Opin. Plant Biol* **1**: 480–485.

Röhring, C., 1998. Induction of pathogenesis-related proteins of group 1 by systemic virus infections of *Nicotiana tabacum* L. Beitr. *Tabakforsch. Internat* **18**: 63-67.

Rossi, M., F. L. Goggin, S. B. Milligan, I. Kaloshian, D. E. Ullman, V. M. Williamson, 1998. The nematode resistance gene Mi of tomato confers resistance against the potato aphid. *Proc. Natl. Acad. Sci. USA* **95**: 9750–9754.

Ryals, J. A., U. H. Neuenschwander, M. G. Willits, A. Molina, H. Y. Steiner, M. O. Hunt, 1996. Systemic acquired resistance. *Plant Cell* **8**: 1809–1819.

S.C. Williams, S.S. Dincher, D.L. Widerhold, D.C. Alexander, P. Ahl- Goy, J.P. Métraux, J.A. Ryals, 1991. Coordinate gene activity in response to agents that induce systemic acquired resistance. *Plant Cell* **3**: 1085-1094.

Sarowar, S., Y.J. Kim, E.N. Kim, K.D. Kim, B.K. Hwang, R. Islam, J.S. Shin, 2005. Overexpression of a pepper basic pathogenesis-related protein 1 gene in tobacco plants enhances resistance to heavy metal and pathogen stresses. *Plant Cell Reports* **24**: 216-224.

Schaller, A., P. Roy, N. Amrhein, 2000. Salicylic acid-independent induction of pathogenesis-related gene expression by fusicoccin. *Planta* **210**: 599-606.

Scheel, D., 1998. Resistance response physiology and signal transduction. *Curr. Opin. Plant Biol* **1**: 305–310.

Schenk, P.M., Kazan, K., Wilson, I., Anderson, J.P., Richmond, T., Somerville, S.C. and Manners, J.M. 2000. Coordinated plant defense responses in Arabidopsis revealed by microarray analysis. Proceedings of the National Academy of Sciences USA 97, 11655-11660.

Schoffelmeer, E.A.M., Klis, F.M., Sietsma, J.H. and Cornelissen, B.J.C. 1999. The cell wall of *Fusarium oxysporum*. *Fungal Genetics and Biology* **27**: 275-282.

Schultheiss, H., C. Dechert, L. Király, J. Fodor, K. Michel, K.-H. Kogel, R. Hückelhoven, 2003. Functional assessment of the pathogenesis-related protein PR-1b in barley. *Plant Sci* **165**: 1275-1280.

Sela-Buurlage, M.B., Ponstein, A.S., BresVloemans, S.A., Melchers, L.S., van den Elzen, P.J.M. and Cornelissen, B.J.C. 1993. Only specific tobacco (*Nicotiana tabacum*) chitinases and β- 1,3-glucanases exhibit antifungal activity. *Plant Physiology* **101**: 857.

Selitrennikoff, C.P., 2001. Antifungal proteins. Appl. Env. Microbiol., 67, 2883-2894.
Siddiqui, I.A., S.S. Shaukat, 2004. Systemic resistance in tomato induced by biocontrol bacte- ria against the root-knot nematode, Meloidogyne javanica is independent of salicylic acid production. *J. Phytopathol* **152:** 48-54.

Sessa, G., Yang, X.-Q., Raz, V., Eyal, Y. and Fluhr, R. 1995. Dark induction and subcellular localization of the pathogenesis-related PRB-1b protein. *Plant Molecular Biology* **28**: 537 547.

Silva, H.S.A., R.S. Romeiro, R. Carrer Filho, J.L.A. Pereira, E.S.G. Mizubuti, A. Mounteer, 2004. Induction of systemic resistance by Bacillus cereus against tomato foliar dis- eases under field conditions. *J. Phytopathol* **152**: 371-375.

Singh, N.K., C.A. Bracker, P.M. Hasegawa, A.K. Handa, S. Buckel, M.A. Hermodson, E. Pfankoch, F.E. Regnier, R.A. Bressan, 1987.Characterization of osmotin. A thaumatin- like protein associated with osmotic adaptation in plant cells. *Plant Physiol* **85**: 529- 536.

Somssich, *I.E.* and Hahlbrock, K. 1998. Pathogen defence in plants - a complex paradigm of biological complexity. *Trends in Plant Science* **3**: 86-90.

Staskawicz, B.J., F.M. Ausubel, B.J. Baker, J.G. Ellis, J.D.G. Jones, 1995. Molecular genetics of plant disease resistance. *Science* **268**. 661-667.

Steffenson, B. 1997. Common root rot and seedling blight. In: Mathre, D.E, (Ed.). Compendium of barley diseases, second edition. The American Phytopathological Society Press, Minnesota, USA. 35-36 pp.

Steiner-Lange, S., Fischer, A., Boettcher, A., Rouhara, I., Liedgens, H., Schmelzer, E. and Knogge. W. 2003. Differential defense reactions in leaf tissues of barley in response to infection by *Rhynchosporium secalis* and to treatment with a fungal avirulence gene product. *Molecular Plant-Microbe Interactions* **16**: 893-209.

Sticher, L., B. Mauch-Mani, J. P. Metraux, 1997. Systemic acquired resistance. *Ann. Rev. Phytopathology* **35**: 235–270.

Stintzi, A., Heitz, T., Prasad, V., Wiedemann-Merdinoglu, S., Kauffmann, S., Geoffroy, P., Legrand, M. and Fritig, B. 1993. Plant "pathogenesis-related" proteins and their role in defense against pathogens. *Biochemie* **75**: 687-706.

Surplus, S.L., B.R. Jordan, A.M. Murphy, J.P. Carr, B. Thomas, S.A.-H.-Macklerness, 1998. Ultraviolet-B-induced responses in *Arabidopsis thaliana*: role of salicylic acid and reactive oxygen species in the regulation of transcripts encoding photosynthetic and acidic pathogenesis-related proteins. *Plant, Cell, Env* **21**: 685-694.

Tahiri-Alaoui, A., Dumas-Gaudet, E. and Gianinazzi, S. 1993. Immunocytochemical localization of pathogenesis-related PR-1 proteins in tobacco root tissues infected *in vitro* by the black root rot fungus *Chalara elegans*. *Physiological and Molecular Plant Pathology* **42**: 69-82.

Takahashi, W., M. Fujimori, Y. Miura, T. Komatsu, Y. Nishizawa, T. Hibi, T. Takamizo, 2005. Increased resistance to crown rust disease in transgenic Italian ryegrass (*Lolium multiflorum* Lam.) expressing the rice chitinase gene. *Plant Cell Rep* **23**: 811-818.

Tamás, L., Ciamporová, M. and Luxová, M. 1998. Accumulation of pathogenesis-related proteins in barley leaf intercellular spaces during leaf senescence. *Biologia Plantarum* **41**: 451-46.

Tamás, L., J. Huttová, Z. igová, 1997. Accumulation of stress-proteins in intercellular spaces of barley leaves induced by biotic and abiotic factors. *Biol. Plant* **39**: 387-394.

Tinline, R.D., Wildermuth, G.B. and Spurr, D.T. 1988. Inoculum density of *Cochliobolus sativus* in soil and common root rot of wheat cultivars in Queensland. *Australian Journal of Agricultural Research* **39**: 569-577.

Tonón, C., G. Guevara, C. Oliva, G. Daleo, 2002. Isolation of a potato acidic 39 kDa β-1,3- glucanase with antifungal activity against *Phytophthora infestans* and analysis of its expression in potato cultivars differing in their degrees of field resistance. *J. Phytopatho* **150**: 189-195.

Van Kan, J.A.L., M.H.A.J. Joosten, G.A.M. Wagemakers, G.C.M. Van den Berg-Velthuis, P.J.G.M. De Wit, 1992. Differential accumulation of mRNAs encoding extracellular and intracellular PR proteins in tomato induced by virulent and avirulent races of *Cladosporium fulvum*. *Plant Mol. Biol* **20**: 513-527.

Van Loon, L. C., 1997. Induced resistance in plants and role of pathogenesis-related proteins. *Eur. J. Plant Pathol* **103**: 753–765.

Van Loon, L.C. and van Strien, E.A. 1999. The families of pathogenesis-related proteins, their activities, and comparative analysis of PR-1 type proteins. *Physiological Molecular Plant Pathology* **55**: 85-97.

Van Loon, L.C., 1999. Occurrence and properties of plant pathogenesis-related proteins. In: Pathogenesis-related proteins in plants. Eds. S.K. Datta, S. Muthukrishnan, CRC Press LLC, Boca Raton, 1-19.

Van Loon, L.C., 2001. The families of pathogenesis-related proteins. 6th International Work- shop on PR-proteins. May 20-24, 2001, Spa, Belgium. Book of abstracts, p. 9.

Van Loon, L.C., A. Van Kammen, 1970. Polyacrylamide disc electrophoresis of the soluble leaf proteins from *Nicotiana tabacum* var. "Samsun" and "Samsun NN" II. Changes in protein constitution after infection with tobacco mosaic virus. *Virology* **40**: 199- 211.

Van Loon, L.C., E.A. Van Strien, 1999. The families of pathogenesis-related proteins, their activities, and comparative analysis of PR-1 type proteins. *Physiol. Mol. Plant Pathol* **55**: 85-97.

Van Loon, L.C., Gerritsen, Y.A.M. and Ritter, C.E. 1987. Identification, purification, and characterization of pathogenesis-related proteins from virus-infected Samsun NN tobacco leaves. *Plant Molecular Biology* **9**: 593-609.

Velazhahan, R., Datta, S.K. and Muthukrishnan, S. 1999. The PR-5 family: thaumatin-like proteins. In: Datta, S.K. and Muthukrishnan, S. (Eds.). Pathogenesis- related proteins in plants. CRC Press. 107-129 pp.

Veronese, P., Ruiz, M.T., Coca, M.A., Hernandez-Lopez, A., Lee, H., Ibeas, J.I., Damsz, B., Pardo, J.M., Hasegawa, P.M., Bressan, R.A. and Narasimhan, M.L. 2003. In defense against pathogens. Both plant sentinels and foot soldiers need to know the enemy. *Plant Physiology* **131**: 1580-1590.

Vigers, A.J., S. Wiedemann, W.K. Roberts, M. Legrand, C.P. Selitrennikoff, B. Fritig, 1992. Thaumatin-like pathogenesis-related proteins are antifungal. *Plant Sci* **83:** 155-161.

Vigers, A.J., W.K. Roberts, C.P. Selitrennikoff, 1991. A new family of plant antifungal pro- teins. *Mol. Plant-Microbe Interact* **4**: 315-323.

Walz, C., Giavalisco, P., Schad, M., Juenger, M., Klose, J. and Kehr, J. 2004. Proteomics of cucurbit phloem exudate reveals a network of defence proteins. *Phytochemistry* **65**: 1795- 1804.

Ward, E.R., Uknes, S.C., Williams, S.C., Dincher, S.S., Wiederhold, D.L., Alexander, D.C., Ahl- Goy, P., Métraux, J.-P. and Ryals, J.A. 1991. Coordinate gene activity in response to agents that induce systemic acquired resistance. *Plant Cell* **3**: 1085-1094.

Wilson, K. and Walker, J.M. 2000. Principles and Techniques of Practical Biochemistry, fifth edition. Cambridge University Press, Cambridge, UK. pp. 592- 595.

Woloshuk, C.P., Meulenhoff, J.S., Sela-Buurlage, M., van den Elsen, P.J.M. and Cornelissen, B.J.C. 1991. Pathogen-induced proteins with inhibitory activity toward *Phytophtora infestans*. *The Plant Cell* **3**: 619-628.

Wubben, J.P., C.B. Lawrence, P.J.G.M. de Wit, 1996. Differential induction of chitinase and β- 1,3-glucanase gene expression in tomato by Cladosporium fulvum and its race- specific elicitors. *Physiol. Mol. Plant Pathol* **48:** 105-116.

Wubben, J.P., Joosten, M.H.A., van Kan, J.A.L. and de Witt, P.J.G.M. 1992. Subcellular localization of plant chitinases and 1,3-β-glucanases in *Cladosporium fulvum* (syn. *Fulvia fulva*)- infected tomato leaves. *Physiological and Molecular Plant Pathology* **41**: 23-32.

Yadav, B.S. 1981. Behaviour of *Cochliobolus sativus* during its infection of barley and wheat leaves. *Australian Journal of Botany* **29**: 71-79.

Yang, Y., Shah, J. and Klessig, D.F. 1997. Signal perception and transduction in plant defense response. *Genes and Development* **1:** 1621-1639.

Yun, D.-J., Zhao, Y., Pardo, J.M., Narasimhan, M.L., Damsz, B., Lee, H., Abad, L.R., Dúrzo, M.P., Hasegawa, P.M. and Bressan, R.A. 1997. Stress proteins on the yeast cell surface determine resistance to osmotin, a plant antifungal protein. *Proceedings of the National Academy of Sciences* USA 94, 7082-7087.

Zhou, J., X. Tang, R. Frederick, G. Martin, 1998. Pathogen recognition and signal transduction by the Pto kinase. *J. Plant Sci. Res* **111**: 353–356.

Zhu, B., Chen, T.H.H. and Li, P.H. 1996. Analysis of late-blight disease resistance and freezing tolerance in transgenic potato plants expressing sense and antisense genes for an osmotin- like protein. *Planta* **198**: 70-77.

Zhu, Q., Maher, E.A., Masoud, S., Dixon, R.A. and Lamb, C.J. 1994. Enhanced protection against fungal attack by constitutive co-expression of chitinase and glucanase gene in transgenic tobacco. *Bio/technology* **12:** 807-812.

Transformation of Indian Agriculture through Innovative Technologies *Pages* **191–198**
Editor: **Dr. Shahid Ahamad & Dr. Jag Paul Sharma**
Published by: **ASTRAL INTERNATIONAL PVT. LTD., NEW DELHI**

12 Management of Vegetable Diseases in Rural Areas through Eco-friendly Techniques

Ranbir Singh, Sachin Gupta and Sunita Rani

Vegetable are important source of dietary, minerals and vitamins. All the developed and developing countries realize the importance of vegetables as an essential diet due to medicinal and nutritional value for human health. There is steady upward trend in vegetable production. China is ranking first in world and currently produces 237 million tons of vegetables. India has a quantum jump in vegetable production securing the second positions in the world. The total production of vegetable is more than 91 million tons in the country.The Indian scenario has changed tremendously during the last decade because of change in the life style and food habits. The people are becoming more aware to eat healthy foods. Hence, protection of plants from pathogens is the preoccupation of agricultural scientist around the world and it is the unifying goal of plant pathology to control plant disease and chemicals play a major role in accomplishing that goal in contemporary agricultural production.Vegetables are an important component of health food and provide nutritional and health security In spite of this, the productivity of vegetable per unit area is very low. Thus the produces at present is approximately half of the requirement as per dietary standard against 250-300 g/day/adult. Vegetable being more succulent and rich in nutrient are more prone to disease infection, thereby incurring high yield losses during pre and post production period. Disease pressure in vegetable crop from seedling stage to harvest caused by mainly fungi, bacteria and virus are the most important constraints for low production. Hence protection of plants from pathogens is the preoccupation of agricultural scientist around the world and it is the unifying goal of plant pathology to control plant disease and chemicals play a major role in accomplishing that goal in contemporary agricultural production. The survey of literature reveals that vegetables either grown directly or through transplanted seedling suffer from a variety of biotic, mesobiotic and abiotic causes. Control methods invariably recommended includes cultural practices, host resistance, chemical control, physical and biological control methods. Individually different methods have been recommended for management of different disease, but among the recommendation application of pesticides is really high and thereby posing problems of residue poisoning.The growers, out of ignorance, and with the motive to get instant results with high yield are at times compelled to switch on to applying toxic chemical pesticides which result in damaging the health of soil with reducing the beneficial microbes as well. The proposed study aims to constitute an ideal, eco-friendly alternative to the conventional, expensive and hazardous chemical-based farming that is still in practice today. In recent years, there has been worldwide swing in the use of organic farming. The striking feature of all the above components is environment friendly and easily biodegradable there by resulting in lower pesticide residue and pollution problems associated with chemical farming. The proposed study is an attempt for evolving bio organic package against pest and diseases as an alternative to the conventional chemical based farming to achieve the mission by including all the sustainable components in the management programme. Under this situation application of integrated disease management (IDM) appears most appropriate.

Guidelines for Developing IDM

- ☆ Vegetables are different from cereals because cereals are harvested only once after sowing therefore enough time is available to apply management strategies. On the other hand most vegetables are harvested several times at different stages of the crop growth and therefore the application of control method particularly fungicide suffer due to shortage of time, also for consideration of safe period.
- ☆ Vegetable are raised repeatedly following the principles of intensive farming. This practices favor survival of primary inoculum and subsequently infection and spread of secondary inoculum in the crops. To ensure success of any IDM schedule the impact and the effect of intensive cultivation on diseases must be the major input in developing the schedule.
- ☆ The vegetable in India is still today are grown on small scale. Cucurbits and beans are grown ever around the house and the such crops are reservoir of inoculum of number of pathogens as no control measure is invariably applied. These major sources of inoculum must also be considered for developing the schedule.
- ☆ The schedule develop must be easy approachable and effective to be used at community or co-operative level.

The commercial production also derives the monitoring benefits to the growers. Hence it is pertinent to grow more vegetables with good quality and clean and green produce.This can be achieved by employing improved agro techniques of precision farming and greenhouse technology.

Eco-friendly Techniques of Disease Management

I. Cultural Management

Infected Host Eradication

Certain pathogens of annual spice crops overwinters only or mainly inperennial wild plants. Eradication of host in which the pathogen overwintersis sometimes enough to eliminate completely or to reduce drastically the amount of inoculum that can cause infection in the following season. It is routinely carried out in nurseries, greenhouses and fields.

Vermicompost

Vermicompost has found to effectively enhance the root formation, elongation of stem and production of biomass, vegetables, ornamental plants *etc.* (Grappelli *et al.*, 1985). One of the unique features of vermicompost is that during the process of conversion of various organic wastes by earthworms, many of the nutrients are changed to their available forms in order to make them easily utilizable by plants. Therefore, vermin-composts have higher level of available nutrients like nitrate or ammonium nitrogen, exchangeable phosphorous and soluble potassium, calcium and magnesium derived from the wastes (Buchanan *et al.*, 1988). These distinctive properties can be utilized in considering vermin-compost as a suitable carrier for the bio-formulation and as a important soil fertilizers.

Biopesticide

With the increased environmental awareness and the pollution potential and health hazards from many of the conventional pesticides, the demand for nature-based biopesticides has been increasing steadily worldwide. As potential antagonistic fungal bio agents, the filamentous Deuteromycetes fungi *Trichoderma sp.*, The species of *Tricoderma* have been evaluated against wilt, root rot fungi and root knot nematode under glass house and field conditions and reported to give results (Chet, 1987, Kaur and Mukhopadhay 1992). Several isolates of *Tricoderma* against the all pathogenic soil borne fungi and root knot nematode showed outstanding results in combating soil borne diseases problems (Shankar and Jeyarajan, 1996).

Oilseed Cakes

Neem (*Aradirachta indica*) widely used for management of root knot nematodes infecting vegetables and pulses (Goswami and Swarup, 1971) and also soil borne fungi *viz.* wilt, root rot *etc.* (Sharma and Bedi, 1988). Neem oil seed cake have been recorded to be fungicide and reported to enhance the proliferation of the spores of *Glomus fasiculatum* (Lingaraju and Goswami, 1995 and Bhattacharya and Goswami, 1988).

Farmyard Manure

Farmyard manure is a varying mixture of animal manure, urine, bedding material, fodder residues, and other components – is the most common form of organic manure applied in the field. The high organic matter content and the active soil life improve or maintain friable soil structures, increase the cat ion exchange capacity, water holding capacity and infiltration rate, and reducing the risk of soil pests building up.

Crop Rotation

Soil borne pathogens that infect plants of one or few species or even families of plants can sometimes be reduced in the soil by planting non-host crops for 3 or 4 years. In this case, crop rotation can reduce population of pathogen (*e.g. Verticillium*). Ginger can be rotated with cruciferous crops.

Sanitation

Sanitation consists of all activities aimed at eliminating or reducing theamount of inoculum present in a plant, field or a warehouse and at preventingthe spread of the pathogen to other healthy plants and plant products. Thus, ploughing under infected plants after harvest, such as leftover infected fruit, rhizomes or leaves, helps cover the inoculum with soil and speed up its disintegration.

Creating Conditions Unfavourable for the Pathogen

In the production of many crops, particularly containerized stock, using decomposed tree bark in the planting medium has resulted in the successful control of diseases caused by several soil-borne pathogens. *e.g. Phytophthora, Phythium* and *Thielaviopsis* causing root rots, *Rhizoctonia* causing damping off and crown rot, *Fusarium* causing wilt and nematode diseases of several spice crops.

Evasion or Avoidance of the Pathogen

For several spice crop diseases, control depends on attempts to evadepathogens. For example, chilli anthracnose, caused by the fungus *Colletotrichum capsici,* and the bacterial blight of coriander, caused by bacteria *Pseudomonas syringae* are transmitted through the seed. They can be successfully controlled by using disease free seed and seed treatments.

Use of Pathogen Free Material/Seeds

Seed may carry internally one or a few fungi such as those causing yellowsand soft rots, certain bacteria causing bacterial wilts, spots and blights andcertain viruses.

II. Physical Methods

The physical agents used most commonly in controlling plant diseases aretemperature (high or low), dry air, unfavourable light wave lengths and thevarious types of radiations. With some crops, cultivation in glass or plasticgreen houses provides physical barriers to pathogens and their vectors and inthat way protects the crop from some diseases.

Soil Sterilization by Heat

Soil can be sterilized in green houses, and sometimes in seed beds and coldframes, by the heat carried in live or aerated steam or hot water. The soil issteam sterilized either in special containers (soil sterilizers), into whichsteam is supplied under pressure, or on the greenhouse benches, in which case steam is piped into and is allowed to diffuse through the soil. At about 50°C, nematodes, some oomycetes, and other water moulds are killed whereas most plant pathogenic fungi and bacteria, along with some worms,slugs, centipedes, are usually killed at temperatures between 60 and 72°C.

Soil Solarization

The concept of managing soil borne pathogens has now changed. In past, control of these pathogens concentrated on eradication. Later it has been realized that effective control could beachieved by interrupting the disease cycle, plant resistance or the microbial balance leading to disease reduction below the economic injury level, rather than absolute control. The integratedpest management concept encompasses many elements. In this context, soil solarization can play a significant role. Soil solarization is a non-chemical soil disinfestation method applied worldwide for the control of soil-borne plant pathogens, weeds and nematodes. Soil solarization has been demonstrated to control diseases caused by many fungal pathogens such as *Rhizoctonia solani*, *Fusarium* spp., *Pythium*spp., *Phytophthora* spp., *Verticillium* spp., *Sclerotium rolfsii etc.* in many crops (Katan *et al.*, 1983; Abdul *et al.*, 1995). Soil solarization has also been shown to significantly decrease the population of disease causing *Agrobacteria* and *Pseudomonas* (Chellemi *et al.*, 1994). Soil solarization has also been used to control many species of nematodes. However, as nematodesare relatively mobile, may survive solarization deeper in the soil profile and recolonize soil rapidly, soil solarization may not always be as effective as in controlling fungal disease and weeds. Diseases caused by *Meloidogyne* spp., *Heterodera* spp. *etc.* have been successfully controlled by soil solarization (Grinstein *et al.*, 1995). Soil solarization has also been successfully combined with biological control. The use of *Trichoderma harzianum* with solarization in fields infested with *Rhizoctonia solani* has been shown to improve disease control while delaying the buildup of inoculum (Chet, 1987). An increase in population of green *fluorescent pseudomonads* along with an increase of *Penicillium* and *Aspergillus* spp. Heat is used as a lethal agent for the control of plant pathogenic organisms through theuse of transparent polyethylene soil mulches (tarps) for capturing solar energy. Polyethylenecovering of soil induces green house effect and raises soil temperature.

The followingrecommendations are made to bring about effective solar heating of soil:

- ☆ Transparent (clear) not black polyethylene should be used since it transmits most of thesolar radiation that heats the soil. Black polyethylene, though it is greatly heated by itself, is less efficient in heating the soil than transparent sheet.
- ☆ Soil mulching should be carried out during the period of high temperatures and intense.

- ✰ solar irradiation.
- ✰ Soil should be kept wet during mulching to increase thermal sensitivity of resting structures such as sclerotia, chlamydospores, *etc.* and to improve heat conduction.
- ✰ The thinnest possible polyethylene tarp (25-30 µm) is recommended, since it is both cheaper and more effective in heating, due to better radiation transmittance, than the thicker one. Polyethylene reduces heat convection and water evaporation from the soil to the atmosphere. As a result of the formation of water droplets on the inner surface of the polythene film, its transmissivity to long wave radiation is highly reduced, resulting in better heating due to an increase in its greenhouse effect. Ideal plastic mulch is that which is 100 per cent transparentto solar radiation and completely opaque to long wave radiation. This ideal mulch can increase soil temp. by 6-80c over ordinary polyethylene.
- ✰ Since temperatures at the deeper soil layers are lower than at the upper ones, the mulching period should be sufficiently extended, usually 4 weeks or longer, in order to achieve pathogen control at all desired depths.

III. Biological Management

Biological control of plant pathogens refers to the total or partial destructionof pathogen population by other organisms. It occurs routinely in nature but manipulations by human being have resulted in enhanced benefits. It is achieved by suppressive soils, reducing amount of inoculum through antagonistic microorganisms or by direct protection by biological control agents.

Suppressive Soils

Many soil borne pathogens, such as *Fusarium oxysporum* (causing yellows), *Pythium* spp. (causing soft rot) develop well and cause severe diseases insome soils, known as conducive soils, whereas they develop much less andcause much milder diseases in other soils, known as suppressive soils. Manykinds of antagonistic microorganisms have been found to increase in suppressive soils; most commonly, pathogen and disease suppression hasbeen shown to be caused by fungi, such as *Trichoderma*, *Penicillium*, and *Sporidesmium*, or by bacteria belonging to the genera *Pseudomonas*, *Bacillus*, and *Streptomyces*.

Reducing Amount of Inoculum through Antagonistic Microorganisms

a) Control of Soil Borne Pathogens

Several non-plant pathogenic oomycetesand fungi, including some chytridiomycetes and hyphomycetes, and somepseudomonad and ctinomycetous bacteria infect the resting spores ofseveral plant pathogenic fungi. Among the most common mycoparasitic fungi are *Trichoderma* sp., mainly *T. harzianum*. It parasitizes mycelia of *Rhizoctonia* and *Sclerotium*, and inhibits the growth of many oomycetes such as *Pythium*, *Phythophthora*, and other fungi, *e.g.*, *Fusarium*

Control of Aerial Pathogens

Many fungi have been shown to antagonizeand inhibit numerous fungal pathogens of aerial plant parts. For example, *Ampelomyces quisqualis* parasitizes powdery mildew fungi. *Darlucafilum,* and *Verticillium lecanii* parasitizes several rust.

Control through Trap Plants

If a few rows of rye, corn, or other tall plantsare planted around a field of peppers or ginger many of the incoming aphidscarrying viruses that attack the peppers, and ginger will stop and feed on theperipheral taller rows of rye or corn. Trap plants are also used againstnematodes which are sedentary endo- or ecto-parasites. For example, *Crotalaria* plants trap the juveniles of root- knot nematodes.

IV. Host Resistance

Use of resistant varieties in crop cultivation provides undoubtedly the mostcost-effective, logistically the easiest, and also the safest of all the methodsused for disease control. Both from the economic point of view and thepossible health hazards involved in some of the methods used for diseasecontrol, this can probably termed as the **"painless method"**. This approach costs little to the farmer and is, therefore, suitable for the developingcountries like India. Use of resistant varieties not only reduces environmentpollution and eliminates hazards to human health, but also checks diseaseepidemics and thus helps to maintain the biological balance in the ecosystem.For many diseases like the vascular wilts and those caused by viruses, whichare difficult to control effectively by some other means, and others like rusts,powdery mildews, and root rots, which do not appear to be economicallypractical to be controlled by other methods, the cultivation of resistantvarieties provides the only means of producing acceptable yields withoutusing toxic compounds.

REFERENCES

Abdul Rahim, M. F., Satour, M. M., Mickail, K.Y. and El Eraki, S. A. (1995). Effectiveness of soil solarization in furrow irrigated Egyptian soils. *Plant Dis.*.**72**: 143-146.

Bhattacharya, D., Goswami,B.K. (1988). A study on the comparative efficacy of neem and groundnut oil cakes against root knot *Meloidoygne incognita* as influenced by microorganism on sterilized and unsterilized soil. *Indian Journal of nematology* 17(1): 81-83.

Buchanan, D., Boddy, D. and McCalman, J. (1988). Getting In, Getting On, Getting Out and Getting Back, In Bryman, A. ed. Doing Research in Organizations, pp. 53-67, London: Routledge.

Chaube, H.S and Ramji Singh, 2001. Introductory Plant Pathology JBD Co. Lucknow 360 pp.

Chellemi, D.O.., Olson, S.M. and Mitchell, D. J. (1994). Effect of soil solarization and fumigation on survival of soil borne pathogens of tomato in Northern Florida. *Plant Dis.* 78: 1167- 1172.

Chet, I. (1987). Trichoderma: application, mode of action and potential as a bio-control agent of soil borne plant pathogenic fungi, p. 137–160.InI. Chet (ed.), Innovative approaches to plant disease control. John Wiley and Sons, New York

Goswami,B.K. and G. Swarup: (1971). Effect of oilcakes amended soil on the growth of tomato and root knot nematode population. *Indian Phytopath.* 24: 491.

Grappelli, A. Tomati, V. Galli E and Vergari, B. (1985). Earthworm casting in plant propogation. *Horticulture Science*, 20: 874-876.

Katan,, J., Fishler, G. and Grinstein, A. (1983). Short and long term effects of soil solarization and crop sequence on Fusarium wilt and yield of cotton in Israel. *Phytopathology*. **73**: 1215-1219.

Kaur, N. P. and Mukhopadhyay, A.N. (1992). Integrated control of chickpea wilt-complex by *Trichoderma* and chemical methods in India. *Trop. pest Management* 38: 20-23.

Lingaraju, S. and Goswami, B.K. (1995). Studies on the effect of neem and mustard oil-seed cakes on cowpea in *Glomus fasciculatum* and *Rotylenchulus reniformis* interaction. *Indian Phytopath.* 51 (1): 33-37.

Shankar P. and Jeyarajan, R. (1996).Biological control of sesamum root rot by seed treatment with *Trichoderma* spp. and *Bacillus subtilis*. *Indian Journal of Mycology and Plant Pathology*. 26: 217-220.

Sharma J. R., and Bedi. P.S. (1988). Effect of soil amendment with oil cakes on wilt of cotton. *J. Punjab agric. Res* 23: 414-416.

Singh, R.S. 1984. Disease of Vegetable Crops. Oxford and IBH Publishing Co. New Delhi, 346 pp.

Transformation of Indian Agriculture through Innovative Technologies *Pages 199–203*
Editor: **Dr. Shahid Ahamad & Dr. Jag Paul Sharma**
Published by: **ASTRAL INTERNATIONAL PVT. LTD., NEW DELHI**

13 Diseases of Saffron and their Management

Sabiya Bashir, Mohammad Najeeb Mughal and Ali Anwar

Saffron (*Crocus sativus* L.) world's most sought after and expensive spice is an important spice cash crop of Kashmir. There are more than 2 lakh people who are directly or indirectly involved with the saffron trade. The crop covers an area of 5000 acres (2023 ha) in Jammu and Kashmir. Saffron is especially grown in uplands and karewa areas of Kashmir valley especially Pampore and adjoining areas. The other place where saffron is grown are Budgam, Pulwama and Anantnag in Kashmir valley and Kishtwar district in Jammu Division. Saffron covers about 4 per cent of total cultivated areas of Kashmir valley and provides about 6 per cent of total agricultural income (Mir, 1992).

The yield of saffron dwindle year after year. The average productivity in J&K reached to 2.7 kg as against 3.29 kg/ha in 1997 (Zargar, 2002). The decline in production continues though the newer areas are being are brought under its cultivation. The intensive cultivation and mono-culturing of saffron in saffron

Figure 13.1. Full Blooming of Saffron Crop in Kashmir.

growing belts of valley together with the continual use of diseased material resulted in frequent occurrence of saffron corm rot diseases incited by pathogens like *Phoma crocrophila, Rhizoctonia crocorum* (sheath blight and corm rot), (*Madan et al.*, 1967), *Fusarium moniliforme* var *intermedium,* non sporulating basidiomycetous fungus (Dhar, 1992), *Macrophomina phaseolina* (Thakur *et al.*, 1992), *Fusarium oxysporum, F. solani, F. pallidoroseum, F.equiseti, Mucor* spp, *Penicillium* spp (Wani, 2004, Ahmed and Sagar, 2006), *Sclerotium rolfsii* (Kalha *et al.*, 2007). Of these diseases, corm rot of saffron caused by *Fusarium oxysporum* and *F. solani* is considered most destructive (Wani, 2004; Ahmed and Sagar, 2006) and takes considerable proportion of the produce every year. Dhar (1992) reported 6.7 to 15.2 per cent corm rot disease incidence and observed that none of the saffron growing areas in Kashmir valley was free from this disease. Thakur (1997), however, reported corm rot incidence to the magnitude of 70 to 85 per cent in saffron growing fields of Kashmir.

Symptoms caused by important fungal corm rot pathogens are briefly described as under:

Rhizoctonia crocorum

The fungus attacks the corm after penetrating through the protective covers and cause the symptoms known as sheath blight.

Phoma crocophila

The fungus attacks the corm transferring their colour from white to yellow. Ultimately turning it black and resulting in destruction of entire corm.

Sclerotium rolfsii

Symptoms appeared as brown to dark brown sunken, irregular patches below corm scales. Lesions are usually 1 mm deep with raised margins. The foliage of severely infected corms start drying from tip downward. White fungal mycelia appear on the bulbs that rot at later stages of disease development. Sclerotia are formatted which are hard, brown to black and about 1 to 2.1 mm in diameter with pseudo-parenchymatous rind.

Fusarium spp.

The disease is characterised by poor growth and wilting of saffron plants besides causing damping off and basal stem rot. The corm rot causing symptoms are mostly localised in root and bud region. The affected corms yield yellow foliage after bloom which gets scratched easily. The morphological and cultural characters of three important *Fusarium* spp. associated with corm rot of saffron are presented in Table 13.1

Disease Management

1. Cultural Practices

Use of healthy and apparently disease free corms procured preferably from disease free area, is recommended besides field sanitation (collection and destruction of rotten corms or plant debris).

Table 13.1. Morphological and Cultural Characters of Three Important *Fusarium* spp.

Fusarium sp.	*Characteristics*	*Description*
Fusarium solani	Growth rate	Fast
	Colony characters	Green to bluish brown with little aerial mycelium
	Micro conodia	Abundant 0-1 septate, 8.5 -16.0 μm oval with thicker wall, hyaline.
	Macro conidia	Moderately curved with short blunt apical cell and pedicillate basal cell, 3 septate, 29.0- 42.0 μm × 4.0 – 6.0 μm
	Conidiophores	Distinct sample elongated narrow slightly towards Apex later sparsely shorter branched vertical borne forming sporodochium.
	Chalamydospores	Produce singly or in pairs in terminal laterally or inter calary portion, hyaline smooth 6.5 to 10 μm.
	Growth rate	Fast
Fusarium oxysporum	Colony characters	white to purplish tint, aerial mycelium reverse of culture, purple to blue.
	Micro conodia	Generally abundant, round, 6.0-10.0 μm ×2.0 -3.5 μm, non septate, ellipsoidal and straight.
	Macro conidia	Fusiform, moderatelty curved, pointed at both ends, basal cell padicellate. 22 -25 μm × 3.0-4.0 μm.
	Conidiophores	Sparsely branched terminal hyaline, smooth 6-12 μm
	Chalmydospores	Sparse, hyaline, smooth walled, both intercalary and terminal 7.75 ×7.31 μm
	Growth rate	Fast
Fusarium moniliforme	Colony characteers	White granulated to purple, reverse in culture
	Micro conodia	Rare,5- 12 μm ×2.5 -3.4 μm, non septate, formed in chains
	Macro conidia	Clavate to non septate with flattened base 4.5 – 11.0 μm × 2.3 μm, formed in chains, basal cell bearing to 3 apical phylloid.
	Conidiophores	Branched.
	Chalmydospores	Absent.

2. Soil Amendments

Well rotten farm yard manure @ 20 tonnes ha^{-1}and application of NPK @ 30: 40: 10kgha^{-1} is beneficial in reducing the per cent corm rot incidence and severity.

3. Soil Solarisation

Soil solarisation for a period of six week coupled with application of FYM and irrigation has been reported most beneficial in increasing the soil competitive mycoflora, increasing the yield and decreasing the corm rot incidence (Wani, 2004).

4. Biological Management

As most of the saffron pathogens are soil borne, the use of biological control agents is desirable for eco-friendly management of saffron diseases. Among various

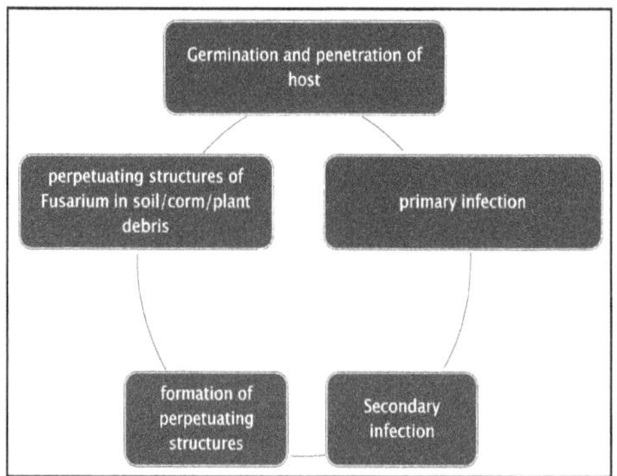

Figure 13.2. Disease Cycle of *Fusarium* sp.

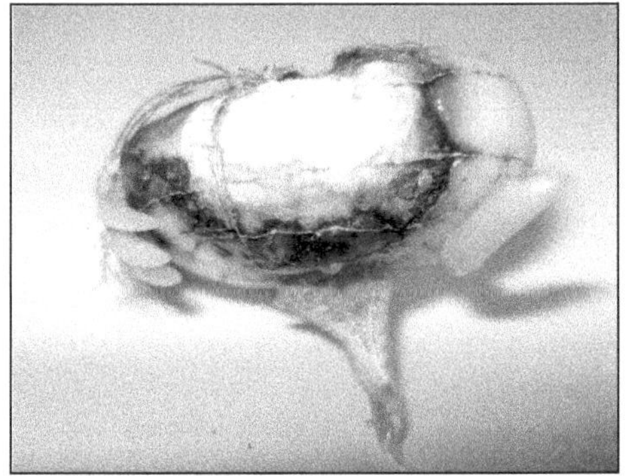

Figure 13.3. Corm Rot of Saffron.

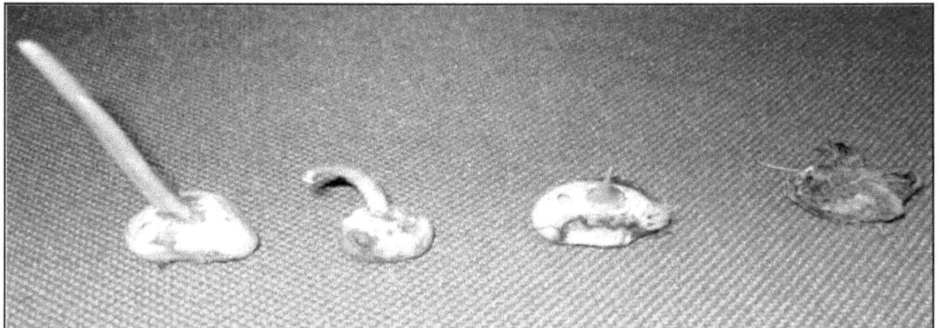

Figure 13.4. Stages in the Development of Saffron.

Figure 13.5. Saffron Flowers.

Figure 13.6. Saffron Field in Kashmir.

fungal biological control agents tested in pot culture experiment (Ahmed and Sagar, 2006) *Tichoderma viride* 2 and *T. viride* 1 reduced corm rot severity (due to *Fusarium oxysporum*) to minimum (15.6 and 11.1 per cent and to 23.7 per cent and 15.9 per cent, respectively) when applied as corm treatment 1×10^{-8}spores ml^{-1} or into soil through FYM (20 tonnes ha^{-1}) as compared to 64.4 per cent, in untreated control.

5. Chemical Management

For immediate and efficient management of saffron diseases, various chemicals have been used. Application of carbendazim (Bavistin) and Tecto (Thiobendazole) applied at 0.2 per cent as corm dip or drenching provide complete disease control (Sud *et al.*, 1999). Corm dip or wet slurry treatment of seed corms with carbendazim (0.2 per cent), Myclobutanil (0.2 per cent) or Mancozeb (0.3 per cent) have also been reported to provide efficient management of corm rot caused by *Fusarium oxysporum* and *F. solani* (Ahmed and Sagar, 2007).

6. Use of Resistant Genotypes

Use of genotypes known to be resistant to corm rot disease is the best way of management of diseases. In a screening study conducted by the authors at SKUAST-K on screening of germplasm against the corm rot disease during 2009 and 2010, 11'Highly resistant' genotypes *viz.*, 0.5Kr, 0.75Kr, SMD-1, SMD-3 SMD-27, SMD-146, SD-147, SD-224 SD-45, SD-52, SD-68, with disease score ranging from 0 to 5.0 were identified while 17 'Resistant Genotypes' *viz.*, SMD-161,0.25 Kr,SMD-152, SMD-93, SMD-13, SMD-157, SMD-47, SMD-11, SMD-76, SMD-52, SD-21, SD-1, SD-35, SD-81, SD-180, SMD-192 and SD-31 with disease score ranging from 5.1 to 10.0 were identified.

REFERENCES

Ahmed, M and Sagar, V 2006. Annual progress report on 'integrated management of corm/tuber rot of saffron and kalazeera (Horticuture Technology Mission-1 project 2.2) p. 10.

Ahmed, M and Sagar, V 2007. Annual progress report on 'integrated management of corm/tuber rot of saffron and kalazeera (Horticuture Technology Mission-1 project 2.2) p. 12.

Dhar, A.K.1992. Bio-ecology and control of corm rot of saffron (*Crocus sativa* L.) *Master's thesis* Division of Plant Pathology, SKUAST-K (J&K) p. 109.

Kalha, C.S, Gupta, V and Gupta, D 2007.First report of sclerotial rot of saffron caused by *Sclerotium rolfsii.*

Madan.C.L., Kapoor. B.M and Gupta, U.S. 1967. Saffron. *Economic Botany*, **20**: 377-385.

Mir, G.M. 1992. *Saffron agronomy in Kashmir*,Gulshan Publishers, Srinagar, p. 163.

Sud, A.K., Paul,Y.S and Thakur.B.R.1999. Corm rot of saffron and its management, *Journal of Mycology and Plant Pathology*, **29**: 380-282.

Thakur R N, Singh C and Koul B L1992. First report of corm rot in *Crocus sativus* L. *Indian Phytopathology* **45**: 278.

Wani, A 2004. Studies on corm rot of saffron (*Crocus sativus* L), Division of Plant Pathology, SKUAST-K, Srinagar (J&K), India, p. xvii+108.

Zargar, G. H 2002. Genetic variation in saffron and importance of quality seed corm. In: Proceeding seminar/workshop on development of saffron, SKUAST-K, pp: 25-36.

Transformation of Indian Agriculture through Innovative Technologies *Pages 205–223*
Editor: **Dr. Shahid Ahamad & Dr. Jag Paul Sharma**
Published by: **ASTRAL INTERNATIONAL PVT. LTD., NEW DELHI**

14 Speciation in Plant Pathogens and Emergence of New Plant Diseases

Arif Hussain Bhat, Mudasirbhat, Ali Anwar, V.K. Ambardar and Saleem Dar

Introduction

Speciation, the splitting of one species into two, is one of the most fundamental problems of biology, being the process by which biodiversity is generated. Understanding how the 1.5 millions of fungal species (Hawksworth, 1991) have arisen is of fundamental interest and has tremendous applied consequences in the cases of agricultural pathogens, emerging human diseases, or fungal species used in industry and biotechnology. Although much progress on the origin of species has been made since the book of Darwin (1859), the subject remains heavily debated. Fungi are excellent models for the study of eukaryotic speciation in general, although they are still rarely included in general reviews on this subject (Coyne and Orr, 2004). First, many fungi can be cultured and crossed under laboratory conditions, and mycologists have long reported numerous mating experiments among fungal species. Second, fungi display a huge variety of life cycles and geographical distributions, allowing the study of which parameters most significantly influence the speciation processes. Third, numerous species complexes are known in fungi, encompassing multiple recently diverged sibling species (Dettman *et al.*, 2003), which allows investigations on the early stages of speciation. Understanding the demographic and ecological factors that lead to the evolution of plant-pathogen species can greatly advance our knowledge of how they adapt to agricultural systems and provide new insights for plant disease management. In addition to these potential applications, fungi and are ideal models for understanding the biological features that lead to new species in general. Fungi comprise a large number of species (About 70,000 species) (Hawksworth, 1997), of which many can cause disease in a wide variety of plant hosts. These factors make them appealing models for evolutionary biologists because pathogen biology provides an extraordinary window into how species interactions shape divergence patterns in different organisms and allows for clear hypotheses formulation. For example, given the dependence of plant pathogens on their host species, it is expected that some of the major mechanisms underlying plant-pathogen species emergence include host-range expansion and host jumps (Broders *et al.*, 2012). We first address the question of species definitions and species criteria and then review the patterns of speciation in fungi, situating them in the general theory as applied to eukaryotes. We focus particularly on the aspects that have seen recent and significant developments, such as sympatric speciation, co speciation, hosts shifts, reproductive character displacement, and the time course of speciation.

Species Definition vs Species Criteria

To study speciation, it seems necessary to first define species. The continual proposal of new species concepts may lead one to think that there is no general agreement about what species are. To the contrary, it has been argued that all modern biologists agree that species correspond to segments of evolutionary lineages that evolve independently from one another (De Queiroz, 1998). The apparently endless dispute about species concepts stems from the confusion between a species definition (describing the kind of entity that is a species) and species criteria (standard for judging or recognizing whether individuals should be considered members of the same species). Many so called "species concepts" actually correspond to species criteria, *i.e.*, practical means to recognize and delimit species (De Queiroz, 2007). The Biological Species Concept (BSC) for instance emphasizes reproductive isolation, the Morphological Species Concept (MSC) emphasizes morphological divergence, the Ecological Species Concept (ESC) emphasizes adaptation to a particular ecological niche, and the Phylogenetic Species Concept (PSC) emphasizes nucleotide divergence. These species criteria correspond to the different events that occur during lineage separation and divergence, rather than to fundamental differences in what is considered to represent a species. One may wonder why there are conflicts over which species criterion we adopt. There are three main reasons why such criteria cannot be universal: (i) speciation is a temporally extended process, but one which varies tremendously in its place among different types of organisms, (ii) several modes of speciation can occur, during which the phenomena used for species recognition do not necessarily appear in the same chronological order, (iii) characteristics of certain organisms render some criteria difficult to apply. Let us take as example the most popular yet the most challenged species criterion, the BSC. For proponents of the BSC, the capacity to interbreed delimits the infra species level, and "Biological Species" are inter sterile groups (Mayr, 1942). This criterion is based on reproductive isolation, but this is only one of the many stages of speciation. Depending on the mode of speciation, inter sterility can occur at early or late stages of speciation, and can constitute the critical stage (in sympatric speciation), or it may be only a by-product of genetic divergence (in allopatric speciation). Obviously, the BSC will be most useful in the first case (sympatric speciation), whereas species criteria based on evidence for lack of gene flow using molecular markers will be more discriminating in the latter case. Inter sterility is the stage at which the process has become irreversible, but this stage may take very long to reach. Until quite recently, the most commonly used species criterion for fungi has been the MSC. However, many cryptic species have been discovered within morphological species, using the BSC (Anderson and Ullrich, 1978), or the GCPSR (Genealogical Concordance Phylogenetic Species Recognition), an extension of the PSC. This latter species criterion uses the phylogenetic concordance of multiple unlinked genes to indicate a lack of genetic exchange and thus evolutionary independence of lineages. Species can thus be identified that cannot be recognized using other species criteria due to the lack of morphological characters or incomplete pre-zygotic isolation. The GCPSR criterion has proved immensely useful in fungi, because it is more finely discriminating than the other criteria in many cases, or more convenient (*e.g.* for species that we are not able to cross), and is currently the most widely used within the fungal kingdom (Dettman *et al.*, 2003; Johnson *et al.*, 2005).

How to Study Speciation in Plant Pathogens

The completion of speciation is signalled by the existence of distinct RIMs. Thus, understanding how species originate necessarily focuses on understanding what biological features prevent gene flow. Although research in plant pathology is usually not characterized as speciation research, a series of approaches has aimed at dissecting the biological basis of reproductive isolation between plant-pathogen species. We divide these studies into two categories and refer to the studies that have identified RIMs between species as the classical studies. The second category, genomic studies, involves genome-scale studies that identify genomic features that are involved in reproductive isolation or in interspecies differences. Classical studies include allopatric speciation, sympatric speciation and ecological speciation while as genomic studies include speciation by hybridization and chromosomal speciation. Each of these studies are discussed below (Restrepo *et al.*, 2014).

Allopatric Speciation

How new species arise in nature is still a highly active field of research. It has long been believed that species originate mostly through allopathic divergence (Mayr, 1963), because extrinsic geographic barriers seemed obvious impediments to gene flow. Fungi could appear as exceptions because eukaryotic micro-organisms have long been considered to have global geographic ranges (ubiquitous dispersal hypothesis; Finlay, 2002), at least for those not dependent on a host having a restricted range. This was in particular true for airborne fungal pathogens because their spores can be dispersed over very long distance. Among the numerous complexes of sibling species recently uncovered using the GCPSR criterion, many however appear consistent with allopatric divergence, because the cryptic species occupy non-overlapping areas separated by geographic barriers (Taylor *et al.*, 2006). Among the recent examples, a multiple gene genealogies approach revealed the existence of cryptic species among the morphological species *Neuro sporacrassa* (Dettman *et al.*, 2003). They had non-overlapping geographical ranges, suggesting allopatric speciation: one phylogenetic species was located in the Congo, another in the Caribbean and Africa (but not Congo), and a third one was restricted to India. In yeasts, Kuehne *et al.* (2007) showed, also using a multiple gene genealogies approach, that *Saccharomyces paradoxus*, a close relative of *Saccharomyces cerevisiae* present in temperate woodlands in the northern hemisphere, was composed of two distinct genetic groups, A and B. The majority of isolates from group A were from Eurasia whereas all isolates from group B had been collected in North America, suggesting a differentiation of these incipient species in separate continents. Another example comes from *Fusarium graminearum*, a fungus responsible for scab on wheat and barley, which had long been considered as a panmictic species with a broad distribution. Recent studies however identified at least nine phylogenetically distinct and geographically separated species (O'Donnell *et al.*, 2004). Four of them were clearly endemic to South America, one was found only in Central America, one in India and one in Australia (O'Donnell *et al.*, 2004). Examples (Table 14.1) can also be found among basidiomycetes, for instance in *Armillaria mellea*, where North American and European strains have been shown to belong to different species(Anderson *et al.*, 1989).

Table 14.1. Speciation Cases Reported to be Allopatric

Sl.No.	Species	Speciation Mode	Comments
1	*Ceratocystis fimbriata*	Allopatric	Fungi and hosts show disjoint ranges
2	*Phytophthora nicotiana*	Allopatric	Speciation occurred after an allopatric period
3	*Venturiaina equalisecotypes*	Allopatric	Wide geographic divergence
4	*Fusarium nepalensis*	Allopatric	Geographic range of all species does not overlap
5	*Fusarium gramineciarum*	Allopatric	Geographic range of all species does not overlap
6	*Pyrenophoratritici-repentis*	Allopatric	Glacial refuges may have provided conditions for speciation
7	*Microbotrymlychnidis-dioicae*	Allopatric	Very little or no recent gene flow was detected

Sympatric Speciation

Compelling evidence for the sympatric divergence is extremely difficult to provide, because excluding a past period of allopatry is almost always impossible (Coyne and Orr, 2004). Evidence consistent with sympatric divergence of fungal populations driven by parasitic adaptation to different hosts has however been reported. An example is provided by *Ascochyta* pathogens, where recent multilocus phylogenetic analyses of a worldwide sample of *Ascochyta* fungi causing blights of chickpea, faba bean, lentil, and pea have revealed that fungi causing disease on each of these hosts form distinct species (Peever, 2007). Experimental inoculations demonstrated that infection was highly host-specific, yet in vitro crosses showed that the species were completely interfertile. The host specificity of these fungi may therefore constitute a strong reproductive barrier, and the sole one (Peever, 2007), following a mechanism of sympatric divergence by host usage. The coexistence in sympatry of interfertile populations specialized on different hosts that remain reproductively isolated cannot indeed be explained currently by models other than the reduced viability of immigrants. This mechanism seems to be able to maintain the species differentiated in sympatry and could similarly have created the divergence in sympatry. It is however difficult to exclude a period of allopatry in the past that would have facilitated specialization, *i.e.*, the accumulation of different alleles beneficial on alternate hosts. An elegant way to demonstrate the sympatric occurrence of speciation is to show that gene flow has occurred after initial divergence (Wu and Ting, 2004). This approach is very promising and has been used so far in fungi only on *Mycosphaerella graminicola*, showing that this wheat pathogen arose recently, most probably during wheat domestication in the fertile crescent, by sympatric differentiation from *Mycosphaerella* species pathogens of natural grasses. Table 14.2 shows some more examples.

Table 14.2. Speciation Cases Reported to be Sympatric

Sl.No.	*Species*	*Speciation Mode*	*Comments*
1	*Botrytis cineria*	Sympatric	Ecological factors favoured speciation
2	*Didy mellarabiei*	Sympatric	Speciation occurred geographic proximity
3	*Mycosphaerella graminicola*	Sympatric	Coalescent analysis show little or no gene flow during the divergence process
4	*Phytophthora infestans*	Sympatric	No change in geographic range since speciation
5	*Botrytis cineria ecotypes*	Sympatric	No geographic barriers separated pathogen populations
6	*Rhizoctonia solani*	Sympatric	Species are currently sympatric in their geographic range
7	*Phytophthora mirabilis*	Sympatric	No geographic barriers separated pathogen populations

Nature of Reproductive Isolation

As seen above, a sine qua non of speciation in sexually reproducing organisms is the decrease of gene flow between incipient species due to the development of reproductive barriers. Two types of reproductive barriers are usually distinguished, prezygotic and post-zygotic, depending on their time of occurrence, before or after fertilization. In fungi having a long dikaryotic stage, nuclear fusion occurs long after individual or gamete fusion, which may render the term post zygotic ambiguous. We will therefore here use the terms pre- and post-mating for fungi, which qualifies time before or after cell fusion. Premating isolation may include different kinds of barriers: for organisms depending on biotic vectors, specialization of these vectors can prevent contact between two populations even if they lie close to one another, yielding ecological isolation. For example the complex *Microbotryum violaceum*, where the insect vectors are different to some extent between host species, leading to a reduction in mating opportunities among strains from different plants, although the barrier is not complete. Specialization may also allow for ecological premating isolation if mating occurs within habitats (hosts for parasites), as discussed above (Giraud *et al.*, 2006). Allochrony, *i.e.*, differences in the time of reproduction, may also be efficient to promote premating isolation. The sister species *Saccharomyces cerevisiae* and *S. paradoxus* exhibit for instance different cell growth kinetics; this allows most individuals of one species to undergo homospecific crosses before or after reproduction of the individuals of the other species. Proportion of inter-specific mating can therefore be significantly reduced without the need of incompatibility factors. As has been invoked in plants, a high rate of selfing may be efficient in limiting inter-specific mating. Selfing has been suggested to act as a reproductive barrier in the anther smut fungus *M. violaceum*. Assortative mating due to mate recognition occurs if individuals or gametes are able to discriminate between conspecifics and heterospecifics. Assortative mating seems to be especially important in the reproductive isolation of *Homobasidiomycota*, where clamp connections between mycelia of opposite types are almost exclusively observed when the tested mycelia belong to the same species. Post-mating isolation refers to barriers associated with hybrid in viability and sterility and is expected to arise as a result of the divergence of incipient species. In the case of post-mating isolation, hetero-specific crosses occur and lead to the production of unfit offspring. Hybrids may be in viable or sterile due to genetic incompatibilities if mutations fixed independently in the diverging lineages display negative epistatic interactions when brought together in the same individual, a phenomenon known as Dobzansky–Muller incompatibilities (Orr and Turelli, 2001). This kind of intrinsic postmating reproductive isolation is responsible for the numerous reported cases in fungi of crosses that initiate and subsequently abort during in vitro experiments. For instance, heterospecific crosses among *Microbotryum* species produce in vitro fewer viable mycelia than conspecific ones, and crosses among *Neurospora* species lead to few or abnormal perithecia or few viable ascospores (Dettman *et al.*, 2003). Post-mating isolation may also be linked to ecological factors. Hybrids are then perfectly viable and fertile in a benign environment, such as in vitro conditions, but unfit in a natural environment. This can be the case if hybrids display intermediate traits between parental phenotypes and, as a result, are poor competitors in either parental environment. Despite its potential importance to reduce gene flow, such ecological, post-mating barriers have rarely been investigated in fungi. In the species complex *M. violaceum*, hybrids between two close species were inoculated onto both parental host species. In one of the host species, hybrids performed as well as the parental species specialized for this host, indicating that there are no genetic incompatibilities in hybrids. However, when inoculated in the reciprocal host, hybrids did not perform as well as the parental species specialized for this host, showing that in other environmental conditions the same hybrids had a lower viability (Le Gac *et al.*, 2007). Mycologists have extensively studied the pre and post-mating reproductive barriers that are accessible via in vitro crosses, namely intrinsic premating mate recognition and intrinsic post-mating barriers. Despite the potential importance of ecological barriers to gene flow, they are still understudied in fungi. Using fungal systems to investigate reproductive isolation both in the lab and in nature would be a great approach to the virtually unexplored question of the relative contributions of the various reproductive barriers to the decrease of gene flow between sibling species (Ramsey *et al.*, 2003) and to understand which barriers arise first during speciation.

Speciation by Hybridization

Many fungal species do not exhibit complete inter-sterility which gives the opportunity for hybridization. Hybrid speciation is classified according to the ploidy level of the resulting individuals: when hybrids have a chromosomal number that sums that of the parental species, the process is called allopolyploid speciation, whereas hybrids with ploidy identical to that of the parents are referred to as allodiploids or homoploids. Allopolyploids have a higher ploidy level than the parental lines, but their karyotypc is interestingly often not the exact addition of the two parental genomes, due to losses of chromosomes (Le Gac *et al.*, 2007). As a consequence, many ancient polyploidy speciation events may have been overlooked. Recent allopolyploid hybrids have however, been identified in diverse genera: *Botrytis allii*, the agent of gray mold neck rot of onion and garlic. The presence of multiple hybrids in some taxa suggests that hybrids could have selective advantages over parental species, at least in some cases. Alloploidy would provide simultaneously instant reproductive isolation, due to triploidy in backcrosses, and a new ecological niche. Evidence for homoploid speciation comes from a ploidy level identical to that of its parents and a broad heterozygosity. Contrary to allopolyploids that are reproductively isolated from their parents, homoploid hybrids are in competition not only with their parents but also with backcrossed individuals, which renders stable allodiploid species much more unlikely than polyploid ones. A well described case of homoploid speciation is that of the rust *Melampsora columbiana* that emerged from hybridization of *M. medusa*, parasite of *Populus deltoides*, and *M. occidentalis*, parasite of *P. trichocarpa* (Newcombe *et al.*, 2000). This hybrid emerged in 1997 when a poplar hybrid resistant to the two parental rust species was widely grown in California, the hybrid rust being able to infect the hybrid poplar. In this case, the homoploid hybrid clearly had a novel ecological niche, a new host. Another question is why many loci actually stay heterozygous despite potential recombination among F1 hybrids? This may be due to a selective advantage of simultaneous heterozygosity at many loci. Interestingly, recent focus on the gene expression in hybrids provides a potential mechanism for such an advantage. Hence, hybridization may allow exploring fitness landscapes outside that of the parental species. This would facilitate the maintenance of hybrids in a new niche and thereby their persistence as a new species.

Chromosomal Speciation

Another mechanism allowing instant speciation is chromosomal speciation. The first model of chromosomal speciation (speciation due to chromosomal rearrangements) considered that if two isolated populations had fixed karyotypic differences, and that recombination between rearranged chromosomes were generating unbalanced gametes that lowered fitness, between-population gene flow could be prevented upon secondary contact (White, 1978). This model was then dismissed on the rationale that rearrangements that cause a sufficient reduction in fitness in heterozygotes could not be fixed in a population precisely because of this reduction in fitness. When at low frequency, the rearrangement will indeed always be in a heterozygous state, and should not be able to increase in frequency. Fungi may however be some of the rare organisms where this speciation scenario could occur because of asexual reproduction and selfing that allow mutants with karyotypic rearrangements to reproduce without loss of fitness. New models of chromosomal speciation consider that the effects of chromosomal rearrangements on recombination rates are more important than those on fitness to explain speciation: a chromosomal rearrangements creates a large region of suppressed recombination where one or more specialization genes can accumulate and lead to the localized restriction of gene flow, which could eventually drive the populations to speciation (Le Gac *et al.*, 2007). In fungi, the small size of chromosomes has long been a barrier to the study of chromosomal rearrangements and chromosomal speciation. The invention of the pulsed field-gradient-gel-electrophoresis allowed the separation of intact fungal chromosomes and revealed that an extremely high proportion of fungal species exhibited chromosome- length-polymorphism (CLP). Chromosomal rearrangements leading to CLP reported in fungi include deletions, reciprocal and insertional translocations, chromosome breakage and fusion or complete chromosome loss, which may in large part be due to transposable elements and other dispersed repetitive sequences (Zolan, 1995). In the ascomycete *Sordaria macrospora* and in the basidiomycete *Coprinus cinereus*, intraspecific sexual crossing of strains harboring different karyotypes resulted in low fertility in the progeny, concordant with the idea that chromosomal rearrangement can play a role in the speciation process (Zolan *et al.*, 1994). However, these karyotypically differentiated strains may also have differed in their genic content. In order to isolate the effect of karyotypic rearrangement, Delneri *et al.* (2003) elegantly constructed strains of *Saccharomyces cerevisiae* differing uniquely by the presence of reciprocal translocations but otherwise completely isogenic. They showed that crosses between such strains had lowered spore fertility and proposed that chromosomal rearrangements, for yeasts at least, are able to provide partial isolation. Chromosomal rearrangements can thus theoretically have a role in speciation in fungi, but showing that the rearrangements were a cause of the divergence, and not only its consequence, remains a challenging task.

Asexual Fungi

In asexual fungi, the theoretical issues of species formation are completely different from those in sexual organisms. There is no recombination to break down combinations of multiple alleles adapted to a given habitat, and the selective pressure on one gene has an effect on the whole genome. Any new allele allowing adaptation on a new niche can thus give rise to a new "species". The difficulty in asexual organisms is rather to understand if, and why, discrete entities exist that we can recognize as species, instead of continuous distributions of phenotypes/genotypes. Asexual organisms in fact seem to form discrete species, and the hypotheses invoked to explain their existence despite lack of homogeneizing gene flow are the existence of discrete ecological niches, random processes of extinctions of intermediate genotypes/phenotypes (Coyne and Orr, 2004), or the recurrent apparition of asexual species from sexual ones. The example is *Magna porthegrisea* complex, many species of which are strictly asexual and host-specific. One of the species of this complex, *M. oryzae*, an important fungal pathogen of rice, has been shown to have arisen recently, possibly in association with rice domestication (Newcombe *et al.*, 2000). Isolates from rice, millet, cutgrass, and torpedo grass appeared also strictly asexual, and to constitute recent host-specific lineages. These patterns in the *M. grisea* complex appear consistent with the idea that acquisition of abilities to infect new hosts in asexual parasitic fungi can readily form new species because recombination will not prevent the differentiation from the ancestral populations. Fungi, with their enormous diversity of modes of reproduction, seem ideal subjects to test the different hypotheses on the nature of species in non-recombining organisms. In some asexual fungi however, recombination can still occur between individuals via somatic recombination (Bos, 1996), which can be considered as equivalent to sex as regards the speciation issue. Hyphal fusions between genetically different individuals is controlled by elaborate vegetative compatibility systems (Bos, 1996), resulting in a condition of heterokaryosis. The exchange of nuclei and organelles can lead to parasexuality via highly transient nuclear fusion and subsequent chromosomal segregation and/or ameiotic recombination. In fungi undergoing such somatic recombination, vegetative compatibility groups (VCG) could be considered as reproductively isolated from each other and therefore as distinct species. This has been suggested in *Aspergillus flavus*, where the different VCGs indeed formed genetically distinct lineages (Newcombe *et al.*, 2000).

Which Factors usually Restrict the Possibility of Speciation?

The existing theory of ecological speciation (Rundle and Nosil, 2005) can be used to better understand factors constraining or promoting the adaptation of pathogens to a new host. According to this theory, alleles providing an advantage on a new host need to greatly increase in frequencies in the local population on that new host. This increase is accomplished by strong selection for local adaptation. The theory also tells us that locally advantageous combinations of alleles need to be protected from being 'diluted' by ancestral alleles brought by immigrants. Such a dilution is expected if mating is random and immigration is recurrent. Therefore, protection of locally adapted allele combinations is required through the evolution of assortative mating (which can be achieved by mate choice, or habitat choice if mating occurs within habitats) or by strongly reduced viability of immigrants (Nosil *et al.*, 2005).The evolution of assortative mating and speciation can be prevented by several factors, including a lack of genetic variation, immigration of ancestral genes, or the costs of being choosy in the selection of mate and/or habitat. Moreover, recombination and segregation will work against the establishment of locally advantageous allele and trait combinations by continuously destroying them. Therefore the overall success of ecological speciation in the presence of gene flow is dependent upon a delicate balance of several factors, as discussed above.

Features of Fungal Pathogens of Plants Promoting Ecological Speciation

Several features of life-history traits in fungal pathogens are conducive to ecological speciation by reducing the constraints that usually impair speciation.

We detail these features below and examine their consequences for the possibility of ecological speciation.

Very Large Numbers of Spores

Population persistence and large mutational input: Pathogenic fungi can produce thousands of spores per lesion per day (Andrivon *et al.*, 2007) and multiple asexual cycles on the same individual plant can yield hundreds of separate infections. This means that billions of spores can be released from a given plant during an infection by a single fungal genotype. Such large numbers of spores can allow the population to persist on a new host even if selection against allele combinations adapted to infection of the ancestral host is extremely strong and the initial degree of adaptation to a new host is very low. Moreover, such large numbers of spores allow the rapid and recurrent creation of genetic variation by mutations. Empirical examples support this logic. For instance, consider evolution of virulent strains able to overcome resistance in crops. New cultivars of plants with resistance genes conferring complete resistance often become susceptible in just a few years if deployed over large areas due to the rapid appearance of new fungal genotypes by mutations that can infect the hitherto resistant plants. Producing a very large number of spores represents an alternative reproductive strategy which can make the evolution of the mechanisms for the choice of host and mate unnecessary.

Mating within Hosts and Outside Hosts

Pleiotropy between host adaptation and assortative mating: Many pathogenic fungi can disperse over large distances after mating or by asexual spores, but they cannot disperse between infection of the host and mating. This is the case for the many ascomycete fungal pathogens (Figure 14.1) responsible for most of the devastating crop diseases. In cases of obligate biotrophs undergoing sex within their host plant, mating occurs only between individuals able to grow on the same host. This means that mutations providing adaptation to a new host will pleiotropically affect local adaptation and mating patterns. This 'magic trait' scenario is one of the most favourable for ecological speciation (Gavrilets *et al.*, 2004). A theoretical model has shown that, because of this characteristic of the lifecycle of some fungal pathogens, adaptation to a new host can significantly restrict gene flow even in sympatry without the need for mate choice or host choice. In this model, the barrier to gene flow is the reduced viability of immigrants With very strong selection this can completely prevent neutral gene flow. Several studies provide empirical support for the generality of this mechanism in natural fungal plant pathogens (Nosil *et al.*, 2005).

Whether a plant pathogen mates within or outside its host has important consequences for the level of gene flow between populations adapted to different hosts. Consider two populations of an obligate biotroph fungal pathogen adapted to two different hosts. Assume there exists a single diallelic haploid locus involved in a gene-for-gene relationship, with allele A1 preventing infection of host 2 (because it codes for an effector recognized by the plant and inducing a defence reaction), and allele A2 preventing infection of host 1. Consider also a neutral locus with alleles B1 and B2. Figures 14.1 and 14.2 depict different types of barriers that could

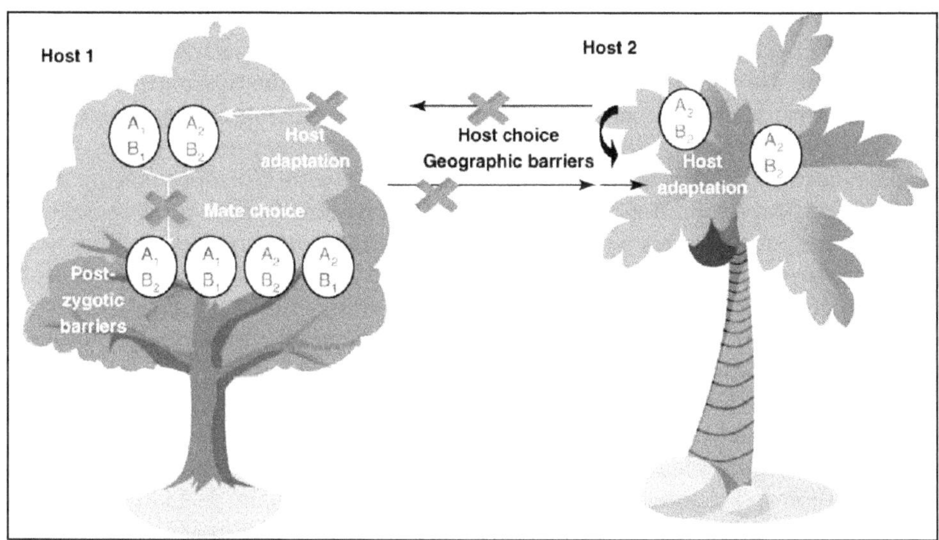

Figure 14.1. Lifecycle of a Pathogen Mating within its Host and the Possibility of Ecological Speciation by Host Shift.

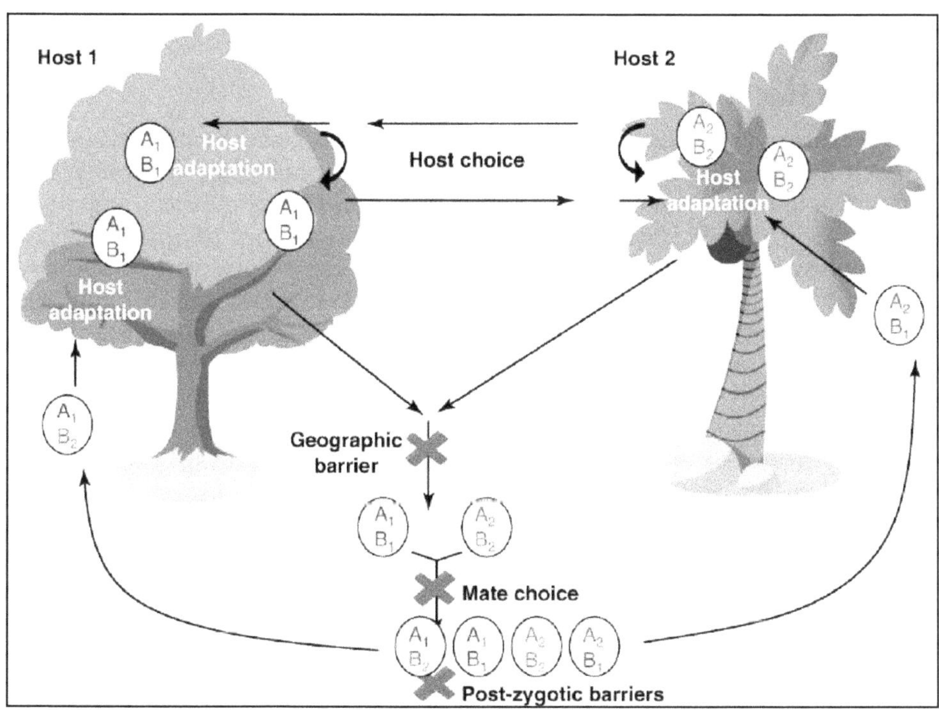

Figure 14.2. Lifecycle of a Pathogen Mating Putside its Host and the Possibility of Ecological Speciation by Host Shift.

act to restrict neutral gene flow between the pathogen populations adapted to the alternative hosts, *i.e.* host choice, host adaptation, geographic barriers between the two hosts, mate choice (*i.e.* assortative mating between the host races), and post-zygotic barriers. The arrows show potential gene flow. The red crosses indicate the steps in the lifecycle where the barriers to gene flow act to prevent gene exchange at the neutral locus. The figures illustrate that, for obligate biotrophs mating within their hosts, host adaptation alone can be a barrier to gene flow (even at neutral loci) because only individuals able to grow on the same host can eventually mate. In contrast, host adaptation cannot substantially reduce gene flow at neutral loci if mating occurs outside the host. A lifecycle of a pathogen mating outside of its host is represented in Figure 14.1. Geographic barriers, mate choice, and postzygotic barriers can prevent gene flow at a neutral B locus (B_1 and B_2 alleles). Host choice and host adaptation can decrease the frequencies of allele A_1 on host 2 and allele A_2 on host 1, but cannot prevent gene flow at the neutral locus B. This is a typical lifecycle of some basidiomycete fungal pathogens such as rusts and smuts. A lifecycle of an obligate biotroph mating within its host is represented in Figure 14.2. Host choice and host adaptation can prevent gene flow at the host choice or specialization loci (A_1 and A_2 alleles), but also at the neutral locus (B_1 and B_2 alleles), as can geographic barriers, mate choice, and post-zygotic barriers. Host adaptation can therefore pleiotropically cause specialization and reproductive isolation if mating occurs within hosts, which is not the case if mating occurs outside the host (Figure 14.1). This is a typical lifecycle of ascomycete plant pathogens.

Special Recognition in Plant Pathogens

Because reproductive isolation is often not easily assessed, fungal and oomycete species have traditionally been defined with morphological and phenotypic traits. Contemporary species description integrates these classical approaches with sequence-based phylogenetic approaches that attempt to find evidence of reproductive isolation (Baum and Donoghue, 1995.). The premise of such approaches is that different gene genealogies of two groups that have ceased interchanging genes should reflect concordant topologies among loci. The most common practice involves the use of gene genealogies to detect long branches, which in turn can be interpreted as discontinuities in population structure that might signal lack of gene exchange. Such studies have revealed that the number of species in fungi and oomycetes tends to be underestimated by a factor of two (Hibbett and Taylor, 2013). In an attempt to unify molecular taxonomy, some general guidelines have been proposed as to what levels of divergence are required to designate species in fungi (Dettman *et al.*, 2003), although how these correlate with reproductive isolation is unknown. Accordingly, species description should rely on multiple, unlinked loci from neutrally evolving genes with strong phylogenetic signals (Avise and Wollenberg, 1997). Phylogenetic species recognition practices and the application of other species concepts are extensively reviewed elsewhere. Identifying species is a challenging task for asexual populations. The majority of species concepts have been developed for individuals that reproduce sexually and recombine with individuals from the same gene pool. Thus, a rigorous framework is required to study species formation in those groups of organisms that do not reproduce sexually or for which no sexual stage is known.

Population genetics and evolutionary theory for asexual taxa provide some guidance and several quantitative extensions of the gene genealogies approach have been proposed. Two of the most comprehensive proposals are the K/θ method (or 4X rule) and the application of branching rates in phylogenetic trees. The K/θ method (Birky and Barraclough, 2009) uses DNA polymorphism and basic coalescent theory to identify clusters of organisms that are diverged enough to be considered separate species. The average number of differences between individuals of two species (after correcting for multiple hits) equals, on average, 8Neμ, where Ne is the effective population size and μ is the mutation rate (Barracloughet *et al.*, 2003). This quantity is known as K. Intraspecific polymorphism, θ, is roughly equivalent to 2Neμ. The K/θ metric reflects divergence scaled by the effective population size and is thought to reveal different species if greater than 4, hence the reference to the 4X rule. This approach identifies putative species only on the basis of reciprocal monophyly using rooted trees and ascribes a metric to delimitate populations from true species (Rosenberg, 2003). An alternative framework for considering whether a clade has diversified into discrete genetic clusters is to consider branching models in phylogenetic trees (Pons *et al.*, 2006).This approach is more complex than the K/θ method but allows for a global test of the relative rates of divergence within a clade. Under a null model, the entire sample derives from a single asexual population, *i.e.*, without divergence into independently evolving or ecologically distinct species. Under this null hypothesis, the pattern of branching is expected to conform to a standard coalescent model for a single population. However, if speciation has taken place, the branching pattern of a phylogenetic tree deviates from the coalescent null hypothesis and has longer branches in the deeper regions of the phylogenies. These long branches are thus a diagnostic that reveals species boundaries. A large body of theoretical work is available to specify the likelihood of a given pattern of branching under a particularmodel (Kaplan *et al.*, 1991), ranging from a neutral coalescent in a population of constant size to populations that increased through time or experienced different forms of selection (Rosenberg and Nordborg, 2002). To our knowledge, this approach has not been applied to plant pathogens but, along with the K/θ method, has the potential for application to the identification of species boundaries and determination of rates of speciation.

Speciation and the Emergence of New Diseases

Understanding the appearance of new pathogenic species can inform control methods and prevention plans (Ahmed *et al.*, 2012, Stukenbrock and McDonald, 2008.). Population genetics inferences and the study of evolutionary processes in fungi and oomycetes can reveal how pathogens emerge to cause new diseases. Emergence proceeds by one of three processes. The simplest scenario does not involve the evolution of new species and occurs when a pathogen colonizes a new geographical area in which it finds the host species that it infected in its ancestral location. *Mycosphaerella fijiensis* is an example of a recent worldwide epidemic that affects banana plantations. Molecular genealogies indicate that the range expansion of this pathogen has occurred through bottlenecks (*i.e.*, severe population reductions) and that the spread of the disease might be caused by movement of plant material infected with ascospores (Halkett *et al.*, 2010). A similar pattern is observed with the sudden oak death pathogen *Phytophthora ramorum*. This pathogen consists of several distinct genetic lineages that appear to have diverged a long time ago, possibly before modern agriculture (Goss *et al.*, 2009). A second process of disease emergence occurs when a pathogen colonizes a new geographic area and develops the capability of infecting new hosts. This process, mediated by adaptation (via host shifts or expansion of the host range), enables the pathogen to thrive in hosts different from its ancestral plants. If the new adaptation leads to reproductive isolation from the parental species, then speciation is correlated with the emergence of the new pathogen and of the new pathogenic syndrome. *Rhynchosporium*, the causal agent of scald disease, seems to be composed of three cryptic pathogen species that have evolved by ecological divergence and host specialization to barley, rye, and *Agropyron* (Zaffarano *et al.*, 2008). Gene genealogies also suggest that *Colletotrichum kahawae* emerged as a pathogen that specializes in green coffee berries (Silva *et al.*, 2012). The recent emergence of these pathogens is correlated with host shifts, suggesting that agriculture has played an important role in the emergence of diseases. A final possibility is that emergent pathogens and diseases originate after the hybridization between two species; the hybrid then how a new suite of traits not observed in the parents (a type of inheritance known as transgressive segregation), allowing it to infect new hosts. Hybridization and admixture after bringing closely related pathogens into contact could favor cross-host-species disease transmission (Gladieux *et al.*, 2011). The demographic inference of gene exchange between plant-pathogen species in an explicitly spatial context can lead to a deeper understanding of the range constraints of plant-pathogen species.

Some Implications of Linking Emerging Diseases and Ecological Speciation

Recognizing that emerging diseases caused by fungal plant pathogens often result from host shift speciation, and that several characteristics of fungal plant pathogens render them conducive to this type of ecological speciation, will improve our knowledge of the mechanisms responsible for disease emergence and the biodiversity of fungi. In addition, this recognition has several important consequences for our understanding of disease dynamics and evolution, as well as for designing more efficient and sustainable control programs. First, if host adaptation alone can be sufficient for speciation, then intersterility, one of the most commonly applied criteria for delimiting species, will not be an appropriate criterion. Interfertility can be retained long after gene flow has ceased between plant pathogens species if the sole reproductive barrier is host adaptation. Failure to recognize that host adaptation can be an efficient barrier facilitating speciation by host shifts means that two distinct pathogenic species could be considered as one. This can lead to the development of control measures that overlook the specificities of each species, such as specific fungicide resistance (Giraud *et al.*, 1997). Accurate delimitation and identification of species is also fundamental for making sound quarantine decisions and policies, and for the implementation of strategies specifically designed to target the right taxa. Furthermore, host adaptation alone can allow for very rapid speciation via host shift, so rapidly evolving markers would be needed to delimit the species. Second, it is important to link emerging diseases with ecological speciation to assess if new diseases are due to spillover, host range expansion, or host shift speciation. These different scenarios affect the control measures to be taken (*e.g.* whether one or multiple hosts should be targeted by fungicides). Also, the dynamics and evolution of the disease will be different if the pathogen is adapted to a single versus multiple hosts (Gandon *et al.*, 2004). or if the newly attacked host is only a reservoir for the second host. Finally, if host adaptation is sufficient for speciation onto a new host via host shifts, then disease emergence can be relatively rapid. This should be taken into account in theoretical models aiming to understand and predict disease emergence. Furthermore, models of the evolution of fungal pathogens characteristics (*e.g.* virulence) should take into account the specificities of the pathogen lifecycle. For instance, the software package Quantinemo (Giraud *et al.*, 1997).in which no dispersal between selection and mating is allowed, can be used for modelling pathogens mating within their hosts. Also, the lifecycle should be considered when predicting which pathogens are the most serious threats: plant fungal pathogens mating within their hosts are expected to cause disease emergence on novel hosts more readily. This feature should be accounted for in quarantine policies, control design, and plant breeding programs.

Conclusion

In conclusion, important advances have been made recently on the speciation in fungi, and they have proved tractable biological models for the general study of speciation. Fungi also exhibit some specific and interesting modes of speciation, and many open questions remain which will be fascinating to explore. Recently developed analytical methods for studying past gene flow and differentiation should be useful to determine in which cases fungal speciation by specialization onto novel hosts has occurred in sympatry. Evolutionary biology provides a powerful framework for understanding key aspects of the natural history of pathogens (Xhaard *et al.*, 2011). The study of evolutionary processes in plant pathogens is a nascent field that promises to deliver new model systems and a comprehensive view of mechanisms involved in speciation and adaptation. On a more practical scale,the development of multilocus databases will allow for the identification of pathogenic lineages in a phylogeographic framework for additional comparative studies to understand the population dynamics of plant pathogens. One of the ultimate goals of plant pathology is to be able to predict the emergence of new pathogens, and this can be done using an evolutionary perspective. The combination of classical approaches and high-resolution genomic studies will reveal demographic and genetic characteristics of how speciation occurs in oomycetes and fungal plant pathogens. The addition of new technological developments will bolster classical approaches and questions. Studying the evolutionary and ecological dynamics of plant pathogens can lead to a better understanding of how these organisms affect agricultural processes over time. The study of recently diverged pathogenic species can also address key questions in evolutionary theory. What is the level of gene flow required to override divergence by selection? What is the role of hybridization in the generation of more virulent pathogens? Do hybrids show new traits or do they always show reduced fitness compared with their parents? Does hybridization (and admixture) lead to the collapse of species boundaries? These questions, among others, are at the very core of evolutionary biology research but also have direct implications in the day-to-day management of plant diseases, as they essentially resolve the origins and mechanisms of emergence of new pathogens.

REFERENCES

C. Darwin(1859). On the Origin of Species. Murray, London.

C.J. Bos (1996). Somatic recombination. In: Bos, C.J. (Ed.), Fungal Genetics, Principles and Practice. Marcel Dekker, New York, pp. 73–95.

C.W. Birky and T.G. Barraclough (2009). Asexual Speciation. In: *Lost Sex: The Evolutionary Biology of Parthenogenesis*, ed. I. Schron, K. Martens and P. Van Dijk, pp. 201–216. Dordrecht, Neth. Springer.

D. Andrivon (2007). Adaptation of *Phytophthora infestans*to partial resistance in potato: evidence from French and Moroccan populations. *Phytopathology*, **97**: 338–343.

D. Delneri, I. Colson, S. Grammenoudi, I.N. Roberts,E.J. Louis and S.G. Oliver (2003). Engineering evolution to study speciation in yeasts.*Nature,* 422: 68–72.

D.A. Baum and M.J. Donoghue (1995). Choosing among alternative "phylogenetic" species concepts.*Syst. Bot., 20*: 560–573.

Dettman, J.R., Jacobson, D.J. and Taylor, J.W. 2003.Amultilocus genealogical approach to phylogenetic species recognition in the model eukaryote *Neurospora. Evolution.* 57: 2703–20.

Felsensein, J. 1981. Skepticism towards Santa Rosalia, or why are there so few kinds of animals? *Evolution* **35**: 124–138.

Finlay, B., 2002. Global dispersal of free-living microbial eukaryote species.*Science* **296**: 1061–1063.

Gandon, S. 2004. Evolution of multihost parasites.*Evolution.* **58**: 455–469.

Gavrilets, S. 2004. Fitness Landscapes and the Origin of Species, Princeton University Press

Giraud, T. 1997. RFLP markers show genetic recombination in *Botrytis cinerea* and transposable elements reveal two sympatric species. *Mol. Biol. Evol.* **14**: 1177–1185.

Giraud, T. 2006. Selection against migrant pathogens: the immigrant in viability barrier in pathogens. *Heredity* **97**: 316–318.

Giraud, T., Gladieux, P. and Gavrilets,S. 20010. Linking the emergence of fungal plantdiseases with ecological speciation. *Trends in Ecology and Evolution* **25**: 387–395.

Gladieux, P., Vercken, E., Fontaine, M.C., Hood, M.E. and Jonot. 2011. Maintenance of fungal pathogen species that are specialized to different hosts: allopatric divergence and introgression through secondary contact. *Mol. Biol. Evol.* **28**: 459–71.

Goss, E.M., Carbone, I. and Grunwald, N.J. 2009.Ancient isolation and independent evolution of the three clonal lineages of the exotic sudden oak death pathogen *Phytophthoraramorum.Mol. Ecol.* **18**: 1161–74.

Halkett, F., Coste, D., Platero, G.G., Zapater, M.F, Abadie, C. and Carlier, J. 2010. Genetic discontinuities and disequilibria in recently established populations of the plant pathogenic fungus *Mycosphaerella fijiensis*. *Mol. Ecol.* **19**: 3909–23.

Hawksworth, D.L. 1991. The fungal dimension of biodiversity: magnitude, significance, and conservation. *Mycol. Res.* **95**: 641–655.

Hawksworth, D.L. 1997. Fungi and biodiversity: international incentives. *Microbiologia***13**: 221–26.

Hibbett, D.S. and Taylor, J.W. 2013. Fungal systematics: Is a new age of enlightenment at hand. *Nat. Rev. Microbiol.* **11**: 129–33.

J.A. Coyne and H.A. Orr(2004).Speciation. Sinauer Associates, Sunderland, MA.

J.B. Anderson and R.C. Ullrich(1978). Biological species of *Armillaria mellea* in North America. *Mycologia,* **71**: 402–414.

J.B. Anderson, S.S. Bailey and P.J. Pukkila (1989). Variation in ribosomal DNA among biological species of *Armillaria,* a genus of root-infecting fungi.*Evolution,* **43**: 1652–1662.

J.C. Avise and K. Wollenberg (1997). Phylogenetics and the origin of species.*Proc. Natl. Acad. Sci. USA,* **94**: 7748–7755.

Johnson, J.A., Harrington, T.C. and Engelbrecht, C.J.B. 2005.Phylogeny and taxonomy of the North American clade of the *Ceratocystis fimbriata* complex. *Mycologia* **97**: 1067–1092.

Johnson, P.A., Hoppensteadt, F.C., Smith, J.J. and Bush, G.L., 1996. Conditions for sympatric speciation: a diploid model incorporating habitat fidelity and non-habitat assortative mating. *Evol. Ecol.***10**: 187–205.

K. De Queiroz (1998). The general lineage concept of species, and the process of speciation. In: Endless Forms: Species and Speciation. Eds., D.J. Howard and S.H. Belocher, Oxford University Press, pp. 57–75.

K. De Queiroz (2007). Species Concepts and Species Delimitation. *Taylor and Francis,* **56**: 879–888.

K.D. Broders,A. Boraks,A.M. Sanchez and G.J. Boland (2012). Population structure of the butternut canker fungus, *Ophiognomonia clavigignenti-juglandacearum,* in North American forests. *Ecology and Evolution,* **2**: 2114–2127.

Kaplan, N., Hudson, R.R. and Iizuka, M. 1991. The coalescent process in models with selection, recombination and geographic subdivision. *Genet. Res.* **57**: 83–91.

Kawecki, T.J. 1998. Red Queen meets Santa Rosalia: arms races and the evolution of host specialization in organisms with parasitic lifestyles. *Nature* **420**: 635–651.

Kondrashov, A. 1986. Sympatric speciation: when is it possible? *Biol. J. Linn. Soc.* **27**: 201–223.

Kuehne, H.A., Murphy, H.A., Francis, C.A. and Sniegowski, P.D. 2007.Allopatric divergence, secondary contact and genetic isolation in wild yeast populations. *Curr. Biol.* **17**: 407–411.

Le Gac, M., Hood, M.E., Fournier, E. and Giraud, T. 2007. Phylogenetic evidence of host-specific cryptic species in the anther smut fungus. *Evolution* **61**: 15–26.

Mayr, E., 1942. Systematics and the Origin of Species. Columbia University Press.

Mayr, E., 1963. Animal Species and Evolution. Harvard University Press, Cambridge, MA.

Newcombe, G., Stirling, B., McDonald, S. and Vradshaw J. 2000. *Melampsora x columbiana*, a natural hybrid of *M. medusae* and *M. occidentalis*. *Mycol. Res.* **104**: 261–274.

Nosil, P. 2005. Perspective: reproductive isolation caused by natural selection against immigrants from divergent habitats. *Evolution* **59**: 705–719.

O'Donnell, K., Ward, T.J., Geiser, D.M., Kistler, H.C. and Aoki, T. 2004. Genealogical concordance between the mating type locus and seven other nuclear genes supports formal recognition of nine phylogenetically distinct species within the*Fusariumgraminearum*clade. *Fung. Genet. Biol.* **41**: 600–623.

Orr, H.A. and Turelli, M. 2001. The evolution of postzygotic isolation: accumulating Dobzhansky–Muller incompatibilities. *Evolution* **55**: 1085–1094.

Peever, T. 2007.Role of host specificity in the speciation of *Ascochyta* pathogens of cool season food legumes.*Eur. J. Plant Pathol.***119**: 119– 126.

Pons, J.D., Barraclough, T.G., Gomez-Zurita, J., Cardoso, A. and Duran, D.P. 2006. Sequence-based species delimitation for the DNA taxonomy of undescribed insects. *Syst. Biol.* **55**: 1–15.

Ramsey, J., Bradshaw, H.D. and Schemske, D.W. 2003.Components of reproductive isolation between the monkey flowers *Mimulus lewisii* and *M. caridinalis* (Phyrmaceae). *Evolution* **57**: 1520–1534.

Restrepo, S., Tabima, J.F., Mideros, M.F., Grunwald, N.J. and Matute, D.R. 2014. Speciation in fungal and oomycete plant pathogens. *Annu. Rev. Phytopathol.***46**: 75–100.

Rice, W.R., 1984. Disruptive selection on habitat preference and the evolution of reproductive isolation: a simulation study. *Evolution* **38**: 1251–1260.

Rosenberg, N.A. 2003. The shapes of neutral gene genealogies in two species: probabilities of monophyly, paraphyly, and polyphyly in a coalescent model. *Evolution* **57**: 1465–77.

Rosenberg, N.A. and Nordborg, M. 2002. Genealogical trees, coalescent theory and the analysis of genetic polymorphisms. *Nat. Rev. Genet.* **3**: 380–90.

Rundle, H.D. and Nosil, P. 2005. Ecological speciation. *Ecol. Lett.***8**: 336–352.

S. Ahmed, D.T. Labrouhe and F. Delmotte (2012). Emerging virulence arising from hybridisation facilitated by multiple introductions of the sunflower downy mildew pathogen *Plasmopara halstedii*. *Fungal Genetics and Biology,* **49**: 847–855.

Silva, D.N., Talhinhas, P., Cai, L., Manuel, L. and Gichuru, E.K. 2012. Host-jump drives rapid and recent ecological speciation of the emergent fungal pathogen *Colletotrichum kahawae*. *Mol. Ecol.* **21**: 2655–70.

Singh, R.P. 2008. Will stem rust destroy the world's wheat crop? *Adv. Agronomy* **98**: 271–309.

Slatkin, M., 1987.Gene flow and the geographic structure of natural populations. *Science* **236**: 787–792.

Stukenbrock, E.H. and McDonald, B.A. 2008. The origins of plant pathogens in agro-ecosystems. *Annu. Rev. Phytopathol.* **46**: 75–100.

T.G. Barraclough, C.W. Birky and A. Burt (2003).Diversification in sexual and asexual organisms.*Evolution*,**57**: 2166–2172.

Taylor, J.W., Turner, E., Townsend, J.P., Dettman, J.R. and Jacobson, D., 2006. Eukaryotic microbes, species recognition and the geographic limits of species: examples from the kingdom fungi. *Philos. Trans. R. Soc. B.***361**: 1947–1963.

White, M.J.D. 1978. Modes of speciation. Freeman, San Francisco.

Wu, C. and Ting, C., 2004. Genes and speciation. *Nat. Rev. Genet.***5**: 114–122.

Xhaard, C., Fabre, B., Andrieux, A., Gladieux, P. and Barres, B. 2011. The genetic structure of the plant pathogenic fungus *Melampsoralarici-populina*on its wild host is extensively impacted by host domestication. *Mol. Ecol.* **20**: 2739–55.

Zaffarano, P.L., McDonald, B.A. and Linde, C.C. 2008. Rapid speciation following recent host shifts in the plant pathogenic fungus *Rhynchosporium*. *Evolution* **62**: 1418–36.

Zolan, M.E. 1995.Chromosome-length polymorphism in fungi. *Microbiol. Rev.* **59**: 686–698.

Zolan, M.E., Heyler, N.K. and Stassen, N.Y.1994.Inheritance of chromosome-length polymorphisms in Coprinuscinereus. *Genetics* **137**: 87–94.

Transformation of Indian Agriculture through Innovative Technologies *Pages 225–248*
Editor: **Dr. Shahid Ahamad & Dr. Jag Paul Sharma**
Published by: **ASTRAL INTERNATIONAL PVT. LTD., NEW DELHI**

15 Women Empowerment and Self Help Groups (SHGs) for Rural Transformation

Banarsi Lal and Shahid Ahamad

India is the home of fourth largest agricultural sector in the world. It has been observed that 84 per cent of the Indian rural women livelihood depends on agriculture. Women farmers contribute significantly to the Indian agriculture. Women are involved in seed selection, sowing, planting, harvesting, and other aspects of agriculture. They are actively involved in every farm activity. But their contribution is always marginalised. This has been proved by various researches. The results of these researches, however, could not get due reconisation in the planning and implementation of agricultural related programmes. Despite their dominance of the labour force women in India still face extreme disadvantage in terms of land rights, wages and participation in agricultural programmes. It is estimated that 52-75 per cent of the Indian women engaged in agriculture are illiterate. Although they are playing a critical role in agricultural growth but even then they face persistent obstacles and economic constraints. Agricultural extension is one such effort taken by the government and non-government agencies that aims at reaching to farmers. The efforts include bringing about a positive change in knowledge, attitude and skills of the farmers by providing training and technical advice and also assisting them in taking decisions in adoption of new research results. Importantly, the clientele of such programmes and efforts is inclusive of both farmers and farm women. Managers of these programmes often consider men as farmers and women as farmer's wife thereby systematically marginalising and underestimating women's productive role in agriculture. The agricultural extension services in India has limited in its operations to a larger extent on male farmers only and it has failed to tackle the great structural problem of invisibility of female farmers. Women farmers are bypassed by male extension workers. It would be correct to state that women farmers in India have failed to get their due share in extension services apropos their contribution to the Indian agriculture. Extension services in India need to be refined, modified and redesigned so as to reach farm women effectively. The purpose of agricultural extension services can be achieved for sustainable rural development only if sincere attempts are made to provide and improve farm women's access to the available extension services thereby leading to their technological empowerment.

The extension machinery in India can be classified in four heads namely(a) Extension services offered by the Indian Council of Agricultural Research(ICAR), (b) Extension services provided by the Ministry of Agriculture and Co-operation, (c) Extension services provided by the Ministry of Rural Development and (d) Extension services offered by Non-Government Organisations (NGOs).Out of four extension systems, training and visit is the major extension system operating in India under the Ministry of Agriculture and Cooperation for more than 30 years. Although this programme is in operation for so long, it still lacks necessary inbuilt structural arrangements for reaching female farmers. Contact farmers involved in this very programme are mostly male farmers and the numbering of female farmers is very low. It was reported that the extension needs of women were often perceived by the extension agents to be in the disciplines of home science, nutrition, childcare, tailoring *etc.* The information regarding to new farm technologies was seldom passed on to them. Various rural development programmes were launched in India from community approach in 1950s' to special target group approach in 1970s'.Non of these programmes addressed to the specific needs of women farmers and remain concentrated on male farmers. In 1980s' integrated approach was started that attempt to integrate women in the mainstream of development by structurally making them beneficiaries up to an extent of 40 per cent. A number of services supportive for women's socio-economic empowerment *viz.*, Support to Training and Employment Programme for Women (STEP), Rashtriya Mahila Kosh, Indira Mahila Yojna, Mahila Samridhi Yojna, Self-Help Groups *etc.* were implemented. These approaches were not directed towards fulfilment needs for agriculture-related services and concentrated mainly on the issue of employment and social empowerment. In 1993-94, a project aimed at gender-gap reduction among women farmers of the Northern India was launched by the government of India. The limited coverage of this project shortens its impact. Such programmes need to be appreciated for being the pioneering one in this regard. ICAR is another important system for transfer of farm technologies all over the country. This system has operated through various frontline extension programmes, all of which now have been merged with the Krishi Vigyan Kendras (KVKs) since April 1996.In KVKs, provision for special training programmes for women are made. These programmes restricted mainly in the areas like home economics and ignored women's productive role in agricultural operations. The efforts made so far in this direction appear to be localized and remain largely invisible. Involvement of women in agricultural development process by ICAR has been further strengthened when the concept of Farming Systems approach to research/extension was institutionalized by several ways including assessment and refinement of agricultural technologies through institutionalizing village linkage programme. Still these efforts are very limited to make a substantial dent on the overall agricultural scenario. From the very beginning Non-Government Organisations (NGOs) have involved women component in their mandate. They too have given more emphasis on issues related to social empowerment of women. They also have given little attention on women's role in agriculture. Therefore, all the major extension systems in India, the participation and benefits accrued to women farmers are minuscule.

There is a need to delineate and discuss the reasons attributable to poor access of farm women to extension services in India. The causative factors found were-lack of approach of farm women to extension centres, less communication channel of farm women, less income to purchase farm inputs by farm women, lack of land, inconvenient time and location of meetings, gender biasness by extension staff, lack of gender-appropriate technology and lack of authority to them. The other factors were low farm women literacy, lack of tendency to innovate and make decisions in farming, less women workers in extension services and lack of structural arrangements for women farmers in extension programmes. Indian women despite playing an important role in agricultural production, processing and storage, generally lack the right to property and the control of resources usually pass on from men to men keeping women out of the chain of inheritance. This makes them the largest group of landless labourers.

Certain structural changes need to be done in the existing extension machinery of India. Changes in attitude of professionals towards women need to be done. In order to improve the access of farm women, it is prime importance to sensitize the concerned extension workers, extension managers, development administrators and policy makers with the realities of farm women, so that they can be considered as an equally strong force for agricultural and rural development on the same footing as men. Attitudinal changes of extension workers, extension managers and all other concerned with agricultural development efforts cannot be ruled out in order to bring significant improvement in the women's access to extension system in India. In India, where these functionaries are mostly males, gender sensitization training can be one of the methods to bring out these attitudinal changes. The efforts for improving the women's access to extension services need to be directed towards bringing out some institutional changes in the present machinery of extension in India. Farm women should be provided with greater access to credit facilities and other inputs by simplifying the existing procedures suiting to the educational levels of women folk. Flow of credit, inputs and marketing facilities to farm women can be done through women's cooperatives and mahilla mandals. Different extension agencies are focusing on a limited scale to integrate farm women in mainstream of development efforts. The efforts of these agencies need to be managed more efficiently so as to have a synergistic effect to solve the problem of poor women access to extension services. The planners should give due reconisation to women farmers in designing the development programmes. There is also the need to recruit more female extension functionaries in all levels of agricultural extension system. Certain institutional adaptations in the present extension system should be made for a positive step in this direction.

Development may be defined as a function of economic, social, educational and cultural betterment of people. In India various programmes have been initiated for women betterment. Women share has been ensured as members and chairpersons in rural and urban local governments. Women empowerment these days has become a buzz word. For the concept of Women's Component Plan which was mooted as far back as in the Seventh Plan, put into practice in the Ninth Plan as one of the important strategies to earmark not less than 30 per cent of funds and benefits in all women related sectors by the Centre and the State Governments. But if we see women empowerment from perspective of their social development, we observe that women are still at margin even after more than five decades of planning and development. The disparity between them and their counterparts is glaring.

The pitiable health and nutrition conditions of women present a gloomy outcome of more than five decades of the planning and development of the country for women in India. This may be seen in the intensity of anaemia among the women. Presently more than 50 per cent of women are anaemic in India. Anaemia in women is highest in Assam and lowest in Kerala. Although the states like Punjab, Haryana and Maharashtra which are known for their prosperity, the anaemia among women is more than 40 per cent. Before 1966 Himachal Pradesh and Harayana were the parts of erstwhile Punjab. If we compare these states we observe that problem is less serious in Himachal Pradesh than the other two states. This could be due to the focus of the leadership on creation of egalitarian society and ensuring people's participation in developmental activities in the state of Himachal Pradesh. More than 75 per cent of children in our country are anaemic. If we observe at state level, we find that this problem is highest in Haryana where about 84 per cent of children are suffering from anaemia. This problem is 44 percent in Kerala. This problem is acute in all over the country. We can estimate the fate of building in terms of national building if the bricks in terms of children are weak. The accesses to antenatal and post-natal care coupled with lack of nutritive food to pregnant women are responsible for this. It has been observed by the Office of the Registrar General, India that 15 per cent of female child below the age of one year, 17.3 per cent in the age group between 1 year to 4 year, 7.9 per cent in the age group of 5 years to 14 years and 59.2 per cent in age group of 15 years and above died due to anaemia. The percentage of male child who died due to anaemia is less than their counterparts. Its main causes are lack of awareness about general health among the people, consumption pattern and level of sanitation. For example, consumption pattern in North Haryana and West Uttar Pradesh is potato and dal as lunch and dinner and chapattis curd with salt mixed with pepper as breakfast. Vegetables rarely form the parts of their consumption basket. This is not in case of poor people but also in case of those who live in the lap of luxury. This fact is hard to swallow but this is the practice which is on this part of the country. It has been revealed that birth-weights of babies born to women in poor income groups are much less than higher groups. The Mid-Term Appraisal of the 9th Five Year Plan commented "Low dietary intake is the most important cause of under-nutrition. Other major factors responsible for under-nutrition in children are poor infant feeding practices, infections due to poor sanitation, lack of safe drinking water, poor access to health care. In spite of higher average dietary intake, under nutrition rates are higher in

MP, UP and Odisha because of lack of equitable distribution of food and access to health care. Prevalence of anaemia among pregnant women ranges between 50 per cent -90 per cent."

The adverse sex ratio is also an indication of deprivation of women in the society. Presently the sex ratio in India is 940 females per 1000 males as per Census 2011 while it should be 1000 because number of males should be equal to females. Many states and Union Territories sex ratio is less than national average. In the sates like Harayana and Punjab which are considered as the most prosperous states of the country sex ratio is abysmally low. Let us highlight some factors responsible for this.

- It may be observed that sex ratio is not adverse in southern states of the country. It is due to practice of kinship which is male -friendly in northern states of the country and female- friendly in southern part of the country.
- The adverse sex ratio particularly in northern part of the country is due to dowry system. It has been observed that dowry is the main reason for female foeticide. It has been found that female foeticide is found more in the families which are prosperous. These families suffer from superiority complex.
- Female participation in the economic activities also determines the sex ratio because the work determines the worth in the society. It has been observed that in those families where women participate in economic activities, this problem is not severe as compared to those families which do not participate in economic activities because in the former families women have not been considered as liability.

The issues of dowry and employment were found to be the main reasons for the social evil. We all know that education to a boy means educating a single man, while educating a girl means educating the whole family. The education among the females is less as compared to males as is evident from the fact that as per the latest Census as against 82.14 percentage literacy of males their counterparts is merely 65.46 per cent. In most of the states the primary education has been put under the domain of the Panchayati Raj and power of supervision and control was also given to Panchayats. It has been observed in the state like Haryana, we can safely say that neither the elected functionaries nor the selected functionaries of the Panchayats are aware about their roles in supervising and controlling the schools located in the villages. They do not have much interest in these kinds of activities.

The social development of women has not been achieved as expected during the planning era, although various measures have been initiated for them. Malini Karkal has worked out a Composite Quality of Life Index (PQLI) based on three measures namely infant mortality rate, life expectancy at age one and literacy for population aged 15 and above using four census periods-1961, 1971, 1981 and 1991 for all states of India and found that the differences for rural and urban areas for all the variants in the four census periods show the poor PQLI for women in the rural areas both as compared to rural males and urban females. These disparities continue for different groups of women, varying with their access to infrastructure and services. Therefore, government should pay more attention on the social development of women.

Women play a vital role at the grass root level for social development. Education is the prime tool for upliftment of society particularly for a girl child who in turn educates the family. The man without woman is like an eye without vision. In our country mythology, the status of women has always adjudged with high honour so much so that women were even worshipped. Even in the Vedic Age, the women were much respected. Women of the Rigvedic period was asked to compose some of the hymns and have risen to the rank of seers like Visvavara, Apala and Ghosha. The women were performing the social duties with full responsibility and were considered as an integral part of the society. We have the evidence that women led a free life in the Vedic period and education was not denied to them. Many Madrasas were set up by the kings especially for girls in Mughals era. Although there were some atrocities took place against women in Mughals era but they promoted the women's status.

Remarkable participation of women in India's freedom struggle throws much light on the potential and strength of women. Nobody can deny the significant role played by the women during the Indian National Movement. Vijayalakshmi Pandit, Sarojini Naidu, Madam Bhikaji Cama, Sucheta Kriplani are to name a few of the women who devoted their whole life for the freedom of the country. Captain Lakshmi Sehgal headed the Rani Jhansi Regiment of the Indian National Army. Sarojini Naidu is said to be the "Nightingale of India", was elected as the Congress President. Our freedom struggle would have been an arduous task without the key contribution of women. There were many undisclosed faces of Indian women who provided moral support for national building.

The International Labour Organization report says that women constitute 50 per cent of the population make up, 30 per cent of the labour force, receives 10 per cent percent of the world's income and own less than 1 per cent of the world's property. According to 2011 Census, female literacy level in our country is 65.46 per cent against the male literacy level of 82.14 per cent.During 2001-2011 decadal period, the growth in female literacy rate (11.8 per cent) was substantially faster than male literacy rate (6.9 per cent).This is a tremendous growth when compared to the ratio in 1951.During the decade 2001-2011, there is a tremendous progress in the field of female literacy in our country. Women are contributing a lot in the Indian economy growth. We made a bold step in respect of the political empowerment of women at the grass root level during Rajiv Gandhi regime.64th Constitutional Amendment Bill for reservation of women in Panchayats was introduced. Though it failed in Rajya Sabha, it was reintroduced during Narasimha Rao's regime. In December 1992, Parliament passed the 73rd and 74th Constitutional Amendments providing for 33 per cent reservation for women in the Panchayats Raj elections. In the last decade, the women have played a significant role in local bodies. Women's Reservation Bill is still being debated. The presence of women in the national politics has grown significantly. Women have proved that if any opportunity is given to them could be completed successfully in any field. Smt. Indira Gandhi was the world's 2nd Woman Prime Minister and Ms. Vijayalakshmi Pandit was the first woman President of the

UN General Assembly. India has many examples of umpteen women in respect of Governors, Chief Ministers, Judges, Civil Servants *etc.* In the entertainment industry also Indian women have proved their talent. In the World Beauty Pageant contests, India is the country which has won the Miss World context more than twice. In sports also the Indian women namely P.T.Usha, Sania Mirza, Anju Bobby George, Mithali Raj, Mary Com *etc.* have gained prestigious positions. Kalpna Chawla and Sunita Williams have proved that they are not lesser than anybody even in the field of Space Technology. Although we are proud of the Indian women in the global arena but we should also look into the atrocities against them in the name of female foeticide, gender inequality, sexual harassment, dowry system and many more evils. The atrocities against women start even before they are born and get manifested in the environment with a new face. The evils against them are man -made that shackles the women's growth from keeping the pace with the development. The plethora of atrocities and denial of basic rights to women, have led the country to the dark development. The sexual harassment cases are increasing day-by-day. The female foeticide cases are alarmingly increasing in our country. A report of World Health Organization reveals that a woman is being raped after 54 minutes. After every four minutes, a woman or minor girl becomes the victim a sexual harassment. Gang rape and murder case of 23 year medical girl in Delhi demanded the strict law to prevent this heinous crime against women. Prostitution is a shame for our country which is spoiling the rich culture and traditional history. As per National Crime Records Bureau there has been increase in women sex harassment cases in India.

Women Recognisation by World Organizations and India:

- 8th of March is celebrated as "International Women Day" every year with great importance.
- In 1998, Food and Agricultural Organization observed the World Food Day and the theme was "Women Feed the World".
- The year 2001 was declared as "Women Empowerment Year".
- Every year, 25th of November has been earmarked as the "International Day against Violence to Women".
- The decade 1991-2001 was declared as "SAARC Decade for Girl Child".
- Landmark legislation for the protection of Girl Child in Goa.
- The Gujarat Government declared 2002-03 as the "Year of Girl Child".

The policy makers and political leaders have clearly defined the code of conduct for the peaceful and prosperous society, while making the Constitution. Till 1947, the violence against women was considered as a social problem but now that is considered as an individual problem. Now many organizations are realizing the emerging scenario of women discrimination. It is really a serious concern and it needs a consistent approach. Man and woman are interdependent on each other and they cannot live individually. It is the responsibility of every male to create a peaceful and conducive atmosphere for the females, who are the emerging stalwarts of the society. Mere slogans, public meetings and wayward announcements are not sufficient. Active participation and the sense of belonging articulated by every

Indian are very important. We should think that women are the social base of the country and through their tutelage we can build a strong nation. We should give more emphasis on gender equality. Each of us should think that men and women are equal and men should treat women as their equal partners with dignity and humanity. Men are born from the womb of women and they should respect them in every sphere of life.

Women constitute around half of the world's population but in fact they are the largest excluded category in all aspects of life. They have only 1/10th of the global income. Around 70 per cent of women are still living below the poverty line. They constitute almost invariably a small minority of those holding elected office. The data reveals that they were around 10 per cent of the world's parliamentarians in 1980 which rose to 14.8 per cent in 1988, came down to 12.7 per cent in 1997 and 19.5 per cent in 2011.The women around the world are striving for gender equality. Women specially in the rural areas are subjected to gender oppression and gender discrimination. The year 2001 was observed as the "Year of Women Empowerment" in India. Poverty, illiteracy, unemployment, ill-health *etc.* are the major problems in our country from the dawn of independence. These factors are impending the socio-economic development of our country. Women are mostly affected by these socio-economic factors. It has been observed that ill-health, unemployment and illiteracy rates are higher among the women as compared to men in our country. It has also been observed that women- headed households suffer more from the poverty as compared to men-headed household. Between 2001 and 2009 the poverty rates for women were double than that of men. They still face injustice from birth to death and discrimination among them is still very common. Every year; three million women die due to gender based violence.

Poverty is the greatest hurdle in the path of development. According to World Bank in 2011, 32.7 per cent per cent of the Indian people fall below the poverty line. The recent observations by the Indian government estimated that around 38 per cent of the Indian population is poor. India still has the world's largest number of poor people in the world. The incidence of poverty customarily is more in rural areas as compared to the urban areas and also the poverty in SC and ST households is more as compared to general castes. The Human Development Report has noted that gender equality is essential for empowering women and eradicating the poverty from the society. From time to time the Indian government has been emphasizing on empowerment of women. The only motive behind these endeavors' is to bring them into the main stream of development. Different policies, programmes and plans are laid to raise the economic status of women. Article 15 of the constitution prohibits any discrimination on grounds of sex while the directive principles of state policy urges that states shall direct its policy for securing an adequate means of livelihood for women and securing equal pay for men and women. In many five years plans, emphasis was given on women empowerment. It has been envisaged to organise women into self-help groups and thus empower them to equip with skills in different trades to make them economically independent and to increase access to credit by setting up development bank for women. It has also been envisaged that the women programmes and policies need to be redesigned. It has also been designed to provide marketing channels for them. The various programmes like Indira Mahila Yojana, Development of Women and Children in Rural Areas, Mahila Samridhi Yojana, Rashtriya Mahila Kosh *etc.* have been launched by the Indian government from time to time.

Prevalence of patriarchy norms, stereotyped gender biased and preference for son's right to inherit parent's property have deprived girl children from their right to property. In general, the son enjoys the inheritance father's property. Women cannot claim equality with men unless she has the same right as men hold and inherit. Ironically, the land reforms measures undertaken by different state governments are also silent about the women's right to property. In present era, growing modernization has escalated social aspirations and family tension, resulting in growing single women and women-headed households in the form of divorce, deserted and separated. The reports indicate that 16.23 million of women headed house-holds are existing in India of which 72 per cent are residing in rural areas. Therefore, empowering the rural headed households would enable them to fight against the income and poverty. The sweets of women welfare benefits are mostly enjoyed by the highly literate women and those who belong to upper strata of urban society. Women specially the rural women are deployed for the unpaid activities. Only around 15 per cent women are working in the organised sectors. Women employment opportunities are characterized with seasonality, unskilled assignments, discrimination in wage *etc.* in unorganized sectors. In the arid and semi-arid areas, the women specially belonging to Scheduled Caste, Scheduled Tribes and impoverished households work hard in order to mitigate hunger and poverty. The data reveals that 43 per cent of working women constitute almost half of total earnings of their families and 18 per cent families entirely dependent on their earnings.

Women accessibility to savings and credit facilities improves their gender economic status. Women accessibility to saving and credits from banking and non-banking sectors is low particularly for the rural women. The empowerment of women largely confines to the few literate and employed urban women and the rural women remain beyond the reach of empowerment. Even it has been observed that the bank officials hesitate to sanction loan to the rural women thinking them to be too poor because the ownership of assets usually rested with men. Women are meagerly represented in the business because they lack of savings, credits and investment facilities. The Self Help Groups formulation is really a stepping stone for the economic empowerment of rural women. These types of schemes really enable the rural women for the small saving and help them to carry out small investments in income generation activities at the rural level. The women Sarpanches, Panches and village level workers of ICDS can further help for strengthening these types of schemes. Credibility among the women, difficulty in regular deposition of money, uncertainty about the benefits of activities *etc.* are the major hurdles in the formation of self-help groups. The income from the small income generating activities would definitely help to substantiate the rural household activities and food security. The women have a decisive role in the eradication of household poverty. Empowering women with the property rights and banking facilities for the loans would contribute for the household income. The income in the hands of women can contribute more in the household economic status than the income in the hands of men. The economic empowerment of women would certainly help for the social upliftment and eradication of poverty specially from the rural areas.

Rural Women and Food Security

India has achieved self-sufficiency in food grains with the green revolution technology. In India, food security was placed as a national objective much earlier than other developed and developing countries. India now has greater share of the world's poor people than thirty years ago. Now, India has one third of world's poor people as recorded by the World Bank. About 35 per cent of the households below the poverty line are headed by women. In India, around 88 per cent of pregnant women suffer with anaemia. Women's role in the production of major food grains and minor millets illustrates their contribution to the food security. Women also ensure supply of food as food vendors and post-harvest processors. As major meal makers, women ensure food security. Women provide the nutrition to the young children. Women are the major decision-makers in ensuring food to the next generation. India today is not only self-sufficient in food grain production but also has a substantial reserve. The progress in agriculture during the last four decades is the biggest success story of independent India. Agriculture and allied activities constitute the single largest contributor to the gross domestic product. In India, about two-third of the work force depends on agriculture. Despite these impressive gains, India at present is in the midst of a paradoxical situation. On one hand there are record food grain stocks and on the other hand over 250 million of India's people are underfed. There is need to bridge this gap. India has a potential to meet these challenges. This potential can be realized through policy and infrastructure support from the government and by strengthening proactive synergies among various sectors. With the following suggestions sustainable agriculture and food security can be achieved.

Agriculture is an integral part of general development system, serving the system as a whole. The sustainability of agriculture is affected by the other sections of the development. Agriculture centres on integrated use of natural resources such as soil, biological diversity, water *etc.* The integration of agriculture with the other aspects of ecosystem conservation is essential in order to promote both environmental sustainability and agricultural production. A state should provide each and every person, adequate and other basic necessasities so that people can live the life with dignity. Food security is an emerging issue. It involves not only the production but access, process, policy, household realities and quality food. The concept of food security should be broadened so as to mean every individual has the physical, economic and environmental access to a balanced diet, safe drinking water, hygienic environment, primary health care and education. Sustainable land and water management should be directly linked with the food security. Food security must focus on diversified food and not just food grains only. Food banks should be maintained at grass root level. The major challenge is to produce additional food while conserving depleting natural resources. Programmes and policies for agriculture and food security should involve public and private institutions. The role indigenous technical knowledge in conserving food and agriculture should be acknowledged. Gender concerns should be mainstreamed. Nutrition security must be placed high on the agenda for development plans and programmes.

Issues Involving the Women in Food Security

- ☆ State imperatives sometimes increase productivity but without ensuring commensurate income and other benefits for women.
- ☆ The increasing drudgery and time spent by women is not compensated by increase in value added. Women lack access to markets.

- ☆ Sustainable development policies impact on agriculture particularly food security on poor women especially female-headed households.
- ☆ Structural adjustments, policies and transition to market economies do not pay adequate attention to their impact on women.
- ☆ Inadequate gender-discriminated data and data gaps with regard to rural women. Women are marginalised in the planning process.
- ☆ Lack of women participation in terms of gender differences in the design, monitoring and evaluation of policies, projects and programmes.
- ☆ Women are treated as welfare recipients in many employment and income generating projects.
- ☆ Inadequate appreciation of the impact of policies on women.
- ☆ Inadequate appreciation of the impact of demographic changes on women.
- ☆ Institutional barriers to women's political participation.
- ☆ Covert and overt policy biases against women due to policies overlooking or excluding gender-equity considerations.
- ☆ Mandates in regard to women are either absent or not enforced.
- ☆ Agricultural policies do not articulate gender issues so they are not considered.
- ☆ Lack of awareness at all levels in all cultures. Social and cultural constraints on women's participation.
- ☆ Low status and disadvantaged position of women resulting in lower education. Traditional knowledge systems become distorted and undermined while new knowledge is often inaccessible to rural women.
- ☆ Complexity of ecological issues, their impact on gender at macro and micro level.
- ☆ Policies should spell out specific legislation and programmes of action to entitle women to productive assets and women full membership in organisations.
- ☆ Efforts to formulate sustainable development policies should recognize the impact of unsustainable practices on women.
- ☆ Agricultural policies and technology should seek drudgery to improve economic efficiency and wages for the time spent by women.
- ☆ Measures to increase productivity should be accompanied by policies to ensure commensurate increase in incomes wages and other benefits for women.
- ☆ Appropriate methodologies and guidelines are needed to recognise and value women's contribution to productive activities in the economy.
- ☆ Planning, project formulation and design, monitoring and evaluation must include women.
- ☆ Agricultural and rural development policies should accord women's access to and control over productive assets.

- Gender issues should be clearly identified and targets stated in policies and programmes accompanied by appropriate methodologies.
- Women's development should be given priority as an indispensable part of human resource development.
- Policy biases against women in agricultural and rural development must be identified and removed.
- Policy must recognise women's right as an integral part of human rights.
- Policy instruments should be applied to encourage an enabling environment for women's participation.
- Agricultural and rural development policy must address the issue of women's low self-esteem through policy measures.
- Women should not be held responsible environmental degradation and restoration of the natural resource base.

Both men and women play their important role in agriculture throughput the world.60-80 per cent of food is produced by the rural women in most of the developing countries. Despite their contribution to global food security they are underestimated in development of strategies. Traditional inheritance and land tenure laws limit women's ownership and use of land. In the developing countries, banks and other credit institutions are less inclined to lend women because without property and land rights, they lack collateral security. Extension activities and cooperatives inputs rarely reach to women. Few of the world's extension agents are women. Needs and priorities of women are rarely considered in the research and development. Women constitute half of our population and play a vital role in the development of the nation. Unless women's potential is properly recognised, no transformation and development is possible.

Technological Empowerment of Women Farmers

Women farmers contribute enormously to the Indian agriculture. This has been proved by various researches. The results of these researches, however, could not get due reconisation in the planning and implementation of agricultural related programmes. Agricultural extension is one such effort taken by the government and non-government agencies that aim at reaching to farmers. The efforts include bringing about a positive change in knowledge, attitude and skills of the farmers by providing training and technical advice and also assisting them in taking decisions in adoption of new research results. Importantly, the clientele of such programmes and efforts is inclusive of both farmers and farm women. Managers of these programmes often consider men as farmers and women as farmer's wife thereby systematically marginalising and underestimating women's productive role in agriculture. The agricultural extension services in India has limited in its operations to a larger extent on male farmers only and it has failed to tackle the great structural problem of invisibility of female farmers. Women farmers are bypassed by male extension workers. It would be correct to state that women farmers in India have failed to get their due share in extension services apropos their contribution to the Indian agriculture. Extension services in India need to be refined, modified and redesigned so as to reach farm women effectively. The purpose of agricultural extension services can be achieved for sustainable rural development only if sincere attempts are made to provide and improve farm women's access to the available extension services thereby leading to their technological empowerment.

The extension machinery in India can be classified in four heads namely(a) Extension services offered by the Indian Council of Agricultural Research(ICAR), (b)Extension services provided by Ministry of Agriculture and Cooperation, (c) Extension services provided by Ministry of Rural Development and(d) Extension services offered by Non-Government Organisations (NGOs).Out of four extension systems training and visit is the major extension system operating in India under the Ministry of Agriculture and Cooperation for more than 30 years. Although this programme is in operation for so long, it still lacks necessary inbuilt structural arrangements for reaching female farmers. Contact farmers involved in this very programme are mostly male farmers and the incidence of female farmers is very low. It was reported that the extension needs of women were often perceived by the extension agents to be in the disciplines of home science, nutrition, childcare tailoring *etc.* The information regarding to new farm technologies was seldom passed on to them. Various rural development programmes were launched in India from community approach in 1950s' to special target group approach in 1970s'.Non of these programmes addressed to the specific needs of women farmers and remain concentrated on male farmers. In 1980s' integrated approach was started that attempted to integrated women in the mainstream of development by structurally making them beneficiaries up to an extent of 40 per cent. A number of services supportive for women's socio-economic empowerment *viz.*, Support to Training and Employment Programme for Women (STEP), Rashtriya Mahila Kosh, Indira Mahila Yojna, Mahila Samridhi Yojna, Self Help Group *etc.* were implemented. These approaches were not directed towards fulfilment needs for agriculture-related services and concentrated mainly on the issue of employment and social empowerment. In 1993-94, a project aimed at gender-gap reduction among women farmers of northern India was launched by government of India. The limited coverage of this project shortens its impact. Such programmes need to be appreciated for being the pioneering one in this regard. ICAR is another important system for transfer of farm technologies all over the country. This system has operated through various frontline extension programmes, all of which now have been merged with the Krishi Vigyan Kendras (KVKs) since April 1996.In KVKs, provision for special training programmes for women are made. These programmes restricted mainly in the areas like home economics and ignored women's productive role in agricultural operations. The efforts made so far in this direction appear to be localized and remain largely invisible. Involvement of women in agricultural development process by ICAR has been further strengthened when the concept of Farming Systems approach to research/extension was institutionalized by several ways including assessment and refinement of agricultural technologies through institutionalizing village linkage programme. Still these efforts are very limited to make a substantial dent on the overall agricultural scenario. From the very beginning Non-Government Organisations have involved women component in their mandate. They too have given more emphasis on issues related to social empowerment of women. They also have given little attention on women's role in agriculture. Therefore all the major extension systems in India, the participation and benefits accrued to women farmers are minuscule.

There is need to delineate and discuss the reasons attributable to poor access of farm women to extension services in India. The causative factors found were-lack of approach of farm women to extension centres, less communication channel of farm women, less income to purchase farm inputs by farm women, lack of land, inconvenient time and location of meetings, gender biasness by extension staff, lack of gender-appropriate technology and lack of authority to them. The other factors were low farm women literacy, lack of tendency to innovate and make decisions in farming, less women workers in extension services and lack of structural arrangements for women farmers in extension programmes. Indian women despite playing an important role in agricultural production, processing and storage, generally lack the right to property and the control of resources usually pass on from men to men keeping women out of the chain of inheritance. This makes them the largest group of landless labourers.

Certain structural changes need to be done in the existing extension machinery of India. Changes in attitude of professionals towards women need to be done. In order to improve the access of farm women, it is prime importance to sensitize the concerned extension workers, extension managers, development administrators and policy makers with the realities of farm women, so that they can be considered as an equally strong force for agricultural and rural development on the same footing as men. Attitudinal changes of extension workers, extension managers and all other concerned with agricultural development efforts cannot be ruled out in order to bring significant improvement in the women's access to extension system in India. In India, where these functionaries are mostly males, gender sensitization training can be one of the methods to bring out these attitudinal changes. The efforts for improving the women's access to extension services need to be directed towards bringing out some institutional changes in the present machinery of extension in India. Farm women should be provided with greater access to credit facilities and other inputs by simplifying the existing procedures suiting to the educational levels of women folk. Flow of credit, inputs and marketing facilities to farm women can be done through women's cooperatives and mahilla mandals. Different extension agencies are focusing on a limited scale to integrate farm women in mainstream of development efforts. The efforts of these agencies need to be managed more efficiently so as to have a synergistic effect to solve the problem of poor women access to extension services. The planners should give due reconisation to women farmers in designing the development programmes. There is also the need to recruit more extension functionaries in all levels of agricultural extension system. Certain institutional adaptations in the present extension system should be made for a positive step in this direction.

Gender Discrimination

The development in any society would be slow if women who constitute about 50 per cent of population are not facilitated to participate in the developmental activities. India with a female population of over 600 million possesses a vast reservoir of women power which exceeds the combined total population of South-East Asian countries. In the 73rd Amendment of the Indian Constitution for the first time in the history of India, minimum numbers of seats were allotted to women in Panchayats. Meager representation of women in the state and national legislatures, reservation not less than one-third of the total number of seats and chairpersons of Panchayats should be considered a significant landmark in the process of political empowerment of women. Clause(3) of Art.243-D inserted in the Indian Constitution by the 73rd Amendment Act provides that not less than one-third of the total number of seats to be filled by direct election in every Panchayat shall be reserved for women and such seats may be allotted by rotation to different constituencies in a Panchayat. Clause (2) ofArt.243-D provides that not less than one-third of the total number of seats shall be reserved for women belonging to the Scheduled Castes or Schedule Tribes. Seats for these marginalised sections of the society should be provided in every Panchayat in proportion of their respective population in the total population in each Panchayat and such seats have to be allotted by rotation to different constituencies in the Panchayat. Clause (4) of the mentioned Art. stipulates that the offices of the chairpersons in the Panchayats at the village level or any other kevel should be reserved for the Scheduled Castes and Schedule Tribes.

The 73rd Amendment to the Indian Constitution has greatly contributed to the political empowerment of women and marginalized sections of society to the political empowerment of women. There were skeptics who were favorably disposed to the proposition of women leadership. Guided by their traditional dominance in a patriarchal society, the males used to cite some of the disabilities of women like illiteracy, family responsibilities, experience, poverty, and communication skills *etc.* as the inhibiting factors for effective participation of women in the decision – making process at the local level. The upper caste males were frantically in search of methods through which their traditional hold in the rural sector could be retained. The women from marginalized communities in the rural areas were not initially very confident of their abilities to assume their leadership in the Panchayats. The male-dominant rural power structure did not like to lose its traditional grip over the rural institutions. This led to the nomination of women members of their families or relatives for the non-SC/ST political seats in the Panchayats. Many of these women who never left their homes had to contest the polls with the support of their husbands. Caste, money and muscle power were also used by the dominant males to ensure their victory in several cases. There were many instances where the elected women in the Panchayats had to depend on their family members to perform their official duties. Most of these women did not know the nuances of Panchayat administration and they used to dependent on their husbands for transaction official business. In many cases husbands or the brothers of elected women presided over the Panchayat meetings and deliberations in absence of the elected women. The elected women in Panchayats were not so literate, aware, experienced *etc.* and in many cases they were depending on their male counterparts in decision-making. In regard to the elected Sarpanches and Panches in the village Panchayats they had to depend on their masters who were the traditional power-holders. The officials working at the village level were not reconciled to work under the control of women

Sarpanches. With the few exceptions, women members of marginalised communities who are relatively literate and have political ambitions or family history of political participation, volunteered to contest elections in the Panchayats. These women had also to depend on their own family members and relatives for electioneering. It should be emphasized that there is nothing wrong if the women seek the support of the traditional male leadership as a learning process. Problems arise if the traditional power holders do not reconcile to the fact they cannot indefinitely continue to have their dominance on popular institutions.

The significant achievements of 73rd Amendment Act concerning reservation of seats and political offices in favour of women and the disadvantaged sections of the rural community is that it had improved their awareness, perceptional levels and rightful share in the decision-making exercise. A brief spell of five to eight years is not enough in the history of nation to judge the rationale of political empowerment of women and other weaker sections of society. Social change in the rural India is already perceptible. Thanks to the mass media and the urge among the weaker sections to improve their educational, social and economic status. The influence of electronic media and the improvement in education, income, knowledge and awareness are affecting the value system, attitude expectations and aspirations of the rural disadvantaged sections in recent years. Political improvement holds the key for their social and economic improvement. There is needed to be cynical about the prospects of the Constitutional safeguards provided to the women and weaker sections to ensure their effective participation in the decentralized democratic decision-making process. The disabilities suffered by these deprived sections of society are bound to disappear in the long run. The 73rd Constitutional Amendment could be considered not only the historic but radical for the first time in the history of India, it has made mandatory provisions for the reservations of a minimum number of seats and offices of chairpersons to women as well as to the marginalised sections of the society. All states have introduced these Constitional imperatives in their respective Panchayat Acts. In the context, many rural women entered the political arena for the time due to persuasion of their family members' caste and political leaders. The male-dominant rural power is not reconciled to their socialisation in politics. This does not deter them from actively participating in the democratic decentralized developmental process at the local level. The male-dominant rural power should desist from applying social pressure on women aspirants who possess the necessary enthusiasm and ability to assume political leadership. The village males should rather encourage and offer support to them. The men folk should develop a positive attitudinal changes and mental make-up in favour of women.

The elected rural women and weaker sections should be educated and trained by which they can get the knowledge and understanding. Special training programmes for the elected women members in Panchayats should be organised. The State Institutes of Rural Development and the Non-Government Organisations may be required not only to prepare appropriate curriculum for them but also to organise the programmes for them. The training programmes for the elected women members in Panchayats should be organised at state, district and block level respectively. The teaching methods for these women's should be simpler as possible. Group discussions, success stories and case studies should be the part of training. Electronic media and audio-visual aids should be utilized in the training programmes. State government should introduce incentives for the Panchayats headed by women or marginalised sections of the society for good performance, attendance rather developmental activities taken up, organisation of gram sabha meetings *etc.* Ultimately the improvements in the literacy among women and weaker sections hold the key factor for their effective participation in decision making process and involvement with the developmental activities in the rural areas.

Self-Help Groups (SHGs) for Women Empowerment

Sufia Begum was a 21 years old villager and a mother of three children when an Economics Professor Mohammad Yunus of Bangladesh (Nobel Peace Prize Winner 2006) met her in 1974 and asked her how much she earned. She replied that she borrowed 5 takas (about $0.09) from a middleman for the bamboo for each stool. About $0.02 of that went back to the lender. "I thought to myself, my God, for five taka's she has become a slave."Then he said "I couldn't understand how she could be so poor when she was making such beautiful things." The following day he and his students did a survey in the woman's village, Jobra, Bangladesh and discovered that 43 villagers owned a total of 856 taka's (about $27). "I couldn't take it anymore. I put the $27 out there and told them that they could liberate them. "An idea of self- help groups came in his mind. His momentary generosity grows into a full fledged concept that came to fruition with founding of Grammeen Bank in 1983.

The term 'Self Help' was coined by Samuel Smiles in his book in 1859 to describe the people whom we might today call entrepreneurs. The critical factor that he found in people was not their individualism but their willingness to help themselves rather than waiting for others to help. The concept of Self- Help Groups gained significance after 1976 when Professor Mohammad Yunus of Bangladesh began experimenting with micro-credit and women SHGs. His planning made a revolution in Bangladesh in poverty eradication.

Self-Help Group is a small, economically homogenous and affinity based group of people who have decided to save and contribute to a common fund to be lent to its members as per the group decisions. SHGs are a way to involve rural men and women actively in the developmental process by increasing the income, technical skills, mutual help, address constraints *etc.* The agenda of SHGs is 'Empowerment' and shift from dormant masses of the rural women to vibrant masses by bringing more income in the hands so that they can lead better social life. The group can be formed by the initiative of a group of people or by the initiation of an NGO, bank, project, govt. programme *etc.* SHGs general objectives are: to encourage small saving habits among the people, to bring an overall change in socio-economic conditions of people living below poverty line, to understand the dynamics of managing and collection of money, to make the people aware of banking procedures, to identify the leadership qualities in the people and also to encourage the people to avoid the traditional source of financing *i.e.* money lenders and also encourage the people to develop cordial environment within and outside the group.

According to the NABARD, at the end of March, 2007, 2.92 million SHG, s cumulatively received bank loans of Rs 180,410 million. 40.95 million Poor households have been provided with credit from formal institutions. Over 200 million poor people have been provided credit by the banking sector through the SHG's. It is estimated that more than 400 women in India join self help groups every hour and one non- governmental Organisation joins NABARD micro-finance programme every day. Over 90 per cent of the bank linked groups are women groups. SHG's are concentrated in south India mainly in Andhra Pradesh.

By forming the self help groups members can increase their income and employment, they can identify their potential to perform, they can develop collective approach, dependence on money lender is reduced,they can gain social recognisation,women can become more assertive in confronting social evils and problems, initiate new ventures through which the group members can be benefitted, facilitates group members to have easy access of information regarding state and central government schemes,programmes, projects *etc.*, speedup developmental efforts by developing the self-confidence and self reliance. Self-Help Groups can promote holistic development of rural India. The states like J&K has lot of scope in Self-Help Groups.

Self-Help Groups (SHGs) especially for women help to start entrepreneurial activities and thus can help to eradicate the poverty. Formation of Self-Help Groups can be an effective tool for the women empowerment. With the help of Self-Help Groups women can be trained for different kinds of skills under different programmes and schemes by the various organizations which further can help them to improve their socio-economic status. The concept of women's empowerment is the result of several important critiques, debates and discussions generated by the women's movement across the globe, especially by the third world countries.

Women constitute around half of population in the social system but still there is a lot of gender discrimination in every sphere of life. Poverty and gender discrimination threaten the well-being of women, deny their choice for education and training, restrict them for their participation in different programmes, curtail political and economic rights *etc.* Women are given secondary importance in every section of society. Right from beginning of her life restrictions are imposed and she is prohibited to express herself properly. It has been observed that large percentage of population of women is unemployed, have poor socio-economic status, least decision making, lack education and awareness on different developmental aspects and so on. So, there is dire need to empower the women through different kinds of schemes or programmes launched by different governmental and non-government organizations.

Self-Help Group (SHG) is a small, economically homogenous and affinity based group of people who decide to save and contribute to a common fund to be lent to its members as per the group decisions. SHGs are a way to involve rural men and women actively in the developmental process by increasing the income, technical skills, mutual help, address constraints *etc.* The agenda of SHGs is 'Empowerment' and shift from dormant masses of the rural women to vibrant masses by bringing more income in their hands so that they can lead a better social life. Self Help

Groups (SHGs) formation in J&K especially in hilly areas can assist in the women's empowerment. The origin of Self-Help Groups is the brain child of Economics Prof. Mohammad Yunus of Chittagong University who founded Gramin Bank of Bangladesh in the year 1976.This was exclusively established for the poor section of the society. The Self-Help Group movement is a silent revolution to uplift the poor people across the globe. World Micro Credit Summit was held at Washingtonin 1997, converged the developed and developing countries of the world to overcome the serious problem of poverty by using micro credit as a tool to empower the poor section of the society. The major objectives of SHGs are: -(1) To inculcate the habit of saving and banking among the rural women. (2) To develop credibility among the rural women and the bankers. (3) To develop group activity so that women can start the entrepreneurships.

There is no doubt that the women of Jammu and Kashmir have suffered a lot during the last 30 years. SHGs formation can be one of the important options to assist these women to come out from economic distress. For this purpose, government has launched several state and centrally sponsored schemes/programmes for the upliftment of women. Some of these are as: (1) Women belonging to minority community *viz.* Buddhists, Sikhs, Christians *etc.* are being financed under the national minorities' development and finance corporation.(2) Swarnjayanti Gram SwarozgarYojana (SGSY) scheme was introduced in 1999 with the major objective of developing micro enterprise in rural areas, thereby developing the potentials of rural poor who belong to below poverty line, financial assistance both in the form of loan and subsidy is provided to the beneficiaries, both individuals as well as Self-Help Groups so that they can set up their own entrepreneurships. As per the reports, the Swarnjayanti Gram Swarozgar Yojana scheme is implemented in all the districts of Jammu and Kashmir. This scheme mainly emphasis on the formation of women's Self-Help Groups. (3) Reports signify that Jammu and Kashmir women's development corporation (JKWDC) has put forth various socio-economic schemes for the upliftment of women belonging to minorities, backward and other classes.(4) Various social welfare departments also implement various schemes for the women. Presently, there are around 150 social welfare centers which impart different kinds of training to the women so that they can start their entrepreneurships. Presently there are 19 Krishi Vigyan Kendras (KVKs) under SKUAST-J, SKUAST-K and CITH in the state which are imparting need based farmers and vocational trainings to the women and the trained women start the entrepreneurships in agriculture and allied sectors. These KVKs assist in the formation, management and strengthening of SHGs.

The size of an ideal Self-Help Group should be of 10 to 20 members. An informal group should not have more than 20 members and the group members should have homogeneity in their socio-economic status. It has been observed that the members in a big group do not actively participate and the chances of conflict also increase. Only one person from a family should become a member of a self-help group and in this way more number of families can join Self-Help Groups. The group should be either of only men or only women. The group leaders *viz.* chairman, secretary and treasurer are decided by the group members. The group leaders should be benevolent in nature and they can be rotated as per the need. It has been observed that the women groups generally perform well and women are considered more credible by the banks for returning the group dues than their counterparts. The group members should follow rules and bye-laws of SHGs. The group should meet at weekly or fortnightly intervals and the participation of all the group members in all the group meetings makes easy to stabilize and develop credibility among the group members. Fine can be fixed for not attending the group meetings and after sometimes interloaning can be initiated in the group. Registers of the group should be kept up to date by the group by making the entries regularly. The funds should be managed and controlled by the group members themselves without any outside interference. Self-Help Groups are mainly based on savings by the group members and credits from the banks. Savings and credit services from local banks are the logical methods of extension of SHGs growth strategy to mitigate the increasing credit demand in the group. The Self-Help Groups' main function is to economically empower the group members. The members join the group on voluntarily basis. They save the money and then link their credit to a nearby bank to start an economic activity. Group members are guided by government or non-governmental organizations (NGOs). In hilly district Reasi women Self-Help Groups on milk processing, food processing, bracelets making *etc.* are assisted by the Krishi Vigyan Kendra, Reasi, local banks and NGOs and are working excellently. The successful SHGs can be a source of inspiration for the other people who want to earn a substantial income for the upliftment of their socio-economic status. The Self-Help Groups can be an effective technique for women empowerment in the society. The group members should have passion to work honestly for their economic improvement. In order to run the SHGs smoothly there is a need of support from the heads of the family of group members, local people and external organizations. Keeping unemployment scenario in view presently the SHG concept has immense scope in J&K. Government and Non Governmental Organizations (NGOs) should involve all the rural as well as urban areas economically vulnerable people to participate in the SHGs.

Self-Help Groups for Rural Development

Over the years a plethora of poverty alleviation programmes has been implemented in India and a huge amount of social and financial investments has been made to achieve the sole objective of poverty eradication. Most of these programmes were based on top-down approach and did not consider the needs of the people. Activities for poverty alleviation programmes in most cases adhered to fund based development. The economic vulnerable groups were forced to remain under-confident and the guarantee for security of their livelihood was found missing in the entire endeavor to eradicate poverty from the country. Considering the large number of people still living below the poverty line, therefore the resources used for poverty eradication and provision of subsidies in the name of poor have not been much effective in achieving the goal of poverty alleviation. In the late 1990s evaluation reports of the Integrated Rural Development Programme (IRDP), a major programme for creating self-employment opportunities in rural areas reflected the flaws in the implementation of the programme. After that holistic programme called Swarnjayanti Gram Swarozgar Yojana (SGSY) was announced by the Government of India. This programme was based on group approach rural development where the rural poor were organized into self-help groups.

Self-Help Groups (SHGs) are small functional groups in rural areas to increase the resource base of the members through the act of thrift and credit among themselves. To form quality groups, rural participation plays a pivotal role in identifying its members who are brought into the SHG-fold through the process of social mobilization. Group functions are assessed and monitored by the external agencies with active support of government, the lead bank of the region and the Panchayats Union. The SHGs developed under various programmes provide a great opportunity for convergence of various programmes of various ministries and organizations. Necessary training can be provided to the SHGs members to create awareness on community health, traditional and modern agricultural practices, micro-credit, veterinary practices, water resource management, Panchayati Raj and other issues. These trainings could be helpful in increasing the abilities and confidence of the rural poor people that may enable them for an effective contribution towards their own community development. Various micro-enterprises like pickle manufacture, agarbati making, honey and food processing, spices production, dairy farming,group leaf plate making *etc.* have proved to be most viable economic activities in the country to drive beneficiaries out of the poverty trap in the rural areas. The products are produced according to the local demand patterns. To sustain the community economic activities, leadership and membership trainings backed by participatory management is a must for the SHGs. The SHGs members if imparted entrepreneurial trainings combined with exposure visits to the successful micro-enterprise of the same nature would have a greater impact on the quality of the products produced by the poor beneficiaries. The success of the economic activities taken by the self-employed persons largely depends on their social influence, their role in decision making process, broader financial base through enhanced thrift and credit activities and widened ownership rights to the assets created by them.

The agricultural extension programmes of the Ministry of Agriculture could be effectively implemented by the help of SHGs members in the rural areas as 66 per cent of the rural people depend on agriculture. Dairy, poultry farming, sheep breeding are preferred activities among rural people. The benefits of the activities can be maximized for the SHGs by ensuring appropriate forward and backward linkages with the activities of the department of animal husbandry and dairy development. The services of SHGs can be utilized for the eradication of diseases like Tuberculosis, Polio, and Acquired Immuno Deficiency Syndrome (AIDS) which are common among the rural people. The SHGs can play an instrumental role in the population stabilization programme of the Ministry of Health and Family Welfare through reaching the communication as well as benefits of these programmes to the BPL families who tend to have large families leading to higher dependency ratios. Maternal and Infant Mortality Rates are higher in these families. Mechanism would need to be developed for the Reproductive and Child Health Programme benefits to reach the SHGs. The members of the SHGs can be involved in the on-going Adult Education programme of the Department of Education where the self-employed members could be considered for training under Total Literacy Campaign continuing Adult Literacy programme activities. In this endeavor, a major chunk of adult illiterate population could be systematically covered and would have multiplier effect in increasing the performance of the literacy drive in the country. The organizations like Khadi and Village Industries Commission, Small Scale Industries, Ministry of Textiles, Development Commissioner, Handloom *etc.* have programmes for development of clusters in places where traditionally some activities have been going on. For instance, carpet weaving, handloom, pottery, wood-craft, stone carving cane and bamboo *etc.* are popular in some parts of the country. There is need to identify all such activities that could be taken up in clusters covering the groups of rural artisans in various areas. Cluster approach has the advantage of bringing in economies of scale as well as developing backward and forward linkages. The National Programme of Nutritional Support to Primary Education popularly known as Mid-Day Meals Scheme is under implementation in the country since 1995 with an objective to ensure universalisation of primary education by increasing enrolment and attendance and reducing school drop-out rates and simultaneously impacting on nutritional status of children in primary classes. There is possibility of utilizing the services of the SHGs for implementing the on-going Mid-Day Meals Scheme in different states. Quality SHGs can be identified and a flexible decentralized approach be adopted for involving the members of these groups in cooking and supplying the mid-day meals to the school going children. This endeavor would ensure successful implementation of the centrally sponsored mid-day meals scheme. The role of people effected by drought and various other calamities has to be properly outlined while planning and executing calamities mitigation activities. An atmosphere encouraging community participation through SHGs has to be created where access to information, knowledge on effects of drought and possible ways of relief from droughts like situation can be ensured through the government and non-government machineries. The SHGs can be trained to analyze and understand the social, economic, political and environmental consequences of regular scarcities of food, water and fodder. The SHGs can be inspired to interact with the government and non-government agencies to initiate various relief measures taken up by the central and state governments. The SHGs can also be given responsibility of running fair price shops under the Public Distribution System in the rural areas. Community organizations like SHGs could be asked to take up the activity of dissemination of necessary awareness on sanitation procedures in rural areas. Various ministries, departments, organizations *etc.* should look towards the SHGs for targeting their programmes which ultimately would help in improving the quality of life in rural areas.

Transformation of Indian Agriculture through Innovative Technologies *Pages* **249–263**
Editor: **Dr. Shahid Ahamad & Dr. Jag Paul Sharma**
Published by: **ASTRAL INTERNATIONAL PVT. LTD., NEW DELHI**

16 Integrated Pest Management in Winter Vegetables

Hafeez Ahmad, Suheel Ahmad Ganai and Thanlass Norboo

During the last quarter of a century, the concept of IPM has overshadowed the general principles of pest control. Since vegetable crops are consumed both raw as well as cooked, the pest control strategy on these crops should includes only safer insecticides. Hence the use of bio –pesticides and biological control agents becomes the most acceptable strategy against the insect pest of vegetable crops. Besides these, the employment of various cultural practice *viz.* use of resistant varieties, clean cultivation, inter-cropping, use of trap crops, *etc.* also play an important role in reducing the incidence of insect pest as well as the use of chemical insecticides on the target crop to combat them. The important insect pest attacking the major winter vegetable crops have been listed here under along with the commonly followed management practices against the same. Besides, some of the latest practices reported effected/promising in combating these pests in different areas have also been discussed.

Tomato

Fruit Borer: *Helicoverpa* (=*Heliothis*) *armigera*

Whitefly: *Bemisia tabaci*

Hadda beetle: *Henosepilachnadodecastigma*

Leafhopper: *Amrasca biguttula biguttula*

Fruit Borer

Management

1. Deep ploughing before transplanting helps to kill the pupae hiding in the soil.
2. Grow pest resistant varieties.
3. Early transplanting crop escapes the damage of this pest.
4. Inter-cropping with marigold helps in checking this pest.
5. Pick up the caterpillars and kill them.
6. Provide perching places for birds in the field so that they may pick up the larvae from the crop.
7. Use pheromone and light traps to monitor and collect male moth population.
8. Spray biopesticides like *Bacillus thuringiensis* (Bt) formulation *viz.*,

Biolep, Bioasp, Delfin, Dipel(@1.0kg per hectare) or NPV preparations @750 LE per hectare or chemical insecticides as cypermethrin 10EC(375ml)or deltamethrin 2.8 EC (500ml) or fenvalerate 20 EC(185ml) or endosulfan 35 EC (1.251) or carbaryl 50 WP (1.25kg) per hectare in 500-600 litre water.

Note: The above mention insecticides should be used alternately.

Ten days interval should be given between spray application and fruit picking.

Biological Control

Two egg parasitoids have been reported to check the incidence of this pest in different parts of India. In Himachal Pradesh, the release of *Trichogramma brasiliensis* and *T. pretiosum* has been reported to reduce the larval population of *H.armigera* on tomato by 55.9 per cent (Rawat and Pawar, 1993).According to Krishnamurthi and Mani (1996) in Karnataka,5-6 inundactive releases (2.5 lakh adult/ha) of these two egg parasitoids, reduced the damage by this pest to 9.92 and 7.72 per cent as compared to 23.6 and 13.72 per cent, respectively, in control.In another experiment four releases (at 10 days interval) of *T. pretiosum* @ 50000 Corcyra eggs/ha resulted in 58,93,28 and 37 per cent parasization of *H. armigera* eggs as against 22-35 per cent in control (Gupta *et al.*, 1998). Comparing of *T. pretiosum*/ha NPV and endosulfan in Maharashtra revealed that infestation levels of tomato were 19.2 and 7 per cent, respectively (Mehetre and Salunkhe, 1999).

The microbial pesticides *viz. B. thuringiensis* formulation and Ha NPV preparation have also been shown to be effective against this pest. While the former have not been too effective, the latter alone as well as in combination with chemical insecticides have been reported to be effective. In Haryana, Spraying of half dose of endosulfan (350g a.i/ha) mixed with NPV 300 LE/ha also has been reported to check this pest effectively (Satpathy *et al.*, 1999).

Though the studies on these aspects are widely in progress, but no concrete recommendations have emerged till date.

Use of Resistant/Tolerant Varieties

A lot of studies have been conducted on this important aspect of the IPM against tomato fruit borer, *H. armigera*. Though none of these cultivars/varieties have been identified/developed which can be claimed to have true resistance against it, yet the studies have been conducted to identify the varieties/cultivars which have comparative resistance and the same have been discussed here under.

In the studies conducted by Mishra and Mishra(1993), var. BT -1 was identified to be the least susceptible. In another study they identified BT6-2,BT-10,BT-17, T-30 and T-32 as resistant cultivars (Mishra and Mishra, 1995). Studies conducted by Brar *et al.* (1995) revealed that species *Lyopersicon chimeliwshi, L.hirsutum, L. pimpinellifolium* and *L. peruvianum* were highly resistant. Accession LA-2992, La-2449,LA-2531 of *L. esculentum* proved to be the most promising as far as resistance to *H. armigera* was concerned.

Sivaprakasam (1996) observed that the month of this pest lays more eggs on glabrous leaves;cv. Paiyur with low trichome density and long calyx harboured less eggs. On the contrary the vars. with hairy pedunkles *viz.* Pusa Early Dwarf, Arka Vikas and Pusa Gaurav were found to be less susceptible in M.P. (Gajender *et al.*, 1998).While working in Himachal Pradesh, Thakur etal.(1998) observed var.S-12 to be the most susceptible whereas Sabkhya and Verma (1997) recorded wild spp. *L. glandulosum, L. pimpinellifolium* and *L. peruvianum* to be less susceptible alongwith EC-27921,EC-23575, EC-28769 and EC-170791.

Intercropping

Intercropping of tomato crop with African tall Marigold, *Tagetes erecta* cv. Golden Age has been reported to check the incidence of *H. armigera* on tomato crop. Marigold has been observed to attract the month of this pest for egg laying. The caterpillars of this insect are also reported to prefer the flowers of marigold. According to Srinivasan *et al.*(1994) planting of one row of marigold after every 10 rows of tomato reduced the no. of eggs laid on tomato plants, no. of larvae per plants and no. of bored tomato fruits.Marigold seedlings should be 40 days old and those of cabbage 25 days.These should be transplanted in the ratios of 1: 20 rows. Marigold should not be treated with insecticides so that it remains attractive to *H. armigera* and natural enemies are also conserved.It is estimated that attractive to *H. armigera* and natural enemies are also conserved. It is estimated that only 10 per cent population of the pest is attracted to tomato crop which should be taken care of by spraying NPV @250 LE/ha or any other conventional insecticide.Grown up larvae of this pest can be picked up from the marigold flowers and destroyed manually.

In another study, Patil *et al.* (1997) while working on the intercropping of different crops with tomato, observed that the lowest damage of tomato by this pest was observed when it was intercropped with radish.

Other Cultural Practices

Staking of tomato plants was reported to reduce the fruit damage by *H. armigera* in Himachal Pradesh (Bhardwaj *et al.*, 1995) and in Assam the least damage by this pest occurred when the crop was transplanted in the month of October (Borah, 1995).

Whitefly

Management

1. Grow whitefly resistant varieties of tomato.
2. Protect the young plants in nursery by applying furadan 3G or phorate10 G@ 1.0k ga.i/ha (calculate the nursery area and granular insecticide accordingly).
3. In Haryana Kharif season tomato crop is heavily damaged by viral diseases, hence rabi season crop should be grown.
4. Spray oxy-demeton methyl 25 EC or dimethoate 30EC @ 75ml or phosphamidon 85 WSC @ 175ml per hectare in 500-600 litre water before fruit in initiation.

Besides these, intercropping with sweet pepper, use of colour mulches, brushing, drought conditioning. Use of vinyul films which absorb UV rays also have been reported to reduce this pest along with other soft bodies insects like thrips, aphids, *etc.*

Potato

Aphids: *Myzus persicae, Aphis gossypii*

Whitefly: *Besmisia tabaci*

Cutworm: *Agrotis ipsilon, A. segetum*

Leafhopper: *Amrasca biguttula A. biguttula*

Green stink bug: *Nezara viridula*

Potatol tuber moth: *Phthorimaeao perculella*

Epilachna beetle: *Epilachnao cellata*

White grub: *Holotrichia longipennis*

Aphid

Management

1. Grow resistant varieties.
2. Spray 750 ml dimethoate 30 EC or oxy-demeton methyl 25 EC or 175 ml phosphamidon 85 WSC in 500-600 litres water per hectare at an interval of 10 days.

Besides these, date of sowing and method of irrigation of the crop have been reported to affect aphid population/multiplication on potato crop. Karimullah *et al.* (1995) observed less incidence of aphids on potato crop sown late. In another study Parihar and Singh (1995) reported incidence of aphids on potato to be the least when the crop was irrigated with sprinkler system and maximum in case where furrow irrigation was applied.

Whitefly

Management

1. Early sown crops are severely damaged by this pest.
2. Chemical control mentioned for potato aphids, takes care of whiteflies, too.

Cutworm

Management

1. Frequent raking of soil helps in exposing the hidden larvae to birds and sun.
2. Spray the crop and drench the soil with 1.5 l endosulfan 35 EC in 750 litre water per hertare.

Besides, Parihar *et al.* (1995) observed that predatory birds account for a substantial toll of cutworms and white grab population.The former's population was reported to reduce from 30 to 4.8 per 6 m^2 area and that of the latter from 3 to 1.8 per 6 m^2.

Potato Tuber Moth

Management

1. Earthing up should be done so that tubes are not exposed to egg laying.
2. Infested tubers in the godown should be destroyed and only good and clean tubers should be stored.
3. Spray the crop with 1.25 l endosulfan 35 EC in 600 litre water per hectare.

Host Plant Resistance

In a study conducted in Himachal Pradesh, Parihar and Chandla (1995) found that none of the cultivars evaluated against potato tuber moth was free from its incidence, though Kufri Sindhuri, KufriDewa and Kufri Badshah were classified into moderately tolerant group. Studies conducted in Australia reveal that wild cultivars of potato, *Solamum pinnatisectum* and *S. chakoense*, having foliar pubescence were preferred less for egg laying.

Use of Pheromone Traps

Studies conducted by Lal (1994) reveal that Raman funnel Trap proved to be most effective in trapping the moths. Moreover, relative humidity (RH) was reported to be positively correlated with male orientation (Chandramohan, 1995). It was further revealed that every per cent increase in RH reduced the no. of males caught in the trap by 2.04. In Maharashtra, comparison of dry and water traps revealed that more moths were trapped in water traps as compared to dry trap (Tamhanker and Hawalker, 1994).

Use of Cultural Practices

Date of sowing and method of irrigation have also been reported to affect the incidence of pest. Nandihalli *et al.* (1993) observed 26th May to 5th July sowing affording the least damage by this pest. In another study irrigation of potato crop with sprinkler system, twice a week, proved effective in checking the incidence of potato tuber moth (Parihar and Ramkishore, 1998).

Use of Biopesticides

Studies conducted in Maharashtra revealed that spray of *B. thuringiensis, Grannulosis Virus* and endosulfan proved equally effective against potato tuber moth (Kurhade and pokharkar, 1997) Plantmix I and II (Neemrich +oil of Salvadoraoleoids, II -Neemrich +Neem extract-) both prove effective for 13-18 days as oviposition deterrent and anti-feedant.

Onion and Garlic

Thrips: *Thrips tabaci*

Onion fly: *Delia antique*

Cutworm: *Agrotis* spp.

Tobacco caterpillar: *Spodoptera litura*

Gram caterpillar: *Helicoverpa armigera*

Thrips

Management

1. Grow resistant varieties. Spanish white varieties of onion are are comparatively tolerant to this pest
2. 2.Spray 185ml fenvalerate 20 EC or 450ml deltamethrin 2.8 EC or 375ml cypermethrin 10 EC or 925 ml endosulfan 35 EC or 750 ml malathion 50 EC in 600 litre of water per hectare at an interval of 10- 15 days.

Note. 1) Add 500 ml sticker (*viz.* washing powder,Agrovat, Selvat, *etc.*) in each tank of spray solution.Use insecticides of different groups for subsequent sprays.

Chemical insecticidesshould be used only when the pest population attains Economic Threshold (ET).In Haryana, Bhardwaj, *et al.* (1995) worked out ET against this pest to be 5 thrips per plant (4 leaf stage). Elsewhere, 0.9-2.2 thrips were reported as ET for different stages of the crop (Fournier *et al.,* 1995). Studies conducted by Sinha *et al.*(1993) revealed that vars./cvs. Pusa Red and N-53 proved resistant against this pest. According to Walters and Eckurode (1996) Neem and Tobacco leaf extracts proved equally effective against thrips as that of methemadophos.

Onion Fly

Management

Crop rotation has been reported to help in reducing the plant damage by this fly pest. Studies conducted by Walters and Eckurode (1996) suggested that it has less than 2 per cent as compared to more than 10 per cent where crop rotation was n ot followed.

Tobacco caterpillar (*Spodoptera litura*): Height of sex pheromone trap and amount of pheromone in the dispenser effective in moth catch werecalibrated by Rao and Subbaratnam (1998).An amount of 5mg of pheromone per dispenser fitted in a trap fixed at a height of 125cm was reported to be most effective.

Pea

Leaf miner: *Phytomyza horticola*

Pod borers: *Etiellaz inckenella, Maruca* sp., *Helicoverpa armigera, Poliomatus (Lampiticus) boeticus*

Stemfly: *Ophomyia phaseoli*

Aphids: *Acryrthosiphum pisum*

Leaf Miner

Management

1. Grow resistant varieties, leafless varieties should be preferred.
2. Collect and burn infested leaves.
3. Spray 1.01 dimethoate 30 EC or 1.251 oxy- demeton methyl 25 EC or formothion 25 EC in 600 litre water per hectare, before pod formation.

Host-Plant Resistance

A lot of work has been undertaken in Himachal Pradesh on this aspect. Of the 57 cvs. screened by Bhatia *et al.*(1995), none could be classified as highly resistant and cvs. JP-826 and DPH-65having up to 2 damaged leaves per plant in class named resistant. According to Mehta *et al.* (1997), all the entries screened had >20 cent infestation; entries IP-3, FCI and Sel -82 were termed as mildly infested. These workers in another study, evaluated 40 genotypes and found none to be immune (Mehta *et al.*, 1998) genotype NDVP-1, KS- 136,KS- 225, KS-226, VG-5 and FCI could be place in resistant group on the bases of no. of infested leaves and no. of larvae per plant.

Studies conducted in Punjab revealed that none of the 25 cvs, screened was free from leaf miner, however 7 entries showed tolerance (Brar *et al.*, 1995).However, any variety resistant to this pest is yet to be developed. Results obtained by Mehta and Sharma (1999) suggest that cvs. having greater leaflet area, thin leaves,low total sugars, low amino acids and high phenols show resistance to this pest.

Stem Fly

Management

Date of Sowing

Late sowing of pea crop has been suggested, in general, to reduce the incidence of this pest. In Punjab crop sown on 6th October had less attack of this pest as compared to those crops which were sown on 16th and 22nd Sep.(Singh, 1996). Likewise, Geeta and Qureshi (1998) while working in Rajasthan observed that pea crop sown in 1week of Nov. afforded minimum incidence of stem fly.

Chemical Control

Application of carbofuran @0.75kg per hectare and phorate @ 1.0 kg per hectare was observed by Brar *et al.*(1993) to give effective control of this pest.

Gram blue pod butterfly: *Euchrysopscnejus*

Screening of different varieties against this pest revealed that the lowest incidence was recorded on var. Kinnauri (10.97 per cent) and Bonville (12.16 per cent) and the highest on var. Lincoln (32.46 per cent) (Thakur and Kashyap, 1995). No other information on this pest is available in the literature.

Pod Borers

Management

1. Spray 150 ml cypermethrin 25 EC or 1.251 endosulfan 35 EC in 600 litre water per hectare and repeat, if needed, after 15 days.

Cole Crops

1. Cabbage aphid: *Lipaphis erysimi, Brevicoryne brassicae*
2. Diamondblack moth: *Plutella xylostella*
3. Tobacco caterpillar: *Spodoptera litura*
4. Cabbage semi-looper: *Tricoplusiani*
5. Leaf webber: *Crocidolomia binotalis*
6. Cabbage Butterfly: *Pieris brassicae* (L.), *P.canidia, P. rapae*
7. Mustard sawfly: *Athalialug ensproxima* (Klug)
8. Cabbage borer: *Hellulaundalis* Fab.
9. Painted bug: *Bagrada cruciferarum* Kirkaldy. *B. hillaris*
10. Leafminer: *Phytomy zahorticola* (Goureau)

Termites, Brown ants, *etc.* are some of the soil dwelling insect pests of these crops.

Cabbage Aphid

Management

1. Regular surveillance of the crop is important.Remove the infested plant parts, early in the season, and destroy along with aphid. Whenever 5 per cent or more plants/leaves are found infested with this pest, remedial measures should be initiated as suggested below.
2. Spray 1.01 malathion 50 EC or 1.01 endolsulfan 35 EC in 600 litre water per hectare.

Common Predator of Aphids

Coccinella septempunctata (L.), *Chilomenus* (*Menochilus*) *sexmaculantus* (Fab.), *HIPPOdemia* (*Adonia*) *variegate*, *Ichiodans cuteliralis* (Fab.) *Syrphus balteatus*, *S. issaci*, *S. serarius*, *S. catractus*, *Chrysoper lacariana*, *C. scelestes*, *Leocopis* spp. (Family Chamaemyiidae), *Lasiopticus selenticus*, *Erisatalisqin quelineatus*.

Common Parasitoid of aphids: *Diaeriellarapa* (Curtis)

Fungi infecting aphids: *Entomophthora coronara*, *Cephalosporium aphidicola*.

Diamond Black Moth

Management

1. Before planting new crop, the plant debris of the previous crop must be destroyed, since their leaves are good source of infestation in the new crop.
2. Spray 1.0 kg Bioasp (*Bacillus thuringiensis* var. *kurstakti* formulation) and other such foumulations or 750 ml diazinon 20 EC or 150 ml dichlorvos 76EC or 1.01 malathion 50 EC or 1.01endosulfan 35 EC in 600 litre water per hectare. Repeat the application with some alternate insecticide, at 7-10 days interval, if needed.

Besides these, neem seed kernel extract has been reported to be effective in Tamil Nadu (Moorthy *et al.*, 1999) found firpronil 5 per cent to be better than cartap hydroxide and endosufan in controlling this pest on cole crops. Whenever NpV formulations are used against DBM, it is suggested that mixing of 1 per cent Indian Ink in the spray solution negates the effect of sunlight (Padmavathamma and Veeresh, 1995).

Biogical Control

Common Parasitoids: *Cotesia* (*Apanteles*) *plutellae*, *Diadegma semiclausum*.

In the Philippines release of > 7000 cocoons/ha of *Cotesiaplutellae* at weekly intervals+ 1-2 sprays of Bt (based on ET *i.e.* 2 larvae/plant; 5 larvae/plant) provided 123 per cent increase in yield of cabbage over control (Morall- Rejesus *et al.*, 1996). Studies conducted by Usha Chauhan *et al.* (1997) in Himachal Pradesh, India recorded parasitization of this pest up to 70 per cent during the month of April.

Use of Cultural Practices

Date of sowing: In Karnataka lowest damage by *P.xylostella* (16.87 per cent) and highest yield was recorded by Shashidhar *et al.*(1994) when the crop was planted in week of October.

Trap crop: Studies conducted by Pawar and Lawande (1995) revealed that among the treatment cabbage + mustard, cabbage+cartap hydrochloride and cabbage alone; cabbage with mustard as trap crop afforded best Cost Benefit Ratio of 1: 6.

Intercropping

In China this pest has been under check by combining different practice *viz.* cultural practice, biological control, judicious use of safer insecticides, *etc.* Under the expert guidance of AVRDC and supported by Asian Development Bank, Collaborative Vegetable Research Network of S.E. Asia (AVNET) was established in 1989. Besides this South Asian Research Network (SAVERNET) has linked some S. Asian countries since 1992. Both these networks have IPM of DBM as one of their sub networks(Dhaliwal and Arora, 1998). They developed the master plan which involves release of parasitoids,Bt or 'neem' kernel extract sprays, pheromone traps and growing mustard as trap crop. In Malaysia, the release of *Diadegma semiclausam* has reduced the use of chemicals against DBM to the extent of 86 per cent. Adoption of IPM technology in the Philippines has resulted in $10.5 m cost reduction over 7000 ha of cabbage (Shanmugasundram, 1994).Similarly in Thailand, according to Chiang and Rumakan (1993), as high as 145 per cent increase in net profits were reported from IPM adopted fields as compared to non IPM adopted fields. In India, paired rows of bold seeds Indian mustard (first row sown 15 days before transplanting of cabbage, second row 25 days after transplanting)attracted 80-93 per cent DBM colonization.

Further, spray of Neem Seed Kernel Extract (NSKE)4 per cent at primordial stage followed by 2-3 sprays at post heading stage, at 10-15 days interval, check this pest effective.NSKE is also safer to the natural enemies *viz. Cotesiaplutellae*. Mustrad rows should be treated with dichlorvos 0.1 per cent at 10-15 days intervals.

Intercropping of cabbage/cauliflower with tomato (Srinivasan and Veeresh, 1986)or carrot (Varela and Guharay, 1998) has been observed to reduce the incidence of this pest by making the micro- climatic conditions less favorable for DBM. Of late, Kandoria *et al.* (1999) observed less incidence of DBM on cauliflower planted 30 days after planting tomato in alternate rows.

Host Plant Resistance

Though not much information is available pertaining to this aspect of IPM of DBM, yet glossy plants have been observed to reduce the survival of this pest (Eigen brode *et al.*, 1995).

Use of Sex Pheromone Traps

(Z)- 11-hexa decenal;(z)- 11-hexa decenal acetate and(Z)- 11-hexa decen -1-ol in the ratio of 50: 50: 1.5 mecro gram per septum mixed with some anti –oxidant has been used effectively in mass trapping of the pest (Reddy and Ure, 1997).They also reported that I and III of these chemicals used in the ratio of 70: 30 effectively disrupted the mating of DBM.

Different types of traps have been tried as sex pheromone traps; of these sticky delta trap proved to be best followed by CPI trap (Reddy and Urs, 1995). Water trap was the least effective. Sex pheromone traps have been to disseminate the pathogen (Zoophthoraradicans) of this pest (Pell *et al.*, 1993).

Methods of Irrigation

Methods of irrigation has been reported to affect the incidence of DBM. According to McHugh and Foster (1995) incidence of *P.xylostella* was observed to be less when intermittent overhead irrigation was applied to cabbage as compared to when the crop was irrigated by drip irrigation system.

REFERENCES

Bhatia, R.,Gupta, D., Pathania, N.K.1995.Screening of pea germplasm against the leafminer, *Chromatomyia atricornis* Meigen in the foothills of Himachal Pradesh (India). *Entomological Res.*, 19;57-59.

Bhardwaj,C.L.,Thakur, D.R.,Tomwal, R.S.1995.Effect of fungicide spray and staking in diseases and disorder of tomato (*Lycopersicon esculentum*). *Indian J. agric. Sci.*, 65: 148-151.

Bhardwaj, B. S., Srivastava, P.K., Gupta, R.P.1995.Economic threshold value of thrips population on the yield of Kharif cotton crop. Newsletter National Horticulcural Research and Development Foundation,15: 10-12.

Borah,R.K.1995.Incidence of fruit borer, *Heliothis armigera* in relation to the date of planting of tomato in the hill zone of Assam. *Horticultural Journal*, 8: 94-94.

Brar,K.S.,Dhillon, G.S., Singh,Dhillon,T.S.1995.A note on the resistance in pea genotypes to leafminer, *Chromato myiahorticola* (Goureau). *Haryana J. Hort. Sci.*, 24: 156-158.

Brar,K.S., Raman,H.,Dhaliwal,H.S.,Cheema,D.S.1995.Field screening of different species of *Lycopersicon* against tomato fruit borer, *Helicoverpa armigera* (Hubner). *J. Insect Sci.*, 8: 196-197.

Butani,D.K., Jotwani, M.G. Insects in Vegetables. Periodical Expert Book Agency, VivekVIhar, New Delhi.

Dhaliwal,G.S.,Arora,R.1998.Principles of insect pests Management. Kalyani Publishers, Ludhiania, pp.297.

Dhaliwal,G.S.,Arora,R.2001.Integrated Pest Management: Concept and Approaches. Kalyani Publisher, 1/1, Rajender Nagar, Ludhiania.

Eigenbrode, A.D., Moodie,S., Castagnola, T.1995.Predator mediated host plant resistance to phytophagous pests in cabbage with glossy leaf wax. *Entomologia Experimental etapplicata*, 77: 335-342.

Fournier, F., Bovin, G., Stewart, R.K. 1995.Effect of *Thripstabaci* (Thysanoptera: Thripidae) on yellow onion fields and economic threshold for its management. *J.Econ. Entomol.*, 88: 1401-1407.

Gajendra, Chandrakar, Ganguli, R.N., Kaushik, V.K.,Dugey,V.K.1998. Resistance to the fruit borer, *Helicoverpa armigera* in tomato. In: Advances in IPM for Horticultural crops. Proc. 1 Nat. Symp. On Pest Management in Horticultural crops: Enviromental implication and thrusts, Bangalore. India, 15-17 Oct., 1997.

Geeta, Bali, Qureshi, Q.G. 1998.Date of sowing on the incidence of stem fly. *Melanagro myzaphaseoli* (Coquillett). In: Advances In IPM for Horticulture crops Proc. 1 Nat. Symp. On Pest management in Horticultural crops environmental implications and thrusts, Banglore, India, 15-17 Oct., 1997.

Gupta,P.R., Babu, B.R.R.M.1998. Management of *Helicoverpa armigera* on tomato with *Trichogramma pretiosum* and *Bacillus thuringiensis* var. *kurstaki*. In: Advances in IPM for Horticultural crops. Proc. 1 Nat. Symp.on Pest Management in Horticulturalcrops: environmental implications and thrusts, Bangalore, India, 15-17 Oct., 1997.

Karimulla, Ahmad, S., Parcha, A.M.1995. Integrarted control of potato aphid, *Myzus persicae*(Sulz.) in Peshwar Sarhad *J. Agriculture*, 11: 165-171.

Kandoria, J.L., Singh, Gurdip, Singh,Labh.199. Effect of inter-cropping cauliflower with tomato on the incidence of diamond back moth. *Insect Environment*, 5: 133-138.

Krishanamurthy, A., Mani,M. 1996. Biosuppression of *Helicoverpa armigera* (Hubn.) on tomato using two egg parasitoids, *Trichogramma brasiliensis* (Ashm.) and *T. pretiosum* (Riley). J.*Entomological Res.*, 20: 37-41.

Kurhade, V,P., Pokharkar D.S. 1997. Biological control of potato tuber moth, *Phthorimaea operculella* (Zeller) on potato. *Journal of Maharashtra Agricultural Universities*, 22: 187-189.

Lal, L. 1994. Potato tuber moth sex pheromone trap design suitable for high rain fall area of India. *J. Indian potato Association*, 21: 234-236.

McHugh, J.,J Foster, R.E. 1995. Reduction of diamondback moth (Lepidoptera: Plutellidae) Infestation in cabbage heads by overhead irrigation. *J. Econ.Entomol*, 88: 162-168.

Mehetre, S.T., Salunkhe, G.N. 1999. Comparative efficacy of *Tichogramma pretiosum* Riley and *HaNPV* with endosulfun in controlling *Helicoverpa armigera* Hub. infesting tomato. Journal of Maharashtra Agricultural Universities, 23: 175-176.

Mehta, P. K., Chandel, Y.S., Chandel, R.S. 1997. Reaction of pea varieties against leafminer, *Chromoto myiahorticola* (Goureau). *Insect Environment*, 3: 118.

Mehta,P.K., Sharma, T.N.1999. Resistance in pea genotypes to agromyzidleafminer, *Chromato myiahorticola* Goureau in relation to biophysical and biochemical charcters of the plant. Pest Management and Economic Zoology, 5: 127-132.

Mehta, P. K., Sharma, T.N., Chandel, R.S. 1998. Resistance in pea (*Pisum sativum* var. *arvense*) genotypes to leafminer (*Chromato myiahorticola*). *Ind. J. Agric, Sci.*, 68: 271-273.

Mishra, N. C., Mishra, S.N.1993. Performance of tomato varieties against wilt, *Fusarium oxysporum* F. spp. Lycopersicae and fruit borer, *Helicovarpa armigera* in the North Eastern Ghats of Odisha. *Ind.J. Plant Prot.*, 21: 183-186.

Mishra, S.N., Mishra, N. C. 1995. Genetic parameters and varietal performance of tomato in NE Ghat zone of Odisha. *Environment and Ecology*, 13: 182-187.

Moorthy, P.N. K, Kumer, N.K. K., Selvaraj, G., Daniel, J.S. 1998.Neem seed kernel extract application for diamondback moth management: transfer of technology for mechanized farming. *Pest Management in Horticultural Ecosystem*, 4: 128-130.

Morallo- Rejesus, B., Inocencio, E.L.Eusebio, J.E.1996. Comparative effectiveness of IPM –DBM technology versus farmers' control practice for diamondback moth, *Plutella xylstella* (L.) control. *Philippines Entomologist*,10: 57-65.

Nandihalli, B.S., Hingappa, S. 1993. Role of planting time and insecticides in the management of Potato tuber moth. Karnataka. *J Agric. Sci.*, 6: 66-68.

Padmavathamma, K., Veeresh, G.K.1995 Effect of sunlight protectant and time of application on the virulence of NPV of *Plutella xylostella*. *Current Science*, 24: 92-94.

Panda, S. K., Behera,U. K., Nayak, S. K. 1999. Bioefficacy of fipronil 5 per cent SC against diamondback moth, *Plutella xylostella* (Linn.) (Yponomeutidae: Lepidoptera) infesting cabbage. *Insect Environment*, 5: 104-105.

Parihar, S. B.S., Chandla, V.K. 1995. Reaction of potato varieties/hybrids to tuber moth *Phthorimaeao perculella*.J. *Indian Potato Association*, 22: 90-91

Parihar, S. B. S., Sharma, C. K., Ram Kishore. 1998. Birds as predators of insects in potato. *Insect Environment*, 4: 95-96.

Parihar, S.B.S., Ram Kishore. 1998. Irrigation methods-a component in the management of *Helicoverpa armigera* Hub. on potato. *Insect Environment*, 4: 20-21.

Parihar,S.B. S., Singh, N.1999. Influence of irrigation methods on the aphid, *Myzus persicae* (Sulzer). *Insect Environment*, 5: 25-26.

Patil, S., Kotiyal, Y. K., Revanappa, Patil, D.R. 1997. Effects of inter-cropping tomatoes (*Lycopersion esculentum* Mill.) on the infestation of tomato fruit borer, *Helicoverpa armigera* Hub. Advances in Agricultural Research in India, 8: 141-146.

Pawar, D.B., Lawande,K.E.1995. Effect of mustard as a trap crop for diamondback moth on cabbage. Journal of Maharashtra Agricultural Universities, 20: 185-186.

Pell, J. K., Macaulay, F. D. M. and Wilding, N. 1993. A pheromone trap for dispersal of the *Zoophthora radicans* Brefeld.(Zygomycetes: Entomophthorales) against populations of diamondback moth, *Plutella xylostella* L. (Lepidoptera Yponomeutidae). *Biocontrol Science and Technology*, 3: 315-320.

Pokharkar,S.S.,Chaudhary, S.D.,Verma, S.K.1999. Utilization of Nuclear Polyhedrosis Virus in the integrated control of fruits borer (*Helicoverpa armigera* Hub.)on Tomato (*Lycopersicin esculentum*). *Indian J. Agric. Sci.*, 69: 185-188.

Rao, D.V.S.,Subbaratnam, G.V.1998.Sex pheromone monitoring of the ragicutworn, *Spodoptera exigua* (Hubner) in onion. *Pest Management and Economic Zoology*,6: 21-25.

Rawat, U. S., Pawar, A. D. 1993. Biocontrol of tomato fruit borer, *Heliothis armigera* (Hubner) in Himachal Pradesh, India. *Plant Prot. Bull. (Faridabad)*, 45: 34

Reddy,G.V.P., Urs, K.C.D.1995. Comparative performance of five types of traps for sex trapping of diamondback moth in cole crops. *J. Insect Sci.*, 8: 24-26

Reddy, G.V.P., Urs. K.C.D.1997. Mass trapping of diamondback moth, *Plutella xylostella* in cabbage fields using synthetic sex pheromones. *International Pest Control*, 39: 125-126.

Sankhyan,S., Verma, A.K.1997. Fields screening of tomato germplasm for resistance against the fruits borer *Helicoverpa armigera* (Hubner) (Lepidotera: Noctuidae). *Pest Management and Economic Zoology*, 5: 107-111.

Satpathy, S., SamarjitRai, Chattopadhyay, M.1999. Field evaluation of NPV and insecticides against tomato fruits borer *Helicoverpa armigera* (Hubner). *Insect Environment*, 5: 117-118.

Sharma, R. N. Vrushali, T., Deshpande, S.G.1998. Improvedplant based formulation for preventation *Phthorimaea operculella* damage on stored potatoes – Plantmixand Plantmix. *Pesticide Res. J.* 10: 214-218.

Shashidhar, Viraktamath, Shekarappa, Reddy, B.S., Patil, M.G.1994. Effect of date of planting on the extent of damage by diamondback moth, *Plutella xylostella* on cabbage. *Karnataka J. Agric. Sci.*, 7: 238-239.

Singh, J., Bhalla, J.S., Brar,K.S. 1996. Effect of dates of sowing on the incidence of pea stem fly, *Ophiomyia phaseoli* (Tryon) in early sown pea crop. *Ind. J. Ecology*, 23: 64-66.

Sinha,A. K., Sinha, R.B.P., Kumer, A.1993. Rection of onion cultivars to thrips (*Thripstabaci*). *J.Applied Biology*,3: 104-105.

Sivaprakasam, N. 1996. Ovipositional preferences of *Helicoverpa armigera* to tomato cultivars. *Madras Agricultural Journal*, 83: 306-307.

Srinivasan, K., Moorthy, P.N.K., Raviprasad, T.N.1994. African marigold as trap for the management of thefruit borer, *Helicoverpa armigera* on tomato. *International J. Pest Management*, 40: 65-63.

Srinivasan, K., Veeresh, G.K. 1986. Economic analysis of promising cultural practices in the control of moth pests of cabbage. *Insect Science and its Application*, 7: 559-563.

Srivastava, K.P., Butani, D. K. 1998. *Pest Management in Vegetables*. Research Periodicals and Book Publishing House, India. p. 589.

Talekar, N.S., Hau, T.B. H., Chang, W.C.1999. *Solanum viarum*-a trap crop for *Helicoverpa armigera*. *Insect Environment*, 5: 142.

Tamhankar, A.J., Harwalkar, M.R.1994. Comparison of a dry and water trap for monitoring potato tuber moth, *Phthorimaea operculella* Zuller. Entomom, 19: 163-164.

Thakur, S.S.,Kashyap, N.P.1995.Screening of pea germplasm against gram blue pod butterfly, *Euchrys opscnejus* Fab. and its chemical control . *Himachal J. Agricultural Res.*, 21: 93-97.

Thakur, S.S., Chandel, K.S., Kashyap, N.P.1998.Field evaluation of tomato varieties against *Helicoverpa armigera* (Hubner) in the higher hill of Himachal Pradesh. *Insect Environment*, 4: 51-52.

Usha Chauhan, Bhalla, O. P., Sharma, K.C. 1997.Biology and seasonality of diamondback moth, *Plutella xylostella* (L.) (Lepidoptera: Yponomeutidae) and its parasitoids on cabbage and cauliflower. *Pest Management in Horticultural Ecosystems*, 3: 7-12.

Varela, Ochoa Guharay, F. 1988.The use of multiple cropping (cabbage-carrot),as a component of integrated pest management of cabbage defoliators. Revista Nicaraguense de Entomologia, 4: 49-50.

Walter, T.W. and Eckenrode, C.J. 1996. Integrated management of onion maggot (Diptera: Anthomyiidae). *J. Econ. Entomol.*, 89: 1582-1586.

Transformation of Indian Agriculture through Innovative Technologies *Pages* **265–276**
Editor: **Dr. Shahid Ahamad & Dr. Jag Paul Sharma**
Published by: **ASTRAL INTERNATIONAL PVT. LTD., NEW DELHI**

17 Weed Management in Pulses for Higher Productivity and Profitability

R. Puniya, B. R. Bazaya and Anil Kumar

Pulses are the important component of Indian agriculture as well as in human diet for it nutrition value. The major pulse crops of the country are bengal gram, lentil, peas, red gram, greengram and blackgram. Besides other constraints, weeds are also one of the major cause which leads considerable loss to pulses. Critical period for crop-weed competition in pulses varies from 20-60 DAS. Depending on weed type and crop-weed competition, weeds reduce yield to the tune of 20-87 per cent in pulses. Weeds in pulses are managed by preventive, cultural and mechanical and chemical methods. The major preventive measures are use of weed free seed, use of well decomposed FYM or compost, proper cleaning of farm machinery before sowing and keeping farm bund and irrigation/drainage channel free from weeds. Stale seedbed techniques, crop rotation, increase the competitive ability of the crop, time of seeding and irrigation, inclusion of cover crops and intercropping are important cultural practices for manageging weeds. The weeds can be effectively managed by mechanical methods including the use of hand hoe and wheel hoe etc. The pre-emergence herbicides like pendimethalin, oxyfluorfen, metolachlor, pendimethalin+imazethapyr, alachlor, oxadiargyl and oxadiazon and trifluralin as pre plant incorporation herbicide applied in pulses/pulse based cropping systems. However, imazethapyr, quizalofop-ethyl, imazethapyr+imazamox, fluazifop, fenoxaprop and clodinafop-propargyl are used as post-emergent herbicides in pulses.

Introduction

Pluses are the second most important group of foodgrain crops after cereals in India. These are essential for nutritional security, soil health and sustainable agriculture. The India is one of the largest producer, consumer and importer of pulses in the world. Pulses in India are grown in an area of about 25.23 million hectares with an annual production of nearly 19.27 million tones (Agricultural Statistics at a Glance 2014). The India accounts for over one third of the total world area and over 20 per cent of total world production. About 90 per cent of the global pigeonpea, 75 per cent of chickpea and 37 per cent of lentil area falls in India. Large population and low pulse yield in the India compared to other counties is attributed to lower per capita availability of pulses in the country. The current per capita net pulse availability is around 50 grams against recommendation of 65 grams of pulses per capita per day acoording to ICMR. This shortfall has serious nutritional implications especially to children and women in rural area. Pulses are important food crops due to their high protein content 20 to 25 per cent, carbohydrates 55 to 60 per cent, rich in calcium and iron also. All pulses play a key role in improving of soil fertility through biological nitrogen fixation with the help of rhizobium bacteria found in their root nodules.

Presently the realized productivity of pulses in the country is less than 1 t/ ha. The low production of pulses has been attributed by a number of factors, the major being ever-increasing population, abrupt climatic changes, complex disease-pest syndrome, socio-economic conditions of the farmers, stagnation in cultivated area and cropping intensity due to dearth of irrigation facilities over the years and non availability of HYV seeds, non availability of fertilizers, non availability of plant protection chemicals at the time, lack of knowledge about seed rate, seed, treatment, herbicide dose and method of fertilizer application *etc.* (Kumar *et al.*, 2009 and 2010; Narayan and Kumar, 2015). Weeds are also one of the major cause which leads considerable loss to pulses. Weed management is critical to maintaining productivity of crops. Weeds are an important constraint in agricultural production systems, acting at same tropic level as the crop; weeds capture a part of the available resources that are essential for plant growth (Smith *et al.*, 2010). Weeds compets for light, water, space and nutrients and reduce crop yield and quality and responcible for billions of dollars in global crop losses annually (Das, 2008).

Weeds in Pulses

The kharif/summer pulses (pigeonpea, greengram, blackgram and cowpea) and rabi pulses (chickpea, lentil, peas and rajmash) were infested with grassy, broad-leaved weeds and sedges. The dominant weeds of pulses in country are given in Table 17.1.

Crop-Weed Competition and Yield Loss due to Weeds

Weeds compete with crop plants mainly for nutrients, moisture light and space. Many weeds secrete toxic allele-chemical which adversely affect the growth and development of crops. The competition becomes severe due to smothering effect when weeds emerge earlier than the crop. The initial growth of some pulses like pigeonpea, chickpea and lentil is slower then weeds. Thus severe problem of crop-weed competition is observed in these pulses. Sometimes total crop failure is also observed due to heavy infestation of diverse weed flora in the rainy season. Maintaining weed free environment throughout the growth period is difficult and uneconomical. In fact, near maximum yield can be achieved through effective weed control during critical period of crop-weed competition. However, the weeds which emerge later may not reduce the yield but produce large number of seeds

Table 17.1. Major Weeds in Pulses

Crops	*Associated Weeds*	*Critical Period of Competition*	*Reference*
	Kharif/summer pulses		
Pigeonpea	*Cynodon dactylon, Bracharia* sp., *Cyperus rotundus, Alternathera triandra, Acalypha indica, Digeria arvensis, Amaranthus viridis, Phyllanthus niruri, Cyanotis axillaris, Commelina benghalensis* and *Parthenium hysterophorus.*	60-70 days	Goud and Patil, 2014
Greengram	*Dactyloctenium aegyptium, Digitaria sanguinalis* L. *Digera arvensis, Trianthema portulacastrum, Mollugo distachya, Cleome viscosa, Cucumis callosus, Corchorus tridens, Corchorus aestuans, Tribulus terristeris, Cyperus rotundus* L., *Cyperus compressus* and *Bulbostyllis barbata.*	20-40 DAS	Punia *et al.*, 2013
Summer greengram	*Trianthema portulacastrum, Eleusine aegyptiacum, Digitaria sanguinalis* and *Cyperus rotundus.*	10-40 DAS	Singh, 2011
Blackgram	*Ageratum conyzoids, Boreria hispida, Commelina banghalensis, Echinochloa colona, Cynodon dactylon, Paspalum scrobiculatum, Digiteria sanguinalis* and *Cyperus rotundus.*	15-45 DAS	Das *et al.*, 2014
Summar Blackgram	*Cyperus rotundus, Sorghum halepense, T. monogyna* and *C. benghalensis.*	10-40 DAS	Kumar and Tewari, 2004
Cowpea	*Commelina benghalensis, Cyperus rotundus, Berrovia hispida, Cynodon dactylon, Digera arvensis* and *Echinochloa colona.*	40-45 DAS	Hanumanthappa *et al.*, 2012
	Rabi pulses		
Chickpea	*Cynodon dactylon, Cyperus rotundus, Melilotus alba, Melilotus indica, Anagallis arvensi, Convolvulus arvensi, Rumax dentatus, Asphodelus tenuifolius* and *Chenopodium* *Album*	20-40 DAS	Sharma, 2009
Lentil	*Cynodon dactylon, Cyperus esculentus, Mimosa pudica, Cyanotis axillaris, Commelina bengalensis, Ipomoea aquatica* and *Vicia sativa.*	20-30 DAS	Aktar *et al.*, 2013
Pea	*Anagalis arvensis, Fumeria parviflora, Melilotus indica, Cynodon dactylon, Convolvulence arvensis, Avena fatua, Vicia sativa, Cornopus didymus, Trianthema monogyna* and *Medicago denticulata. Euphorbia helioscopia, Cannabis sativa* and *Chenopodim album.*	30-60 DAS	Kumar *et al.*, 2009
Rajmash	*Anagallis arvensis, Melilotus alba, Melilotus indica* and *Phalaris minor, Cynodon dactylon, Alternanthera* sp., *Cyperus iria etc.*	-	Panotra *et al.*, 2012

for next season (Kumar *et al.*, 2016). Critical period for crop-weed competition in kahrif pulses (Table 17.1) is varies from 60 to 70 DAS in pigeonpea (Goud and Patil, 2014), 20 to 40 DAS in greengram, (Punia *et al.*, 2013), 15 to 45 DAS in blackgram (Das *et al.*, 2014) and 40-45 DAS in cowpea (Hanumanthappa *et al.*, 2012). Critical period for crop-weed competition in rabi pulses (Table 17.1) is varies from 20 to 40 DAS in chickpea (Sharma, 2009), 20 to 30 DAS in lentil (Aktar *et al.*, 2013), 30 to 60 DAS in pea (Kumar *et al.*, 2009). The work done in different soil types at different parts of India weeds reduces yields by 20.3-87.4 per cent (Table 17.2). The total economic losses will be much higher, if indirect effects of weeds on health, losses of biodiversity, nutrient depletion, grain quality, *etc.* are taken into consideration.

Table 17.2. Losses in Pulses Yield Due to Weeds

Crops	*Yield Losses (per cent)*	*Soil Type*	*Reference*
Pigeonpea	20.3	Clayey	Goud and Patil, 2014
Greengram	48.7	Sandy loam	Nandan *et al.*, 2011
Blackgram	59.2	Sandy-loam	Patel *et al.*, 2015
Cowpea	66.4	sandy clay loam	Hanumanthappa *et al.*, 2012
Chickpea	68.3	sandy loam	Khope *et al.*, 2011
Lentil	37.9-87.4	Loamy soil	Shyam *et al.*, 2008 and Yadav *et al.*, 2013
Peas	43.4-72.5	clay loam	Verma *et al.*, 2004 and Singh and Angiras, 2004
Rajmash	41.4-50.0	Sandy loam	Panotra *et al.*, 2012

Methods of Weed Management in Pulses

Most of the methods employed by farmers as part of their production system are designed to create and environment that allow the crop growth to the greatest extent possible than weeds. Besides herbicides, preventive methods, cultural practices and mechanical methods played an important role in this regard. Enhancing crop competitiveness against weeds could provide a low cost and safe tool for weed management.

Preventive and Cultural Methods

For successful weed management, it is most important to prevent the distribution of weed seeds from one field to another, and from infested to uninfested area. Weed seeds are dispersed by several mechanisms like wind, water, farm machinery, animals *etc.* Weed prevention is an important step in avoiding competition with pulses for resources is to prevent the presence of weed species in new places. Use of clean crop seed, use of well decomposed FYM or compost, proper cleaning of farm machinery before sowing and keeping farm bund and irrigation/ drainage channel free from weeds are the most effective preventive method of weed management (Verma and Singh, 2008). Deep *burial* of weeded plants is necessary to prevent the reproduction and dissemination of seeds. In fact the basic principle of managing annual weeds, which reproduce mostly from seeds, lies in killing them before flowering, so as to prevent seed setting and their infestation in the next season. Cultural methods provide competitive advantage to crop against weeds by reducing weed establishment and through facilitating faster crop growth to smother weeds. Cultural methods are stale seedbed techniques, crop rotation, increase the competitive ability of the crop, time of seeding and irrigation, inclusion of cover crops, and intercropping (Kaur *et al.*, 2015), use of cropping pattern, and tillage systems, employing time, method, rate of fertilizer, inter and mixed cropping, and spacing (Verma and Singh, 2008); smother crop, summer ploughing (Dubey, 2014) have carried out for successful weed management.

Selection of good crop rotation is must to prevent development of diverse weed population in pulses. Growing of sesame (*Sesamum indicum* L.) during the rainy season reduces the weed population including *Cyprus* in winter pulses. Puddling in rice also help in minimizing *Cyprus rotundus* in the following chickpea. Similarly, *Phalaris minor* and *Chenopodium album* in chickpea can be minimized by following puddling in rice (IIIPR, 2009). Inclusion of pulses in the rainy season can reduce the infestation of *Phalaris minor* and *Avena fatua* in winter crops. Crops like mungbean and urdbean which grow fast and compete with the weeds should be included either as sole cropping or intercropping (Kumar *et al.*, 2016).

Pulses are normally gwown under inter-or mixed cropping in rainfed agro-ecosystems of India. Intercropping has been found to suppress the weeds through formation of good canopies due to competitive planting pattern. The suppression of weed growth in intercropping system is mainly due to increased leaf to increased leaf area and light interception. Inclusion of short-duration and quick-growing intercrops in between the rows of long-duration and tall-growing crop has been found to suppress the weed infestation (Kuamr *et al.*, 2016). As a result of considerable reduction in weed weight due of inclusion of intercrops in long-duration wide-spaced crops, the weeding requirements have been found to be reduced substantially (Ali, 1988).

Mechanical Weed Control

Mechanical methods involve removal of weeds with various tools and implements, and include tillage, hoeing, mulching, burning *etc.* Intercultural practices are performed with implements used by hand, bullocks or tractor to create favourable conditions for the growth of crops. One or two hand-weedings provide satisfactory weed control in all pulses if done at critical stages. Although herbicides are fast replacing other weed-management practices, mechanical methods are very much needed to make weed control more effective, manageable and economical (Kumar *et al.*, 2016). Localtion-specific small hand tools which enhance the working efficiency, and reduce the drudgery and cost have been developed in different crops and regions. The old practice of stale seedbed (Gopinath *et al.*, 2009) and summer deep ploughing are also followed for weed management in pulse-production system. Since the competition is more in the initial growth stage, two hands weeding (15 to 30 DAS and 40-50 DAS weeks after sowing) should be done in rabi pulses (Buttar *et al.*, 2008). Also two hand weedings at 20 and 40 DAS is sufficient to control weeds in mungbean, kharif blackgram, summer blackgrma (Nandan *et al.*, 2011; Khot *et al.*, 2015).

Chemical Weed Control in Pulses

Herbicides are chemical used to kill or inhibit the growth of weeds without affecting crop plants. Weed management through herbicides is gaining popularity due to scarcity of labour for weeding on time. Further, manual weeding is also expensive, less efficient and cannot be performed under adverse soil and weather condition. Therefore, herbicides can be used as an alternative of manual or mechanical weeding. The efficiency of these herbicides depends largely on their nature and agro-climatic conditions in which they are used. There are different categories of herbicides used in pulses to manage weeds based on the time of application of herbicides viz; pre-emergence herbicides and post emergence herbicides. Usage of pre-emergence herbicides assumes greater importance in the view of their effectiveness from initial stages. Pendimethalin, oxyfluorfen, metolachlor, pendimethalin+imazethapyr, alachlor, oxadiargyl and oxadiazon are some of the mostly used pre emergence herbicides applied in pulses. However, imazethapyr, quizalofop-ethyl, imazethapyr+imazamox, fluazifop, fenoxaprop and clodinafop-propargyl are used as post emergent herbicides in pulses (Table 17.3).

Integrated weed management relies on weed management principles that have proved to be suitable for long term weed management by combining the use of cultural, mechanical, thermal, biological and chemical means based on ecological approaches (Singh, 2014; Kewat, 2014), that will prevent weed reproduction, emergence, promote weed seed bank depletion and minimize weed competition (Malviya and Singh, 2007), which is the key component of sustainable agriculture. In pulses different herbicides are used in combination with hand weedning or hoeing or ridging or intercultural operations to control of composite weed flora.

Pulses are commonly grown under inter/mixed cropping systems. Several pulse based intercropping systems are prevalent in India. The herbicides which are working in sole crop may not be work in intercropping systems due to diverse nature of intercrops. Hence, different strategies for weed management need to be followed in pulses-based intercropping. Hence different herbicides are given in Table 1.7.4 for different pulse-based inter cropping systems.

Conclusions

Weeds reduce yield to the range of 20-87 per cent in pulses. The major preventive measures are use of weed free seed, use of well decomposed FYM or compost, proper cleaning of farm machinery before sowing and keeping farm bund and irrigation/drainage channel free from weeds. Stale seedbed techniques, crop rotation, increase the competitive ability of the crop, time of seeding and irrigation, inclusion of cover crops and intercropping are important cultural methods for manageging weeds in pulses. The weeds can be effectively managed by mechanical methods including the use of hand (khurpi), hand hoe and wheel hoe *etc.* The pre-emergence herbicides like pendimethalin, oxyfluorfen, metolachlor, pendimethalin+imazethapyr, alachlor, oxadiargyl, oxadiazon herbicides applied in pulses/pulse based cropping systems. While imazethapyr, quizalofop-ethyl, imazethapyr+imazamox, fluazifop, fenoxaprop and clodinafop-propargyl are used as post-emergent herbicides in pulses.

Table 17.3. Chemical Weed Control in Pulses

Crops	*Herbicide*	*Dose (kg/ha)*	*Time of Application*	*WCE (Per cent)*	*WI (Per cent)*	*Reference*
Kharif pulses						
Pigeonpea	Pendimethalin *fb* paraquat	0.75-1.0 fb. 0.40	PE *fb*. 6-8 WAS	74.2-79.7	2.0-11.6	Goud and Patil, 2014 and Sharma *et al.*, 2014
	Imazethapyr *fb* paraquat	0.075 fb. 0.40	15-20 DAS *fb*. 6-8 WAS	69.0-82.0	13.9-19.0	
	Pendimethalin *fb* HW	1.0	PE *fb* 60 DAS	94.2	7.8	Malik and Yadav, 2014
	Imazethapyr *fb* HW	0.10	15 *fb* 50 DAS	86.1	5.1	Rao *et al.*, 2015
Greengram	Imazethapyr	0.075	17 DAS	58.9	20.6	Singh *et al.*, 2015a
	Pendimethalin	1.0	PE	51.6	29.8	
	Imazethapyr + imazamox	0.040-0.060	20 DAS	99.4-99.5	14.6-15.4	Komal *et al.*, 2015
Blackgram	Oxyfluorfen	0.200	PE	79.3	16.1	Patel *et al.*, 2015
	Pendimethalin	1.5	PE	68.2	10.6	Malliswari *et al.*, 2008
	Alachlor	2.0	PE	69.7	9.4	Chand *et al.*, 2004
	Metolachlor	1.0	PE	58.5	21.1	
	Pendimethalin + imazethapyr (pre-mix)	1.0	PE	96.6	10.1	Yadav *et al.*, 2015
	Pendimethalin *fb* HW	0.75	PE *fb*. 45 DAS	92.6	-	Kumar *et al.*, 2006
Cowpea	Pendimethalin	0.75	PE	90.1	6.9	Hanumanthappa *et al.*, 2012
	Imazethapyr	0.075	POE	84.4	15.0	Ram and Verma, 2015

Crops	*Herbicide*	*Dose (kg/ha)*	*Time of Application*	*WCE (Per cent)*	*WI (Per cent)*	*Reference*
Rabi pulses						
Chickpea	Oxadiargyl	0.075	PE	71.5	1.8	Patel *et al.*, 2006
	Imazethapyr	0.063	15 DAS	65	24.3	Ratnam *et al.*, 2011
	Pendimethalin	0.75	PE	-	6.9	Pedde *et al.*, 2013
	Pendimethalin + imazethapyr	1.0	PE	58.6	20.6	Poonia and Pithia, 2013
	Pendimethalin *fb* Clodinafop	1.0 fb 0.060	PE *fb* 45 DAS	81.3	4.1	Kumar *et al.*, 2014a
	Pendimethalin *fb* Pinoxaden	1.0 fb 0.050	PE fb 45 DAS	82.3	4.1	
Lentil	Pendimethalin	1.0	PE	65.2-91.2	5.9-26.9	Lhungdim *et al.*, 2014 and Yadav *et al.*, 2013
	Pendimethalin + imazethapyr	1.0	PE	68.4	16.4	Lhungdim *et al.*, 2014
Peas	Pendimethalin	0.75	PE	41.0	11.7	Verma *et al.*, 2004
	Oxadiazon	0.50	PE	53.5	0.0	
	Alachlor+pendimethalin	1.00+0.90	PE	87.4	0.0	Rana *et al.*, 2004a
Rajmash	Pendimethalin	1.0	PE	78.4	1.9	Panotra *et al.*, 2012
	Alachlor	1.5	PE	53.7-85.5	1.3-6.7	Sharma *et al.*, 2004
	Metolachlor	1.5	PE	86.6	11.2	

Table 17.4. Weed Management in different Pulse Based Intercropping Systems

Intercropping	*Weed Management*
Pigeonpea + mungbean	Pendimathalin 1.0 kg/ha
Pigeonpea + Soybean	Fluchloralin 0.5-0.75 kg/ha or alachlor 2.0 kg/ha
Groundnut + pigeonpea	Fluchloralin 1 kg/ha
Maize + Urdbean	Butachlor 1.25 kg/ha
Chickpea + linseed	Pendimethalin 1.0 kg/ha or oxadiazon 0.50 kg/ha
Chickpea + Mustard	Pendimethalin 1.25-1.5 kg/ha fb quizalofop-ethyl 60 g/ha
Lentil + linseed	Pendimathaline 1.0 kg/ha
Wheat + Chickpea	Pendimathaline 1 kg/ha fb quizalofop- ethyl 100 g/ha
Mungbean + Pearlmillet	Butachlor 1kg/ha
Pigeonpea + Maize	Alachlor 2 kg/ha fb hand-weeding
Sorghum + Pigeonpea	Pendimethalin 1.25 kg/ha
Urdbean + sesame	Pendimethalin 1.0-1.25 kg/ha, metalachor 1.0-1.5 kg/ha and oxyflurofen 0.1-0.2 kg/ha
Pigeonpea + Cowpea	Pendimethalin at 1.0 kg/ha or fluchloralin at 1.0 kg/ha fb hand-weeding

Source: Kumar *et al.*, 2016.

REFERENCES

Agricultural Statistics at a Glance 2014. Directorate of Economics and Statistics, Ministry of Agriculture. Directorate of Economics and Statistics. http: //eands. dacnet.nic.in

Aktar, S., Hossain, M.A., Siddika. A., Naher N. and Amin M.R. 2013. Efficacy of Herbicides on the Yield of Lentil (*Lens culinaris* Medik.). *The Agriculturists* **11**(1): 89-94.

Ali, M. 1988. Weed suppressing ability and productivity od short-duration legumes intercropped with pigeonpea under rainfed conditions. *Tropical Pest Management* **34** (4): 384-387.

Buttar, G.S., Aggarwal, Navneet and Singh, Sudeep 2008. Efficacy of Different Herbicides in Chickpea (*Cicer arietinum* L.) under Irrigated Conditions of Punjab. *Indian J. Weed Sci.* **40** (3 and 4): 169-171.

Das, Rajib, Patra, B.C., Mandal, M.K. and Pathak, Animesh 2014. Integrated weed management in blackgram (*Vigna mungo L.*) and its effect on soil microflora under sandy loam soil of West bengal. *The bioscan* **9**(4): 1593-1596.

Das, T.K. 2008. Weed science: basics and application. Jain Brothers Pub, New Delhi, First edition p. 901.

Dubey, R.P. 2014. Integrated weed management- an approach. In Training Manual Advance Training in Weed Management, held at DWSR, Jabalpur, India on 14-23 January, pp. 19-21.

Gopinath, K. A., Kumar, N., Mina, B. L., Srivastva, A. K. and Gupta, H. S. 2009. Evaluation of mulching, satle seedbed, hand-weeding, and hoeing for weed control in organic garden pea. *Archives of Agronomy and Soil Science* **55** (1): 115-123.

Goud, V.V. and Patil, A.N. 2014. Increase in growth and yield of pigeonpea with weed management. *Indian Journal of Weed Science* **46**(3): 264–266.

Hanumanthappa, D.C., Mudalagiriyappa, R., Veera, Kumar, G.N. and Padmanabha K. 2012. Effect of weed management practices on growth and yield of cowpea (*Vigna unguiculata* L.) under rainfed conditions. *Crop Res.* **44** (1 and 2): 55-58.

IIPR. 2009. 25 Year of pulses Research at IIPR. ICAR-Indian Institute of Puleses Research, Kanpur, UP, India.

Kaur, R, Raj, R., Das, T.K., Shekhawat, K., Singh, R. and Choudhary, A.K. 2015. Weed management in pigeonpea-based cropping systems. *Indian Journal of Weed Science* **47**(3): 267–276.

Kewat, M.L. 2014. Improved weed management in Rabi crops. National Training on Advances in Weed Management pp. 22-25.

Khope, D., Kumar, S. and Pannu, R.K. 2011. Evaluation of Post-emergence Herbicides in Chickpea (*Cicer arietinum*). *Indian J. Weed Sci.* **43** (1 and 2): 92-93.

Khot, D.B., Pagar, R.D. and Munde, S.D. 2015. Effect of different weed management practices on yield and economics of summer black gram (*Vigna mungo* L.). *Agriculture for Sustainable Development* **3**(1): 21-22.

Komal, Singh, S.P. and Yadav, R.S. 2015. Effect of weed management on growth, yield and nutrient uptake of greengram. *Indian Journal of Weed Science* **47**(2): 206–210.

Kumar, A. and Tewari, A.N. 2004. Crop-weed competition studies in summer sown blackgram (*Vigna mungo* L). *Indian J. Weed* Sci. **36** (l and 2): 76-78.

Kumar, A., Sharma, B.C., Nandan, Brij and Sharma, P.K. 2009. Crop-weed competition in field pea under rainfed subtropical conditions of Kandi belt of Jammu. *Indian Journal of Weed Science* **41**(1 and 2): 23-26.

Kumar, N. K., Nath, C. P., Hazra, K. K. and Sharma, A. R. 2016. Efficient weed management in pulses for higher productivity and profitability. *Indian Journal of Agronomy* 61 (4th IAC Special issue): S93-S105.

Kumar, N., Nandal, D.P. and Punia, S.S. 2014a. Weed management in chickpea under irrigated conditions. *Indian Journal of Weed Science* **46**(3): 300–301.

Kumar, P., Peshin, R., Nain, M.S. and Manhas, J.S. 2009 and 2010. Constraints in pulses cultivation as perceived by the farmers *Raj. J. Extn. Edu.* **17** and **18**: 33-36.

Kumar, Suresh, Angiras, N.N. and Singh, Rupinder 2006. Effect of planting and weed control methods on weed growth and seed yield of blackgram. *Indian J. Weed Sci.* **38** (1 and 2): 73-76.

Lhungdim, J., Singh, Y., Singh, O.N. and Chongtham, S.K. 2014. Efficiency of different weed control methods on yield and economics of rainfed lentil (*Lens culinaris* Medikus). *Journal of Food Legumes* **27**(1): 32-36.

Malik, R.S. and Yadav, Ashok 2014. Effect of sowing time and weed management on performance of pigeonpea. *Indian Journal of Weed Science* **46** (2): 132–134.

Malliswari, T., Reddy, P.M., Karuna, G.S. and Chandrika, V. 2008. Effect of irrigation and weed management practices on weed control and yield of blackgram. *Indian J. Weed Sci.* **40** (1 and 2): 85-86.

Malviya, A. and Singh, B. 2007. Weed dynamics, productivity and economics of maize as affected by integrated weed management under rainfed condition. *Indian J. Agronomy*. **52**(4): 321 24.

Nandan, Brij, Kumar, A., Sharma, B.C. and Sharma, N. 2011. Chemical and cultural methods for weed control of mung bean under limited moisture conditions of Kandi belt of Jammu. *Indian Journal of Weed Science* **43** (3 and 4): 241-242.

Narayan, Prem and Kumar, Sandeep 2015. Constraints of growth in area production and productivity of pulses in India: An analytical approach to major pulses. *Indian J. Agric. Res.* **49** (2): 114-124.

Panotra, N., Singh, O.P. and Kumar, A. 2012. Effect of chemical and mechanical weed management on yield of French bean–sorghum cropping system. *Indian Journal of Weed Science* **44**(3): 163–166.

Patel, B.D., Patel, V.J., Patel, J.B. and Patel, R.B. 2006. Effect of fertilizers and weed management practices on weed control in chickpea (*Cicer arietinum* L.) under middle Gujarat conditions. *Indian J. Crop Science* **1**(1-2): 180-183.

Patel, K.R., Patel, B.D., Patel, R.B., Patel, V.J. and Darji, V.B. 2015. Bio-efficacy of herbicides against weeds in blackgram. *Indian Journal of Weed Science* **47**(1): 78–81.

Pedde, K.C., Gore, A.K. and Chavan, A.S. 2013. Integrated weed management in chickpea. *Indian Journal of Weed Science* **45**(4): 299.

Poonia, T.C. and Pithia, M.S. 2013. Pre- and post-emergence herbicides for weed management in chickpea. *Indian Journal of Weed Science* **45**(3): 223–225.

Punia, S.S., Hooda, V.S., Duhan, Anil, Yadav, D. and Amarjeet 2013. Distribution of weed flora of greengram and blackgram in Haryana. *Indian Journal of Weed Science* **45**(4): 247–249.

Ram, S. and Verma, H.P. 2015. Weed growth, yield and quality of summer cowpea [*Vigna unguiculata* (L.) Walp.] as influenced by weed management. *Ann. Agric. Res.* **36** (3): 304-308.

Rana, M.C., Kumar, N., Sharma, A. and Rana, S.S. 2004a. Management of complex weed flora in peas with herbicide mixtures under Lahaul valley conditions of Himachal Pradesh. *Indian J. Weed Sci.* **36** (1 and 2): 68-72.

Rao, P.V., Reddy A.S. and Rao, Y.K. 2015. Effect of integrated weed management practices on growth and yield of pigeonpea (*Cajanus cajan* (L.) MILLSP.) *International Journal of Plant, Animal and Environmental Sciences* **5** (3): 124-127.

Ratnam, M., Rao, A.S. and Reddy, T.Y. 2011. Integrated weed management in chickpea (*Cicer arietinum* L.). *Indian J. Weed Sci.* **43** (1 and 2): 70-72.

Sharma, G.D., Sharma, J.J. and Sood, Sonia 2004. Evaluation of alachlor, metolachlor and pendimethalin for weed control in rajmash (*Phaseolus vulgaris* L.) in cold desert of North-Western Himalayas. *Indian J. Weed* Sci. **36** (3 and 4): 287-289.

Sharma, J.C, Prakash, Chandra, Shivran, R.K and Narolia, R.S. 2014. Integrated weed management in pigeon pea (*Cajanus cajan* (L.) Millsp.) *IJAAS* **2** (1 and 2) 57-62.

Sharma, O.L. 2009. Weed management in chickpea under irrigated conditions of western Rajasthan. *Indian Journal of Weed Science* **41**(3 and 4): 182-184.

Shyam, R., Mishra, O.P. and Singh, R. 2008. Evaluation of post-emergence herbicides for weed control in lentil *Indian Journal of Weed Science* **40** (1 and 2): 100-101.

Singh Harinder and Angiras N. N. 2004. Weed Management Studies in Garden Pea (*Pisum sativum* sub sp. hortens L.) *Indian J. Weed Sci.* **36** (l and 2): 135-137.

Singh, G. 2011. Weed management in summer and kharif season blackgram [*Vigna mungo* (L.) Hepper] *Indian J. Weed Sci.* **43** (1 and 2): 77-80.

Singh, G., Kaur H., Aggarwal, N. and Sharma, Poonam 2015a. Effect of herbicides on weeds growth and yield of greengram. *Indian Journal of Weed Science* 47(1): 38–42.

Singh, R. 2014. Weed management in major kharif and rabi crops. National Training on Advances in Weed Management. pp. 31-40.

Smith, R.G., Mortensen, D.A. and Ryan, M.R. 2010. A new hypothesis for the functional role of diversity in mediating resource pools and weed-crop competition in agro-ecosystems. *Weed Research* **50**: 37-48.

Verma, R., Nepalia, V. and Kumawat. S.K. 2004. Influence of Weed Control and Sulphur Nutrition on Weed Dynamics and Productivity of Pea (*Pisum sativum* L.). *Indian J. Weed* Sci. **36** (3 and 4): 285-286.

Verma, S.K. and Singh, S.B. 2008. Enhancing of wheat production through appropriate agronomic management. *Indian Farming* **58**(5): 15-18.

Yadav, K.S., Dixit, J.P. and Prajapati, B.L. 2015. Weed management effects on yield and economics of blackgram. *Indian Journal of Weed Science* **47**(2): 136–138.

Yadav, R.B., Vivek, Singh, R.V. and Yadav, K.G. 2013. Weed management in lentil. *Indian Journal of Weed Science* **45**(2): 113–115.

Transformation of Indian Agriculture through Innovative Technologies *Pages* **277–296**
Editor: **Dr. Shahid Ahamad & Dr. Jag Paul Sharma**
Published by: **ASTRAL INTERNATIONAL PVT. LTD., NEW DELHI**

18 Symptomatology and Etiology of *Alternaria* Leaf Spots and Blight of Oilseed Crops

Udit Narain, Alka Kushwaha, Vandana Krishna and Ved Ratan

Eight oleiferous crops are grown in diverse agro-climatic conditions in India in all the seasons, Rabi, Kharif and Zaid. The oil extracted from their seeds are not only utilized for cooking purposes but some of them have medicinal values also (castor, linseed, sesame).

Eleven species of Alternaria cause leaf spots and blight on eight oil yielding crops in India. Some of them are very specific to parasite only a particular oilseed plant like A. helianthi (sunflower), A. brassicae and A. brassicicola (rapeseed and mustard), A. linicola (linseed), A. sesami (sesame), A. ricini (castor) and A. carthami (safflower). Alternaria alternata is of wide spread occurrence causing leaf spots and blight of sunflower, mustard, groundnut and soybean. A. tenuissima is next to A. alternata in causation of disease. The disease symptoms on all the oil yielding crops alongwith morphological characters of causative Alternaria spp. have been given in greater detail. On rapeseed/mustard and sunflower there is the involvement of four species of Alternaria on each crop so, a comparative study of symptomatology and etiology has been made and a very simple and feasible key has been framed for ready and correct identification of Alternaria spp. associated with them.

keywords: *Olieferous crops, Alternaria spp., Alternaria leaf spots and blight, Symptomatology, Etiology.*

Oilseeds play an important role in the national economy due to their oil contents which form the essential constituent of human diet. Oils from rapeseed and mustard, groundnut, sunflower, sesame and safflower are mainly utilized for cooking purposes whereas oils from castor and linseed are used for industrial and medicinal purposes. The oilcakes of these oil yielding crops are good quality of animal feed.

The diseases mainly caused by the fungal origin constitute one of the most serious constrains for successful cultivation and production of oilseed crops. Alternaria leaf spots and blight of eight oleiferous crops caused by eleven species of *Alternaria* are one of them which not only deteriorate the quality of seeds but also reduce the oil content considerably.

Out of eleven species of *Alternaria,* the most common and destructive diseases are those caused by four species of the pathogen parasitizing on sunflower and rapeseed and mustards. Groundnut being cultivated in larger areas has a prominent place among oilseed crops, and is attacked by two species of *Alternaria* and similar situation is in case of linseed and soybean and on rest of oilseed crops like safflower, castor and sesame there is the involvement of a single species of *Alternaria* in causation of disease (Table 18.1).

Table 18.1. *Alternaria* spp. Associated with Oilseed Crops

Sl.No.	*Oilseed Crops*	*Alternaria spp.*
1.	Sunflower	*Alternaria alternata*
		Alternaria helianthi
		Alternaria tenuissima
		Alternaria zinniae
2.	Rapeseed and mustard	*Alternaria alternata*
		Alternaria raphani
		Alternaria brassicae
		Alternaria brassicola
3.	Groundnut	*Alternaria alternata*
		Alternaria tenuissima
4.	Linseed	*Alternaria lini (syn. A. alternata)*
		Alternaria linicola
5.	Sesame (Til)	*Alternaria sesami*
6.	Safflower	*Alternaria carthami*
7.	Castor	*Alternaria ricini*
8.	Soybean	*Alternaria alternata*
		Alternaria tenuissima

Elliot (1917) described first the taxonomic characters of genus *Alternaria.* A great deal of diversity and parasitism occur in *Alternaria* (Joly, 1964; Narain and Kant, 2008; Narain *et al.,* 2016) and they also differ in their biology, epidemiology and pathogen city (Rotem, 1994). Some of the diseases caused by *Alternaria* spp. were considered to be of minor importance in the past but due to newer agriculture technology, cultivation of newer varieties of crops and climatic change, they are occupying the prominent place in their disease situation.

The symptoms of Alternaria leaf spots and blight on all eight oil yielding crops have been described alongwith their pathogens. Due to involvement of four species of *Alternaria* with sunflower and rapeseed and mustard, the symptom differentiation, comparative morphological variation in pathogens and the suitable key for species identification have been given.

Sunflower

Symptoms

1. The disease caused by *Alternaria alternata* initially appears as numerous small spots on the upper surface of leaves. Later, these spots extent along with whole leaf surfaces due to the coalescence of the adjacent spots. The spots are usually dark brown, circular to oval and vary from 0.2 to 1 cm in diameter. In several instances in later stage of disease development, the affected areas are perforated irregularly due to the development of shot holes.

   A. alternata was for the first time reported by Bose (1942) on sunflower in India. A report of *A. alternata* causing head rot in sunflower is also available from Australia (Middleton, 1971).

2. The symptoms of disease caused by *Alternaria helianthi* are observed throughout the growing season of the crop during *Rabi, Kharif* and *Zaid*. The characteristic symptoms are more prevalent and prominent during the rainy season. The symptoms produced by the fungus appear on leaves, petioles, stems, sepals and petals also. The disease initially appears in the form of small, scattered brown spots on the leaf lamina. The spots following the rains increase in size and a number of them coalesce involving larger area of leaves. The spots are dark brown with paler margin and "yellow halo" and indistinct zonations, usually irregularly circular, 0.5-2.0 cm size.

   Linear, dark brown necrotic lesions also appear on the petioles and stems. On sepals and petals, brown to brown black spots, irregularly circular are produced. Irregular dark brown lesions are also seen on the base of the heads.

   Leaf spot disease of sunflower caused by *A. helianthi* was first time reported by Tubaki and Nishiharn (1969) from Japan. In India, it was recorded by Narain and Saksena in 1973 at Kanpur (U.P.). Later Anil Kumar *et al.* (1974) and Mukewar *et al.* (1974) also reported the occurrence of leaf spot of sunflower caused by *A. helianthi* from Tamilnadu and Delhi, respectively.

3. The symptoms due to *Alternaria tenuissima* are confined to leaves only. The disease appears in the beginning as straw coloured to light brown small spots scattered on leaf lamina. Later, they become circular to irregular, pale to dark brown spots measuring 2 mm to 1.5 cm in diameter with cracked centre. The affected leaves dry up and shred. Very rarely the stems also get infected. The disease was reported for the first time from India by Narain and Chauhan (1981).

4. In case of *Alternaria zinniae* the disease initially appears on the foliage in the form of small, scattered light brown spots which increase in size following the rains and become dark brown and a large number of them coalesce to form larger patches on leaves. Mature lesions on leaves are irregularly or roughly circular, 0.5-2.0 cm in diameter, dark brown with paler margins and with distinct or indistinct zonations.

   Linear necrotic lesions also appear on petioles. The lesions on stems begin as small flecks which gradually enlarge longitudinally to form streaks. They are large, dark brown, 0.5-5 cm in the length and sometimes these lesions completely girdle the thin stems. Zelle (1932) for the first time in Russia observed on sunflower brown leaf spots with concentric rings suggestive of target like spottings caused by *Alternaria zinniae* infection. Nattrass (1950) listed *A. zinniae* as the cause of leaf spot of sunflower in Kenya and this was the first record of its occurrence on sunflower.

   McDonald and Martens (1963) reported leaf and stem spots of sunflower caused by *Alternaria zinniae* from Canada. The disease produced the dark, circular leaf spot and dark brown longitudinal stem lesions (Middleton, 1971). Takano (1963) also reported the occurrence of *Alternaria zinniae* on sunflower from Japan. From India *A. ziniae* has been reported on sunflower (*Helianthus annuus* L.) by Narain and Srivastava (1996).

   Leaf spots and blight and also head rot of sunflower (*Helianthus annuus* L.) caused by different species of *Alternaria* have been also reported by Zelle, 1932; Hansford, 1943; Sackston, 1957; Martens *et al.*, 1970; Hedjaroude, 1973 from different parts of the world.

Pathogens

1. *Alternata alternata* (Fr.) Keissler (*Beih. Bot. Zbl.*, 29: 434, 1912)

Colonies moderately fast growing, usually black or olivaceous black, sporulation abundant; *Mycelium* septate, branched, olive buff, 2.5-9.5 mm wide; *Conidiophores* arising singly or in groups, simple, septate, straight or curved, 24.5-68.5µm in length and 3.5-7.0µm in width; *Conidia* formed in long and often branched chains, obclavate, obpyriform, ovoid or ellipsoidal, dark brown, smooth or verrucose with 0-10 cross and 0-8 longitudinal septa, 8.50-53.50 x 5.30 x 21.50µm in size and often with short conical or cylindric beaks, *Beaks* short, mostly half of spore body, 2.50-45.50µm in length and 1.5-6.5µm in width and 0-3 septate (Figure 18.1).

The morphological and cultural characters of the fungus is identical to those as described by Simmons (1967) for *Alternaria alternata*, the type species of the genus *Alternaria* and that of Bose (1942) and Narain and Srivastava (1996) on sunflower causing the leaf spot disease.

2. *Alternaria helianthi* (Hansf.) Tubaki and Nishihara (*Trans. Br. mycol. Soc.*, 53: 147-149, 1969)

Colonies grey to dark blackish brown; *Mycelium* immerged and hyphae septate and branched, pale olivaceous brown, 3.5-7.5µm wide; *Conidiophores* arise singly or

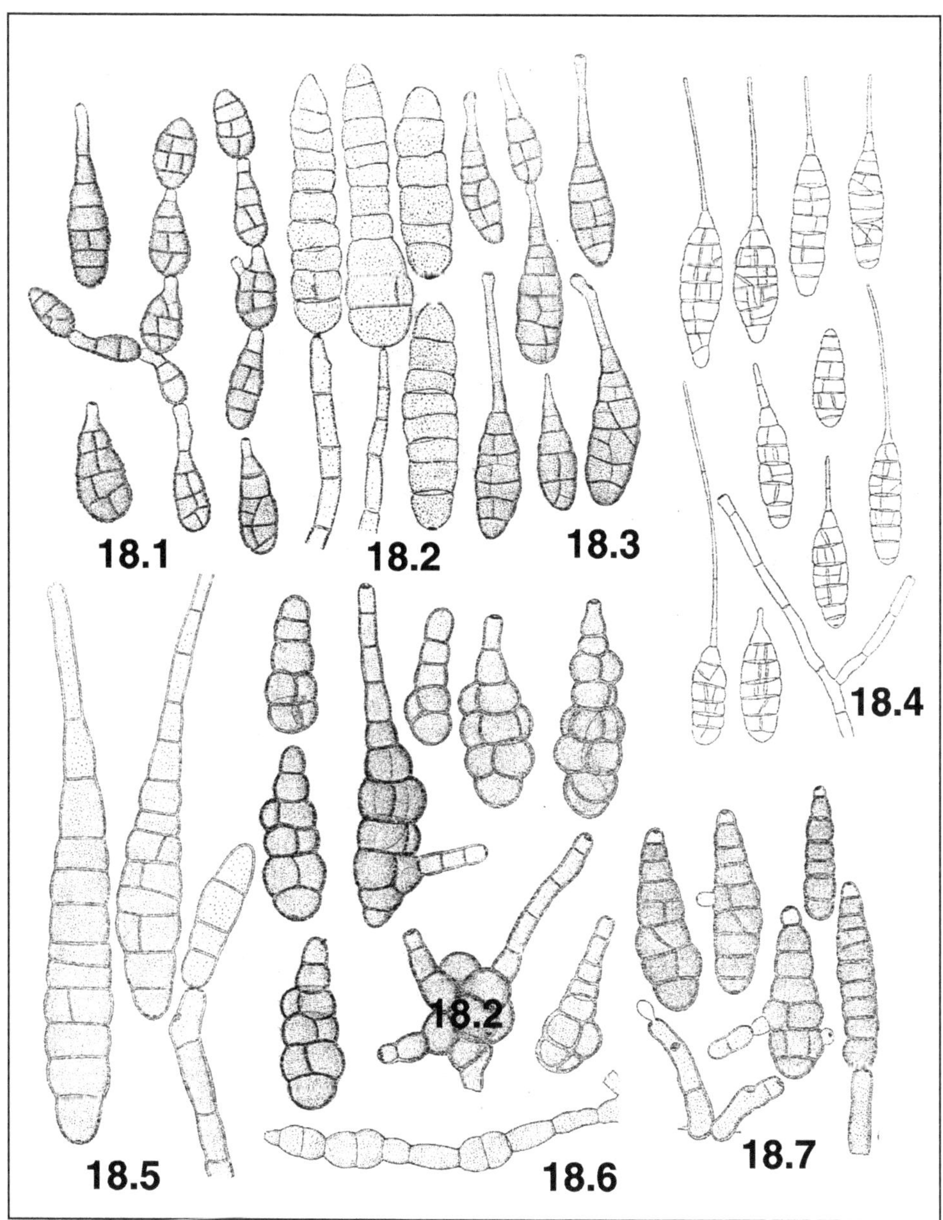

Figures 18.1–18.7. Morphology of *Alternaria* spp. Parasitic on Oilseed Crops.

18.1: *A. alternata*, 18.2: *A. helianthi*, 18.3: *A. tenuissima*, 18.4: *A. zinniae*, 18.5: *A. brassicae*, 18.6: *A. raphani*, 18.7: *A. brassicicola*.

in groups of 3 to 6, cylindrical, straight or curved and septate, geniculate with scars, 35-100 x 4.0-8.0µm in size; *Conidia* formed singly, cylindrical to ellipsoid, mostly straight with rounded ends and without beaks, often slightly constricted at septa, pale yellow to pale brown, smooth, with many cross septa (2-12) an occasional longitudinal septa (0-3), 35.0-100 µm in length and 10.0-30.0µm in width (Figure 18.2).

The morphological and cultural characters of this fungus fit with the description of Tubaki and Nishihara (1969), first described from Japan and of Narain and Saksena (1973) from India on sunflower leaves causing leaf spots.

3. *Alternaria tenuissima* (Kunz ex Pers.) Wiltshire (*Trans. Br. mycol. Soc.*, 18: 157, 1933)

Colonies growing faster, centre raised and deep greyhish; *Mycelium* septate, branched olive buff, 2.8-10.7 µm in width, *Conidiophores* produced either as side branches from cells of hyphae or terminally to hyphae, septate and branched, dark brown, measuring 19.5-75.0µm in length and 2.7-6.20 µm in width; *Conidia* in short chains (3-6), conical to oval, dark olive buff, tapering gradually into beaks, with 3-9 cross and 0-8 longitudinal septa, measuring 12.0-66 µm in length and 7.0-15.5µm in width, *Beaks* usually smaller to spore body in culture but in nature upto same length, sometimes forming a terminal swelling, 3.50-70.0µm in length and 3.50-5.50µm in length (Figure 18.3).

Characters of *A. tenuissima* observed in culture and in nature agree with that of described by Wiltshire (1954) and Narain and Srivastava (1996).

4. *Alternaria zinniae* Pape (*Ang. Bot.* 24: 16-22, 1942)

Colonies amphigenous, *Mycelium* septate and branched, olive buff to deep olive buff, 2.5-8.5 µm in width; *Conidiophores* usually cylindrical, erect to slightly bent, simple, septate, 42.5 - 205 µm x 6.0 – 9.0 µm in size; *Conidia* produced singly or rarely with a secondary conidium, oval to cylindrical in shape, dark olive buff, constricted at septa with 3-12 cross and 0-10 longitudinal septa and measure 20.5-95.0 µm in length and 8.5 – 26.5 µm in width; *Beak* abruptly formed from conidia, long, filiform, septate, simple, straight to curved, 10.5-175.0 µm in length and 2.0-3.5 µm in width and 0-3 septate (Figure 18.4).

The morphological and cultural characters of this isolate are similar to those originally described by Pape (1942) and by Neergaard (1945) on other hosts and by McDonald and Martens (1963) on sunflower.

The symptoms of disease in sunflower caused by four species of *Alternaria* have been presented in Table 18.2.

For identification of *Alternaria* spp., a simple and feasible key has been framed as follows:

Key to *Alternaria* spp. Associated with Sunflower

I. Conidia formed in long chains consisting of even 20 spores

Conidia usually polymorphic, often with short conical or cylindrical short beaks .... *A. alternta*

Table 18.2. Comparative differentiation of *Alternaria* spp. Occuring on Sunflower

Morphological Characters	*A. alternata*	*A. helianthi*	*A. tenuissima*	*A. zinniae*
Conidiophores	Olive brown, erect, simple, septate	Pale-grey yellow, simple or rarely branched, erect, septate	Dark brown, erect, simple of branched, septate	Dark brown, erect, simple, septate
Length	24.5-68.5	35.0-110µm	19.50-75.0µm	42.5-205.5µm
Width	3.50-7.25	4.0-8.0µm	2.70-6.20µm	6.0-9.0µm
Conidia				
(a) No. in chains	10 or even more	Single	3-6	Single
(b) Shape	Polymorphic	Cylindrical to ellipsoid	Conical to oval	Oval to cylindrical
(c) Colour	Dark brown	Pale-grey yellow to pale brown	Dark olive buff	Olive brown
(d) Septa : cross	3-10	2-12	3-9	3-12
Longiludinal	0-8	0-3	0-8	0-10
(e) Length	8.50-53.50µm	35-100µm	12.0-65.0µm	20.5-95.5µm
(f) Width	5.30-21.50µm	10.0-30.0µm	7.0-15.50µm	8.50-26.5µm
Beaks	Short, Conical or cylindrical	Absent	Cylindric, some times swollen terminally	Long, filiform and formed abruptly
(a) Length	2.5-45.5µm	-	3.50-70.0µm	10.5-17.5µm
(b) Width	1.5-6.5µm	-	3.50-5.50µm	1.75-3.50µm
(c) Cross septa	0-3	-	0-8	0-4

II. Conidia formed in short chains consisting of about 3-6 spores

Conidia conical to oval, tapering gradually in to beak, of same length of spore body with terminal swelling.. *A. tenuissima*

III. Conidia formed singly

(a) Conidia cylindric to ellipsoid, mostly straight with occasional longitudinal septa and without beaks................................. *A. helianthi*

(b) Conidia oval to cylindrical with several longitudinal septa and long filiform beaks .... *A. zinniae*

Rapeseed and Mustard

During the course of investigations on the taxonomic studies of phytopathogenic fungi, particularly of the genus *Alterneria* associated with oil yielding Brassicaceous crops, four species *viz., A. brassicae* Ellis (1968a), *A. brassicicola* (Ellis 1968b), *A. raphhani* (Gorves and Skolko, 1944) and *A. alternata* (Prasad and Narain, 2007) were found to be parasitic on them, causing the characteristic leaf spots and blight, lesions on stems, branches and petioles, inflorescence and siliquae. So, a comparative study of disease symptomatology and etiology was made with ultimate aim to develop a most suitable and feasible key for ready identification of *Alternaria* spp. associated with oleiferous crops of Brassicaceae.

Symptoms

The disease appears initially from the seedlings. Symptoms caused by *A. brassicae* start from margin of leaves or at the tip and spread inward and downward accordingly. The dark brown, irregular spots are accompanied by characteristic yellow halo appearance and have distinct zonations, which gradually increase in size in concentric manner. Two or more such spots may coalesce to form larger area. In well advanced stage, some of the spots are covered with dark olive green mass of the fungal spores with shot-hole appearance.

The leaf spot due to *Alternaria brassicicola* are circular to elliptical and olivaceous in colour which vary from 4-12 mm in size with indistinct or distinct zonations. In severe infection, several lesions coalesce to cover large area and cause death of leaves. The young and actively developing leaves are almost free from the disease or may have the small necrotic areas. The severely infected leaves show burning effect, which ultimately dry up and may fall prematurely.

In case of symptoms of Alternaria leaf spot due to *Alternaria raphani*, the spots are brown to dark brown in patches starting from margin of leaves, developing inwards, in irregular fashion. The spots are circular to irregular in appearance, dark brown with indistinct zonations on both surfaces of leaves. The affected areas soon turn blighted and wither. The severely infected leaves ultimately dry up and may fall prematurely.

Alternaria alternata initially appears to produce minute, dark brown patches which start from the margin or centre of leaf and at later stage of disease development, a number of spots coalesce to form the irregular dark brown spots.

On plants grown for seed, dark brown lesions occur on the main axis, inflorescence, branches and on the siliquae. These lesions coalesce rapidly and cause premature ripening and splitting of siliquae and may result in high level of seed infection. Infected seeds are shrunken, discoloured or covered with fungal growth and have low quality. Cankers may form just below the inflorescence which may result in premature death of plants.

In general, the symptomatology reveals that the disease appears initially as small brown spots on the leaf lamina or on margin of the leaves which later increase rapidly in size or inward and become mostly circular or irregular in shape and turn light to dark brown in colour with slight distinct zonations with or without halo (Chahal *et al.*, 1977; Sharan *et al.*, 2005). It has been reported that in severe cases, the spots coalesce and give the leaf a blighted appearance with dark brown mass of spores. On some hosts symptoms develop concentric rings on the leaf (Kadian and Saharan, 1983 and 1984). The leaf spots later enlarge, become nearly circular, 4-12 mm in diameter, olivaceous in colour with zonations and profuse sooty sporulation in case the infection is specially caused by *A. brassicicola* (Ellis, 1971). The symptoms caused by different species of *Alternaria* on different Brassicaceous hosts have close resemblance with that described by Changsri and Weber (1963), Walker (1952), Ellis (1968a and b), Narain and Saksena (1975), Atkinson (1950), Narain (1986), Gupta and Basuchaudhary (1992) and Chand *et al.* (2007).

Alternaria brassicicola has been reported to be pathogenic on white cabbage, cabbage, knol-khol, broccoli and radish (Rao, 1977; Ellis, 1968b and Sangwan *et al.*, 2002; Sangwan and Mehta, 2007 and Chand *et al.*, 2007). *A. raphani* isolated from radish was pathogenic to a number of Brassicaceous hosts and parasitized the vegetable crucifers as well as oil crops (Atkinson, 1950; Khalid *et al.*, 2004).

Changsri and Weber (1963), Narain and Saksena (1975), Narain (1986) and Chand *et al.* (2007) have also reported more or less the same type of symptoms caused by *Alternaria raphani.*

The comparative symptom differentiation of *Alternaria* spp. occurring on rapeseed and mustard has been presented in Table 18.3.

Pathogens

1. *Alternaria brassicae* (Berk.) Sacc. (*Michelia*, 2: 129, 1880)

Colonies fast growing, amphigenous, usually in beginning pale olive and later turn into dark olivaceous in colour with abundant sporulation; *Mycelium* septate, branched, yellowish brown, smooth, 4-8 µm thick; *Conidiophores* arising in groups of 2-10 or more from hyphae, emerge through stomata, usually simple, erect, geniculate, septate, grayish to olive in colour, swelling at its base, measuring 5-150 x 5-10 µm in size, bearing one to several small but distinct conidial scars; *Conidia* solitary or occasionally in chains upto 4, straight or slightly curved, obclavate, rostrate with 5-10 transverse and 0-5 longitudinal or oblique septa, olive or grayish olive in colour and 40-150 x 15-30 µm in size; *Beak* usually pale brown, cylindrical, 10-130 µm in length and 3-8 µm in width (Figure 18.5).

Table 18.3. Comparative Symptom differentiation Due to *Alternaria* spp. on Oil Yielding Brassicaeous Crops

Symptoms on	*Alternaria brassicae*	*Alternaria brassicicola*	*Alternaria raphani*	*Alternaria alternata*
Leaves	Spots dark brown to black with halo chlorotic tissues, 0,5 to 2.0cm in diameter and in concentric circle	Numerous, dark brown to black spots, 4-12mm in diameter, zonate, lesions with sooty black appearance	Spots dark brown or black, 3-8mm in size, raised with distinct or indistinct zonations	Spots minute to large, dark brown in patches, often colescing to form irregular dark brown lesions
Stems	Dot like linear or elongated lesions	Spots numerous, dark brown to black	Lesions dark brown to black, in linear manner	Spots/lesions dark brown in patches
Siliquae	Linear, elongated or dot like lesions, without concentric rings	Appearance of minute dark brown to black spots	Irregular brown to black lesions	Lesions coalescing and splitting type

2. ***Alternaria brassicicola* (Schw.) Wiltshire (*Mycol. pap.*, 20: 8, 1947)**

Colonies growing as dark olivaceous to dark blackish brown, velvety; *Mycelium* septate, branched, hyaline at first, later brown or olivaceous brown, inter and intracellular, smooth, 25-75 x 1.5-7.5 µm in size; *Conidiophores* produced either singly or in groups of 2-12, emerging through stomata, usually simple, erect or curved, occasionally genticulate, more or less cylindrical but often slightly swollen at the base, septate, pale to mild olivaceous brown, smooth, 20-75 µm long and 4-8 µm thick; *Conidia* mostly in chains of upto 20 or more, usually tapering slightly towards apex, obclavate, basal cell round, the apical cell being more or less rectangular or resembling a truncate cone with 2-10 cross and 0-5 longitudinal septa, often constricted at the septa, pale to dark olivacecous brown, smooth or becoming slightly verriculose with age and measuring 20-120 µm length and 7-25 µm width; *beak* usually non-existent (Figure 18.7).

3. ***Alternaria raphani* Groves and Skolko (*Can. J. Res., Sect.* C. 22: 227, 1944)**

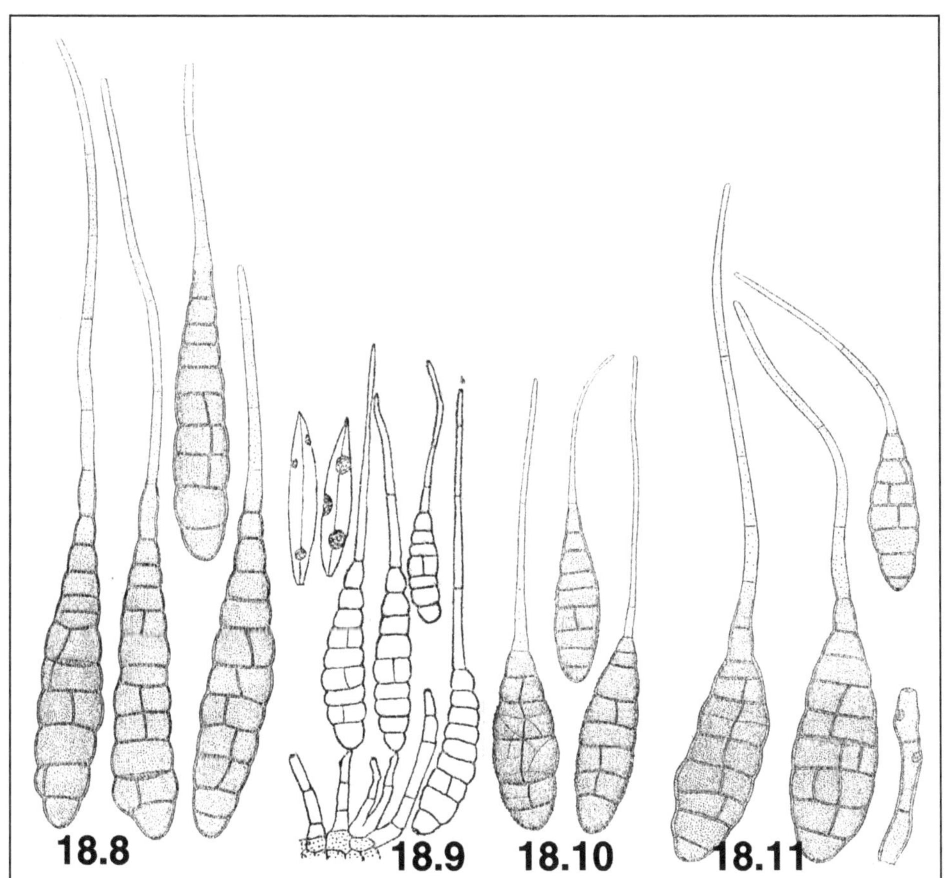

Figures 18.8-18.11. Morphology of *Alternaria* spp. Parasitic on Oilseed Crops. 18.8: *A. carthami*; 18.9: *A. linicola*; 18.10: *A. ricini*; 18.11. *A. sesami*.

Colonies grow faster, centre sometimes raised and light brown to dark brown in colour; *Mycelium* septate, branched, hyaline at first later become olive buff in colour, 2-7 µm thick, sometimes swollen to form chlamydospores in chains or in groups; *Conidiophores* simple or occasionally branched, septate, olivaceous brown, 2.5-8.0 x 2.5-150 µm in size with short beak, some times swollen slightly at tip and usually with a single conidial scar; *Conidia* commonly in chains consisting of 3-4, obclavate or ellipsoidal in shape, dark golden brown to olivaceous brown, smooth or some times minutely verruculose with 3-7 transverse and often a number of longitudinal or oblique septa, constricted at septa and measuring 20-60 µm long and 10-30 µm thick; *Beaks* of conidium smaller than that of *A. brassicae*, but longer than that of *A. alternata* (Figure 18.6).

4. *Alternaria alternata* (Fr.) Keissler (*Beih. Bot. Zbl.*, 29: 434, 1912)

Colonies fast growing, usually black or olivaceous black with abundant sporulation; *Mycelium* septate, branched, olive buff in colour, 2.5-9.5 µm in width, sometimes swollen ranging from 12.5-23.5 µm to form chlamydospores; *Conidiophores* usually arise single or sometimes in groups, mostly simple, septate, straight or curved and geniculate, pale to mid olivaceous brown, smooth and measure 24.5-68.5 µm in length and 3.50-125 µm width; *Conidia* formed in long and often branched chains, varying in shape and sizes, dark brown in colour, smooth to verruculose and 8.50-53.50 x 5.0-21.50 µm in size, *Beaks* of conidia usually short, conical or cylindrical but never equal in the length of conidium, often one third or half of the spore body, usually lighter in colour, 10-45.5 µm in length and 1.5-6.5 µm in width and 0-3 septate (Figure 18.1).

All the four species of the genus *Alternaria* were observed to be parasitic on rapeseed and mustard and differ from each other on the basis of their mode of formation of spores (Neergaard, 1945) and conidial morphology (Ellis, 1971; Khalid *et al.*, 2004; Prasad and Narain, 2007). In *A. alternata* and *A. brassicicola* long chains of conidia are formed but latter is without beak. In *A. brassicae*, conidia are solitary or seldom in chains with comparatively larger conidia and beaks but in *A. rapliani* conidia are produced in short chains (3-4) and there is frequent formation of chlamydospores in culture (Narain and Saksena, 1975; Verma and Saharan, 1994).

Morphological observations of all the four species of *Alternaria* were made in culture (PDA) and in nature (host) and their comparative differentiation has been presented in Table 18.4 and based on their distinguishing characters, a very simple and feasible key has been framed for ready identification of *Alternaria* spp. associated with rapeseed and mustard.

Key to *Alternaria* spp. Parasitic on Rapeseed and Mustard

I. Conidia solitary or occasionally upto four, obclavate, rostrate, tapering gradually into thick cylindrical beaks.. *A. brassicae*

II. Conidia in long chains consisting of twenty or even more in a chain:

(a) Conidia usually cylindrical, basal cell rounded and apical cell more or less rectangular and beak usually almost non-existent.. *A. brassicicola*

Table 18.4: Comparative differentiation in *Alternaria* spp. Occurring on Rapeseed and Mustard

Morphological Characters	*Alternaria brassicae*	*Alternaria brassicicola*	*Alternaria raphani*	*Alternaria alternata*
Conidiophores	Mild-pale grayish to olive, simply erect, swelling at base, geniuculate, septate	Pale mild olivaceous brown, simple, septate, erect or curved	Olivaceous brown, simple or occasionally branched, septate	Olive brown erect, simple and septate
Length	15-150 µm	25-75 µm	25-150 µm	24.5.68.5 µm
Width	5-10 µm	1.5-7.5 µm	3.5-7.5 µm	3.50-7.25 µm
Conidia				
Conidial chain	Solitary or occasionally upto 4	Upto 20 or more	Upto 3-4	Upto 20 or more
Shape	Obclavate rostrate	Obclavate	Obclavate or ellipsoidal	Polymorphic
Colour	Olive or grayish olive	Pale to dark olivacious brown	Golden brown to olivaceous	Dark brown
Cross septa	5-16	2-10	3-7	3-10
Longitudinal septa	0-8	0-5	0-5	0-8
Length	40-150 µm	20-120 µm	20-60 µm	8.5-35.5 µm
Width	15-30 µm	7-25 µm	10-30 µm	5.0-21.5 µm
Beak				
Length	10-130 µm	Non-existent	5-75 µm	10-45.5 µm
Width	3-8 µm	-	2.5-6.0	1.5-6.5 µm
Cross speta	0-5	-	0-3	0-3

(b) Conidia usually polymorphic, often with short conical or cylindrical short beaks.. *A. alternata*

III. Conidia 3-4 in chains, straight or curved, obclavate, generally with short beak, chlamydospores in abundance in culture......... *A. raphani*

Soybean

Soybean (*Glycinc max*) although a leguminous crop is also utilized for oil purposes. This crop also suffers from the leaf spot diseases caused by two species of *Alternaria viz.,* A. *alternata* and *A. tenuissima* (Misra and Prakash, 1975). The symptoms of the disease and causal organisms almost are similar to those as described earlier for *Alternaria alternata* and *A. tenuissima* causing leaf spots and blight in other oleiferous crops.

Groundnut

Alternaira Leaf Spots and Blight

In groundnut, two species of *Alternaria* are involved. Lesions produced by *Alternaria* spp. are brown in colour and irregular in shape, surrounded by yellowish halo, while in case of *A. tenuissima* are characterized by blighting of apical portions of leaflets, which turn light to dark brown in colour. In the later stages of infection, blight leaves curl inward and become brittle. Lesions produced by *A. alternata* are small, chlorotic that spread over the surface of the leaf. The lesions become necrotic and brown and are round to irregular in shape (Ghewande *et al.*, 1987).

In all the two cases, affected leaves show chlorosis and in severe attack become pre-maturely senescent. Lesions (spots) can coalesce, give the leaves a ragged and blight appearance. Profuse sporulation occurs on the affected portion of mature lesions in case of *Alternaria alternata* (Narain *et al.*, 1987).

Descriptions of morphological characters of both the causative fungi have already been given in case of sunflower causing leaf spots and blight (Figures 18.1 and 18.3).

Linseed

Symptoms

All the above ground parts of the plants are attacked by the pathogen(s). The first symptoms appear on cotyledons as black lesions. Then the lower leaves get infected and small dark brown spots appear which later increase in size and pass over to stems and floral parts. In case of severe infection, the buds fail to open. If the infection takes place after fertilization, the development of ovule is normal but seeds become undersize, shriveled and capsules give burnt appearance.

The species of *Alternaria viz., A. lini* Dey [syn. *A. alternata* (Fr.) Keissler] was first reported from Kanpur (U.P.), India (Dey, 1933).

A. linicola Groves and Skolko was first reported by Groves and Skolko (1944), subsequently by Nergaard in 1945. It is more pathogenic and common on linseed causing leaf spot. Narainn and Koul (1982) reported *A. linicola* for the fist time from India on *Linum grandiflorum* L. Garg (1982) has also mentioned about infection of *A. linicola* in the cotyledons of *Linum usitatissimum* L. (linseed).

Pathogen

Alternaria linicola Groves and Skolko (*Can. J. Res., Seet, C.* 12: 217-234)

Conidiophores deep olive buff to dark olive buff, septate, 15-75 µm long and 6-8 µm wide, unbranched (simple), erect, often geniculate with one or more scars, formed singly or in small branches in culture; *Conidia* normally formed singly, conical to ellipsoid to obclavate, hardly constricted at septa, 22.5-130 µm in length and 7.5 to 28.5 µm in width with 4-15 cross and 0-4 longitudinal septa; gradually tapering into a long filiform often bifurcated *beak* which is 16.5-110 µm in length and 3-4.5 µm in width with 0-3 transverse septa (Figure 18.9).

Description of this species exactly resembles with those described by Neergaard (1945) and Singh (2007) on linseed (*Linum usitatissimum*) and Narain and Koul (1982) on *Linum grandiflorum* causing leaf spots on both the hosts.

Sesame

Symptoms

The fungus attacks seedlings, stems of young plants, leaves and pods. In the leaves the spots are brown, round to irregular in shape varying from 1 mm to 8 mm in diameter. The spots later become darker in colour with concentric zonations demarcated with brown lines inside the spots on the upper surface. In severe infections, several spots coalesce together involving a major portion of the leaf blade and the affected leaves dry and usually drop off (Mohanty and Behera, 1958).

Pathogen

Alternaria sesami (Kawamura) Mohanty and Behera, (*Curr. Sci.*, 27: 493, 1958)

Conidophores arising singly or in small groups, usually simple, straight or flexuous, almost cylindrical, septate rather pale brown or yellowish brown, *Conidia* solitary or sometimes in chains of 2, straight or slightly curved, obclavate or ellipsoidal, tapering to the *Beak* which is usually the same length or upto twice as long as the spore body, with 6-11 cross and 0-8 longitudinal or oblique septa, *Beak* simple or branched, 2-4 µm thick and 24-210 µm in length (Figure 18.11).

Safflower

Symptoms

In the beginning the spots appear as minute spots scattered on the leaf surface. Later, they increase in size, become brown in colour, about 1 cm in diameter; concentric zonations on spots are also deserved. The spots frequently coalesce into large, irregular lesions. Infected flower buds often shrivel without opening (Narain *et al.*, 1988).

Pathogen

Alternaria carthami **Chowdhury, (*J. Indian Bot. Soc.*, 23: 65, 1944)**

Conidiophores erect, simple or flexuous sometimes geniculate, septate, brown or olivaceous brown, upto 90µm long, 5-8µm thick, *Conidia* solitary or in very short chains, straight or curved, obclavvate, rostrate, 60-110µm long, 15-25µm thick with 7-11 cross and up to 7 longitudinal septa, *Beak* 25-160µm long, 4-6 µm thick at the base, tapering to 2-3µm with upto 5 transverse septa (Chowdhary, 1944) (Figure 18.8).

Exceptionally in *A. carthami,* the conidia instead of tapering into true beaks form a few to innumerable chlamydospores. The beak length greatly increase (upto 385µm) due to the formation of a quite longer continuous chain of chlamydospores (Narain, 1982).

Castor

Symptoms

The symptoms of disease initially appear as minute dark brown lesions which start form margin and centre of leaves and at later stage of disease development, a number of spots coalesce to form the irregular dark brown patches covering larger parts of foliage, often zonate and surrounded by yellow halo. The disease appears throughout the crop life *viz.*, seedlings, foliage, inflorescence and capsules also (Pawar and Patel, 1957).

Rainfall, atmospheric temperature and R.H. play important role in the disease development. Average temperature 30.5 to 33.5°C (max.) and 25.5 to 26.5°C (min.), continuous rainfall for several days and average R.H. of 81.0 to 92.23 per cent (max.) and 70.26 to 87.33 per cent (min.) have been found congenial for infection and disease development (Singh *et al.*, 1994).

Pathogen

Alternaria ricini **(Yoshii) Hansford (*Proc. Linn. Soc. Lond.*, 1942-43: 52, 1943)**

Colonies when on leaves amphigenous. *Conidiophores* arising singly or in groups, erect, simple, straight or flexuous, almost cylindrical or rather thicker towards the base, septate, rather pale, brown, smooth, with 1 or a few conidial scars, up to 80µm long,, 5-9µm thick. *Conidia* solitary or occasionally in chains of 2, straight or curved, obclavate or with the body of the conidium ellipsoidal, tapering rather abruptly to a very narrow beak which is equal in length to or up to twice as long as the spore body, pale to mid golden brown or reddish brown, smooth, with 5-10 transverse and several longitudinal or oblique septa, sometimes constricted at the septa, overall length 70-170 (140) µm, 13-27 (19) µm thick in the broadest part; *Beak* simple, pale, 1-2.5µm thick along most of its length (Figure 18.10).

REFERENCES

Anil Kumar, T.B., Urs, S.D., Seshadri, V.S. and Hegde, R.K. (1974). Alternaria leaf spot of sunflower. *Curr. Sci.*, **43**: 93-94.

Atkinson, R.G. (1950). Studies on parasitism and variation of *Alternaria raphani*. *Can. J. Res. Sect. C.*, **28**: 288-317.

Bose, A.B. (1942). *Alternaria* on leaves of sunflower in India. *J. Indian bot. Soc.*, **21**: 179-184.

Chahal, A.S., Kang, M.S. and Chauhan, J.S. (1977). Alternaria blight: A serious disease of rapeseed and mustard. *Progressive Farming*, **12**: 21-22.

Chand, Gireesh, Narain, U., Kumar, M. and Verma, Shilpi (2007). Symptomatology, etiology and eco-friendly management of Alternaria leaf spots and blight of broccoli. In: *Eco-friendly management of plant diseases*. Eds. Ahamad, Shahid and Narain, U., Daya Publishing House, Delhi, pp. 461-472.

Changsiri, W. and Weber, G.P. (1963). Three *Alternaria* spp. pathogenic on certain cultivated crucifers. *Phytopathology*, **53**: 643-648.

Chowdhury, S. (1944). An Alternaria disease of safflower. *J. India bot. Soc.*, **23**: 65.

Dey, P.K. (1933). An Alternaria blight of linseed plant. *Indian J. Agric. Sci.*, **3**: 881-896.

Elliott, J.A. (1917). Taxonomic characters of the genera *Alternaria* and *Macrosporium*. *Am. J. Bot.*, **4**: 439-476.

Ellis, M.B. (1968a). *Alternaria brassicae*. Commonwealth Mycological Institute. Description of pathogenic fungi and bacteria, No. 162, CMI, Kew, England.

Ellis, M.B. (1968b). *Alternaria brassicae*. Commonwealth Mycological Institute. Description of pathogenic fungi and bacteria, No. 163, CMI, Kew, England.

Ellis, M.B. (1971). *Dematiaceous Hyphomycetes*. Commonwealth Mycological Institute, Kew, England, 464-497.

Garg, S.K. (1982). Studies on leaf spot and black point disease of linseed caused by *Alternaria* sp. Ph.D. Thesis, C.S.A.U. of Agric. and Tech., Kanpur, 165 pp.

Ghewande, M.P., Nandagopla, V. and Reddy, P.S. (1987). *Plant protection in groundnut*. Tech. Bull., National Research Centre for Ground nut, Junagarh, Gujarat, p.7.

Groves, J.W. and Skolko, A.J. (1944). Notes on seed borne fungi II. *Alternaria*. *Can. J. Res. C.*, **22**: 217-234.

Gupta, D.K. and Basuchaudhary, K.C. (1992). Occurrence and prevalence of *Altrenaria* species in crucifers grown in Sikkim. *Indian J. Hill Farming*, **5**: 129-131.

Hansdford, C.G. (1943). Contributions towards the fungus flora of Uganda. V. Fungi Imperfecti. *Proc. Linn. Soc. Lond.*, **1**: 34-37.

Hedjaroude, G.A. (1973). Brown spot of sunflower (*Helianlhus annuus* L.) in Iran. *Phytopath. Zeitschrif*, **78**: 274-277.

Joly, P. (1964). *Le Genre Alternaria*. Editions Paul Lechevalier, Paris, 150 pp.

Kadian, A.K. and Saharan, G.S. (1983). Symptomatology, host-range and assessment of field loss due to *Alternaria brassicae* infection.

Kadian, A.K. and Saharan, G.S. (1984). Studies on spore germination infection of *Alternaria brasicae* of rapeseed and mustard. *J. Oilseed Res.,* **1:** 183-188.

Khalid, Abdul, Akram, Mohammad, Narain, U. and Srivastava, M. (2004). Characterization of *Alternaria* spp. associated with Brassicaceous vegetables. *Farm Sci. J.,* **13:** 195-196.

Martens, J.W., Ravagon, G. and McDonald, W.C. (1970). Diseases of sunflower in Kenya. *E. Afr. Agric. For. J.,* **35:** 389-395.

McDonald, W.C. and Martens, J.W. (1963). Leaf spot of sunflower caused by *Alternaria zinniae, Phytopathology,* **53:** 93-96.

Middleton, K.J. (1971). Sunflower diseases in South Queensland. *Qd. Agric. J.,* **97:** 597-600.

Misra, B. and Prakash, Om (1975). Alternaria leaf spot of soybean from India. *Indian J. Mycol. Pl. Pathol.,* **5:** 95.

Monanty, N.N. and Behera, B.C. (1958). Blight of sesame (*Sesamum orientale* L.) caused by *Alternaria sesami* (Kawamura) n. comb. *Curr. Sci.,* **27:** 492-493.

Mukewar, P.M., Lambat, A.K., Nath, R., Majumdar, Indira Rani and Chandra, K.J. (1974). Blight disease of sunflower casued by *Alternaria helianthi. Curr. Sci.,* **43:**.

Narain, U. (1982). Chlamycospore formation in conidial beak of *Alternata carthami. Indian Phytopath.,* **35**: 172-173.

Narain, U. (1986). Studies on *Alternata* spp. associated with Crucifers in India. *Adv. Biol. Res.,* **4** (1 and 2): 187-191.

Narain, U. and Chauhan, L.S. (1981). Leaf spot of sunflower caused by *Altenaria tenuissima. F.A.O. Plant Prot. Bull.,* **29**: 29.

Narain, U. and Kant, S. (2008). Diversity of species and parasitism in the genus *Alternaria. Applied Botany Abstracts,* **28:** 272-281.

Narain, U. and Kaul, A.K. (1982). Alternaira leaf spot of *Linum grandiflorum* from India. *Indian J. Mycol. Pl. Pathol.,* **12:** 343-344.

Narain, U. and Saksana, H.K. (1975). A new leaf spot of turnip. *Indian Phytopath.,* **28**: 98-100.

Narain, U. and Saksena, H.K. (1973). Occurrence of Alternaria leaf spot of sunflower in India. *Indian J. Mycol. Pl. Pathol.,* **3**: 115-116.

Narain, U. and Srivastava, M. (1996). Taxonomy and parasitism of *Alternaria* spp. occurring on sunflower in India. *Ann. Pl. Protec. Sci.,* **4** (2): 95-98.

Narain, U., Banerjee, A.K., Singh, Mohit and Swarup, J. (1988). *Alternaria* species associated with Compositae plants. *Farm Sci. J.,* **3:** 164-169.

Narain, U., Chauhan, L.S. and Swarup, J. (1987). Occurrence of two foliar diseases of groundnut – new to Uttar Pradesh. *Farm Sci. J.,* **2** (2): 202-203.

Narain, U., Kant, S. and Chand, Gireesh (2016). Characterization of species and parasitism in the genus *Alternaria*. In: *Crop diseases and their management: Integrated approaches.* Apple Academic Press Canada, pp. 385-404.

Nattras, R.M. (1950). Annual report of the Senior plant pathologist, 1948. Dept. Agr. Kenya, 1948: 95-98 (R.A.M., **30**: 308-309).

Neergaard, P. (1945). *Danish species of Alternaria and Stemphylium*: Taxonoly, Parasitism and economic significance. Oxford University Press. London, 560 pp.

Pape, H. (1942). *Alternaria zinniae* n. sp. *Ang. Bot.*, **24:** 61-79.

Pawar, U.H. and Patel, M.K. (1957). Alternaria leaf spot of *Ricinus communis* L. *Indian Phytopath.*, **10:** 110-114.

Prasad, R. and Narain, U. (2007). Integrated management of Alternaria blight of rapeseed and mustard: An overview. In: *Eco-friendly management of plant diseases.* Eds. Ahamad, Shahid and Narain, U. Daya Publishing House, Delhi, pp. 201-2014.

Rao, B.R. (1977). Species of *Alternaria* on some Cruciferae. *Geobios*, **4:** 163-166.

Rotem, J. (1994). The genus *Alternaria*: biology, epidemiology and pathogenicity. *American Phytopath. Society*, St. Paul, MN.

Sackston, W.E. (1957). Diseases of sunflower in Uruguay. *Pl. Dis. Reptr.*, **41:** 885-889.

Saharan, G.S., Mehta, N. and Sangwan, M.S. (2005). *Diseases of oilseed crops of India.* Indus Publishing Co., New Delhi, 673 pp.

Sangwan, M.S. and Mehta, Naresh (2007). Eco-friendly management of diseases of rapeseed and mustard. In: *Eco-friendly management of plant diseases,* Eds. Ahamad, Shahid and Narain, U., Daya Publishing House, Delhi, pp. 416-427.

Sangwan, M.S., Mehta, N. and Gandhi, S.K. (2002). Some pathological studies on *Alternaria raphani* causing leaf spot and pod blight of radish. *J. Mycol. Pl. Pathol.*, **32**: 125-126.

Simmons, *e.g.* (1967). Typification of *Altenaria, Stemphylium* and *Ulocladium. Mycologia,* **59:** 73.

Singh, J. (2007). Alternaria blight of linseed (*Linum usitatissimum* L): An overview. In: *Eco-friendly management of plant diseases.* Eds. Ahamad, Shahid and Narain, U. Daya Publishing House, Delhi, 222-230.

Singh, S.K., Narain, U. and Singh, Mohit (1994). Effect of rainfall, atmospheric temperature, R.H. on Altrnaria leaf spot of castor. *Ann. Pl. Prot. Sci.,* **2:** 81-82.

Takano, K. (1963). A disease of sunflower caused by a species *Alternaria. Ann. Phytopath. Soc. Japan,* **28:** 300.

Tubaki, K. and Nishihara, N. (1969). *Alternaria helianthi* (Hansf.). Comb. nov., *Trans. Br. mycol. Soc.,* **53:** 147-149.

Verma, P.R. and Saharan, G.S. (1994). Monograph on *Alternaria* diseases of crucifers. *Research Branch Technical Bulletin 1994,* Saskatoon Research Centre, Canada, 95 pp.

Walker, J.C. (1952). Disease of crucifers. In: *Diseases of vegetable crops*. Mc-Graw Hill Book Co., New York, **6**: 150-152.

Wiltshire, S.P. (1954). New records of sunflower diseases. *F.A.O. Plant Prot. Bull.*, **2:** 188.

Zelle, M.A. (1932). Sunflower diseases. Pan-Soviet State Assoc. Control of pests and diseases in Agric. Sylvic. Leningrad, Publ. 6, 32 p. (in Russian) *R.A.M.*, **12**, 570-571.

Transformation of Indian Agriculture through Innovative Technologies *Pages* **297–310**
Editor: **Dr. Shahid Ahamad & Dr. Jag Paul Sharma**
Published by: **ASTRAL INTERNATIONAL PVT. LTD., NEW DELHI**

19 Entrepreneurship Development through Modern Vegetable Production Technologies for Growers of Mid Hill Regions of J&K

Anil Bhushan and J.P. Sharma

Introduction

India is the 2nd largest producer of the vegetables in the world after China producing about 162.90 million tons vegetable in an area of 9396'000 ha in the year 2013-14 (NHB, 2015). It has made tremendous progress in terms of production since 1950'sespecially during the post green revolution period. However, the productivity level is still low in comparison to other developed countries and still it is difficult to provide requisite quantity of vegetables per capita per day. At present, the per capita availability is approximately 200 g per day as against the requirement of 300 g/capita/day recommended for balanced diet. Development of improved vegetable varieties/hybrids/technologies through systematic research coupled with their adoption by the farmers and developmental policies of the government culminated in tremendous increase in area, production and productivity in the country. Compared to area (2.84 million ha), production (16.5 million t) and productivity (5.8 t/ha) in 1950-1951, there had been phenomenal increase in area (2.99 folds), production (8.88 folds) and productivity (2.98 folds) of vegetables in our country during the last 6 decades (Vanitha *et al.*, 2013). The production and marketing of vegetables crops are undergoing continuous change globally. This is mainly due to the growing demands of consumers for safe and healthy vegetables, increased urbanization of societies, and the growth in scale and influence of super- markets chains. On national scale, J&K ranks very low in terms of area and production of vegetable crops with an estimated area of 355.5'000 ha and 2073.9 '000 MT production (NHB, 2015) out of which Jammu province accounts for 31795 ha area and 686498.56 MT production (Anonymous, 2015). However, high average productivity of many important vegetables like brinjal, chillies, okra *etc.* have as compared to national average has been reported. Rapid growth in mean per capita incomes of the emerging educated middle class in developing countries has enabled consumers to purchase a broader range of relatively expensive vegetable commodities such as off-season produce, new or exotic vegetables and organic produce resulting in the expansion of the consumption and domestic vegetable markets especially in urban and peri-urban areas (Dias, 2014).The state of Jammu and Kashmir is divided into three main agro-climatic zones *viz.*, subtropical plains, intermediate hills and temperate zone. The intermediate zone includes lower to mid altitude hilly regions of Rajouri, Poonch, Bhaderwah, Kishtwar, Ramban, Reasi and some parts of Kathua and Udhampur districts.. The diverse nature of agro-climate of state of Jammu and Kashmir presents not only a great challenge but also a huge opportunity for agriculture in general and vegetable production in particular (Bhushan, 2011).It is necessary to explore all possible ways of increasing the sustainable productivity and carrying capacity of the farming systems in this region in order to improve the livelihoods of marginal mountain households (Partap, 1999).

Current Scenario of Agriculture in Mid Hills of J&K

Agriculture is the basis of the livelihood of over 80 percent of the rural population in the countries of the Hindu Kush-Himalayan (HKH) region. However, more than 90 percent of the farmers in the hill and mountain areas are marginal or small land-holding families, cultivating less than one hectare of land each (Koirala and Thapa, 1997; Partap, 1999).Cultivation of subsistence cereal crops is the main occupation of the people of the intermediate zone of Jammu and Kashmir state, which is based upon the centuries old traditional practices and carried out on the narrow patches of the terraced fields. The economic viability and capabilities, for smooth running of livelihood, of these crops are considerably insufficient even to meet the daily food requirement. Farming systems, which include the cultivation of agricultural crops, horticultural crops and rearing of livestock is traditional. Cultivation of these crops is carried out mostly in the mid-slopes and to a certain extent in the valley regions, where concentration of population is high. Under the traditional farming systems, the main crops grown are maize, paddy, wheat, and horticultural crops in which livestock play an integral role. Possibility of using modern innovation applicable in the plain areas is negligible due to smallholdings and undulating and precipitous slopes. Even the seeds sown in the cropped lands are not improved. Thus, they negatively influenced the production and per ha yield of crops. Manure is used as a fertilizer, which is carried from the cowsheds to the cropped land by the women. Though, manure is highly fertile but its proportion in the cropped land is very less, which is not enough to get sufficient production of crops in a season. Furthermore, the upper layer of soil is washed away due to monsoon rain; thus, the soil fertility remains considerably low. Agriculture is based upon the monsoon rainfall, as 90 per cent cropped land is rain fed. The possibility of using various means of irrigation is very less except in the few places of the valley regions. Two main reasons are behind with this: (i) the slope of the land is very steep and its fragility is high and (ii) the mid-slopes and uplands obtain high proportion of humidity throughout the year, therefore, the requirement of irrigation for cropped land is less. Crops are generally sown by broadcasting/ spreading method. Horticultural production is low to moderate in terms of cultivation of fruits and vegetables and is consumed domestically. However, the agro-ecological conditions of the region are quite suitable for growing these crops. Under the traditional system of horticultural farming, fruits and vegetables are grown in the fallow spaces near the home as kitchen garden. Commercial vegetable cultivation is undertaken in a very few pockets/areas. The major questions are: (i) whether the traditional farming system is enough to provide a base for sustainable livelihood? (ii) How far the modern innovations can be implemented in the field of agriculture and (iii) what are the potentials of cultivation of vegetables as an economic enterprise? Remoteness, inaccessibility, fragility of terrain, erosion of soil and consequently infertility, and frequent environmental risks in mountains resulted in comparatively low production and productivity. Technological innovations such as chemical fertilizers, pesticides and high yield crop varieties which transformed low land agricultural areas, could not change the farming system to a great extent, on account of specific constraints of the regions. Dependence on forests for maintaining soil fertility in croplands or expansion of agricultural land itself thus was not substituted by the new technologies. Mountain areas are highly diverse in renewable natural resources and environmental services. This diversity is helping reduce internal competition in mountain areas and partially offsetting the physical vulnerability of the fragile mountain environment. The mountains' biodiversity provides important values to agriculture, medicine, food security, and industry, besides spiritual, cultural, aesthetic, and recreational values. Diverse landscapes and climatic variation throughout the region result in rich biodiversity, fostering multiple livelihood options. These niche-based mountain products and services hold some of the keys for helping mountain people, diversity and enhance their livelihood options while reducing pressure on traditional occupation and the environment. This, however, cannot be done by emphasizing the cultivation of cereal crops alone. If the poor mountain farmers are going to compete favourably in the modern world, they must be given options and alternatives that are not already captured by the competition. Development efforts tend to focus on exploring farming approaches to increase the productivity and carrying capacity of farms (Partap, 1999). Cash crops farming–fruit and vegetable crops suitable to specific agro-climatic conditions – is one comparative advantage that can be exploited by these farmers.

Components of Modern Vegetable Technologies

There are so many means and ways to go for developing a sustainable model of entrepreneurship by adapting modern vegetable technologies. These technologies can be the guiding force for establishing a viable commercial vegetable production venture to the rural youth and women in the hilly areas. Main components of modern agriculture technologies are:

Table 19.1. Predominant Cropping System and Districts Covered under Intermediate Zone of J&K

Irrigated	Rice based (including wheat/fodder/ vegetables/Ginger/Turmeric/pulses/ oilseeds growing)	Areas of Ramban, Doda, Kishtwar, Kathua, Reasi, Udhampur, Rajouri, Poonch
Unirrigated	Maize based (including wheat/fodder/ vegetables/pulses/oilseeds growing)	Areas of Ramban, Doda, Kishtwar, Kathua, Reasi, Udhampur, Rajouri, Poonch

Source: Deparment of Agriculture, 2014-15.

1. Crop Diversification for Enhancing Productivity and Production

Over the years the traditional farming system model has lost its relevance not only in terms of sustaining productivity as well as production but also due to deteriorating soil health and thus making it unsustainable in terms of ecology and profitability. Therefore, crop diversification is being advocated by the scientists, researchers and policy makers working in these areas. Many new options which are complimentary to their traditional farming system have been encouraged and refined catering to their local conditions and needs.

Vegetable cultivation is one such option which is highly suitable and most profitable avenue for the growers of hill regions. The farmers can take vegetable cultivation on commercial scale for increasing their socio economic status for the following reasons:

I. The agro-climate of intermediate hills is most suited for cultivation of most of the vegetables. The summer season as well as temperate zone vegetables can be easily grown as main season and offseason vegetables.

II. Two basic requirements for successful vegetable cultivation are water and farmyard manure which are found abundant in these areas coupled with most conducive environments make the hilly zone as most appropriate for making vegetable growing as important component in crop diversification.

III. Vegetable are short duration crops and easily fit in cropping system in between major cereal crops.

IV. In horticultural orchards, vegetables can be grown as intercrops and thus utilizing the space and earn additional revenue especially during early establishment period of the fruit orchard.

V. Vegetables are high yielding as compared to field crops and more crops can be taken in a season or year.

VI. Vegetables are source of vitamins, minerals and proteins besides being source of various neutraceuticals and other phytochemicals essential for humans and thus reduce the risk of various deficiency diseases of people of hilly regions especially women and children and provide nutritional security.

VII. The mild agro-climates of hilly regions produce high quality produce free of diseases and pests and thus can fetch premium price in the market.

2. Vegetable Nursery Raising: An Enterprise for Rural Youth and Women in Hills

Quality seedling is pre-requisite for obtaing higher vegetable production in the fields. Scientific nursery raising is one such area in vegetable cultivation which can be taken up in a limited area and time but has potential to provide income to the rural youth and women. Due to the prolonged winters and heavy rainfall during monsoons, the production of seedlings for summer as well as winter season crops becomes cumbersome and farmer has to delay the sowing for 15-30 days until climate improves. This delay proves to be costly affair most of the times as tapping of early market becomes very difficult. Therefore, scientific nursery raising using plasticulture (poly house, low poly tunnels, poly bags, portrays *etc.*) can be taken up by the rural youth and women for selling the vegetable nursery for the local growers. The technologies are standardized and refined for the specific areas and agro-climates. An initiative in this direction from all the stake holders *i.e.* farmers, line departments, SAU's and even NGO's can improve the socioeconomic status of the people of the hilly regions.

The plug-tray nursery raising technology is better in comparison to the traditional methods/polybags due to the limitations of the existing traditional methods of nursery raising summarized as under:

I. It is nearly impossible to raise virus free nursery during rainy and post rainy season.

II. Development of soil borne fungus and nematodes mainly during high temperature period, which adversely affect crop growth.

III. Duration of nursery raising during winter is long (50-60 days) due to poor germination in low temperature.

IV. Since there is no competition among plants, therefore greater uniformity in seedlings is obtained under this system.

V. It is suitable for raising the nursery of sexually and asexually propagated vegetables and also of ornamental plants.

VI. Economizes hybrid vegetable nursery by raising by way of reducing the seed rate by 30-40 per cent compared to traditional system.

VII. Easy for transportation after packing for long distances.

VIII. Nursery of vegetables like cucurbits is not at all possible under traditional system.

IX. Off season nursery cannot be raised due to abiotic factors

3. Protected Vegetable Cultivation

Protected cultivation on commercial scale is undertaken in over 50 countries across the globe. In the past half century, tremendous progress has been made both in popularizing and improving greenhouse or protected cultivation technology. Protected cultivation, which encompasses glasshouse, greenhouse, polyhouse, low tunnel, high tunnel, row cover, shade house, screen house, net house and plastic mulch *etc.*, will face an increasing competition of the produce and therefore, now-onwards, crop diversification (product innovation), reduction in production cost, less dependence on energy, yield increase, product safety and minimal environment pollution will be the major issues, which need immediate attention of the green house growers for their long term sustainability in the domestic and foreign markets. Under protected cultivation, higher productivity with high quality in shorter cycle leads to better farm incomes to improve rural livelihoods (Choudhary, 2016).

The protected cultivation of high value crops has become irreplaceable both from economic and environment points of view. It offers several advantages to grow high value crops with improved quality even under unfavourable and marginal environments. However, due to high training needs of the green house growers and trainers especially in a few developing countries, poor quality produce with pesticide residues has been a matter of great concern. These issues can easily be addressed by integrating various production and protection practices including location specific efficient designing and construction of the polyhouses. Therefore, adoption of systematic approach based on the involvement of greenhouse engineers, breeders, crop production and protection scientists and post-harvest technologists will help in increasing the sustainability and relevance of protected cultivation across the globe. Creating awareness among the greenhouse growers about safe food production and judicious use of safe pesticides can be instrumental in providing quality products without polluting the environment.

Greenhouse vegetable production has tremendous scope in hilly regions of the country because of continuous increase in availability of up-markets and liking of consumers towards off-season and high quality produce of different horticultural crops. Greenhouses have tremendous potential in increasing production and productivity of vegetables like normal tomato, cherry tomato, colored sweet peppers, green sweet peppers, slicing cucumbers, muskmelon, summer squash including some rare vegetables even under adverse agro-climatic conditions. There is immense utility of greenhouses and portable low tunnels for raising virus free healthy seedlings of various vegetables during rainy and post rainy season and off-season nursery raising of cucurbits during winter months for out of season production of these crops. The design of the polyhouse (as per location) and its construction by reputed companies is the major factor for successful growing of various potential crops in different geographical conditions. The prevailing climate in hilly regions of North Himalayas is very cold in winter and hot in summer and rainy season. The design of the polyhouse should be such that it meets out both the requirements of providing very good ventilation during summer and rainy season and conservation of heat energy during winters.

Major Advantages of this Technology

I. Raising of vegetable nursery or transplants under protection.

II. The potential of polyhouse production technology to meet the demand of producing good nutrition and healthy foods and quality vegetables free from pesticides can be fully exploited.

III. Vegetable crops can be grown under adverse weather conditions round the year and off-season.

IV. The vegetables can be produced with higher productivity and uniform quality of produce than open field cultivation.

V. Management and control of insect-pests, diseases and weeds is easier.

VI. There is efficient resource management.

VII. In the hilly terrains, the farmers generally have small land holdings and this technology provided a useful impetus to their farming livelihood by more productivity and more money from less land.

Future Needs

Polyhouse vegetable production in the country is still in infancy and for its rapid commercialization, there is urgent need to redress the following issues related to this technology:

I. Standardizing proper design of construction of polyhouses including cost effective and indigenously available cladding and glazing material.

II. Developing cost effective agro-techniques for growing of different vegetable crops in the different types of polyhouses and lowering energy costs of the green house environment management.

III. Major research activities on growing of vegetables under protected covers should be launched by ICAR and SAU's.

IV. Import of planting materials, structural designs and production technologies which are not relevant under Indian conditions should be stopped and in turn emphasis should be given to develop own F_1 hybrid varieties so that seed are made available to the growers in time and at cheaper rates.

4. Vegetable Seed Production: A Viable Option for Improving Socio-economic Status in Hills

Vegetables occupy an important place in diversification of agriculture and have played a pivotal role in nutritional security. With the changing paradigms of food and nutritional securities, the consumption of vegetables have attained tremendous importance. To meet the ever increasing demand of burgeoning Indian population, production and productivity of vegetables has to be increased manifolds. Due to increasing pressure on land through urbanization and industrialization, it is not feasible to increase the area under vegetables commensurate to our requirements hence the preciousness of high quality vegetable seeds becomes much more significant than it has ever been to increase the yield per unit area. Hybrid vegetable technology has made significant impact on most vegetable crops but a major limitation to vegetable production is unavailability of high quality seed. Although use of quality seeds of improved varieties of different vegetable crops has witnessed tremendous growth in vegetable production and productivity, however, the availability of quality seeds in time and at affordable price is still a matter of great concern (Singh *et al.*, 2010). In this direction, the congenial mild climate in hills during summer months with maximum temperature ranging between 30-35°C is single most important factor for taking up vegetable seed production as a viable option in hills. There is at least a gap of 10°C when compared with summer peaks in Jammu plains. Moreover, extended winter season allows the winter crops grown for seed production to mature properly and attain full seed size without facing the difficulties of pod shattering and shrivelled seed due to sudden rise in temperature. Therefore, there is a huge advantage in terms of climate for seed production of temperate vegetables especially Cole crops *i.e.* cauliflower, knolkhol, broccoli and temperate radish and carrot. Valley areas and cool temperate pockets need to be developed for seed production of cabbage, snowball cauliflower and brussel sprout. Besides this there are many vegetables of temperate nature, which require chilling temperature to get transformed into reproductive stage from vegetative stage, seed of such vegetables can be successfully produced in the hilly areas due to natural low temperature prevailing in such areas. Besides favourable climate, there are many other factors like good quality soils, natural Isolation distance, natural pollinators, natural water resources like rivers, nallahs, springs *etc.*, less attack by the insect pests and diseases and availability of cheap labour

The seed production of winter as well as summer vegetables would widen the scope for expanding vegetable seed industry not for indigenous consumption only but also for export to European and Western countries where seed production is becoming a costly affair day by day with the increase in costs of labour. This profitable venture shall not only lead to cheaper cost of seed production but shall also provide employment opportunities to the hilly people. Identification of potential pockets with suitable climate conditions in these areas will also overcome the problem of many diseases and insect pests which otherwise have become quite predominant due to continuous growing of vegetable seed crops in the same tract/ land.

5. Off-Season Vegetable Cultivation

Off-season vegetable cultivation is defined as production of fresh vegetable after or before their normal season by

- ☆ Availing and using different agro climatic conditions
- ☆ Adjusting the planting time
- ☆ Selecting and improving the varieties
- ☆ Creating the controlled environment by making plastic tunnels, polythene house,

The objective is to produce and supply vegetables to the market during their lean period of supply. Off-season vegetable production and marketing is the most profitable farm business giving very high production and income to farmers per unit area of land. These off-season vegetables have a definite market advantage and provide assured better returns to the farmers.

In order to promote this enterprise, niche areas for off-season vegetable cultivation need to be identified and efforts to tap irrigation potential in those areas should be enhanced. Education of farmers for scientific management of crops and provision of improved tools for efficient use of labour have also been undertaken to lower production costs and make the vegetable cultivation more beneficial to farmers, particularly to the small and marginal farmers in the mid hill regions of the state.

Suitability of Hills for Off-Season Production

I. Wide range of climatic and topographical situations. There is -12°C temperature for every rise of 1000m, making possible variation in growing periods of off-season vegetable crops.

II. The peculiar classification of agro-climatic zones on the basis of topography, climate, rainfall *etc. viz.* sub-tropical zone(plains), sub-temperate zone(Midhills), temperate, wet and cold and dry temperate Zone (high hills) is not possible in plains.

III. Act as natural green house of the country for the uninterrupted production of vegetables, which become off-season for the neighbouring plains.

IV. Planting season is short in high snow bound hills (April –Oct) and suitable planting period in different regions are ascertained accordingly. Vegetables grown are short duration in nature *e.g.* Radish variety White Icicle, French Break Fast *etc.* get ready in 30-40 days, Pea varieties Arkel in 60-65 days and Azad Pea in 60-70 days. Due to mild climate 2-3 or more crops in a year can be grown at same piece of land and better fit in cropping system.

V. Due to cool climate maturity is delayed in hills enabling easy post harvest operations and methods of harvesting.

VI. Soils are fertile, shallow, rich in organic matter, potassium, slightly acidic, well drained and free from any residual effect of chemical fertilizers.

VII. The hill tops, forest areas and rivers catchment areas provides natural barriers between valley, pockets or off-season growing areas for obtaining better quality uniform produce with high yields.

VIII. Pollinating agents (pollinators) especially bees and bee flora is available in abundance naturally in the hills which play significant role in crosspollination and provide scope for adopting apiculture as an ancillary activity.

IX. Irrigation water for quality production of vegetables is available in plenty from various sources *viz.* perennial rivers and their tributaries, springs, kuhl, tube wells and well distributed rainfall/snowfall *etc.*

X. Due to chill climate some pest and disease attack less on the crop thus, saving the expenditure on chemical and harvest residue free product.

XI. In hills the women number exceeds men and they are engaged in all types of agriculture work and play significant role from land preparation up to marketing of vegetable produce.

XII. Labour needed from time to time is easily available throughout the year to carry out various operations. Off-season vegetable produce higher yields as compared to cereals and cost of production is comparatively less.

XIII. Various developmental schemes, both long and short terms funded by State Govt. or Govt. of India, operating in hill states facilitate adoption of offseason vegetable cultivation by providing assistance to the farmers. Schemes like Horticulture Technology Mission, Rashtriya Krishi Vikas Yojna, National Horticulture Boards, NABARD provide funding at nominal interest rates and subsidy at higher rates in hilly States.

XIV. Rashtriya Krishi Bima Yojna is also extended to hilly states to provide insurance cover to vegetables from natural calamities.

XV. The transportation, marketing and distribution system is rapidly developing with the road connectivity and establishment of vegetable mandis by the state government in concerned districts for setting up of terminal market network in hilly regions of the state.

Advantages

I. The farmers can learn specific techniques of vegetable production.

II. The farmers can develop confidence and make vegetable production as their main profession.

III. The farmers can get higher profits from off-season production.

IV. The consumers can get fresh vegetables during off-season.

V. Sometimes it is possible to export fresh vegetables and earn foreign currency.

VI. It creates employment for farm labourers all the year around.

Disadvantages

I. It requires highly specialized techniques of vegetable production.

II. Sometimes it becomes a highly risky operation due to incidence of pests and diseases.

III. It is possible on commercial scale only in areas where marketing is not a problem.

IV. It needs regular supervision, support and facilitation from concerned government agencies.

6. Contract Farming: A Potential Yet Unrealized in Hilly Regions

The production of vegetables as commercial enterprise on larger areas demand not only superior seed, modern scientific production technologies but a viable demand and supply chain which ensure the remunerative returns for the grower by timely disposal of his produce in the market at premium prices. In this direction, contract farming has emerged as one among several possible interfaces between vegetable cultivation and its processing and consumption. It is a powerful means of introduction of new crops or farm technologies, especially when both marketing and production uncertainties predominate. Contract farming helps when markets do not exist or are underdeveloped; conversely, contracts diminish in importance with development of competitive markets. Contract farming works when specific quality requirements must be met. Contracts are effective when there is no zero-sum game (one party's gain at the expense of the other). They are ideal for a win-win situation, since they represent a natural mutual dependency. Other factors helping adherence to contracts include exclusiveness, provision of planting materials or inputs, and a strong, self-regulatory social systems among growers. Contract farming is an exciting way of marrying small-farm efficiency to scale economies of processing and marketing. It is an acceptable *via media* for corporate ownership and could be a boon to processing industry, if properly designed and implemented.

The concept of contract farming of vegetable crops is fast catching up and many states like Punjab, Maharashtra, Karnataka and Andra Pradesh have taken a lead in this direction. A list of private companies involved in contract farming of various vegetables in different states of the country is given in the Table 19.2.

Table 19.2. List of Vegetable Contract Farming Companies in India

Sl.No.	*State*	*Crop*	*Company*
1.	Maharashtra	Several vegetables and spices	Ion Exchange Enviro Farms Ltd. (IEEFL)
		Potato	Pepsico (India) Ltd., Mahindra Sulabh
2.	Karnataka	Gherkin	20 Pvt. Companies
		Caprica chillies	AVT Natural Products Ltd.
3.	Madya Pradesh	Several vegetables and spices	Ion Exchange Enviro Farms Ltd. (IEEFL)
		Garlic and White onion	M/s Garlico Industries Limited.
4.	Punjab	Tomato and Chilly	Nijjer Agro Foods Ltd.
		Green vegetables and exotic vegetables	Punjab Agro Foods Park Limited, a joint venture of Punjab Agro Export Corporation and IDMA, a corporate body.
5.	Tamil Nadu	Gherkins	M/s Mahindra Sulabh
6.	Chhattisgarh	Tomato	BEC Co.
7.	Haryana	Turmeric, Mentha	HAFED
8.	Andra Pradesh	Several vegetables	Horticulture Department
		Gherkins	Aduri Natural Products, ACE Agrotech, Copricorn Natural Products and Mahendra and Mahendra
9.	Orrisa	Vegetable seed	Odisha Seeds Production Corporation
		Tomato	M/s Bilati Odisha
10.	West Bengal	Potato	M/s FritoLay India Limited

Source: www.indiastat.com.

Many factors also limit the utility of contract farming. Thousands of contracts are needed for securing even modest quantities. The large, quality-indifferent Indian market for fresh commodities acts as a main roadblock, leading to a "Feast-or-Famine" syndrome; shortages cause supply contracts to be flouted, and surpluses cause gluts at the buyer's doorstep. This also causes problems in meeting stringent SPS requirements, which will become an increasingly important consideration. Legal procedures governing such contracts are seen to be one-sided, with defaults difficult to pin. Little if any recourse to courts or other legal avenues is seen as feasible. These reasons explain the relative absence of international players in this segment of the Indian economy. Contract farming in general helps reduce market risk associated with crop cultivation and is thus a possible instrument of credit deepening. The presence of a third party could lead to outsourcing some preliminaries of credit disbursal and help reduce the cost. The buyers' commitment to purchase substantially reduces the risk of default to the lender. Therefore, credit institutions would be interested in using contract farming as a means of improving credit disbursal and meeting the mandatory targets for priority sector lending. Yet their actual involvement in contract farming is limited because there are not many well-planned contract farming schemes with adequate safeguards for all concerned, backed by credible sponsors, and previous experience has not been all positive.

7. Women in Vegetable Production: Role and Importance

Rural Indian women are extensively involved in agricultural activities. However, the nature and extent of their involvement differs with the variations in agro-production systems. In the vegetable growing, Although the males are mostly taking decisions regarding which and how much area to be choose for vegetable growing, which type of vegetables or varieties to be cultivated? However, women perform a variety of tasks both in cultivation as well as marketing. They are the actual ones who raises nursery, plant seedlings in the field, don most of intercultural operations, do harvesting and picking and packaging the produce for marketing. Practices related to vegetable cultivation have now been considered as a family enterprise in which husband and wife participate to share work and pleasure both. Thus, it is expected that all decisions related to practice of vegetable cultivation are also taken jointly. Realizing the importance of ascertaining the nature and extent of intensity of role of farm women in decision making related to vegetable operations and looking to the major role that women play in rural area. As far as constraints faced by farm women in decision making on vegetable cultivation is concerned, male dominance, no knowledge about improved technology and lack of education were the top most constraints reported by majority of the farm women. Women's participation in the decision making process has a significant impact on their improved status and greater role in society (Musokotwane *et al.*, 2001). Their participation is potentially important to bring equality between women and men in order to achieve sustainable development. Regarding association between decision making and different attributes of farm women, eight variables *viz.* education, land holding, social participation, farm power, economic motivation, scientific orientation, market orientation, mass-media exposure were found to be significantly associated with decision making of farm women, whereas age, family size, material possession, innovativeness and extension contact had showed non- significant association with decision making of farm women (Chouhan *et al.*, 2014).

Conclusion

Vegetables play a major role in Indian Agriculture by providing food, nutritional and economic security and produce higher returns per unit of area and time. In addition, vegetables have higher productivity, shorter maturity cycle, high value and provide greater income to improve the livelihood of farmers of hilly regions. The hilly areas possess a vast potential and serve as a greenhouse for growing off season vegetables on commercial scale to bridge the season gap. Also to bridge the gap between demand and supply, more emphasis need to be given on extension of its cultivation in unexplored valleys/pockets in hilly areas of the country. Western Himalayan region has a great potential due to wide range of congenial agro-climatic conditions enabling the hill farmers to grow off season vegetables in one pocket or other in each zone, this besides maintaining the uninterrupted supply of off season vegetables having typical flavor, better taste, attractive color and appeal and better quality seed with lucrative returns to the farmers, also provide scope for producing exotic vegetables which are in great demand both in the domestic market as well as in neighbouring/Gulf countries. Modern vegetable production technologies like scientific nursery raising, use of plasiculture, micro irrigation, INM/IDM, hybrid technology, protected cultivation, new/exotic vegetable cultivation, seed production of temperate vegetables, reducing pre and post-harvest losses can go a long way in changing the face of hill agriculture from a cereal based subsistence farming to a diversified sustainable commercial enterprise capable of meeting the nutritional requirements of the whole country besides earning remunerative returns to the farmers of the hilly regions.

REFERENCES

Anonymous, (2015). Annual Area and Production Data. *Directorate of Agriculture*, J&K, Jammu.

Anil Bhushan, Susheel Sharma, Sanjeev Kumar and B. Singh (2013). Vegetable cultivation in mid hills of Jammu region: Problems, Prospects and Future strategies. In: *Hill Agriculture: Prospects, Constraints and Mitigations*. Edited by Shahid Ahamad. Daya Publishing House, New Delhi. pp. 130-144.

Choudhary, A.K. (2016). Scaling up of protected cultivation in Himachal Pradesh, India. *Current Science*. 111(2): 272-277.

Chouhan, K.S., Bisht, K., Raghuwanshi, S. and Singh, S.P. (2014).Farm women participation in vegetable cultivation. *Technofame*, 3(2): 19-25.

Deshpande, C.S. (2005). Contract farming as means for value added agriculture-Occasional paper-42. Department of Economic Analysis and Research. NBARD, Mumbai.

Dias, J.C.S. (2014). Guiding strategies for breeding vegetable cultivars. *Agricultural Sciences*, 5(1): 9-32.

Indian Horticulture Database (2015). Data on area, production and productivity on brinjal in India. National Horticulture Board, New Delhi.

Koirala, G.P.and Thapa, G.B. (1997). Food security challenges: where does Nepal stand? (Research Report Series no.36).Kathmandu: HMG Nepal and WINROCK International.

Musokotwane, R.; Siwale, R.M. and Nkhata, B. (2001).Gender awareness and sensitization in basic education. Paris: People's Action Forum UNESCO Basic Education Division.

Partap, T.(1999). Sustainable land management in marginal mountain areas of the Himalayan Region: In Mountain Research and Development. 19: 251-260.

Singh, P.M., Singh, B., Pandey, A.K. and Singh, R.(2010). Vegetable Seed Production - A Ready Reckoner. Technical Bulletin No.37, IIVR, Varanasi.

Vanitha, S.M., S.N.S. Chaurasia, P.M. Singh and Prakash S. Naik. 2013. Vegetable Statistics.Technical Bulletin No. 51, IIVR, Varanasi, p. 250.

Transformation of Indian Agriculture through Innovative Technologies *Pages 311–313*
Editor: **Dr. Shahid Ahamad & Dr. Jag Paul Sharma**
Published by: **ASTRAL INTERNATIONAL PVT. LTD., NEW DELHI**

20 Boost the Agriculture for Rural Development

Banarsi Lal and Shahid Ahamad

India as a country is one of the biggest food consumers in the world and there is dire need to keep our agricultural sector vibrant and alive for our food security. More than half of our country population owes its social, economic and even cultural life to the state of this sector. Governments after governments gave the importance to agriculture but when it came to policy making then it is left to struggle. Agriculture is the sole source of livelihood for more than two-third of the population and there is dearth of basic infrastructure in agriculture for instance irrigation, post-harvest facilities, research centres, loan facilities *etc.* which have impact on farmer income levels. India was facing the formidable famine in 1960s and then the Prime Minister Sh.Lal Bahadur Shastri successfully led the country to Green Revolution. With green revolution India became self-dependent in food production and leaped miles forward in staple food production. The sole objective of green revolution was to increase in productivity and acreage. This led to excessive use of fertilizers, pesticides and other agro-chemicals. Now the states like Punjab, the biggest beneficiary of green revolution has started facing the problem of soil health deterioration, reduction in crop productivity and high cost of cultivation. For the first time in 2003 Govt. of India thought over post harvest aspect of the Indian agriculture and formulated Agriculture Produce Market Committee (APMC).There was lot of debate on APMC and farm insurance. Now the government is contemplating on agriculture not merely as a tool to feed the country but also as a means to uplift the socio-economic status of the farming community of the country. The government has initiated a number of developmental schemes and programmes which have the potential to immensely benefit the farming community by strengthening the roots of agriculture. On 19th February, 2015 the Prime Minister of India launched the nationwide Soil Health Card Scheme from Suratgarh, Rajasthan. Agriculture is a tool for poverty eradication. The govt. has sanctioned 100 mobile soil testing laboratories across the country and approximately 2.53 crore soil samples would be collected during 2015-16.The government has announced that 14.5 crore farmers would get the soil health cards within three years across the country. Soil Health Card Scheme is a national movement across the country. Under this scheme the soil sample is taken by the experts from the farmer's field and tested in a soil health laboratory. Then the soil health card is issued to the farmers regarding the ingredients and deficiencies in the soil. On the basis of basis of the results of the soils of respective farmer field, he can add the nutrients in the soil accordingly. This scheme may not only maintain the health of the soil but will also reduce the cost of cultivation. This will also help to identify the best crop suited in the respective field.

The government also launched another scheme called as Prime Minister Krishi Sinchayee Yojana (PMKSY) with an objective to irrigate the every farmer field and also to increase the water use efficiency. Under this scheme there is immense support to the micro-irrigation and watershed development. The irrigated agriculture accounts for 56 per cent of all food grains production in the country. It is estimated that the total area under agriculture rose only 5.6 per cent during 2000-1 to 2011-12. The net irrigated area reached 46.34 per cent from 40.5 per cent during this period. The major objective of Prime Minister Krishi Sinchayee Yojana (PMKSY) is to achieve convergence of investments in irrigation at the field level, expand cultivable area under assured irrigation, reduce wastage of water, increase water use efficiency, promotion of water conservation practices, reusing treated municipal based water and attract private investment in precision irrigation system. This scheme will also help in bringing the ministries, departments, financial institutions, various agencies under a common platform so that a comprehensive view of the entire water cycle is taken into account. It is hoped that this scheme will go a long way for the upliftment of the socio-economic status of the farming community.

Keeping the agricultural information of the Indian farmers in view the government of India has launched a new TV channel namely DD Kisan which is dedicated to the farmers. This will help to disseminate the agricultural information to the farmers. Now the govt. is working on a new agricultural insurance scheme that would cover all the agricultural inputs of the farmers at his farm. If the govt. succeeds in providing a robust loan to the farmers, it may create a strong social security to the farmers. We have national insurance then we have modified national insurance and now we are thinking for the weather-based insurance cover and also there can be income insurance of the farmers. From the last many years the prices of onions and potatoes have marred the farmers. These are perishable agricultural crops which require proper storage facilities which the farmers lack. So, the farmers are compelled to sale these commodities at throwaway prices and the middlemen earn lot of profits. In order to check this govt. of India has launched a price stabilization fund scheme. These agricultural crops will be procured from the farmers directly from the farm gate and made available to the consumers at the reasonable rates. We daily hear of the farmers suicidal cases because of the farmers fail to pay the credits. It is a herculean task for the government to provide loan to the farmers at easy terms. Keeping realities of the financing to the farmers in view the govt. has increased the agricultural funding. Effective warehousing is primarily required for the storage of farm produce. It has been observed that 2.1 crore tonnes of wheat is wasted every year in India which is around 22 per cent of the total production due to lack of proper warehousing facility. In fruits and vegetables this proportion is 40 per cent of the total national production. National Bank for Agriculture and Rural Development (NABARD) has aimed to make scientific storage of 2.23 lakh tonnes by constructing godowns and warehouses. The government of India has diagnosed the major problem of the farmers and trying to solve it through various policies.

The Cabinet Committee on Economic Affairs approved a Central Sector Scheme for Promotion of National Agricultural Market through Agri tech Infrastructure Fund on 2nd July, 2015. The Government plans to integrate 585 wholesale markets across India by setting up an online platform. There will be electronic auctions for price discovery. The plan is to cover 250 mandis in current fiscal, 200 mandis in 2016-17 and 135 mandis in 2017-18.After the completion of National Agriculture Market (NAM) transfer of agricultural commodities can take place. The market size for the farmers would be limited to a captive market. Through this scheme the farmers will get better prices, improve supply chain and reduce wastages by unifying the markets at state and the central level. This will also solve the genuine problem of the middlemen who eat away the major chunk of profits due to lack of market availability to the farmers. In order to get the employment the new generation from rural India is migrating towards the urban areas. The new schemes and programmes launched by the government may reverse this trend.

Transformation of Indian Agriculture through Innovative Technologies *Pages* **315–320**
Editor: **Dr. Shahid Ahamad & Dr. Jag Paul Sharma**
Published by: **ASTRAL INTERNATIONAL PVT. LTD., NEW DELHI**

21 Transformation of Rural India through Agricultural Technology

Anil Bhat, Jyoti Kachroo, Sudhakar Dwivedi and S.P. Singh

Rural industrialization contributes to the development of agriculture and urban industries. Without rural industrialization it would be considerably more difficult to solve the problem of agricultural unemployment and widespread underemployment. Further it promotes agricultural development and agricultural development promotes rural industry. The development of rural industries increases the level on income in rural areas, and trends to break down the old self-sufficiency of the family and to lessen its cohesiveness, creating opportunities for youth, women and the able bodies as well in changing the pattern of leisure and work. This should be looked upon not merely as a way of containing the rural workers and stopping them from migrating to urban areas by providing them some kind of remunerative employment in the villages, but as a dynamic element in the process of rising to productivity and income levels of the workers in rural areas. Rural industrialization has taken roots in the rural economy in India. This is so because simple forms of manufacture typical of consumer goods industries and varied service industries, are everywhere developed before the more complex process involved in the production of capital goods, and because the size of the establishment of optimum-sized plants in the production of certain capital goods. It is a key to rural development and rural prosperity. It constitutes a significant link in the process of socio economic transformation of rural areas. Primarily, it provides additional opportunities of employment, income, better standard of living and thereby enriches the cultural heritage of the varied social structure in rural areas.

Agro-industries which are considered to be rural industries if established in rural areas are the corner stone of a developing economy and have well established roots in the indigenous environment. The widespread unemployment and under-employment associated with the problem of migratory labour in developing economics can only be met by creation or extension of rural industries in general and ago-based industries in particular. They serve as a mean for providing better employment opportunities to the labour during off-season. These industries also help in the mechanization of agriculture. Agro-industries may reduce the migration of people in rural areas to urban areas. These may be used as an effective tool in bringing prosperity in rural areas. Their establishment may lead to improve the infrastructural facilities in backward areas. Further, agro-industries may help in tackling the problem of exploitation of farming community by traders and middlemen. Transport cost can be minimized. In order to keep rural economy at par with urban economy, rural industrialization especially through agro-industries is one of the important means. Income generation through agro-industries may improve the purchasing power of rural people, thereby creating potential for demand-based industries and increasing the standard of living of rural people. Thus, the rural industrialization has capacity of revitalization of the rural masses and assists the urban sector. Even the United Nations Report suggested that greater effort should be made to get economic and social betterment of the neglected rural areas where most of the population lives and will continue to live for many decades to come.

Rao commented that rural industrialization will not only bring prosperity to rural sector but will bring about equilibrium in the structure. Rural industrialization is the only way of removing poverty of rural masses and enabling them to stay in their homes. There should be coordination at the district level, with local markets, traditional technology not discarded but improved to save the raw material and finances made available. It may encourage the cooperative way. In the recent years, rural markets have acquired significance in developing countries as well as the overall growth of the economy has resulted into substantial increase in the purchasing power of the rural communities. India with majority of its population living in villages is having rural poverty a major problem. The disparity between the rural and urban in terms of income is also on rise. This major reason leads to migration from rural to urban areas. Therefore it is very much necessary to transform rural India through modern agricultural technology and that too in terms of production, marketing, processing, agro-based industries *etc.* All these activities related with agriculture must be given due importance at rural level also so that income of the farmer can be increased. On account of technological change, rural areas are consuming a large quantity of industrial and urban manufactured products. In this context, a special marketing strategy has taken place. Sometimes, rural marketing is mystified with agricultural marketing the later denotes marketing of agricultural commodities of the rural areas to the urban consumers or industrial consumers, whereas rural marketing involves delivering manufactured or processed inputs or services to rural producers or consumers.

Transforming Rural India with the Help of Technologies

1. Modern Markets to be Set Up in Rural India

The markets of rural India are increasingly attractive in relation as compare to urban India because competition is at its peak in urban India. Most of the companies dealing in processed food products are busy in selling products in small packs to rural people as per the demand and capacity which is an example of use of technology in rural India. Processing industries equipped with all modern facilities are now a days established in rural areas which is clear indication of development of rural India.

2. Kisan Credit Card for Farmers as a Technology

The government started Kisan Credit Card through banks with minimum formalities to help the farmers in terms of money for purchase of emergency inputs. The farmer had a choice to take short or medium term loans through these credit cards to buy good quality seeds, fertilizers, pesticides, *etc.* This enabled him to produce more and thereby increase his income.

3. Communication Approach

The method of promotion needs to be modified to suit the outlook of the market and conventional as well as recent media should be used as a marketing strategy. Mass Media and print media has created increased demand for goods and services in rural areas. Smart marketers are employing the right mix of conventional and non-conventional media to create increased demand for products.

4. Information Technology Dissemination in Rural India

Today's rural children and youth will grow up in an environment where they have 'information access' to educational opportunities, career counseling, job opportunities, government schemes and services, health and legal advice and services, worldwide news and information, land records, mandi prices, weather forecasts, bank loans, livelihood options and moreover well awareness about modern agricultural technology. As the digital India and Information technology culture moves into rural India, the possibilities of change are becoming visible.

5. Improvement of Agricultural Marketing System

Government of India has adopted a number of measures to improve agricultural marketing, the important ones being - establishment of regulated markets, construction of warehouses, provision for grading, and standardization of produce, standardization of weight and measures, daily broadcasting of market prices of agricultural crops on All India Radio, improvement of transport facilities, *etc.* These must be implemented at rural level also.

i. Marketing Surveys

In the first place the government has undertaken marketing surveys of various goods and has published these surveys. These surveys have brought out the various problems connected with the marketing of goods and have made suggestions for their removal. This will definitely help to transform rural India.

ii. Grading and Standardization

The government has done much to grade and standardize many agricultural goods. Under the Agricultural Produce (Grading and Marketing) Act the Government has set up grading stations for commodities like ghee, flour, eggs, *etc.* The graded goods are stamped with the seal of the Agricultural Marketing Department -AGMARK The «Agmark" goods have a wider market and command better prices. A Central Quality Control Laboratory has been set up at Nagpur and eight other regional laboratories in different parts of the country with the purpose of testing the quality and quality of agricultural products applying for the Government's "Agmark" have been created The Government is further streamlining quality control enforcement and inspection and improvement in grading.

iii. Organization of Regulated Markets

Regulated markets have been organized with a view to protect the farmers from the malpractices of sellers and brokers. The management of such markets is done by a market committee which has nominees of the State Government, local bodies, arhatiyas, brokers and farmers. Thus all interests are represented on the committee. These committees are appointed by the Government for a specified period of time. Important functions performed by the committees can be summarized as follows:

a. Fixation of charges for weighing, brokerages *etc.*,
b. Prevention of unauthorized deductions, underhand dealings, and wrong practices by the arhatiyas,
c. Enforcing the use of standardized weights,
d. Providing up to date and reliable market information to the farmers, and
e. Settling of disputes among the parties arising out of market operations.

It is the policy of the government to convert all markets in the country including those in rural areas into the regulated type with modern facilities.

iv. Provision of Warehousing Facilities

To prevent distress sale by the farmers, particularly the small and marginal farmers, due to prevailing low prices, rural go downs have been set up. The government has done much to provide modern warehousing facilities in towns and villages so that product can be stored there during excess supply and less demand and then distributed during favourable time.

v. Dissemination of Market Information

The government has been giving attention to the broadcasting of market information to the farmers through radio sets, newspapers, magazines, electronic boards placed at KVKs *etc.* These electronic/print media publish/announce agricultural prices either daily or weekly accompanied by a short review of trends which help them to take decision where to sell and when.

vi. Directorate of Marketing and Inspection

The directorate was set up by the Government of India to co-ordinate the agricultural marketing of various agencies and to advise the Central and State Governments on the problems of agricultural marketing. Activities of this directorate includes the following.

a. Promotion of grading and standardization of agricultural and allied commodities;
b. Statutory regulation of markets and market practices;
c. Training of personnel;
d. Market extension;
e. Market research, survey and planning and

vii. Government Purchases and Fixation of Support Prices

In addition to the measures mentioned above, the Government also announces minimum support price for various agricultural commodities from time to time in a bid to ensure fair returns to the farmers. These prices are fixed in accordance with the recommendations of the Agricultural Price Commission.

viii. Empowering Farmers

With respect to empowerment e-choupal comes up as fine example. It is efficient supply chain system empowering the farmers with timely and relevant information enabling them to get better returns for their produce

Factors Contributing to the Change in the Rural India

1. Green Revolution

Green revolution has helped in improving agricultural productivity. Adoption of new agronomic practices, selective mechanisation, multiple cropping, inclusion of cash crops and development of allied activities like dairy, fisheries and other commercial activities have helped in increasing disposable income of rural consumers. It was a period when agriculture in India increased its productivity due to improved agricultural technology.

2. Infrastructural Facilities

Rural India is in developing phase with growth of infrastructure facilities such as construction of roads, communication network, transportation, PDS *etc.* These factors helps to transform rural India in terms of marketing. It will provide more income in the hands of the farmers.

3. Government's Initiative

The initiatives taken by the government for self-sufficiency has made a tremendous change in rural India. The activities or schemes which fall under these initiatives are white revolution, blue revolution, integrated rural development programme, Kisan credit cards, contract farming *etc.* In addition to this, government has also invested massively to change a shape of rural India in terms of development, agricultural technology, machinery, infrastructure, input subsidies.

4. Agricultural Research

Research in the field of agriculture has resulted in increase in production and quality product which in turn increases scope of agricultural marketing. Use of new scientific methods/technologies has remained successful in increasing the yields to manifolds. With the increase in yield, the income of farmer is also increased which is a clear example of change in rural India.

REFERENCES

Acharya, S.S., and Agarwal, N.L. 2011. Agriculture Marketing in India. Oxford and IBH Publishing Company Pvt. Ltd. New Delhi.

Kumar, K. P. and Swamy, S. 2013. Indian Rural Market – Opportunities and Challenges. *Asian Journal of Marketing and Management Research.* 2(2).

Pratiyogita Darpan, 2011. Indian Economy, Upkar Prakashan, Agra.

Priya, Lakshmi and Bajpai, V. Impact of Rural Marketing on Indian Economy, Kanpur Institute of technology, Kanpur.

Rani, M. 2013. "Rural market: the core of Indian market" G.J.C.M.P. 2(6): 123-125.

Singh, H., Goel, M. K. and Singhal, A. K. 2012. New Perspective in Rural and Agricultural Marketing. *VSRD- IJBMR,* 2 (1): 17-24.

Transformation of Indian Agriculture through Innovative Technologies *Pages 321–334*
Editor: **Dr. Shahid Ahamad & Dr. Jag Paul Sharma**
Published by: **ASTRAL INTERNATIONAL PVT. LTD., NEW DELHI**

22 Rajmash: Cultivation Technology in Hills Regions for Transforming Farmers' Economy

Rohit K. Sharma and Shahid Ahamad

Bhaderwah Rajmash is famous for its unique taste and flavour. It is known for its good cooking qualities, sweetness, easy digestibility and colour, and is in high demand in market. *Rajmash* of other regions are sold in the market on the tag name of Bhaderwah *Rajmash*. Bhaderwah *Rajmash* is a local pole type trailing variety of the *Phaseolus vulgaris* L. Bhaderwah *Rajmash* is an herbaceous annual plant grown for its edible dry bean.

Bhaderwah *Rajmash* is grown locally as an intercrop with maize since ages. The local tall white maize variety gives it the ideal staking for this twining pole type variety. It is harvested in 2-3 pickings at 5-7 days interval as the pods at lower end mature earlier and seeds are dispersed by over maturity into the field. The *Rajmash* harvesting is completed in the of last week of August to September before the maize harvesting. This intercrop combination of nitrogen fixing legume crop and maize is an ideal combination to generate high economy in the hilly regions and is a sustainable soil enriching practice. Maize – *Rajmash* combination also helps to reduce soil erosion of the sloppy lands in hills, which is a major problem of this area particularly in solo maize growing fields. This is a traditional agricultural practice adopted since ages in this region. Bhaderwah *Rajmash* is a pole or trailing type variety 3-4 m long. It bears alternate, green or purple leaves, divided into three oval, smooth-edged leaflets, each 6–15 cm long and 3–11 cm wide. The flowers are creamy white to pinkish white about 1 cm long,, 1–1.5 cm wide, green, yellow or light purple in color, and give way to pods 8–14 cm long each containing 4–7 beans.

The beans are smooth, plump, kidney-shaped, up to 0.75cm –1.5 cm long, ranges widely in colour(light red, medium red to maroon), and are often mottled in two or more colours.

Area Under Production

Rajmash is grown in district Doda, Ramban and Kishtwar of Jammu division. Some of the main Rajmash growing areas of district Doda are Bhaderwah, Sungli, Sartangal, Nalthi, Chinta, Kakol Manthal, Kansar, Lanchan, Thathri, Gandoh and Premnagar. Besides it is also cultivated in Paddar, Chatroo, Warwan, Marwah, Dakshin, Chingam and Dehrna area of Kishtwar district and Gool Gulabgarh area of Ramban district.

Climate

Rajmash commercially grown in temperate areas having elevation between 1500-2500 m of sea level. The optimum temperature for better growth is 16-24 !. If temperature falls below 10 !, plant growth ceases and temperature above 35 ! results into buds and flower drop. The crop is generally raised in areas receiving 50-150cm annual rainfall. Water logging at any growth stage adversely affects its yield. Excessive rains cause flower drop and spread of leaf spot diseases. The optimum relative hunidity is 60-70 per cent.

As it is grown with maize as intercrop, the hampered light availability increases the growth of Rajmash plant. It requires bright sunny locations for its cultivation.

Strong winds with high velocity cause lodging of supportive crop (maize) and damage the Rajmash plant. The strong winds also tend to increase the flower drop.

Soil

Rajmash is grown on a wide range of soils ranging from light sand to heavy clay, but for optimum growth and yield, well drained clay loam soil is best. The crop is sensitive to salinity. The crop is grows successfully in soils whose pH ranges from 5.2-5.8. Soils rich in organic matter promote better vegetative growth.

Production Technology

Pre-harvest Operations

Seeds Collection

Farmers collect their own seeds from better genotypes among population and use it for next generation. Mostly good seeds from previous year's product are obtained and preserved for sowing in the next season.

Preparation of Field

To achieve the desired tilth, the land is ploughed with MB plough (soil turning plough) 3 to 4 times. Each plouging should be followed by planking to ensure fine tilth and moisture conservation. The soil should be well drained, as Rajmash (maize both) are sensitive to water stagnation.

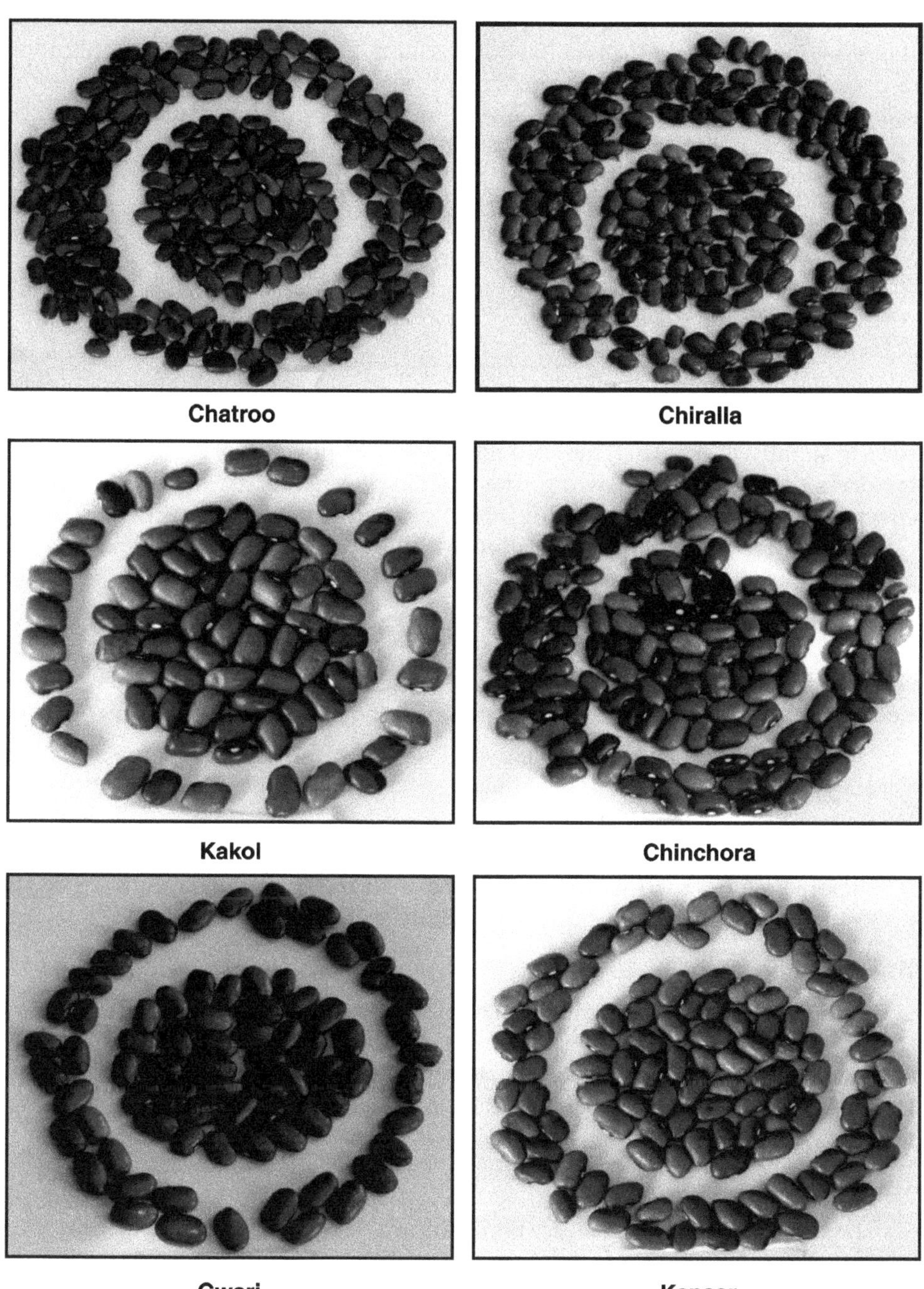

Figure 22.1. Few Selected Germplasms from Bhaderwah Region of Doda District.

Nutrition

Manures and fertilizers both play an important role in *Rajmash* cultivation. It is slow in biological nitrogen fixation because of poor nodulation. Application of organic manure improves the general physical conditions and structure of the soil and water holding capacity. A liberal quantity of bulky manures needs to be applied in the field. About 10-15 tones/ha of well rotten farm yard manure or compost should be applied one month before sowing and worked well into the soil. Additional requirements of crop can be met by applying inorganic fertilizers.

- Nitrogen: 20 kg/ha
- Phosphorus: 60 kg/ha
- Potash: 30 kg/ha
- Zinc sulphate: 10 kg/ha (once in three year)

Half dose of N and full doses of P and K should be applied at the time of sowing slightly deeper than the seed. Remaining half dose of N is applied through top dressing 30 days after sowing. The foliar application of $ZnSO_4$ (0.1 per cent) is effective in enhancing seed yield if it has not been applied at the time of sowing. Generally there is no need of additional fertilizer or nutrient to Rajmash when grown as mixed crop with maize.

Sowing time in Jammu region: April - May

Seed rate: 3-4 kg/ha when taken as an intercrop with maize (30-35 kg/ha). (8: 1)

Seed and Spacing

- Row to row distance: 30 cm
- Plant to plant distance: 15-20 cm
- Sowing depth: 5-8 cm

Before sowing, seed treatment is done with Bavistin or Thiram @ 2g/kg to protect the crop from various diseases at early seedling stage.

Water Management

As a rainy season crop, it does not require irrigation, when rainfall distribution is even throughout the crop cycle. Being a shallow rooted crop, moisture stress at any stage is detrimental to its performance. It is essential to maintain available soil moisture near field capacity during critical period of growth *viz.*, pre-blooming, flowering and pod filling

Thinning and Weeding

To obtain maximum yield, the optimum plant population per unit area is maintained by thinning and thus recommended plant to plant distance is maintained. *Rajmash* suffers severe competition from weeds in initial stages. First 30-45 days after sowing is the critical period for crop weed competition. Thinning should be accompanied with hoeing and weeding after 25-30 DAS. Chemical weed control in *Rajmash* can be done by the application of following chemicals. The application should be done in bright sunny days and avoided in cloudy weather or when wind is blowing.

Pre-emergence

Pendimethalin @ 0.75-1.0 kg/ha for annual grasses and few broad-leaved weeds.

Mixed Cropping System in *Rajmash*

Maize and *Rajmash* are grown mixed in hilly area of Jammu division. Taking twining type *Rajmash* as a pure crop is not economical because returns from the pure crop are meager. The Bhaderwah *Rajmash* varieties require arrangement for support during the crop growth period. So when sown mixed, it gets natural support from maize plants.

Figure 22.2. Staking in *Rajmash*.

Pollination

It is highly self pollinated crop.

Disease and Pest Management

1. Root Rot and Web Blight (Seed and Soil Borne)

Casual Organism- *Rhizoctonia solani* Kuhn

Symptoms

Disease occurs in two phases:

- ☆ **Root Rot-** Reddish brown cankers appear on complete rotting of the root system takes place

Figure 22.3. Root Rot of *Rajmash*.

- ☆ **Web blight**- One leaf and pods, circular slimy water-soaked, light tan in colour spots appear and tan brown to reddish brown discolouration appear on infected seeds.

Management

Treat the seed with Bavistin @ 2g/kg. Follow the soil application of bio-agents *viz. Trichoderma harzianum* (2.5kg/ha).

2. Anthracnose (Seed and Soil Borne)

Casual Organism- *Colletotrichum lindemuthaianum* (Sacc. And Mang.)

Symptoms

Prominent symptoms appear on pods as black, sunken cankers. Spots are usually depressed with dark centers bright red, yellow or orange margins.

Management

Treat seed with Bavistin (0.2 per cent) with 0.4 per cent talc formulation of *T. viride*. Use resistant cultivars.

3. Ashy Stem Blight (White Mould) (Soil Borne)

Casual Organism- *Sclerotinia sclerotiorum*

Symptoms

First symptom of the disease is yellowing of the leaves. In the next 2-3 days, they may drop off. Wilted plants show typical ashy stem blight symptoms. Infected seeds turn blank or brown.

Figure 22.4. *Rajmash* Anthracnose.

Figure 22.5. Ashy Stem Blight.

Management

Follow deep ploughing, use ammonium-type fertilizers and calcium compounds. Seed treatment with Bavistin (0.2 per cent). Drench the infected plants/ soil with copper oxy chloride (0.1 per cent).

4. Angular Leaf Spot (Seed and Soil Borne)

Casual organism: *Phaeoisariopisis griseola* (Sacc.) Ferr.

Symptoms

Small angular, dark brown leaf spots chuck board appearance. Favourable conditions lead to complete defoliation. Infected pods either bear no seeds or produce only shriveled seeds.

Management

Apply seed soaking with Bavistin (0.1 per cent). Spray Benlate (0.1 per cent) or Dithane M-45 (0.2 per cent).

Figure 22.6. Angular Leaf Spot.

5. Common Blight (Seed Borne)

Casual organism: *Xanthomonas axanopodis pv. Phaseoli* (Smith) Dye (XAP).

Symptoms

Leaves: Small, water soaked areas develop into larger reddish- brown area.

Stems and pods: Deeply sunken lesions surrounded by reddish- brown tissue.

Seeds: Discolouration appears especially near the hilum region. Plants arising from infected seeds exhibit characteristic wilting.

Management

Use pathogen free seed, crop rotation. Mixed intercropping of bean with maize. Give the seed treatment with hot water at 50°C for ten minutes followed by diping.

Figure 22.7. Common Blight.

in streptocycline (10 gu/ml) solution. Three foliar spray of copper oxychloride (0.1 per cent) + zineb (0.1 per cent) at 15 days interval.

6. Halo Blight (Seed Borne)

Casual Organism: *Pseudomonas savastanoi pv. Phaseolicola*

Symptoms

Brown to red zonate lesions surrounds water- soaked spots on affected pods. The bacterium leads to shriveling, discolouration and even rotting.

Management

Use disease free seed and follow 2-3 year crop rotation. Burn the debris. Give the seed treatment with streptomycin as hot water soak for 60 minutes with successive tap water rinse and a further 30 minutes soaking in streptomycin.

7. Bean Common Mosaic (Seed and Vector Transmitted)

Casual Organism: Bean common Mosaic virus (BCMV).

Vector: Aphids

Symptoms

Leaves affected with mosaic have irregular shaped, light-yellow and green areas. The characteristic mottling, there may be considerable puckering and malformation of leaves. Infected plants remain stunted. Downward curling of the leaves takes place.

Figure 22.8. Halo Blight.

Figure 22.9. Bean Common Mosaic.

Management

Eradication of infected plants should be done. Use healthy seeds. Apply carbofuran 3G @ 1.0 kg per kanal at the time of sowing for aphid control. Use yellow sticky traps @ 2.0 per kanal.

Insece Pests

1. Black Bean Aphid (*Aphis faba*)

Nature of Damage

Vectors of various virus diseases.

Sucks sap from plant parts.

Management

Spray NSKE @ 5 per cent or Imidacloprid @ 0.30 ml or Metasystox @1.5 ml per litre of water.

Figure 22.10. Black Bean Aphid.

Figure 22.11. White Fly.

2. White Fly (*Bemisia tabaci*)

Nature of Damage

White flies sucks sap from under surface of leaves and vector of bean mosaic disease.

Management

Delta traps or sticky traps @ 2.0 per kanal are effective in catching whiteflies. Rogue out and burn the virus infected plants. Metasystox 25EC or Dimethoate 30EC @ 1.0 ml or Imidacloprid @ 0.30 ml per lit of water at 15 days interval for vector management.

3. Thrips (*Scirtothrips dorsalis*)

Nature of Damage

Nymphs and adults suck sap from the flower and in flower drop results.

Management

Install delta or sticky traps @ 2.0 per kanal. Spray of Metasystox @ 1 ml or Imidacloprid @ 0.3 ml per litre of water.

Figure 22.12. Thrips.

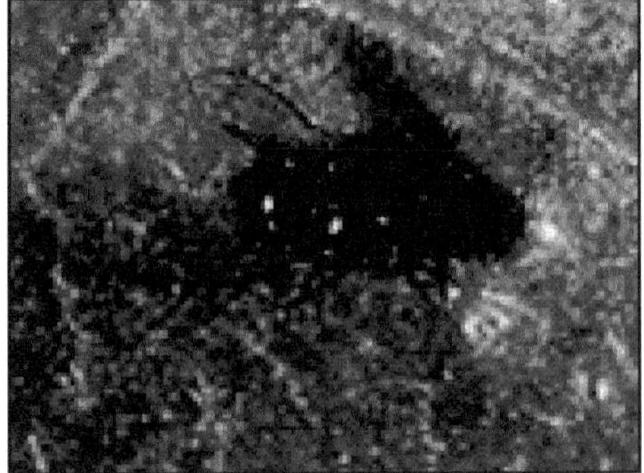

Figure 22.13. Stemfly.

4. Stemfly (*Ophiomyia phaseoli*)

Nature of Damage

Maggots bore the stem causing yellowing and wilting of plants.

Management

Carbaryl 50 per cent WP @ 2kg/ha carbofuran 3G@ 1.0 kg/kanal.

5. Leafminer (*Lariomyza trifolii*)

Nature of Damage

Causes loss of chlorophyll through mining in leaves.

Management

Spray Imidacloprid (0.3 ml per litre of water when the attack begins and repeat at 15 days interval or spray NSKE @ 5 per cent.

6. Leaf Hopper (*Amrasca bigutulla*)

Nature of Damage

Suck the sap from the underside of leaves produce hopper burn symptoms.

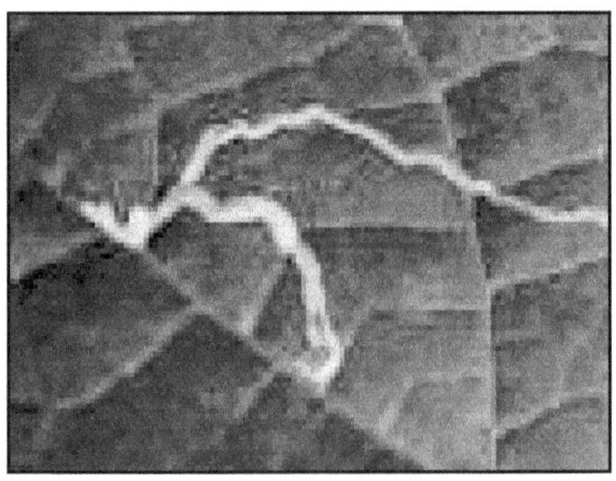

Figure 22.14. Leaf Miner.

Figure 22.15. Leaf Hopper.

Management

Remove crop debris and and spray Malathion @ 2 ml per litre of water or cypermethrin 1.0 ml per litre of water.

7. Pulse Beetle (*Callosobruchus chinensis*)

Nature of Damage

Grubs feed on developing grains.

Management

Spray Quinalphos @ 2ml/l water or Carbryl @ 2ml/l water

8. Bean Gall Weevil (*Alcidodes signatus*)

Nature of Damage

Weevils lay eggs at the node axil and grubs feed on the tissue at the axil resulting in the gall formation at internodes.

Management

Spray Chlorpyriphos @ 1.25 ml or Cypermethrin 1.0 ml or Carbryl @ 2.0 ml per litre of water.

9. Blister Beetle (*Cyaneolylta coerculea*)

Nature of Damage

Grubs and adults feed voraciously on leaves and flowers.

Management

Hand picking of adult beetles and destroy them, spray Malathion @ 2 ml per litre of water.

Figure 22.16. Blister Beetle.

10. Hairy Caterpillar (*Spilosoma oblique*)

Nature of Damage

Larvae completely defoliate the plants.

Management

Spray Malathion @ 2 ml per litre of water.

11. Cutworm (*Agrotis ipsilon*)

Nature of Damage

Serious problem in temperate areas on hills, stoutly built larvae cut the plant at ground level at an early stage of crop growth.

Management

1. Land plouging and irrigation in early evening hours expose the hidden larvae for bird predation.
2. Chloropyriphos (seed treatment) @ 2 gm/kg of seed. Soil application of carbofuran 3G @ 1.0 kg per kanal.

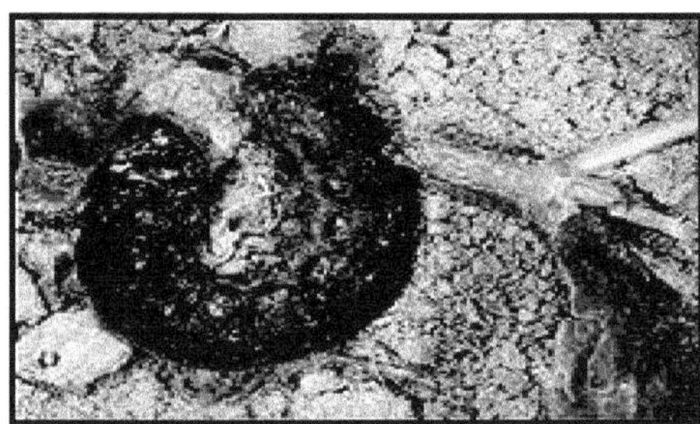

Figure 22.17. Cutworm.

Harvesting and Storage

Harvest is done after 120-130 days after sowing when leaves and pods turn yellowish brown. The upper leaves remain green till final harvest. Delay in harvesting may cause pod shattering. The harvested crop is kept for sun drying for a week and thereafter threshed. After threshing, seeds are further cleaned and dried to moisture content below 9 per cent before storage.

Transformation of Indian Agriculture through Innovative Technologies *Pages 335–348*
Editor: **Dr. Shahid Ahamad & Dr. Jag Paul Sharma**
Published by: **ASTRAL INTERNATIONAL PVT. LTD., NEW DELHI**

23 Walnut Production in J&K for Livelihood Security to the Farmers

Rajesh Kumar, Vikas Sharma and Shahid Ahamad

Introduction

Walnut is the one of the most important temperate nuts grown in India. In India, walnuts are grown in the states of Jammu and Kashmir, Uttaranchal, Himachal Pradesh and Arunachal Pradesh with Jammu and Kashmir occupying the largest share in total area and production. The total area under walnut production is increasing due its high economic returns.

Indian walnuts are categorized in to 4 categories *viz.*, paper-shelled, thin-shelled, medium-shelled and hard-shelled.

Major varieties of walnut grown in different states of India are:

- **Jammu and Kashmir**: Lake English, Drainovsky, Bulbul, Sulaiman, Hamdan, K-5 and Opex Caulchry
- **Himachal Pradesh**: Gobind, Eureka, Placentia, Wilson, Franquetfe and Kashmir Budded
- **Uttaranchal**: Chakrata Selections

The area of walnut in India has increased from 123 thousand hectares with the production of 233 thousand metric tonnes in 2012-13 to 125 thousand hectares and production of 206 thousand metric tonnes in 2014-15 indicating the decline in the production (Horticulture Statistics Division, DAC and FW, 2015). In the state of Jammu and Kashmir also, the area under walnut increased from 93.64 thousand hectares with the production of 209 thousand metric tonnes in 2012-13 to 96.40 thousand hectares and production of 181.44 thousand metric tonnes in 2014-15. Overall in 2014-15, there has been a decline in the production as compared to the production of walnut recorded in previous years.

Figure 23.1. Walnut.

As far as the export of walnut is concerned, there has been a sharp decline in the quantity of the walnut exported from India to major destinations *viz.*, Vietnam Social Republic, Egypt, Arab Republic, Netherlands, United Kingdom, Spain, United States and Germany. In 2013-14, the total walnut export was 6,726.36 MT which sharply declined to 2665.87 MT in 2014-15.

Despite increase in area and advancement in the production technology in different fruit crops grown in the Jammu and Kashmir state, very little attention has been given to increase the production of walnut thereby the average productivity and quality of this valued crop has remained much below the expected potential.

Therefore following data was collected and compiled so that issues will be examined critically.

a. Total Area and Production of Walnut in the State for the Last Ten Years

Table 23.1. Area and Production of Walnut in the State for the Year 2010-11 to 2014-15

(Area in hectares, Production in Metric Tonnes)

Kind of Fruit	*2010-2011*		*2011-2012*		*2012-2013*		*2013-2014*		*2014-2015*	
Walnut	*Area*	*Prod.*	*Area*	*Prod.*	*Area*	*Prod.*	*Area*	*Prod.*	*Area*	*Prod.*
	89788	163745	83613	208738	93641	209051	95620	251930	96397	181443

Source: Economic Survey J&K, 2013-14, Directorate of Horticulture, Jammu, J&K.

Table 23.2. Area and Production of Walnut in Jammu Division for the Year 2012-13 and 2014-15

(Area in hectares, Production in Metric Tonnes)

Kind of fruit	*2012-2013*		*2013-2014*		*2014-2015*	
Walnut	*Area*	*Prod.*	*Area*	*Prod.*	*Area*	*Prod.*
	38081	49506	38816	56640	39349	58263

Source: Directorate of Horticulture, Jammu.

b. Quantity of Walnut being Exported Outside the State/Country for the Last Ten Years

Table 23.3. Export of Walnut Outside Jammu and Kashmir

Year	*Quantity Exported (in Metric Tonnes)*		*Foreign Exchange Earned (Rs. in crores)*
	Shell	*Kernel*	
2010-11	235	3124	87.89
2011-12	142	4321	177.91
2012-13	309	4986	199.82
2013-14*	Not Available	5274	334.89
2014-15*	Not Available	1424	99.90
2015-16*	Not Available	2091	136.72

Source: Directorate of Economic and Statistics Govt. of Jammu and Kashmir (Digest of Statistics, 2012-13).

* Directorate of Horticulture, Planning and Marketing.

As per the data, maximum walnut from J&K is exported to North East, Europe, South America, European and middle ports, Spain, Netherland, Greece, U.K, Norway, Germany, Canada, USA, Sweden and Denmark. It has been also find out from the research studies that share of J&K in walnut export at national level is 85-90 per cent.

(Source: International Journal of Multidisciplinary Research and Development, 2015. 2 (9): 572-73)

Table 23.4. Export of Walnut Outside Country (India)

Year	*Quantity Exported (in Metric Tonnes)*	*Foreign Exchanged Earned (Rs. in lakhs)*
2010-11	5762.34	16629.25
2011-12	5841.56	23108.40
2012-13	5295.47	19983.57
2013-14	6726.36	32453.50
2014-15	2665.87	13645.00

Source: APEDA Website.

c. Price Fetched by the Walnut in the Local Market at least for the Last Ten Years as far as the Availability of Data is Concerned

Table 23.5. Price Fetched in the Local Market of Jammu and Kashmir (Rs per kg)

Year	Price at the Local Market in Jammu		Wholesale Price in Kashmir
	Walnuts	Walnut Kernels	Walnuts
2011-12	NA		100-150
2012-13	162	447	130-175
2013-14	166	492	Not Available
2014-15*	223	679	Not Available
2015-16	203	531	Not Available

* As per the observations of exporters of Jammu division, the increase in prices during the year 2014-15 was due to unprecedented floods in Kashmir valley and repeated damage to walnut crop which later turned out to be a hoax.

d. Price of the Walnut Outside the State/Country (as far as possible)

Table 23.6. Price of Walnut Outside State for the Year 2013-14 and 2014-15. (Rs per quintal)

Year	Price Outside J&K			
	Maharashtra	Punjab	Utter Pradesh	Delhi
2013-14*	62229	27875	26462	26958
2014-15**	25466.67			

Source: * indiaagristat.com; **Trees Nut annual, India.

e. Quantity of Walnut Imported in the State/Country for the Last Ten Years

Table 23.7. Arrival and Dispatch Walnut in different Markets of Jammu Region for the Year 2010-11 to 2012-13. (qty in tonnes)

Sl. No.	Name of the Market	Arrivals (tones)			Dispatches (tones)			Major Destinations where Despatched
		2010-11	2011-12	2012-13	2010-11	2011-12	2012-13	
1.	Narwal	4523	4221	3539	-	-	-	-
2.	Doda	4560	4670	7550	3192	3269	5285	Narwal Mandi Jammu
3.	Batote (local)	4707	5348	6010	3740	4156	4670	F and V Market Jammu and Udhampur

Source: Directorate of Horticulture, Planning and Marketing, Jammu.

f. Total Walnut Imported in the Country

Table 23.8. Quantity of Walnut Imported in the Country from the Year 2010-11 to 2013-14

Year	*Walnut Quantity Imported (in Metric Tonnes)*	*Value (in lakh rupees)*
2010-11	165.24	219.06
2011-12	72.87	214.2
2012-13	341.35	784.99
2013-14	178.00	431.08

The state is a major exporter of walnut and its international market share is about 7 per cent.

(Source: Econmic Survey 2014-15 Vol-1, pp. 66, Government of Jammu and Kashmir, Directorate of Economics and Statistics, J&K)

Local walnut consumption in India: 50 per cent to 60 per cent

(Source: USDA Foreign Agricultural Service, GAIN Report No. IN4081 dated 12/09/2014. India. Tree Nuts Annual, 2014)

g. Quality Parameters of State Walnut *viz-a-viz* the Quality of Imported Walnut

- ☆ In-Shell Walnuts - In-shell walnuts shall be clean, creamish brown in colour. They should be dry.
- ☆ Shelled Walnuts -These shall be obtained by shelling selected walnuts. Kernels may be white, light cream or brown in colour; they shall be dry. The kernels shall possess the characteristic taste and should be free from foreign taste, including rancidity and mustiness.
- ☆ Freedom from Moulds, Insects, etc. - Walnuts in-shell or walnut kernels should be free from living insects and moulds and should be practically free from dead insects, insect fragments and rodent contamination visible to the naked eye (corrected if necessary, for abnormal vision), with such magnification as may be necessary in any particular case. If the magnification exceeds x 10, this fact should be stated in the test report.
- ☆ Extraneous Matter - Walnuts in-shell or walnut kernels shall not have extraneous matter more than 0'25 percent by weight. Entraneous matter would include wood-splinters, husk dirt, walnut meal, shell pieces and any other foreign matter.
- ☆ Grades - Grade designations and their quality characteristics of walnuts shall be given.
- ☆ Packing - Walnuts in-shell or walnut kernels should be packed in new, clean, B-twill jute bags.
- ☆ Marking-Each bag should be marked or labelled giving the following particulars:

a) Name of the materials;
b) Name and address of the packer/producer;
c) Net mass;
c) Grade of the material;
d) Name of the country, where packed; and
e) Any other marking required by the purchaser.

h. Scope of the Replacement of Traditional Walnut Trees by the High Density Walnut with the Comparable Production Level and Period in which the Whole Area under Walnut can be Covered

Sl.No.	*Item*	*Walnut under Traditional Plantation*	*Walnut under HDP*
1.	Plant type	Seedling	Grafted/Budded
2.	Planting distance	10m×10m	3m×6m
3.	No. of plants/ha	100	555
4.	Average nut yield/plant at full bearing stage	50kg	50kg
5.	Total nut yield/ha at full bearing stage	5000kg	27750kg

Total area under walnut plantation in Jammu Province: 38081 hectares (2012-13).

Period for converting whole area of Jammu Province into HDP: 8 years.

As far as the production of walnut is concerned, there is no glut/surplus in the market as inputs received from suppliers, stakeholders, exporters and as well as farmers. The demand is still alive in the country and also in world for walnut but the only concern is of quality.

Definitely increase in production and productivity due to high density plantation will play a major role in export of walnuts. As by high density plantation we will be able to grow quality walnuts of same variety which is having more demand in international market.

In case the imports of walnut in the state are too meager, how can any increase in import duty give protection to the state walnut. In case, state exports walnuts, then how would any increase of importing duty by the Government of India help in enhancing the prices. Once, we are exporting the walnuts, would it not mean that it competes in the international market.

The problems faced by Walnut Industry are not new, they are in fact perpetual. Problem of export of walnut is emerging quickly now a day. Therefore, we would like to enumerate and discuss all above problems briefly under different headings given below:

1. Premature Harvest

Harvest of Walnuts is done by producers at their sweet will. They don't bother whether the nuts are fully matured or not. Their focus is to reach the market before anybody else does that. They are then artificially ripened to remove the hull, resulting in very bad quality. In many cases harvest starts as early as by the third week of August, which should be around second week of September. There is an urgent need to regulate the harvest time.

2. Post-harvest Management

After Harvest and dehulling the Walnuts are required to be dried completely to remove the excess moisture from the product. Many a times growers/traders send walnuts to the market which become rancid very soon.

3. Extraction of Kernels

To extract "Halves" from walnuts, the shellers soak walnuts in the water for Twenty four to forty eight hours and many a times the water is not even potable. This results in bacterial growth in the kernels and thus they are loaded with microbiological problems. We are getting complaints from our overseas buyers and many a times consignments are rejected or sent back. There is a urgent need to stop this practice.

4. Hygienic Environment

Walnuts are being cracked in most unhygienic conditions. Usually there is no separate area embarked for this process. It's an edible product and for this we need very hygienic conditions, which are not there. Therefore there is a need to educate the crackers to maintain proper hygiene. Whilst all the processors/Exporters are registered under FSSAI Act, no cracking unit in the valley is registered.

5. No Single Variety

The biggest problem with our raw material is that we do not have any specified variety. There is wild growth of Walnuts in our State. Product of one tree does not match with the produce of other tree. There are as many varieties as there are trees, whereas other counties have specified varieties and they offer their produce on that basis. Many small countries which were unheard of like Ukraine, Maldives, Hungary, Romania, Italy, Chile, have started giving stiff competition. Previously we were facing competition from USA and China but now these new producers are formidable competitors.

6. Hoarding for High Prices

Our growers/traders tend to hoard the goods to fetch a higher price. They create an artificial scarcity, which thus result in price rise. Higher prices results in lower demand even in the domestic markets. The example is of 2014 crops. During September, 2014, there were unpredicted rains and floods in valley. A shortage psychosis was created that was more than 70 per cent of Walnut crop had perished in rain with the result, prices sky rocketed during the Diwali festival season but demand plunged. As everybody knows that later 40 to 50 per cent crop remained unsold.

7. International Pricing

There was a time when the industry used to export 7000 to 8000 tons of walnut kernels. Over the years, it has come down to 100 to 1500 tons per year. Basic reason is we are being outcompeted by other counties on pricing. Our prices are much higher than the International market prices. In the present scenario, when the world has become a global village we cannot dictate our prices.

The industry has invested crores of rupees in creating infrastructure. The processors have State of art plants with international standards but unfortunately they are working at 40 per cent capacity, as they don't get export orders due to aforesaid reasons of high prices and hoarding.

The government of India has declared Jammu and Kashmir as Agriculture Export Zone for Walnuts but nothing has been done by state government for export of Walnuts.

8. High Cost Due to Multi-stage Middlemen

By the time goods arrive to the exporters, the cost escalates due to multi stage traders and middlemen. Both growers as well as exporters are suffering due to this.

9. Absence of Reliable Production Data

We do not have any access to the reliable production data for Walnuts, whereas other countries give advance forecast of production figures with almost 98 per cent of accuracy. In absence of dates we are unable to plan our Export Sales.

10. Vat Refunds to Exporters

Exporters have not received the refund of VAT on Export of Walnut Kernels. Since the year 2005-2006, which roughly amount to Rs. Four Crores, resulting in serial cash crunch to Exporters.

11. Suggestions

First of all there is an urgent need to develop varieties which can compete with other countries.

There is a wild growth of hard walnuts commonly known as "bata". These walnuts do not contain any kernel and are useless. Many a times these are mixed with good quality nuts by some unscrupulous traders, thus bringing a bad name to industry as a whole. Such trees should be replaced immediately with new varieties.

Growers should be made aware that this is an edible produce and every care should be taken to maintain hygiene around the working area. As explained earlier wet cracking should be discouraged. If at all necessary, clean potable water should be used. The water should be changed every day. The wet kernels should be dried immediately to save them from development of mould, yeast etc.

The government should sanction and amount for development of walnuts. This amount should be allocated for the development of private nurseries, which can provide best planting material to phase out old trees.

After the due consultation and meetings with fruit growers association of Narwal mandi Jammu, some points were suggested by them also for the recovery of dying walnut industry.

The detailed list of the suggestions/recommendations is given below:

- Awareness camps for the proper production of walnut to be organised at district level.
- Pre harvesting of walnut should be avoided and growers should be aware for the consequences of pre harvesting.
- Cracking of nuts must be done in neat and clean condition/hygienically.
- Dry cracking to be promoted for competing with other countries.
- Wet kernels should be dried to stop bacterial growth.
- Batta varieties of walnut to be replaced with new hybrid variety.
- Growers to be educated at village level.
- Imposing of anti-dumping duty on import of walnut.
- Loan facility on horticulture crops.
- Quality parameters to be considered and scientific recommendations should be given.
- In foreign countries well organized and planned scientific orchards are established but we are lacking in that.
- Growers to be made aware about the ill effects of pre matured harvesting walnut crop. It can be done through print media. Govt should take initiative.
- Quality degraded due to above points.
- Grading is not proper at grower's level.
- Department should guide the growers regarding production of walnut.
- Growers are holding the produce which creates hype in market,
- Washing of walnut with acid for its whiteness is another problem.
- R&D sector is lacking in India as compare to foreign countries.
- Commercial variety of walnut is not grown in India.
- Wild growth of walnut can not be run for long time.
- Why should buyer buy from us as we are having wild walnut.
- Export of kernels reduced to 600-700 tonnes in this year as compared to 8000 tonnes few years before.
- Nuts are not exported as now because of quality parameters.
- Like almond, walnut is also losing its fame in J&K state.
- Doda and Kishtwar is best as per walnut cultivation is concerned.
- For export we are not having same quality produce.
- Problem is in mindset of growers.
- Walnut board should be established who should take care of all problems.
- Foreign countries released prices for walnut one year advance.
- Board should promote the things.

- ☆ Bulgarian variety of walnut sponsored by APEDA in J&K state. But now it is vanished.
- ☆ Posing of duty on import of walnut will not help in long run.
- ☆ Growers holds the Walnut and does not sell in the market. Product is available but not as per quality of export.
- ☆ Growers expect more price for their produce but that is not possible because of poor quality.
- ☆ Estimated production and demand of walnut crop before festive season to be required by traders.
- ☆ Remote sensing for the actual production and area under walnut crop.
- ☆ Systematic and scientific survey for walnut is required.
- ☆ Now shell is not exported because of poor quality.
- ☆ Variety of J&K is good.
- ☆ Wrong Production figure and variety is main cause.
- ☆ Quality kernel is required.
- ☆ In 2014 prices were at peak then declined. Peak is not flat.
- ☆ Variety should be as per climate of our state.
- ☆ With the quality of walnut available in J&K state we can compete at particular level only.
- ☆ Promotion and development of walnut can be taken up with board.
- ☆ Mindset that our walnut is best to be changed.
- ☆ Turnover of exporter is reduced to $1/3^{rd}$ or $1/4^{th}$.
- ☆ Exporters are at stake.
- ☆ Personal nurseries to be developed with major stress on variety development.
- ☆ In Uttarakhand private partners are importing walnut and rootstock are prepared by them.
- ☆ Our walnut is by default organic so more emphasis should be given for organic certification to the walnut industry which in turn will fetch more prices. We can take advantage of it.
- ☆ Only 2 persons in Jammu are able to manage organic certification for their kernels.
- ☆ Govt. initiates a scheme for traders "you can import walnut duty free and than export kernel but this process does not fetch much prices.

No doubt all above problems are the reasons for dying walnut industry in J&K and we are supposed to create awareness among farmers, stakeholders and whosoever is involved in this prestigious industry of J&K. All the persons involved should come forward and work in the same platform for saving this industry from any debacle furthermore. The productivity of walnut is another major problem which is because of certain important factors.

The major contributing factors to low productivity in walnut and researchable issues which needs to be discussed are as:

1. Lack of Ideal Varieties in Walnut

Most walnut trees in the state of Jammu and Kashmir are originated from seed, so there is considerable variability in their nut and kernel characteristics. The high genetic variation in walnut trees is due to their seed based propagation, high heterozygosis and dichogamy. High variability in phenological and nut traits has also been reported in walnut trees from different walnut growing regions of India.

The produce obtained from the trees is not uniform and as such, plantation cannot be geared to various market outlets where uniform crop of better quality is in demand. In general, nuts produced from the present plantations belong predominantly to the thick shell group, where shelling percentage is low.

2. Lack of Suitable Methods of Propagation and Availability of Adequate Vegetatively Propagated Planting Material

The availability of grafted plant material is a major limiting factor in the cultivation of walnut. Walnut seedlings grow slowly and require 1 ½ to 2 years to attain desirable size for budding/grafting. During this period, their tap roots grows very deep and pose problem in uprooting and transplanting in the field resulting in a poor survival rate of transplanted saplings. Low and variable success obtained with different methods of vegetative propagation is also one of the constraints for its vegetative propagation.

3. Lack of Clonal Rootstocks

Clonal rootstocks are not available and generally seedlings of the same or related species are used. These seedlings exhibit much variation among the plants. The seed germination in walnut is also a problem due to its hard shell. A large number of factors such as nut type, size and position of seed, planting depth, soil moisture, temperature and seed treatment influence the germination of seeds.

4. Predominance of Old and Senile Orchards

In walnut, there is predominance of old and senile orchards. Their productivity has declined drastically. Rejuvenation of old and declining walnut orchards has not been standardized as in case of other perennial fruit crops.

5. Problem of Re-establishment of Nursery Plants in the Orchard

The expansion of the area under walnut is very limited mainly because of insufficient grafted plants and their poor establishment under field conditions due to a high rate of mortality when nursery plants are planted in the orchard, especially in due to the tap root system which is disturbed while uprooting the plants from the nursery.

6. Lack of Suitable Pollinizers

The walnut tree has monoecious flowers and the problems in pollination and fruit set are due to dichogamy. There are several contrasting features between staminate and pistillate flower development in the walnut. Once a seedling reaches sexual maturity, staminate flowers are produced one to two years before pistillate flowers. Production of staminate flowers first on a given seedling is not advantages as walnut is normally cross-pollinated. Cross-pollinated of walnut ensures heterozygous progeny and increased survival of seedling with sufficient genetic diversity to endure the selective pressure imposed throughout a wide climatic range. Cross-pollination is promoted by dichogamy. When selfing occurs, fruit abortion is increased and kernel development is suppressed. Improvement in productivity can be achieved with the planting of suitable pollinizers cultivars.

7. Lack of Efficient Use of Irrigation Water

Water is the critical input in fruit crop production. Its efficient use is inevitable for achieving higher production from shrinking ground water resources. Micro-irrigation, popularly known as drip irrigation is an efficient way of irrigation with high frequency applications of water close to feeder roots. It economizes water use by 40-60 percent, added with enhanced yield, reduces weed population and quality production. Such techniques have not been fully exploited by orchardists.

8. Lack of Integrated Plant Nutrient Management (IPNM)

IPNM refers to maintenance of optimal soil fertility and plant nutrient supply for sustaining desired crop productivity, through optimization of the benefits from all possible sources of plant nutrients in an integrated manner. It is a holistic approach coupled with use of bio-fertilizers including growing of green manure, cover and intercrops. Fertilizer application should be restricted to need based at required time, right place in soil/plant tissues.

Researchable Issues

1. Development/Selection of Ideal Varieties of Walnut

- ✰ There is a need to identify suitable trees from native seedlings populations or to introduce cultivars from other countries which are suitable for the different climatic conditions prevailing in the state of Jammu and Kashmir.
- ✰ Till date, no walnut breeding programme has been undertaken in India to develop walnut cultivars adapted for use in the agro-climatic conditions suitable for its cultivation. Walnut varieties that have a higher level of phyto-chemicals for fresh consumption, and tolerance to major insect and disease pests.

2. Production of Quality Planting Material

- ✰ Number of quick and efficient plant multiplication technologies has been developed in the recent years. But still plant multiplication is being done with age old techniques. Standardization of proper propagation technique and establishment of a separate mother block propagated from elite walnut plants and their proper maintenance is the need of the hour.

3. Rejuvenation of Old and Senile Orchards

- ☆ Many old trees with declining production and nut quality can be stimulated to more active growth and increased nut production by pruning. Standardization of time, severity for pruning back old/senile walnut trees and standardization of calendar of operation for encouragement of new growth, as well as the tree architecture is one of the researchable issue which requires immediate attention so that there will be increase in the production of nuts with improved quality.

4. Refinement of Production Technology

- ☆ There is no updated technology available to the growers for taking up walnut cultivation on scientific lines under Indian conditions. This is due to a lack of well planned orchards and no systematic trials have been laid out to study the various aspects viz., canopy management (training, pruning), placement of pollinizers, growing of intercrops, possibility of use of growth retarding chemicals, water and nutritional management have not been thoroughly been worked out for walnut nut cultivation which needs due attention for making walnut culture more profitable and productive.

Initiatives to be taken for Increasing Quality and Production in Walnut in Jammu Province – Strategies

1. Cluster approach area expansion will be given emphasis. This can facilitate in post harvest management and in developing proper marketing infrastructure.
2. Production of planting material of elite germplasm with an objective to increase yield production.
3. Imparting training to farming community and extension workers for acquiring the knowledge and know-how of latest technologies at SKUAST-J, SKUAST-K and CITH and other Research Institutions outside the State for intensifying the yield gains through judicious use of modern technologies.
4. Similarly women shall be given know-how of technologies for active participation in production besides ensure their independence for earning income by imparting training in the art of fruit processing preservation and value addition of fruit crops.
5. 20-30 per cent plantation of pecan nut have become unproductive due to age, attack of pest/diseases and changing climatic scenario.. In addition to such plantation there is huge count of wild/inferior pecan trees, whose produce fetch negligible returns to the farming community as the produce is of very inferior quality.. Therefore, rejuvenation of senile/old plantation including top working of wild/inferior fruit cultivars will be given due attention by following corrective pruning, replanting and graft-age of inferior/wild fruit cultivars into superior and commercial viable ones.

This programme activity will ensure increase in productivity of quality and commercially viable fruit crops and employment generation.

6. To reduce post harvest losses of fruits due to poor handling at Farm level, on farm handling units will be constructed in bearing orchards.
7 Mechanization will also be a main priority as such Power Tiller/Tractors, Power Sprayers, Water lifting pumps and pruning tools shall be given due consideration.

Transformation of Indian Agriculture through Innovative Technologies *Pages* **349–369**
Editor: **Dr. Shahid Ahamad & Dr. Jag Paul Sharma**
Published by: **ASTRAL INTERNATIONAL PVT. LTD., NEW DELHI**

24 Pecan Nut (*Carya illinoensis*) Cultivation Techniques in the Intermediate Zone of J&K

Rajesh Kumar, Vikas Sharma and Shahid Ahamad

Introduction

Pecan (*Carya illinoensis*) is one of the most important nut fruits of the world ranking fifth in production. It has excellent nutty flavour and tree characteristics. In the USA, pecan is considered the 'queen of nuts' because of its value both as a wild and as a cultivated nut and acceptable quality of meats. Pecan has been under cultivation only for about one hundred years, but within this short period it has gained immense popularity. It is planted in the form of orchards as well as singly or in groups around the farm building or as road-side shade trees. The Pecan is a highly nutritive and energy giving nut fruit. It has nutty flavor and superb tree characteristics. Pecan nut (Carya illinoensis) belonging to family Juglandaceae, is a valuable and favourite edible nut of the people. It has excellent nutty flavour and tree characteristics.The most urgent need today is to enhance the production, productivity and quality of nutritious food in eco-friendly manner, so as to improve farm income to ensure household food and nutritional security. The topography and agro-climatic conditions of the hill limits the scope for production of field crops, but offer most suitable conditions for horticultural crops. The research evidence shows that a large untapped potential of pecan nut production in J&K state of India, which could benefit particularly weaker section and women through creation of employment opportunities besides export earnings.

Major varieties of pecan nut grown in different states of India are:

- **Jammu and Kashmir**: Nellis, Mahan, and Burkit
- **Himachal Pradesh**: Nellis, Mahan, Clero, Burkit
- **Uttaranchal**: Nillish, Mahan

Pecan nut is the one of the most important temperate nuts grown in India. In India, it can be grown in Jammu and Kashmir and Himachal Pradesh. In Jammu and Kashmir State, pecan nut is being successfully grown in Kathua, Udhampur, Reasi, Doda, Kishtwar, Ramban, Rajouri and Poonch districts of Jammu division and Baramula and Kupwara districts of Kashmir division of Jammu and Kashmir State of India. The total area under pecan nut in Jammu and Kashmir is 412.80 hectares and production is 81.48 MTs during 2015-16 (Source: Directorate of Horticulture, Jammu, J&K). Most plantations are of seedling origin in scattered form and produce nuts of variable quality. Exotic varieties are found fruiting in the country and seedling trees do not yield true to type nuts. Therefore, there is a need to raise high quality nursery through vegetative propagation. Despite increase in area and advancement in the production technology in different fruit crops grown in the Jammu and Kashmir, very little attention has been given to increase the area and production of pecan in Jammu and Kashmir thereby the average productivity and quality of this valued crop has remained much below the expected potential. Potential for bringing additional area under Pecan nut is enormous. As per preliminary estimates about 1392 hectares are still available only two districts *viz.*, Rajouri and Poonch on which Pecan nut cultivation can be undertaken successfully.

As far as the export of pecan nut is concerned, the entire produced are consumed with the state and there was no report of export of pecan nut.

Despite increase in area and advancement in the production technology in different fruit crops grown in the Jammu and Kashmir state, very little attention has been given to increase the production of walnut thereby the average productivity and quality of this valued crop has remained much below the expected potential.

Figure 24.1. Pecan Nut.

Therefore following data was collected and compiled so that issues will be examined critically.

Production

World production of pecans totalled more than 108,000 metric tons (kernel basis), that is, 59 percent of increase in the last ten years. Production of pecans is clearly led by the United States and Mexico, which account for 93 percent of world production. Their production in 2014 was 60,185 and 40,823 metric tons respectively, followed by South Africa with 5,724 metric tons and Australia with 1,080 metric tons. The total area under pecan nut in Jammu province is *412.80* hectares and production is *81.48* MTs during 2015-16 (*Source:* Directorate of Horticulture, Jammu, J&K).

a. Total Area and Production of Pecan Nut in the State for the Last Two Tears

Table 24.1. District-wise Area and Production under Pecan in Jammu Division of J&K State for the Year 2014-15.

Name of the Districts	*Kathua*	*Udh.*	*Reasi*	*Doda*	*Kshtr*	*Rmbn*	*Rajouri*	*Poonch*
Area (HA)	59	47	6	49	7	5	163	304
Production (M T)	1	1	0	0	0	0	3	7

Source: Directorate of Horticulture, Jammu and Kashmir.

Table 24.2. District-wise Area and Production under Pecan in Jammu Division of J&K State for the Year 2015-16.

Name of the Districts	*Kathua*	*Udh.*	*Reasi*	*Doda*	*Kshtr*	*Rmbn*	*Rajouri*	*Poonch*
Area (HA)	43.8	50.0	5.75	25.0	0	0.25	163.0	125.0
Production (M T)	11.47	1.69	1.00	0.12	0	0	60.20	7.00

Source: Directorate of Horticulture, Jammu and Kashmir.

The production of fruit particularly apple and walnut is increasing every year in Jammu and Kashmir, however, potential dry fruit like pecan nut has been given very little attention. Studies revealed that a large untapped potential of pecan nut cultivation exists in Jammu and Kashmir, which could benefit particularly weaker section and women through creation of employment opportunities besides export earnings. Therefore, Jammu and Kashmir government has identified pecan nut production particularly in intermediate zone of the state as a thrust segment for over all development of the state with the view to create the new avenue for income and employment opportunities through vertical and horizontal expansion of production and post harvest management. Pecan nut is the one of the most important temperate nuts grown in India. In India, it is mainly grown in Jammu and Kashmir, and Himachal Pradesh. In Jammu and Kashmir, pecan nut is grown successfully grown in Kathua, Udhampur, Reasi, Doda, Kishtwar, Ramban, Rajouri and Poonch districts of Jammu division. In Rajouri it is grown in Thanamandi, Dhangri,

Manjakote, Gambhir Mughala *etc.* and in Poonch, it is grown in Surankote, Poonch city, Mehander. Presently, total area under pecan nut in Jammu and Kashmir is 656 hectares and production is 13.0 metric ton. After evaluation, improved cultivars like Western Shelley, Desirable, Nellis, Mahan, Burket, Stuart and Wichita have been recommended for its cultivation. Development of appropriate arrangement for providing elite planting material, technological package and integration of various line departments for production, marketing and post harvest management are important to harness the potential of pecan nut fruit cultivation in this zone. The block wise potential area for pecan nut cultivation of Rajouri and Poonch districts of Jammu and Kashmir are presented in Figures 24.2 and 24.3.

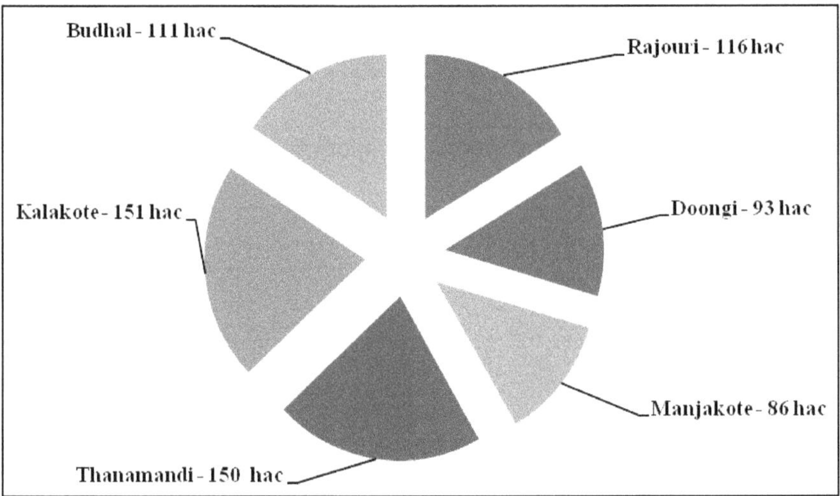

Figure 24.2. Block-wise Potential Area for Pecan Nut Cultivation in District Rajouri of Jammu Province.

Export

As far as the export of pecan nut is concerned, the entire produced are consumed within the state and there was no report of export of pecan nut. In the world, Pecans' exports accounted almost 35,500 MT in 2013, 2 percent down than last year but 7 percent up compared with 2004. Mexico and USA accounted the 97 per cent of all pecans exports in 2013, being USA the main destination of Mexico's exports and Canada, the principal destination of USA's exports.

b. Quality Parameters of State Walnut *viz-a-viz* the Quality of Imported Walnut

- ✰ Freedom from Moulds, Insects, *etc.* - pecan nut in-shell or pecan nut kernels should be free from living insects and moulds and should be practically free from dead insects, insect fragments and rodent contamination visible to the naked eye (corrected if necessary, for abnormal vision), with such magnification as may be necessary in any particular case. If the magnification exceeds x 10, this fact should be stated in the test report.

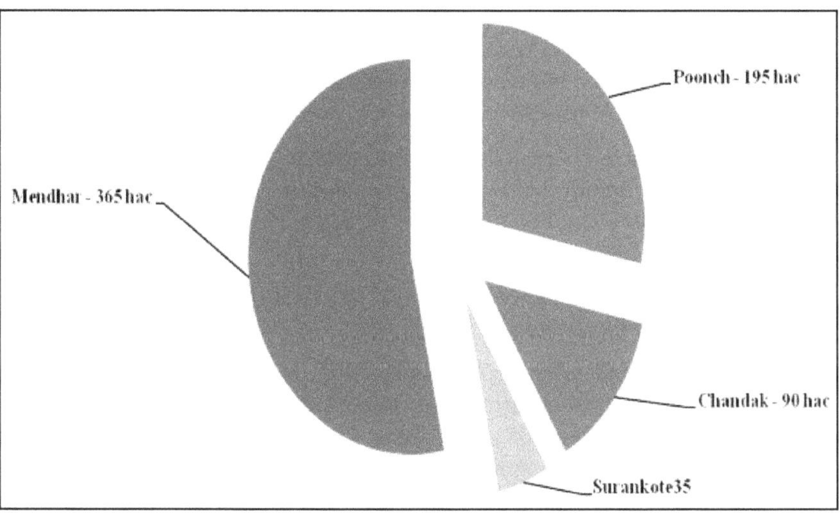

Figure 24.3. Block-wise Potential Area for Pecan Nut Cultivation in District Poonch of Jammu Province.

- ✰ Grades - Grade designations and their quality characteristics of pecan nut shall be given.
- ✰ Packing - pecan nut in-shell or pecan nut kernels should be packed in new, clean, B-twill jute bags.
- ✰ Marking-Each bag should be marked or labelled giving the following particulars:
  - a) Name of the materials;
  - b) Name and address of the packer/producer;
  - c) Net mass;
  - d) Grade of the material;
  - e) Name of the country, where packed; and
  - f) Any other marking required by the purchaser.

c. Scope of the Replacement of Traditional Walnut Trees by the high Density Pecan Nut with the Comparable Production Level and Period in which the whole Area under Pecan Nut can be Covered

Sl.No.	*Item*	*Pecan Nut under Traditional Plantation*	*Pecan Nut under HDP*
1.	Plant type	Seedling	Grafted/Budded
2.	Planting distance	10m×10m	3m×6m
3.	No. of plants/ha	100	555
4.	Average nut yield/plant at full bearing stage	50kg	50kg
5.	Total nut yield/ha at full bearing stage	5000kg	27750kg

As far as the production of pecan nut is concerned, there is no glut/surplus in the market as inputs received from suppliers, stakeholders, exporters and as well as farmers. The demand is still alive in the country and also in world for pecan nut but the only concern is of quality.

Definitely increase in production and productivity due to high density plantation will play a major role in export of pecan nut. As by high density plantation we will be able to grow quality pecan nut of same variety which is having more demand in international market.

The major contributing factors to low productivity in walnut and researchable issues which needs to be discussed are as:

1. **Lack of ideal varieties:** Almost all pecan plantations in Jammu division owe their origin to un-descriptive seedling and are extremely heterogeneous in quality attributes. The existing plantation is of seedling origin. The produce obtained from the trees is not uniform and as such, plantation cannot geared to various market outlets where uniform crop of better quality is in demand. In general, nuts produced from present plantations belong predominantly to the thick shell group, where shelling percentage is low.
2. **Lack of suitable methods of propagation and availability of adequate vegetatively propagated planting material:** The availability of grafted plant material is a major limiting factor in the cultivation of pecan. Pecan seedlings grow slowly and require 2-3 year to attain desirable size for budding/grafting. During this period, their tap roots grow very deep and pose problem in uprooting and transplanting in the field resulting in a poor survival rate of transplanted saplings. Low and variable success obtained with different methods of vegetative propagation.
3. **Lack of Clonal rootstocks:** Clonal rootstocks are not available and generally seedlings of the same or related species are used. These seedlings exhibit much variation among the plants. The seed germination in pecan is also a problem due to its hard shell. A large number of factors such as nut type, size and position of seed, planting depth, soil moisture, temperature and seed treatment influence the germination of seeds.
4. **Predominance of old and senile orchards:** In pecan there is predominance of old and senile orchards. Their productivity has declined drastically. Rejuvenation of old and declining pecan and walnut orchards has not been standardized as in case of other perennial fruit crops.
5. **Problem of re-establishment of nursery plants in the orchard:** The expansion of the area under pecan is very limited mainly because of insufficient grafted plants and their poor establishment under field conditions due to high rate of mortality when nursery plants are planted in the orchard, especially in Poonch and Rajouri districts of Jammu due to the tap root system which is disturbed while uprooting the plants from the nursery.

6. **Lack of suitable pollinizers:** The pecan tree has monoecious flowers and problems in pollination and fruit set are due to dichogamy. There are several contrasting features between staminate and pistillate flower development in the pecan. Once a seedling reaches sexual maturity, staminate flowers are produced one to two years before pistillate flowers. Production of staminate flowers first on a given seedling is not advantages as per pecan is normally cross-pollinated. Cross-pollinated of pecan ensures heterozygous progeny and increased survival of seedling with sufficient genetic diversity to endure the selective pressure imposed throughout a wide climatic range. Cross-pollination is promoted by dichogamy. When selfing occurs, fruit abortion is increased and kernel development is suppressed. Improvement in productivity can be achieved with the planting of suitable pollinizer's cultivars.

Climatic Requirement

Pecan need warm temperate climate and extremes of both temperate and sub-tropical climates are undesirable. It needs adequate chilling hours ranging from 400-600 hours below 7.2°C for different cultivars. Mean monthly temperature ranging from 24.0 to 29.5°C, with little diurnal variation during growing period is best for growth and fruiting. High humidity prevents pollination, increases the incidence of diseases and premature splitting of nuts. The tree can be grown successfully at elevations from 914 to 1829 meters above mean sea levels. It requires a moderate rainfall of about 75 to 100 cm and therefore, heavy rainfall areas are not suitable for its cultivation. Pecan need warm temperate climate and extremes of both temperate and sub-tropical climates are undesirable. Pecan tree is deep rooted and requires a deep fertile, well aerated and well drain sandy loam soil rich in organic matter. The soil should have capacity to retain water and have a depth of six meters. The pH range for pecans should be in between 5.0 -8.0. If the soils are acidic, they should be reclaimed through dolomite limestone application. Planting on eroded hills or bottom lands should be avoided as such sites do not permit free air movement.

Nutritional Importance

Pecan nut (*Carya illinoensis*) is one of the most important nut crop of the family Juglandaceae, having high nutritional value containing high content of proteins (12.5 per cent), fats (71.42 per cent), phosphorus (0.46 per cent) and potassium (0.23 per cent) and is rich in oil content wherein some varieties have shown as high as 76 per cent oil. Pecans are a good source of protein and unsaturated fats. like walnuts, pecans are rich in omega-6 fatty acids, although pecans contain about half as much omega-6 as walnuts. The antioxidants and plant sterols found in pecans reduce high cholesterol by reducing the "bad" LDL cholesterol levels. It is usually taken roasted or salted to supplement normal diet. The nuts are commonly used to add aroma, flavour, crispness, meatiness, tenderness, a rich colour or to garnish a large number of dishes. It is also used in baking products such as fruit cakes, custard pies, cookies, nut bread and cake fillings. Pecan ice cream is a top selling one after vanilla and chocolate. Pecan shell (a byproduct) is used for tannin, charcoal and abrasives in hand soap, flour as a filler for plastic wood, filler for feeds, insecticides, fertilizers and fuel for heating. The shell flour of various sizes is used in adhesives, dynamite and polishing metals. Pecan tree is also a beautiful ornamental tree. The wood has great value in timber market due to its strength and hardiness. The composition of pecan nut is given below in the Table 24.2.

Table 24.2. Composition of Pecan Nut (per 100 g kernel)

Sl.No.	*Components*	*Amount*
1	Fat	71.2 g
2	Protein	9.2 g
3	Carbohydrates	14.6 g
4	Fibre	2.3 g
5	Water	3.4 g
6	Calcium	73 mg
7	Phosphorus	289 mg
8	Potassium	603 mg
9	Magnesium	142 mg
10	Iron	2.4 mg
11	Vitamin A	130 IU
12	Thiamine	0.86 mg
13	Riboflavin	0.13 mg
14	Niacin	0.9 mg
15	Ascorbic acid	2 mg
16	Food value	Calories

Propagation/Crop Improvement

Pecan can be propagated both by seed as well as vegetative methods. However, it is commercially propagated by budding or grafting on seedling rootstock.

Seed Propagation

Pecan seeds (nuts) lose their viability very soon and so they should be stored at 0°C immediately after harvest. The seeds are stratified at low temperatures 1-5°C for about 12 to 16 weeks for obtaining good and rapid germination. To ensure good germination and to facilitate stratification treatments at low temperatures, the seeds should be sown under natural field conditions in December-January. The seedlings can also be raised in polythene bags which can be planted in the desired site with earthen balls without causing any damage to their roots. Deep, fertile and well aerated sandy soils are used for nursery purpose. Pecans do not stand wet-feet condition of the soil. The lilts are sown in well prepared soil in rows and covered with 10 cm thick soil layer. The seeds should be soaked in water for about 10 hours prior to their planting. Inferior small seedlings are removed during the first year of growth. The plants remain in the nursery for 2-3 years in order to attain proper size for budding and grafting and subsequently growth but they are transplanted after one year by shortening the tap root to 25-30 cm length which enhances vigorous fibrous root system and high survival rate. In order to avoid damage caused by rodents and birds, the seed should be painted with a paste made of red lead oxide and linseed oil.

Vegetative Propagation

Pecans can also be propagated vegetatively through budding or grafting selected scion cultivars on pecan seedling rootstocks. Among the vegetative the vegetative methods, patch budding and whip grafting are the most successful. Patch budding can be done in the month of of June by double blade knife when bark peels easily. Rootstocks of more than 2 cm diameter are usually patch budded. The buds of the previous season are used for spring for budding while well developed mature buds of the current season are patch budded in late autumn season. The bud is inserted more than 20 cm high on the stock from the ground and tied with polyethylene or waxed cloth strips. Whip/tongue grafting is also done on 2-3 years old rootstock. A tongue cut is of 3-3.5 cm length is given both on the scion (at base) and rootstock (on top) with one single stroke of the knife. Scion and rootstock each should have a girth of 3.5-4.5 cm. One each of these surfaces, a reverse cut is made. The stock and scion are inserted into each other. The union is tied with polythene sheet. From the successful grafts, polythene strips should be removed at later stages of the growth. The operation should be performed during early March. Hard wood cuttings (50 cm long and 20 mm thick) with IBA @ 10000 ppm have given excellent rooting.

Planting

Pecan trees are generally planted in square system at 8-10 m spacing. For planting, pits of 1m x 1m x 1m size should be dug and filled properly with a mixture of top soil, 30-50kg well rotten farmyard manure, 500g single superphosphate and 30-50g of chlorpyriphos dust. The filled pits are irrigated to settle the soil. This operation is done in October/November. The planting should be done in December-January. Deep ploughing and leveling is done before planting in late winter or early spring. Because of slow growth of fibrous root system, a considerable mortality is experienced at transplanting which can be minimized by removal of top 50 per cent of plant and keeping 50 cm tap root at planting time. In situ grafting of rootstock also gives good results.

Manuring and Fertilization

Fully grown up pecan trees should be applied with 100 kg FYM every year in December. In addition, apply 500 g NPK mixture (15: 15: 15) per year age of the tree and 8 kg mixture to the full bearing trees of sixteen years age and above. The fertilizer should be applied in early spring. Pecan are prone to zinc and manganese deficiency which can be prevented with foliar application of 0.5 per cent zinc sulphate and manganese sulphate.

Irrigation

Water requirement of pecans is higher than other nut fruits. The trees should be irrigated regularly during dry period after every two weeks or so, for the survival of the pecan tree throughout the first year. The watering should be given fairly deep so as to soak the whole area around the roots. Mulching the ground around the tree with dry leaves or straw helps in economising irrigation water as. well as in the establishment of the trees. Occasional irrigations should also be applied in subsequent years as and when required. The common method of applying water are flooding, furrows, sprinklers and drip irrigation. Drip irrigation can be effective for establishing an orchard. Over watering of a pecan tree should be avoided as excessive moisture damages it.

Training and Pruning

The pecan trees require training and pruning during the first few years of their planting in order to develop them in well shaped and strong trees. At the time of planting, one-third or one-fourth of the top portion should be cut so as to balance the shoot and root ratio. The lowest branch on the main trunk is kept at a height of one metre from the ground level. Training begins when about one-third to one-half of the top is removed. From the second year, the plant is trained to central leader: system and branches are allowed to emerge as high as 1.5 to 2m, from the ground on the trunk. The space between the main limbs' should be maintained at 35 to 40 cm. Pecan requires only light to moderate pruning as it is a terminal bearer however, during the initial 5-6 years, light pruning is essential in order to remove the weak, diseased and interlocked branches. Once the framework is established, dried and broken branches should removed and overcrowding branches be thinned.

Soil Management

Soil management is an important consideration in pecan to ensure good plant growth and productivity. Pecan has a long growing season, therefore, adequate soil moisture is essential during the rapid enlargement and filling of nuts. Clean cultivation (periodic soil tillage to remove all vegetation other than the crop being grown) has been found better in pecan than the other soil management system. Periodic disking is also useful to increase tree height, 'diameter and survival by making available more moisture.

Cover Crops

The cover crops are grown during summer and winter seasons, especially on rolling topography to check soil erosion and leaching of the nutrients. The common summer cover crops are soybean, cowpea and Crotalaria spp, whereas arrow leaf and clovers are winter cover crops. It must be ensured that soil has enough calcium and magnesium with a pH around 6.5.

Weed Control

Preemergence and contact herbicides are generally used for controlling the weeds. Diuron and Simazine (2-5 kg/ha) are widely used to control grasses and broadleaved weeds. However, they should not be used before the pecan plants have established for three years. Contact herbicides such as Parquat (1.0 kg/ha) or Glyphosate should be used to check the perennial weeds. Therefore, mixture of two herbicides can be used in rotation.

Flowering and Fruiting

Pecan trees begin fruit production in a period of 2-6 years after budding or grafting depending on the cultivar and vegetative growth. The male (catkins) and female flowers are borne on the same tree separately. The male flowers are in the form of hanging three-branched catkins and are produced in the axils of two-year old wood. The female flowers are borne in clusters of 2-8 at the terminals of the first spring flush of growth. Blooming periods in pecan differ from variety to variety which greatly affects the fertilization and pollination. Pecan tree is wind pollinated.

Figure 24.4. Flowering and Fruiting in Pecan Nut.

Sometimes, poor fruit setting arid premature losses of nuts occur due to inadequate fertilization either for want of pollen or due to un receptiveness of stigma. In order to ensure regular cropping year after year, a row of pollinizing variety should be planted for every nine rows of the main variety. The rows should be at right angles to the prevailing spring winds. Nuts resulting from cross-pollination are better filled, larger and less prone to shedding before maturity.

Maturity

The whole fertilization process takes about 5-7 weeks after pollination. Maturity of nuts begins with the development of shell size and kernel. The weight and specific gravity of the fruit also increase rapidly. Water content decreases from about 75 per cent to 30 per cent. The amount of fat also becomes seven times greater in mature nuts as compared' to the initial stages of fruit development. The hull turns from pale green to greenish black and cracks at full maturity.

Alternative Bearing

Alternate or irregular bearing is more pronounced in pecan than any other nut fruits. There are many factors responsible for this irregular bearing pattern. Late spring freeze can damage the entire crop in an 'on' year resulting in a heavy crop in the next year which in turn will bear a poor crop in the third year. Defoliation due to prevailing drought or due to the incidence of pests and diseases and likewise unfavourable climate induce the irregular bearing pattern. Inadequate fertilization

Figure 24.5. Low Cost Polyhouses Installed at RARS, Rajouri, SKUAST-Jammu for Production of Quality Planting Materials of Nut Crops.

Figure 24.6. Pecan Nut Seeds Sowing under Low Cost Polyhouse for Raising of Rootstocks.

is also one of the main causes of alternate bearing in pecan. When trees set a large nut crop and nutrients and moisture are not enough for the nuts to mature and for the tree to store enough plant food, production will be low in the following year. To prevent alternate bearing, use sound cultural practices. These include disease and insect control, adequate use of fertilizers, pruning and an extra application of fertilizer in late Mayor June in years when nut set is heavy. Use of early ripening cultivars and following proper management practices that maintain healthy foliage are also useful. Such practices reduce alternate bearing by increasing carbohydrate production and maintaining source of flower promoting substances.

Figure 24.7. Pecan Nut Rootstocks Ready for Grafting at RARS, Rajouri.

Fruit Thinning

Thinning of fruits (removal of a few young fruits from healthy bearing clusters or fruiting branches) becomes essential in pecan when in an 'on' year of the alternate bearing pattern, over production reduces nut weight and return crop. This can be overcome by hand thinning of excess fruits. Some chemicals such as maleic hydrazide (0.04 to 0.08 per cent), Ethephon (25 to 200 ppm), 2,4,5- T (2,4,5-trichlorophenoxy acetic acid) @ 10 to 25 ppm and CIPC (3-chloroisoprophyl-N-phenyl carbamate) @ 220 ppm have also been successful in fruit thinning.

Plant Protection Measure

A large number of pests and diseases are found to attack pecan. They are described below.

A. Insect Pests

1. Pecan Weevil (*Curculio caryae*)

The weevil damages the nut both in the orchards and storage. The adults feed on the nuts in late July and August puncturing the nuts and laying eggs about the same time when kernels begin to fill up. The grubs feed on kernels. The nature of damage caused by weevil may be premature nut drop, black spots on the kernel, shrivelled kernels and destruction of kernel by larvae.

Control

Give soil applications of Carbofuran 3G @ 20 kgs per hectare during early spring season.

2. Aphids

Many aphids cause injury to pecans. They feed on both sides of leaves and cause large yellow blotches which turn brown and cause premature leaf fall.

Control

A single application of Imidacloprid can manage the pest effectively.

3. Mites (*Tetranychus hicoriae*)

Mites may cause serious injury to foliage. They feed mainly on the lower side of leaflets along the midrib forming a brown discolouration which covers the entire leaflets. The most effective acaricides (chemicals used to control mites) is phosmite.

4. Serpentine Leaf Miner (*Stigmeta juglandifoliella*)

The leaf miner creates characteristic designs by feeding between the lower and upper surfaces of leaflets and causes defoliation. It can be controlled by the application of Dimethoate, Diflubenzuron or Fenvalerate.

5. Birds

Crows and blue jays are well known to cause considerably big losses to pecans. Crows damage the nuts soon after kernel filling either on the tree or by carrying them away. The damage is more serious to small, well filled and thin shelled nuts. A dead crow should be hanged high on the top of a pecan tree to scare away the crows.

B. Diseases

The diseases attacking pecan nut are described here.

1. Pecan Scab (*Cladosporium coryigenwn*)

This is one of the most serious fungal diseases of pecan. The fungus attacks primarily the growing leaves, shoots and fruits. The damage is accentuated with rains during spring and early summer. The symptoms are irregular, light green to black spots on leaves or young shoots and small dark brown to black circular lesions on the nut.

Control

1. Burn all diseased leaves, twigs and nuts.
2. Spray fungicides such as Bordeaux mixture, Zineb, Cyprex and Benlate in middle of April and at three weeks interval.

2. Leaf Blotch (*Mycospharella dendroides*)

The disease is only serious on neglected, declining trees and nursery plants. The symptoms are the appearance of live green, velvety tufts on the under surface in early summer and yellow spots on the upper surface of leaflets in late season. The leaves may fall on severe infection. The black pimples like structures give a black shiny blotched look to the leaflets in the mid summer. The scab control measures are also applicable to this disease.

3. Vein Spot (*Gnomonia nerviseda*)

It is a foliage disease and has been found to be more serious on several cultivars in some regions. The fungus attacks the vascular tissue at the junction of petiole to the rachis and base of rachis, causing premature leaf fall.

Control

Benomyl has been found to be the most effective fungicide against the vein spots.

4. Brown Leaf Spot

The disease is found mostly in humid regions. Circular, reddish brown disease spots occur on the underside of mature leaflets in June and July. Later, these spots attain irregular shape and on severe infection leaflets may fall.

Control

Repeated sprays of Bordeaux mixture, Zineb and Dodine are quite effective in preventing these spots.

5. Crown Gall (*Agrobacterium tumefaciens*)

It is a bacterial disease and occurs worldwide in distribution. Tumours or wort like growth develops on the collar and adjoining roots of the trce often protruding out of the soil. The disease kills the nursery plants and mature trees.

Control

1. All diseased small plants should be destroyed.
2. The injury to plants by cultivation should be avoided.
3. The affected tissues after removing the gall may be painted with the Elgetol-wood alcohol mixture.

6. Bunch Disease

It is caused by a virus which is transmitted by sucking type of insects. The characteristic symptom of bunch disease is the bushy growth of slender shoots. The branches of severely affected plant get clustered with thick bloomy sucker growth.

Control

1. The affected shoots should be pruned and burnt.
2. Grow resistant cultivars such as Stuart.
3. The diseased wild plants growing in the vicinity of the orchard should be removed.

4. Highly susceptible cultivars should not be grown in the bunch disease prone regions.

Harvesting and Yield

The pecan nut is ready for harvesting in September- October. The nuts which are enclosed in an outer covering or husk splits open at the time of maturity exposing the nut inside. At this stage the nuts are picked by hand and those that are out of reach are knocked down with the help of long poles. For large scale plantations, mechanical harvesting is adopted which involves tree shaking, picking and cleaning of nuts. Harvesting should be done in dry weather conditions as pecan nuts lose quality very quickly on the ground and sprouting occurs during wet weather conditions. The commercial bearing in pecan begins by 7th to 10th year after planting. The yield ranging from 20 to 100 kg per tree is expected from fully grown trees. Well cared orchards of mature trees should produce around 2,500 kg fruits per hectare per year. Some large trees in good condition have been known to produce as much as 150 kg per tree. In Thana Mandi area of district Rajouri, farmer harvested 108kg nuts form one tree of pecan nuts.

Drying and Storage

A. Drying

When pecans are harvested during wet weather or under wet soil conditions, they get discolored and infected with mould and meat quality 'deteriorates. In order to prevent molding, discoloration and rancidity, pecan nuts are dried soon after harvesting to bring the moisture level of kernels to around 4.5 per cent. The drying helps in giving color, texture, flavor and stability to the nuts which occurs naturally under field conditions if the weather and soil remain dry at harvest. The nuts should be dried at room temperature by constant circulation of low moisture air for about three days.

B. Storage

Pecan nut is semi-perishable in nature and thus requires proper drying, packing and storage temperature. Only high quality pecans having good size, color, flavor and oil content should be stored to fetch high prices in the off -season market. The kernel moisture content and storage temperature are two important factors affecting the storage life of the nuts. The storage life of pecan nuts can be extended, beginning from few months at room temperature to many years at 7.6 °C. Unshelled and shelled pecan nuts can be kept at room temperature (21.1°C) for six and three months, respectively. A number of insects and fungi attack the nuts in storage at temperatures above 7° C. The insects can be controlled by fumigation and fungi growth is prevented by maintaining the low moisture level of kernels (3 to 4 per cent).

Specific Problems in Pecan Cultivation

Despite the fact that huge potential for pecan nut cultivation exists in the state, the adoption of pecan nut remained restricted due to the following constraints:

- **Lack of ideal varieties:** Almost all pecan and walnut plantations in Jammu division owe their origin to non-descriptive seedling and are extremely heterogeneous in quality attributes. The existing plantations are of seedling origin. The produce obtained from the trees is not uniform and as such, plantation cannot geared to various market outlets where uniform crop of better quality is in demand. In general, nuts produced from present plantations predominantly belong to the thick shell group, where shelling percentage is low.
- **Lack of suitable methods of propagation and availability of adequate vegetatively propagated planting material:** The availability of grafted plant material is a major limiting factor in the cultivation of pecan and walnut. Pecan seedlings grow slowly and require 2-3 year to attain desirable size for budding/grafting. During this period, their tap roots grow very deep and pose problem in uprooting and transplanting in the field resulting in a poor survival rate of transplanted saplings. Low and variable success obtained with different methods of vegetative propagation.
- **Lack of clonal rootstocks:** Clonal rootstocks are not available and generally seedlings of the same or related species are used. These seedlings exhibit much variation among the plants. The seed germination in pecan is also a problem due to its hard shell. A large number of factors such as nut type, size and position of seed, planting depth, soil moisture, temperature and seed treatment influence the germination of seeds.
- **Predominance of old and senile orchards:** In pecan, there is predominance of old and senile orchards. Their productivity has declined drastically. Rejuvenation of old and declining pecan orchards has not been standardized as in case of other perennial fruit crops.
- **Problem of re-establishment of nursery plants in the orchard:** The expansion of the area under pecan is very limited mainly because of insufficient grafted plants and their poor establishment under field conditions due to high rate of mortality when nursery plants are planted in the orchard, especially in Poonch and Rajouri districts of Jammu due to the tap root system which is disturbed while uprooting the plants from the nursery.
- **Lack of suitable pollinizers:** The pecan tree has monoecious flowers and problems in pollination and fruit set are due to dichogamy. There are several contrasting features between staminate and pistillate flower development in the pecan. Once a seedling reaches sexual maturity, staminate flowers are produced one to two years before pistillate flowers. Production of staminate flowers first on a given seedling is not advantages as pecan is normally cross-pollinated. Cross-pollination of pecan ensures heterozygous progeny and increased survival of seedling with sufficient genetic diversity to endure the selective pressure imposed throughout a wide climatic range. Cross-pollination is promoted by dichogamy. When selfing occurs, fruit abortion is increased and kernel development

is suppressed. Improvement in productivity can be achieved with the planting of suitable pollinizer cultivars.

Strategies to Improve the Established Orchard

Due to lack of proper attention towards this valued crop, the production has remained much below the expected potential. Under different experimental trials, introduced pecan nut cultivars like Western Shelley, Desirable, Stuart and Wichita have performed well, however, location specific information is still lacking. Success of vegetative propagation which is very low also creates barrier in large scale production of quality planting material. Therefore, there is a need to raise high quality true to the type nursery through vegetative propagation techniques. As per preliminary estimates, about 1392 hectares are still available in only two districts *viz.* Rajouri and Poonch on which pecan nut cultivation can be undertaken successfully (Figures 24.2 and 24.3). This will help to create green cover in hilly areas and conserve soil from erosion, besides, providing high value wood for wood carving industry, as well as nuts for export purposes. New species/kinds/ cultivars are being introduced suiting to the agro-climatic conditions of the state to boost horticulture sector. Therefore, for the popularization and adoption of pecan nut cultivation in the intermediate zone of Jammu and Kashmir, following thrust areas should be given attention.

- ☆ **Top working:** The top working with promising cultivars on the low yielding inferior plantation of seedling origin and rejuvenation of old and senile plantations are the main thrust area of pecan nut cultivation.
- ☆ **Bud wood bank:** Establishment of bud wood/gene bank for ensured quality of planting material and further mass multiplication is required.
- ☆ **Improving orchard efficiency:** The other thrust areas are improving orchard efficiency through proper orchard management, farmers need to be educated on the importance of irrigation, nutrient management and plant protection, planting of adequate number of pollinizers and use of pollinators.
- ☆ **Increasing input use efficiency:** Increasing input use efficiency through proper management.

Future Strategies

Expansion of area of cultivation, evolving varieties resistant to major pests and diseases by non conventional approach and use of high density planting and development of efficient canopy architecture through proper training and pruning. Increase the area under organic farming. Therefore, there is a need to standardize different agro-techniques for profitable cultivation of pecan nut in Jammu province which includes introduction, evaluation and selection of high yielding cultivar, standardization of propagation techniques, standardization of establishment techniques for plants in the orchards, pollination/pollinizers studies and standardization of irrigation and nutritional requirement for Jammu province.

The work on these aspects will help to get some good strains of pecan nut which will boost this industry in Jammu province in general and J&K state in particular.

Researchable Issues

- **Development/selection of ideal varieties of pecan nut:** There is a need to identify suitable trees from native seedlings populations or to introduce cultivars from other countries which are suitable for the different climatic conditions prevailing in the state of Jammu and Kashmir. Till date, no pecan nut breeding programme has been undertaken in India to develop pecan nut cultivars adapted for use in the agro-climatic conditions suitable for its cultivation. Pecan nut varictics that have a higher level of phyto-chemicals for fresh consumption, and tolerance to major insect and disease pests.
- **Bud wood bank:** Establishment of bud wood/gene bank for ensured quality of planting material and further mass multiplication is required.
- **Production of quality planting material:** Number of quick and efficient plant multiplication technologies has been developed in the recent years. But still plant multiplication is being done with age old techniques. Standardization of proper propagation technique and establishment of a separate mother block propagated from elite pecan nut plants and their proper maintenance is the need of the hour.
- **Rejuvenation of old and senile orchards:** Many old trees with declining production and nut quality can be stimulated to more active growth and increased nut production by pruning. Standardization of time, severity for pruning back old/senile pecan trees and standardization of calendar of operation for encouragement of new growth, as well as the tree architecture is one of the researchable issue which requires immediate attention so that there will be increase in the production of nuts with improved quality.
- **Refinement of production technology:** There is no updated technology available to the growers for taking up walnut cultivation on scientific lines under Indian conditions. This is due to a lack of well planned orchards and no systematic trials have been laid out to study the various aspects *viz.*, canopy management (training, pruning), placement of pollinizers, growing of intercrops, possibility of use of growth retarding chemicals, water and nutritional management have not been thoroughly been worked out for pecan nut cultivation which needs due attention for making walnut culture more profitable and productive.
- **Improving orchard efficiency:** The other thrust areas are improving orchard efficiency through proper orchard management, farmers need to be educated on the importance of irrigation, nutrient management and plant protection, planting of adequate number of pollinizers and use of pollinators.

Initiatives to be taken for Increasing Quality and Production in Walnut in Jammu Province–Strategies

1. Cluster approach area expansion will be given emphasis. This can facilitate in post harvest management and in developing proper marketing infrastructure.
2. Production of planting material of elite germplasm with an objective to increase yield production.
3. Imparting training to farming community and extension workers for acquiring the knowledge and know-how of latest technologies at SKUAST-J and other Research Institutions outside the State for intensifying the yield gains through judicious use of modern technologies.
4. Similarly women shall be given know-how of technologies for active participation in production besides ensure their independence for earning income by imparting training in the art of fruit processing preservation and value addition of fruit crops.
5. 20-30 per cent plantation of pecan nut have become unproductive due to age, attack of pest/diseases and changing climatic scenario. In addition to such plantation there is huge count of wild/inferior pecan trees, whose produce fetch negligible returns to the farming community as the produce is of very inferior quality.. Therefore, rejuvenation of senile/old plantation including top working of wild/inferior fruit cultivars will be given due attention by following corrective pruning, replanting and graft-age of inferior/wild fruit cultivars into superior and commercial viable ones. This programme activity will ensure increase in productivity of quality and commercially viable fruit crops and employment generation.
6. To reduce post harvest losses of fruits due to poor handling at Farm level, on farm handling units will be constructed in bearing orchards.
7. Mechanization will also be a main priority as such Power Tiller/Tractors, Power Sprayers, Water lifting pumps and pruning tools shall be given due consideration.

REFERENCES

Economic Survey, *Directorate of Economics and Statistics, J&K* 2013-14.

Economic Survey, *Directorate of Economics and Statistics, J&K* 2015.

Goh K.M. and R.J. Haynes. 1986. Nitrogen and agronomic practice. In Mineral Nitrogen in the Plant-Soil System written by Haynes, R.J., K.C. Cameron, K.M. Goh and R.R. Sherlock. Academic Press, Inc. Harcourt Brace Jovanovich Publishers. New York. pp. 379-468.

Government of Jammu and Kashmir. (2012-13), Economic survey, Directorate of Economics and Statistical Planning.

Grant D. Hall. "Pecan Food Potential in Prehistoric North America". *Economic Botany* (New York Botanical Garden Press). *JSTOR* 4256253.

Joolka NK, Singh RR, Sharma MM (2004). Influence of biofertilizers, GA3 and their combinations on groth of pecan seedlings. *Ind. J. Hort.* 61: 226-228.

Package of practices of Fruit Crops. Division of Fruit Science, SKUAST-Jammu.

Pathak R.K.(2004) Rejuvenation of senile Horticulture Plantation for improvement productivity and quality. Invited lecture in First Indian Science Congress held at New Delhi, 6-9 Novemebr, 2004.

Pecans: Cholesterol Lowering Source of Antioxidants, Fiber, Vitamin E, Protein". Ilovepecans.org. *Retrieved* 2010-06-03.

Ravindran C., Sharma M., and Kher R. (2008). Present status and problems of pecan nut (*Carya illinoinensis* (wang.) k. koch) cultivation in jammu and Kashmir. *Acta Horticulturae,* 2: 179-182.

Sharma and K.K.Paranick (eds), pp. 19-26.

Transformation of Indian Agriculture through Innovative Technologies *Pages 371–388*
Editor: **Dr. Shahid Ahamad & Dr. Jag Paul Sharma**
Published by: **ASTRAL INTERNATIONAL PVT. LTD., NEW DELHI**

25 Transforming Rural Agriculture Production Scenario through Latest Pulses Production Technologies

Brij Nandan, B.C. Sharma, Rakesh Kumar, Ranjeet Kour, Monika Banotra and Kapilashiv

Technological interventions in agriculture makes the nation self sufficient in food grain production from 51MT in 1950-51 to a ever time record production of more than 259 MT in the year 2012-13 since Green Revolution. This phenomenal change in agriculture scenario of the Nation is widely supported by Rainfed agriculture. But know this sector currently faces a host of challenges and fresh constraints like ever growing population, increasing food and fodder demands, degradation in natural resources, higher cost of inputs and more importantly the concern of changing behavior of climate The phenomenal enhancement in the food grain production is the direct impact of production technologies that were introduced time and again in these areas. Even though the Nation would requires about 345 MT of food grains by 2030 (GOI, 2009) implying that we have to ensure an increase of about 5.5 MT of food grains every year to achieve this target. Whereas the net cultivated area has hover between 140-142 M ha since the last four decades with almost negligible probability of an increase in this area in future. Moreover, the present agriculture is also facing unhidden fact that the productive lands have been diverted from agriculture to infrastructural development, urbanization and other non agriculture developmental activities. Pro-cereal Green Revolution Technology launched in the late sixties undoubtedly made India a proud nation of surplus food with 60 million tons in buffer stock. This transformation, however, marginalized pulses, oilseeds and coarse cereals, and contained the vast crop diversity required for balanced development of agriculture.The recent reports on soil sickness under continued cereal based production system are alarming and call for a radical shift in cropping systems. For crop diversification, pulses have comparative advantage over others due to their *intrinsic ability to fix atmosphere N_2, conserve natural resource (soil, water), and improve soil fertility, low input requirements, short duration and hardiness. Paradoxically, the area under pulse crops is shrinking despite their unique importance in long –term sustainability of production system. In the recent past, development of short duration, disease resistant and high yield varieties with matching production technology have led to introduction and popularization of several new cropping systems such as pigonpea-wheat, rice/maize-wheat-mungbean, maize-potato/mustard–mungbean/urdbean in North West Plains, rice-chicpea/lentil,maize-pigeonpea in North East Plains and rice-urdbean and sorghum- chickpea in Southern peninsula. Further, pulses are also excellent intercrops. Short duration mungbean,urdbean,cowpea,etc.,have been found quite compatible for intercropping with spring planted sugarcane in north India. The photo-thermo insensitive varieties of these crops may also be pre-rabi crop under intensive cropping systems. Development of extra-short duration varieties of chickpea may open a new avenue for their cultivation for green pods during September- December without scarifying major cereal crops. This paper discuss about various constraints and opportunities in new cropping system as well as their introduction in new niches.. In light of that, the only option left with us is to enhance the productivity vertically instead of horizontal expansion in area, production and productivity. Since, rural agriculture is playing a pivotal role in maintaining the production of food grains besides providing support to the animal feed and fodder. Therefore, there is dire need to transforming rural agriculture production scenario through latest pulses production technologies for higher productivity, food and nutritional security to the farming community of the domain. In this chapter an attempt has been made to exploit the use of pulses in the cropping system and to transforming rural agriculture production scenario through latest pulses production technologies.*

Introduction

With the advent of high yield varieties, augmentation of irrigation facility, increased use of fertilizers, adoption of improved agricultural practices, and concerted efforts of researchers, planners, administrators and above all the farming community, the Green revolution was brought in the country during the mid sixties. The country witnessed great enthusiasm from all the segments of food production campaign. This led to the quantum jump in the food grain production from 51 million tons in 1950-51 to a record figure of 209 million tons in 1999-2000.The area under rice and wheat,which was 36.94 and 14.99 million ha in 1967-68, has gone up to 44.97 and 27.43 million ha in 1999-2000.The country became self sufficient in food supply as far as the need for growing population was concerned, because the 79 per cent food supply was made available from the 165 million tons production of mainly two cereals crops *viz.*, rice and wheat. The process of widespread expansion of cereal based cropping systems replaced most of the sustainable cropping systems, which involved pulses, as these crops were not competitive on the economic ground. Consequently, the pulses in particular was hovering around 16 per cent in 1967-68, has declined to 13.2 per cent at present. Constraints particular abotic stresses did not allow the expression of advantages of improved pulses production technology in terms of total production. The production growth rate of pulses could not keep pace with growing population resulting into progressive decline in the per capita availability of pulses from 69 g in 1961 to 32 g at present. Through the revolution made the country a proud nation of sufficient food, the intensification of cereals production caused heavy drain on soil nutrients, siphoning of ground water and buildup of diseases and insect pest pressure. The adverse affects were realized in the late nineties when compound production growth rate declined from 2.74 per cent during 1981-90 to 1.66 per cent 1991-2000.

To alleviate protein energy malnutrition, a minimum of 50 g pulses capita^{-1} day^{-1} is recommended which is quite high from the present availability of 32 g day^{-1}.To make up this shortfall in supply besides, of course, further demand from burgeoning population, at least 19.6 million tons of pulses are required by end of the xth Plan. Therefore, urgency of raising production to maintain self sufficiency in food and environment security warrants immediate corrective steps in the form of crop diversification so as to maintain a balance between crop intensification and sustainability of the production system. Pulses by virtue of several advantages have been considered the best choice for diversification in cereals –based cropping system to sustain the production and also to ensure nutritional security of Indian masses.

Role of Legume for Balancing Human and Livestock Nutrition

Legumes with 2-2.25 per cent content (20-25 per cent protein; exception being soybean,40 per cent) are rich source of amino acids like tryptophan, lysine, *etc.* however, deficit in S-containing amino acids (cysteine, cystine and methionine) forms balanced staple for human beings when combined with cereals.The consumption of protein from all sources of food falls below the requirement of healthy person in India and thus Protein Energy Malnutrition (PEM) has been identified as a major health and nutritional problem. About 23-70 per cent of the rural population.(Table 25.1) in different parts of the country is suffering from PEM (Prasad, 2003) and the highest is in Southern India. As per ICMR, a gram of protein is required/kg body weight andthus 60 g protein/day is needed (across the age group of human beings).

Table 25.1. Protein Energy Malnutrition (PEM) Scenario in Rural India

Region	*Population Affected by PEM (per cent)*
North	34.9-36.9
East and central	23.5-58.2
West	30.2-39.8
South	31.4-70.3

Source: Vyas *et al.*, 2010.

Pulses supply 6.05 g/caput/day of protein (8.91 per cent of total protein consumption) in the world (2007) however, it is slightly higher in India (FAO STAT, 2010) *i.e.* 8.40 g protein (15.5 per cent of total) that indicates the greater importance of pulses as suppliers of protein in developing countries like India. The production of pulses being static in India with tremendous growth in population has continuously decreased pulse supply from the highest of 70.3 g/caput/day (1956) to 35.9 (2007) and this could be still less currently. There is a need to enhance the production of grain legumes if the supplies are to be maintained even at current levels in coming years

Legume fodders and haulms also meet the protein needs of livestock in combination with cereal fodders and stovers. The seed husk of *dal* making units and local *chakkis*, deoiled cakes forms the basic ingredient of compound livestock feeds manufactured in India and thus helps in enhancing healthy animal growth and their production. The livestock products (meat, milk, eggs, *etc.*) in turn aid in better human nutrition.Therefore helps in maintaining a balance between the two.

Constraints in Adopting the Legume Based Technological Interventions in Cropping Systems at Rural Areas

- ☆ The effective area under different cropping systems and their contribution in agricultural production is still under exploited domain for proper expansion of the cropping systems with different legume interventions.
- ☆ Improper or almost non adoption of machinery in pulses cultivation
- ☆ The lack of availability of quality seed of legumes at different site specific cropping systems is also one of the major constraints in adopting legume based cropping systems.
- ☆ Over the last two decades a large number of region-specific and widely adapted legume based cropping systems have been developed but out of them very few were promoted
- ☆ Lack of awareness among the farmers for the exploitation of recent scientific and agronomic interventions in traditional cropping systems like need based intercropping combinations.
- ☆ Lack of government policies in promoting legume based cropping systems.
- ☆ Capacity building programmes is also lacking in generating awareness among the farming communities.
- ☆ Post harvest handling of the pulses at rural level.

Technology Initiatives (Merits of Legume Inclusion in Cropping System)

Over the last two decades a large number of region-specific and widely adapted cropping systems have been developed but less importance is being paid to inclusion of pulses in the existing cropping systems. There are scientific proven technologies that reflects the importance of pulses inclusion in the cropping systems by way of adding soil organic matter, reducing fertilizer application pressure/load, enhancing weed control efficiencies, reducing methane emission (the main culprit of climate change from agriculture sector), enhancing water use efficiency through appropriate replacement of need based crop in the cropping systems. Reducing over mining of soil nutrients with traditional cropping systems, enhancing factor productivity, enhancing value addition opportunities and generating employment avenues and livelihood securities.(Gangwar and Prasad, 2005). The various research based benefits of adding legumes in the cropping system is highlighted here under:

Leaf Litter and Nutrient Addition

Additional benefit of leaf litter addition besides supplementation of N, P and K in passable amount to the soil which otherwise being not added in the traditional system (Like cereal-cereals). Study regarding the leaf litter and nutrient addition is

Figure 25.1. Use of Seed Drill in Pulses for Transforming Pulses Production Technology.

highlighted here which shows that chickpea added about 1.1-1.7 t/ha leaf litter in addition to the 7-14 kg/ha N, 3-5.5 kg/ha and 8-20 kg/ha K whereas lentil added 1.3 t/ha leaf litter and N,P and K to the tune of 8-10,3.5-4.5 and 12.5-19 kg/ha, respectively similarly Pigeonpea added 1.3-2.8 t/ha and 8-16,2.5-5 and 13.5-24 kg/ ha N, P and K, respectively.

Table 25.2. Leaf Litter Fall and Nutrient Contribution through Leaf Litter

Crop	*Leaf Litters (t ha^{-1})*	*N (kg ha^{-1})*	*P (kg ha^{-1})*	*K (kg ha^{-1})*
Chickpea	1.1-1.7	7-14	3-5.5	8-20
Lentil	1.3-1.6	8-10	3.5-4.5	12.5-19
Pigeonpea	1.3-2.8	8-16	2.5-5	13.5-24

Source: Singh, 2011.

Weed Smothering Effect of Legume Addition

Intercropping

Intercropping has been found to suppress the weeds through formation of good canopies due to competitive planting pattern and thus provided an opportunity to utilize intercropping system as a tool of weed management (Kumar, 2011). The suppression of weed growth in intercropping system is mainly due to increased leaf area index and light interaction. Inclusion of short duration and quick growing intercrops in the rows of long duration and tall growing crops has been found to suppress the weed infestation resulting in considerable saving of weeding requirement in base crops. Weed smothering efficiency of important intercropping system are given in Table 25.3. As a result of considerable reductions in weed weight due of exclusion of intercrops in long duration wide spaced crops; the weeding requirements have been found to be reduced substantially.

Table 25.3. Weed Smothering Efficiency of Important Cropping System

Intercropping Systems	*WSE (per cent)*
Pigeonpea + urdbean	32.82
Pigeonpea + mungbean	31.01
Pigeonpea + cowpea	39.06
Pigeonpea + sesame	36.6
Pigeonpea + pearlmillet	50.8
Maize + blackgram	17.3
Maize + pigeonpea	16.4

Source: Kumar, 2011.

Figure 25.2. Seed Drill Sown Crop Shows Weed Smothering Effect for Transforming Pulses.

Legume Interventions in Cropping Systems for C Sequestration

The inclusion of legumes has the great potential of C sequestration in cropland has provided a promising approach to reducing the atmospheric concentration of CO_2 for mitigating climate change. However, this approach depends on cropping systems, which may be defined as an operating system for growers to follow in their practices for crop production. An ideal cropping system for C sequestration should produce and remain the abundant quantity of biomass or organic C in the soil. The major strategies in developing cropping systems are discussed below.

Cultivation of Legumes – Legumes are Climate Resilient Crops

During nitrogen fertilizer production, for every kg of NH_3 produced, there is a 10 kg of CO_2-C emission. On an average legume crops can fix up to 100 kg of N/ha annually. Thus, for each legume crop grown, there is approximately 1 ton of CO_2-C emission that is avoided. For a 5-year rotation with two hay crops per rotation cycle, total CO_2-C emission that is avoided in a 75-year period is 30 tons.

Figure 25.3. Transforming Rural Economy by Successfully Cultivating Climate Resilient Pulse Crops.

Over the same period, soil carbon storage increase with manure application of 2 tons of carbon per year is approximately 30 tons/ha, with a total manure carbon input of 150 tons/ha. Thus, there is clearly a carbon emission benefit in using legume crops. In addition to reduced carbon emission, there is also an environmental benefit in using legume crops resulting from increased plant residue input due to huge leaf fall which increased soil organic carbon content. More importantly, the carbon emission savings from using legume plants is permanent while soil carbon content increase resulting from increased inputs must be maintained continuously. The studies conducted by Reicch *et al.* (2001) on N_2 fixation enhancement in free air CO_2 enhancement (FACE) revealed that resilience of SNF of legumes to changing climates prompting greater scope for legumes. Therefore, such crops are helpful in reducing the pressure of GHG and can be a useful tool for reducing ill effects of climate change.

Alternate Land Use Systems

The various alternate land use systems *viz.*, agro-forestry, agro-horticulture, and agro-silviculture, are more remunerative for SOC restoration as compared to

Figure 25.4. Pulses as Alternate Crops for Summer Season for Transforming Rural Economy.

sole cropping system. In northeast hill state India, where all the above three land use systems exists that reduce soil erosion and SOC loss considerably. In a 6 year study, organic carbon content was about double in agro-horticultural and agro-forestry systems as compared to sole cropping (Table 25.4). In long-term study in Indogangetic alluvial soils with rice based cropping system C- sequestration was maximum in rice-wheat-jute system (535 kg $ha^{-1}y^{-1}$) at Barrackpore followed by Rice-mustard-sesame (414 kg/ha/y) than rice-fallow-rice (402 kg $ha^{-1}y^{-1}$)at CRRI, Cuttack.

Table 25.4. Organic Carbon in Soil after Six Years of Plantation with different Land Use Options

System	*Organic C (per cent)*	
	0-15 cm	*15-30 cm*
Sole cropping	0.42	0.37
Agro-forestry	0.71	0.73
Agro-horticulture	0.73	0.74
Agro-silviculture	0.38	0.56

Source: Das and Itnal (1994).

In India the major cropping systems are cereal-cereal (rice-rice, cereal-cereal-cereal (rice-wheat-maize), cereal-cereal-legume (rice-wheat-greengram), legume-cereal (pluse-wheat) and oilseed-cereal (oilseed-wheat). Conventional agriculture normally reduces the SOC of the surface or plough layer. Conservation tillage systems increase SOC level. Under intensive cultivation two or three crops are grown per year and the grain yield at the rate of 10 Mg ha^{-1} y^{-1} was achieved.The data in Table 25.5. indicated that rice-wheat-green gram crop sequence has greater SOC restoration than rice-wheat followed by rice-green gram and rice-mustard possibly due to inclusion of legume in cereal-cereal cropping system.

Table 25.5. Soil Organic Carbon Status as Influenced by different Cropping System (after 6 years)

Cropping Systems	*Organic C (g kg^{-1})*
Rice-wheat-fallow	5.8
Rice-wheat-fodder	6.1
Rice-wheat-green gram	6.8
Rice-mustard-fallow	5.6
Rice-mustard-fodder	6.0
Rice-mustard-green gram	6.5
Uncultivated soil	5.1
Initial soil status	5.2
L.S.D. at 5 per cent	0.5

Source: Sharma and Bali (2000).

The six years study of Sharma and Bali (2000) on different cropping systems revealed that the cropping system that contains legumes sequestered more carbon than those cropping system which did not contain any legume in the system. Under irrigated situations the cereal based cropping system rice –wheat –green gram sequestered more organic carbon (6.8) than Rice- wheat alone (5.8) whereas the non cereal based cropping rice- mustard-green gram system fixes more organic carbon (6.5) than rice-mustard-fodder (6.0) cropping system.

Nutrient Mobilization

Nutrient uptake by a crop raised alone or in combination is controlled by availability of water in the root zone. The differences in rooting system of component crops in mixture could occur because of the tendency of roots to avoid overlapping in root network. Therefore, the principle of selection of intercrops being that the crops of different rooting system are chosen and thus the associated crops do not compete for extraction of nutrients from the similar soil layers/depth and moreover, a greater volume of soil is exploited in an intercropping system as compared to sole cropping. The crops thus avoid area that has already been depleted of resources (nutrients and water) by the associated crops and exploit a greater total volume of the soil. This was considered as the possible reason of yield advantage in growing some crops in mixtures. Singh *et al.* (2000) while working on lentil and mustard intercropping at Bihar revealed that intercropping of lentil+mustard recorded lower values of uptake of nitrogen and phosphorus as compared to their sole stand.

Figure 25.5. Pulses for Transforming Nutrients in *Rainfed* Areas for Rural Agriculture.

Water Use Efficiency

Water use efficiency is the quantity of dry matter of the economic yield produced per unit quantity of water used and is generally expressed as kg ha^{-1} cm^{-1} of water used. Crops grown under rainfed conditions are prone to water stress, owing to rapid loss of soil moisture and development of mechanical impedance to root growth. The stress can be alleviated by enlarging root volume in the soil or by regulating the supply of soil moisture. Thus, soil moisture is the major limiting factor for crop production under rainfed conditions. The effects of intercropping on water use/expense efficiency have received less attention than the effect on nutrient uptake and, so far, there are little evidences of beneficial effects. Baker (1974) tried to show theoretically that spreading peak water demands by temporal intercropping must give more efficient use of water. In northern Nigeria, Baker and Norman (1975) suggested that better water use efficiency was probably a common cause of yield advantage in intercropping because of mutual avoidance of similar system of root system of the component crops to exploit different soil layers Further, their differential requirements of water and differential absorption abilities are also responsible for compatibility. Chickpea with toria, sarson or mustard was reported to be the common mixture in North and North Western India (Sexsena and Yadav, 1975).Reddy *et al.* (1980) at ICRISAT observed that water use was increased by intercropping.

Figure 25.6. Pulses Useful in Transforming Water Use Efficiency in *Rainfed* Agriculture.

Legumes as Fertilizer Use Reducing Machines

Legumes with an unique trait of fixing nitrogen in symbiosis with rhizobia are part of N management since time immemorial as nitrogen is the most crucial element limiting productivity of crops in arable lands. The advent of nitrogen management has marginlaized the role of legumes in cropping systems gradually. The composition of N fertilizers grew in leaps and bounds to reach current levels of 0.1 billion tone in the world. However, the energy involved and environmental concerns with production and use of fertilizer N gave re-impetus as source of nitrogen. Though all legumes fix nitrogen, the addition of N_2 to the soil is in the order of:

Green manure > forage > vegetable legumes > grain legumes.

The N_2 fixation usually vary from 20-70 kg/ha in grain legumes to 24-90 kg/ ha in forage legumes as source of nitrogen. In forage legumes, the N_2 fixed is not mobilized fully from the roots too economic products. Hence, there is a greater benefit to the succeeding crop.

Nitrogen Acquisition

Effective use of soil nitrate by legumes growing before rice could conceivably reduce subsequent losses of soil nitrogen by leaching and denitrification and benefit the rice by cycling soil nitrate N through readily mineralizable, N-rich residue to the succeeding rice crop. Nitrogen accumulation by legumes in tropical rice-based cropping systems is influenced by water regime, soil fertility, photoperiod, inoculation, and legume growth duration. With an adequate water and nutrient supply, fast-growing, food-tolerant legumes can accumulate more than 100kg above ground N/ha in 50 to 60 days. Slower growing drought tolerant legumes, such as pigeonpea (*Cajanus cajan* [L]Millsp) and *Indigofera tinctoria,* are better suited for sowing after wet season rice with subsequent N accumulation through the dry season and then incorporation immediately before the next wet season rice.

Intercropping a Tool for Securing Food Production in Rainfed Areas

Intercropping of oilseeds and cereals with legumes may not prove useful only in stabilizing yields and returns over space and time but would help in stepping up the production of oilseeds, cereals and pulses. The success of an intercropping system will depend not only on the choice of component crops but also on their symbiotic compatibility. Plant population and their spatial arrangement in an intercropping system have a bearing on the balance of competition between component crops and their productivity. Continuous decline in per capita holding size of cultivable land is an indication that the scope for horizontal expansion of the crops is almost negligible and the only option left so far is the vertical expansion of crops through the adoption of intercropping systems. Amongst the various factors for vertical expansion, intercroping systems that too with inclusion of legumes as intercrops is one of the viable options for production enhancement, economic advantage, soil fertility improvement and reduction in the risk of complete crop failures over space and time per unit land area.

Intercropping Sequence: The Way to Food Security

This is an another emerging researchable issue for food security under rainfed situation; as in most of the cases Scientist/farmers and policy planners used to promote only single season intercropping production technologies but the sequential intercropping systems *that too with the inclusion of legumes is* still almost a untouchable production area especially in Indian context and if this new area is exploited to a greater extent,will leads to be the better option for food and nutritional security.

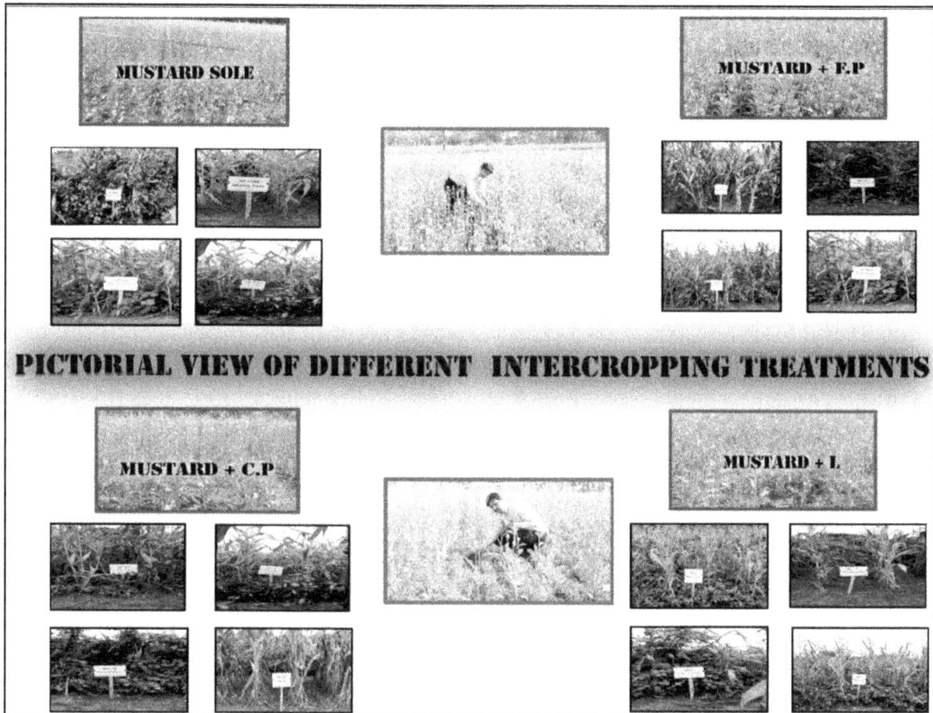

Figure 25.7. Pictorial View of different Intercropping Treatments.

In this line of action, a two years study was undertaken by Nandan *et al.* (2013) on mustard –maize based intercropping systems containing different legumes of both *kharif* as well as *rabi* season under subtropical rainfed *Kandi* areas of Jammu.

The experimental results revealed that there were higher values of system productivity, Land use efficiency, water use efficiency besides improving soil properties and high nutrient mobilization in mustard+fieldpea succeeded by maize+cowpea intercropping sequence than sole mustard –sole maize cropping system (Farmers practices). Sixteen different cropping systems were tested, which includes sole cropping of mustard-maize, one legume intercropped in *rabi* followed by sole maize in *kharif*, sole mustard followed by one intercrop in *kharif* maize and inclusion of one legume each in rabi and *kharif* crops of mustard and maize. The research showed that the intervention of legumes in both *kharif* and *rabi* were beneficial both in terms of monetary as well as ecologically.

Shift in Cropping Systems

Analysis of crop patterns in the past three decades shows that's a remarkable shift took place after the introduction of input responsive high yield varieties of rice and wheat. With expansion in net irrigated area from 31.1 to 54.6 million ha in span of 1967-68 to 1999-2000, rice and wheat emerged as dominant crops with highest total cropped area. In the pre-green revolution period, wide spectrum of diversity in crops and variety was the common feature of Indian Agriculture with major and minor cereals being grown mixed/intercropped with pulses and oilseeds. The king of the cereals, wheat used to be grown in combination with chickpea, lentil, mustard and other oilseed crops. The coarse cereals were cultivation in mixture with short growing pulses like urdbean and mungbean. Each of these crops had wide array of land races/cultivars presenting a wide spectrum of genetic diversity whereas in the post green revolution period, rice and wheat ousted the dominance of several crops by replacing 20 crops each in kharif and rabi seasons. Now both of the cereals share 59 per cent of the total cropped area (Table 26.6). In the states like Uttar Pradesh Punjab and Haryana, these two crops together account for 79 per cent of the total area. The situation in Punjab state is still worse as about 96 per cent cropped area is under rice and wheat. This phenomenon led to the relegation of less productive pulses and coarse cereals to marginal lands. More area under pulses shifted from irrigated tract of North West and North East region to central and southern regions of moisture stress conditions. However, this shift from more productive region to less productive areas somehow kept the total area of pulses almost constant at about 22 million ha without accruing much benefit of the improved technologies in terms of the increased productivity. Expansion of pulses in Andhra Pradesh, Rajasthan, Gujarat, Karnataka, Maharashtra, Madhya Pradesh and Tamil Nadu is the testimony of this phenomenon where area under pulses jumped from 13.92 m ha in 1971-75 to 17.27 m ha at present . The states like Bihar, Haryana Punjab Uttar Pradesh west Bengal and Odisha observed reverse trend from 8.03 m ha to 5.22 m ha during the same period (Table 26.8).

Tables 26.6. Trends in Area under different Crops in Post-Green Revolution Period in India

Crop	*Area under different Crops (m ha)*		
	1967-68	*1999-2000*	*Increases/Decrease*
Rice	36.44	44.97	+8.53
Wheat	14.99	27.43	+12.44
Coarse cereals	47.34	29.34	-17.88
Pulses	22.65	21.19	-1.46
Oilseeds	15.67	24.37	+8.70
Total food grains	121.42	123.06	+1.64
Cotton	8.00	8.76	+0.76
Sugarcane	2.05	4.22	+2.17
Total	147.14	160.40	+13.26
Net sown area	140.27	142.2	+1.93
Total cropped area	165.79	190.76	+24.47
Net irrigated area	31.10	54.57	+23.97
Gross irrigated area	38.19	72.78	+33.88
Kharif	81.49	72.88	-8.61

Rabi	39.93	50.18	+10.25
Cropping intensity (per cent)	118.2	134.3	+16.10

Table 26.7. Major Pulse Based Cropping Systems of the Country

Prominent pulse based Cropping System	*Zone*
Cropping systems (Sequential cropping)	
Pigeionpea- wheat and Urdbean/mungbean-wheat	NWPZ
Maize/Sorghum/peralmillet-chickpea/lentil	Central and South
Rice–Rice -mungbean	South
Intercropping system	
Pigeon pea+ sorghum/groundnut/munbean/urdbean/cotton	Central and South
Pigeon Pea + maize/sorghum (Kharif season)	NWPZ and NEPZ
Chickpea/Lentil+ mustard/sunflower/linseed	NWPZ,NEPZ and CZ

NWPZ: North West Plain Zone; NEPZ: North East Palin Zone; CZ: Central Zone and SZ: South Zone.

Source: Nadarajan, 2010.

The Statistic and Importance Legumes in the Cropping System

More than 250 double cropping systems of primary, secondary and tertiary importance in terms of their spread in the country have been listed. Out of which 30 are of primary importance (Yadav and Prasad, 1997). Among top ten popular cropping systems in the country, only two, *viz.*, rice-chickpea and maize-chickpea contain a pulse crop with less than 6 per cent of the total pulse area.

Table 26.8. Area under Prevalent Cropping Systems and their Contribution to the National Food Basket

Cropping Systems	*Area (m ha)*	*Contribution (per cent)*
Rice-wheat	9.77	25.00
Rice-rice	2.12	5.00
Cotton-wheat	1.39	2.36
Pearl millet- sorghum	1.35	1.68
Maize-wheat	1.29	2.25
Pearl millet- wheat	1.03	1.72
Sorghum-sorghum	0.74	1.65
Rice-chickpea	0.59	0.80
Sugarcane-wheat	0.54	0.86
Maize-chickpea	0.54	0.65

Source: Yadav (1996).

In the pre-green revolution period, legumes (pulses) found significant place in inter/mixed cropping with major and minor cereals. Wheat was used to be generally grown with chickpea, lentil, mustard and other oilseed crops. Similarly, the coarse cereals were grown with short duration pulses like urdbean and mungbean in intercropping/mixed cropping systems. Cropping systems based approach of agricultural research, received little attention, except some considerations for utilizing the beneficial effects of growing crops of dissimilar nature in mixed/intercropping (Aiyer, 1949) or in sequential cropping and role of legumes in green manuring (Singh, 1972). After introduction of high yielding, short stature, photo and thermo-insensitive varieties of wheat and rice in sixties, the entire agricultural systems of country witnessed a change. The low productive, risk prone legumes and oil seed crops were diverted on marginal and fragile land of dry areas, whereas the cereal based multiple cropping systems covered irrigated areas in North. In Andhra Pradesh, Rajasthan, Gujarat, Karnataka, Maharashtra, Madhya Pradesh and Tamil Nadu states, area under pulses increased from 13.92 m ha in 1971-75 to 16.22 m ha in 2005-06, whereas Bihar, Haryana, Punjab, Uttar Pradesh, West Bengal and Odisha witnessed reverse trend, the area declining from 8.0 m ha to 4.6 m ha during the same period. The adverse effect of continued cereal based cropping system in northern India- the Green Revolution belt could be visualized only in the late nineties when compound production growth rate declined from 2.74 per cent during 1981-90 to 1.66 per cent during 1991-2000. Progressive decline in total factor productivity and deterioration in soil health necessitated crop diversification and inclusion of pulses in the system. (Singh *et al.*, 2009).

Adverse Effects of Cereal based Cropping Systems

The ill effect of continuous cereal cultivation was not evident in the beginning as soil had enough reserve of plant nutrients and adequate water resources. complication developed at later stage when the production of wheat and rice in major states like Punjab showed negative pattern. The organic matter content of the soil declined to less than 0.2 per cent.The fertilizer use efficiency reduced from 50.25 kg in 1969-70 to 11.52 kg grains per kg of NPK in 1999-2000.The ratio of N P K fertilizer once being 4 N: 2P: 1K widened to 8.5 N: 3. 1 P: 1 K at the national level with northern states showing maximum distortion (37.1 N: 8.9 P: 1 K). The ground water reserve started depleting. deficiency of micronutrients limited n the expression of yield potential and vulnerability to diseases and pests increased, caused a big threat to Indian Agriculture. The heavy dependence on the use of inorganic fertilizer led to soil sickness and threat to environment and mankind. The diversification of crops involving pulses, which was a feature of traditional agriculture in imparting sustainability, became important for food and nutritional security. The current thinking of diversification in the contemporary agriculture is focused to make them sustainable and profitable besides insurance against the outbreak of diseases, insect epidemics and socio economic advantages.

Prevalent Cropping Systems

The prevalent cropping systems are the outcome of the technological innovations, household needs, reflection of government policies, and availability of production inputs, market forces and socio-economic compulsion. Yadav and Prasad (1998) enlisted 250 double cropping systems of primary, secondary and tertiary importance in term of their spread in the country out of which 30 are of

Figure 26.9. Transforming Rural Post-harvest Handling of Produce through Multi Crop Thresher.

primary importance. Among the top ten popular cropping systems in the country, only two (rice-chickpea and maize –chickpea) contain a pulse crop with less than 6 per cent of the total 20 m ha area whereas reaming cropping systems are purely cereal-cereal with major share of rice-wheat (9.77m ha) and rice-rice (2.12 m ha). There is a very common observation that a large diversity of cropping system exists in rainfed area with an overriding practice of intercropping systems tilted in favour of cereals mainly because of availability of high input responsive and photo insensitive varieties of rice and wheat.

Role of Pulse in Cropping System Diversification

Pulse on account of their short duration and ability to thrive better than other crops under harsh climate and fragile ecosystems on one hand and the national agenda to achieve house –hold nutritional security on the other have great promise for diversification of cropping system. Moreover, pulse in cropping system help in improving the soil fertility. A comprehensive study on sustainability of rice-wheat sequential cropping at pantnagar showed that the organic carbon and total N in the soil declined under rice-wheat rotation after five year whereas significant improvement was noticed in organic carbon and total N when one of the cereal components was substituted with legume crop (Singh *et al.*, 1996) Similar effect was also noticed on the available phosphorus .

REFERENCES

Aiyer, A.K.Y.N. 1949. Mixed cropping in India. *Indian Journal of Agricultural Sciences*, **19**: 439-543.

Baker, E. F. I. 1974. Research on mixed cropping with cereals in Nitrogen farming systems-a system for improvement. *Proceeding of International Workshop on farming Systems*. pp. 287-301. ICRISAT, Hyderabad.

Baker, E. F.I. and Norman, D. W.1975. Cropping systems in northern Nigeria (In) *Cropping Systems workshop, Proceeding IRRI*, Los Banos, pp. 334-361.

Das, S.K. and C.J. Itnal 1994. Capability based land use system: role in diversifying dryland agriculture in dry land area. *Bull Indian Soc. Soil Sci.* 16: 92-100

Kumar Narendra 2011. Issues related to weed management under RCT. In proceeding of " Resource conservation technologies for enhancing input use efficiencies and sustainable pulse production" held at IIPR, Kanpur, September, 8-28, 2011 pp. 73-77.

Nandan, B and Sharma, B.C. and Kumar, Anil (2013): Productivity potential of maize and mustard under different intercropping systems in moisture deficit sub-tropical areas of Jammu and Kashmir, India. *Legume Research an International Journal*, 36(5): 436-441.

Reddy, M. S., Floyal, C. N and Willey, R. W. 1980. Groundnut in intercropping systems. *Proceedings in International Workshop on Groundnut*, pp.13-17. ICRISAT, Patencheru, Hyderabad.

Sexsena, M.C. and Yadav, D.C. 1975. Some agronomic considerations of Pigeonpea and chickpea. In: International workshop on grain legumes held at ICRISAT, Patancheru, Hyderabad, Indian, Jan. 13-16, 1975, 31-61.

Sharma, M. P. and S.V. Bali 2000. Long-term effect of different cropping system on physico-chemical properties and soil fertility. *J. Indian Soc. Soil Sci.*, 48: 181-183.

Singh K.K,. Ali, Masood,.Venkatesh, M.S (2009): Pulses in Cropping Systems,. Technical Bulletin, IIPR, Kanpur, pp. 2-5.

Singh, A. 1972. Conceptual and experimental basis of cropping patterns. In: Proceedings of "Symposium on cropping patterns in India" held at New Delhi, 27-31 January, 1968, ICAR, New Delhi, pp. 271-274.

Singh, K.K. 2011. Residue Management in Pulse Based Cropping System under RCT. In proceeding of "Resource conservation technologies for enhancing input use efficiencies and sustainable pulse production" held at, IIPR, Kanpur, September, 8-28, 2011 pp.57-63.

Stewart, B. A., and Robinson, C. A. (1997). Are agro ecosystems sustainable in semi-arid regions. *Adv. Agron.* 60, 191–228.

Yadav, R.L and Prasad, K. 1997. Identifications of predominant cropping systems in different districts of India. *Annual Report*, Project Directorate of Cropping Systems Research, Modipuram. pp. 30-49.

Yadav, R.L. 1996. Cropping Systems. In: 50 Years of crop science research in India (Eds. R.S.Paroda and K.L.Chadha), ICAR, New Delhi, pp. 117-128.

Transformation of Indian Agriculture through Innovative Technologies *Pages 389–401*
Editor: **Dr. Shahid Ahamad & Dr. Jag Paul Sharma**
Published by: **ASTRAL INTERNATIONAL PVT. LTD., NEW DELHI**

26 Transforming Rural India with Modern Fruit Production Technologies

Parshant Bakshi, Mudasir Iqbal and Kiran Kour

Agriculture has been an integral part of the traditional culture in many parts of Asia particularly in India. There are festivals for worshipping of the land, fruit trees, agricultural tools, water and standing crops, which signify the importance of these resources for the community. Over the years, agriculture has made significant progress with the growth of human civilization. Two major factors, which have contributed to the improvement in fruit cultivation, are the increasing needs of the growing population and initiatives of innovative farmers. With the increasing size of rural families, there was an urge to grow more food and the farmers were compelled to adopt new methodologies. The age old practice of selecting outstanding plants in the field has made a significant contribution in evolving superior genotypes, which are superior in yields and resistant to various pests and diseases. Many tools and implements were developed to increase the speed of tillage operations and reduce strenuous workload. All these innovative technologies like high density planting, micro-irrigation, fertigation, plant propagation techniques, post harvest technologies *etc.*, which were developed by a few elite farmers gradually became popular in the surrounding regions as the people saw the benefits in the field and decided to adopt them. However, in the absence of formal system of extension and efficient mode of communication, the process of transfer of technology has been very slow.

The quality, diversity, and availability of fruits are greater today than at any time in recorded history. Relative wholesale values of most fruits have been stable or declining for decades, higher crop yields are produced on decreasing acreage, and postharvest storage and quality have improved as new technologies have been developed (NRC, 1989). Contemporary fruit growers are more likely to worry about excess production and consequent depression of wholesale prices than crop failures or insufficient supply. Surprisingly, amidst these achievements, there are also increasing doubts about our ability to sustain current production and availability of fruits and other crops. "Sustainability" has emerged as a new paradigm by which fruit production systems and practices are judged (York, 1991). Federal and private funding agencies are reallocating millions of dollars from traditional research into new programs assessing environmental impacts, food safety, and other aspects of agricultural sustainability.

Fruit Production Systems

a) Traditional Fruit Growing and Gathering System

At first glance, traditional fruit production systems appear to provide excellent models of sustainability. They were developed with minimal external inputs, and obviously they have endured for very long periods of time. There are groves where avocado, cashew, olive, apple, grapes, and other perennial fruits have been harvested continuously for many centuries (Simmonds, 1976). A substantial portion of the world's production of several commercially important tropical fruits (*e.g.*, Brazil nuts) still is harvested from semi-domesticated, old forest stands (Smith *et al.*, 1992). Many more fruits gathered in these agro-forestry systems are not commercially marketed, but provide an essential part of the dietary needs of subsistence farmers and indigenous peoples. These traditional fruit production systems are certainly sustainable on their own merits. Nonetheless, rampant deforestation of large tracts for timber, grazing, and conversion to export cash crops threatens these forests, the local people who depend on them for food and income, and the invaluable diversity of flora and fauna that they harbor. In addition, these gathering systems often do not meet the food demands of increasing human populations. If the clear cutting and conversion of tropical forests to pasture or field crops is necessary to provide adequate food for the people of Third-world countries, then traditional forest agro-ecosystems are non-sustainable. However, several studies have indicated that the long-term food and income yields of tropical forests used by low-impact gathering of fruits, nuts, latex, and other resources could surpass the short-term yields of timber or slash-and-burn crops that usually motivate deforestations (Smith *et al.*, 1992). In such cases, it is perhaps the underlying socioeconomic systems that are unsustainable, not the fruit production systems themselves. This fundamental distinction of ten is ignored. In many instances, farming systems are faulted for shortcomings that are caused not by the internal dynamics or integrity of the agro-ecosystem, but by external socioeconomic forces resulting from national policies. It is difficult for farmers to sacrifice voluntarily short-term gains for the sake of intangible future benefits when the cultural, commercial, and political systems in which they must survive are rewarding short-term profits and economies of scale. Traditional fruit and nut gathering systems however intrinsically sustainable they may bear threatened and losing ground everywhere because of external factors unrelated to them agro-ecological integrity of these venerable systems. Without changes in human population dynamics and local/global economic systems, it appears unlikely that many traditional fruit-growing systems can be sustainable (Mervin and Pritts, 1993).

b) Traditional Family Farm Fruit Production

The orchards, vineyards, and berry plantings of 75 years ago were usually just one component of diversified family operated farms. Diversified farms provided some important advantages to growers, insulating them from occasional crop failures and commodity price fluctuations. Smaller plantings and larger families reduced the need for hired help. Integrating livestock with field and fruit crop systems minimized certain off-farm inputs. Organic manures were readily available, and pest-infested fruit could be used for animal feed. On the other hand, the productivity of traditional orchards was generally low, and production/replacement cycles were considerably longer than they are today. Expertise, equipment, and technical support often were limiting. Fruit plantings consisted of many genotypes and were not grafted on clonal rootstocks, resulting in non uniform crop stands that were difficult to manage. Nursery stock was propagated without regard for virus status, and diseased stock severely limited the production of many fruit crops. Pest controls included heavy-metal derivatives such as lead arsenate, or genetic sources of resistance (if they existed). The minor portion of each fruit harvest suitable for fresh consumption was bartered or marketed locally, was relatively expensive, and was available only during a limited season each year.

c) Modern Commercial Fruit Production

Modern production systems differ in many ways from traditional systems. They require more capital to establish, and must be more precocious and productive to be profitable. Fruit trees are usually grafted onto size reducing clonal rootstocks and planted at much greater densities. The harvest index is substantially higher in these "dwarf" orchards; they require less pesticide per acre, and they are easier and safer to prune and harvest because workers do not need to climb ladders. Similar changes have occurred in berry fruit production. For example, most strawberries are now grown in annual production systems. The major cultivars are cloned or micro-propagated, with fewer genotypes suitable for commercial growers dependent on a few varieties of one crop for specialized markets. Intercrops and polycultures are rare, atleast in the industrialized nations. Many of the traditional local varieties or landraces have been replaced by newer cultivars originating from scientific breeding programs. Pest control is accomplished through synthetic chemicals and breeding of resistant or tolerant cultivars. For the twenty odd major economic fruits that are bought and sold internationally, grading standards are high and market competition is fierce. Tropical fruits are consumed increasingly by people in temperate climate areas, and vice versa. Consumers in developed nations are offered an astounding diversity of relatively inexpensive, high quality fresh produce that is available year round from growers throughout the world. Changing consumer demands, coupled with stringent soil and climatic requirements, mandate that today's fruit plantings be re-established continuously in the regions where growing conditions are optimal. The ability to replant fruits successfully in these sites without extensive modification is itself a good test of the sustainability of prior production practices.

In commercial situations today, severe replant problems sometimes develop in previously productive orchards, vineyards, and berry plantations. Correction of nutrient deficiencies and/or soil physical problems such as drainage or compaction will sometimes solve these replant problems, but, in other situations, soil fumigation with broad spectrum biocides such as methyl bromide is necessary to maintain previous levels of production (Mai and Abiwi, 1981).

Modern Fruit Culture in Rural India

India has a wide variety of climate and soils on which large number of horticultural crops are grown. After the green revolution in the sixties, it became clear that the horticulture for which the Indian topography and agro climate are well suited, is an ideal method of achieving sustainability of small holding diversification in horticulture is a best option as there are several advantages of growing horticultural crops. Those crops highly remunerative, have potential for development of waste lands, need less water than food crops, provide higher employment opportunities, are important for nutritional securities, are environment friendly, are high value crops with high potential of value addition, have high potential for foreign exchange earnings and make higher contribution to GDP. A large variety of fruit, are grown in India of these, mango, citrus, guava, sapota, ber, litchi, aonla, peach, plum, pear and grapes are important. A Production technology of these fruits has been developed. Emphasis is given on high density plantation, micro-irrigation (Anonymous, 2015). The various technologies followed in rural farms regarding as modern fruit cultivation are given as:

i. Precision Farming

Precision farming can be defined as cultivation by adopting those technologies which give maximum precision in production of a superior crop with a desired yield levels and quality at competitive production. These include use of genetically modified crop varieties, micro-propagation, integrated nutrient and water management, integrated pest management, protected cultivation, organic farming, use of modern immunodiagnostic techniques for quick detection of viral diseases and high-tech post-harvest technologies including cold chain.

Precision agriculture is the application of technologies and principles to manage spatial and temporal variability associated with all aspects of agricultural production for improving production and environmental quality. The success in precision agriculture depends on the accurate assessment of the variability, its management and evaluation in space-time continuum in crop production. The agronomic success of precision agriculture has been quite convincing in crops like sugar beet, sugarcane, tea and coffee. The potential for economic, environmental and social benefits of precision agriculture is largely unrealized because the space-time continuum of crop production has not been adequately addressed. Precision agriculture is a relatively new area that combines the latest in geographic technology with cropping situations to optimize inputs, reduce waste, and generate the maximum possible yields. The precision farming systems are designed for use in all types of agricultural systems, from row crops to dairies, and the technology has seen widespread adoption across the US and worldwide. Precision farming basically depends on measurement and understanding of variability, the main components of precision farming system must address the variability. Precision farming technology enabled, information based and decision focused, the components include, (the enabling technologies) Remote Sensing (RS), Geographical Information System (GIS), Global Positioning System (GPS), Soil Testing, Yield Monitors and Variable Rate Technology.

ii. High Density Planting

High density planting (HDP) is one of the methods to enhance productivity per unit area both in short duration and perennial horticultural crops. In perennial crops, it is more useful since it permits efficient use of land and resources, realizing higher yields, net economic returns per unit area, easy canopy management suited for farm mechanization and cultural operations, efficient spray and weed control, improvement in fruit quality and easy and efficient harvest of high quality produce, *etc.* There are five important methods to achieve HDP namely

(a) Use of dwarf scion varieties
(b) Adopting dwarfing rootstocks and inter-stocks
(c) Efficient training and pruning
(d) Use of plant growth regulating chemicals and
(e) Suitable crop management practices.

There are several fruit crops where desired success has been achieved using HDP. These crops include apple, peach, plum, sweet cherry and pear among temperate fruits, and banana, pineapple and papaya among tropical fruits. In India, HDP technology has been successfully demonstrated in banana, pineapple, papaya and recently in mango, guava and citrus where two to three times increase in yields are realized. In plantation crops like cashew, coconut and arecanut, such dense plantings can best be adopted using multi-crop species cropping systems.

iii. Improved Varieties

Over hundreds of varieties in different fruit crops comprising of aonla, acid lime, apple, banana, custard apple, grape, guava, litchi, mango, papaya, pomegranate and sapota have been released during the last two decades. A large number of varieties have been identified and released for fresh consumption, culinary purpose, processing, nutritionally rich and those suited for export purposes. Several varieties are also available with resistance to diseases and pests. Varieties have also been developed in various fruit crops for short growing period, resistance to late fungal pathogens, tolerance to viruses and immunity to various diseases and resistance to nematode.

iv. Use of Plastics

Plastics have several applications in commercial horticultural activities. These include drip irrigation; plastic film mulches; greenhouse structures; high and low tunnels; post harvest operations; *etc.* Use of plastics has proved beneficial to promote the judicious utilization of natural resources like soil, water, sunlight and temperature. During the last decade, there has been a substantial increase in area under micro-irrigation and protected cultivation of different horticultural crops. Drip irrigation has got maximum area under fruit crops followed by plantation crops. Among the crops, maximum adoption of drip system has been in fruit crops (35 per cent) followed by plantation crops (18.5 per cent). In fruits, the maximum area is under grape followed by mango, pomegranate and banana. This system saves irrigation water varying from 30 to 70 per cent in different crops. At present, about 2,756.8 lakh ha have been covered under micro-irrigation, which is further extended to 4,584.80 lakh ha with addition of 1,828 lakh ha during last few years (Anonymous, 2007).

v. Propagation

Propagation of plants is defined as a controlled multiplication of plants by man to perpetuate selected individuals/or a group of plants which have specific vale to him. Plant propagation deals with the multiplication and propagation of plants using propagules representing a specific group of genotypes. A propagule is any plant part used to produce a new plant or a population of plants. Specific propagules include seeds, cuttings, layers, buds, scions, explants and various kind of specialized structures such as bulbs, corms and tubers. The propagation of plants has been one of the fundamental occupations of mankind since civilization begins. Later, progress was made on the selection of the desirable types from the wild population, natural hybrids and natural population. Now, every day new dimensions are made in the field of plant propagation in general and fruit plants in particular. Vegetative propagation techniques have been standardized in several fruit and plantation crops. Rapid methods of propagation through micro-propagation are available in banana, several fruit plants. Production of virus-free planting material through shoot tip grafting has been standardized in citrus.

vi. Plant Growth Regulators

Use of plant growth regulators is now successfully exploited for flower production, improvement in fruit set and size, fruit quality and checking fruit drop and effective ripening. One of the most desirable controlling mechanisms of crop regulation in fruit trees. Example: Morphactins, ethephon, NH_3 ions, cytokinins, daminozide and maleic hydrazide (MH) can be used to increase flowering intensity, whereas bromacil could be used to reduce it. Auxin-61 and KNO_3 have also been reported to induce "off" season flowering in fruit crops. Ethephon and NAA have been found to be very effective chemicals for crop regulation in oranges and mandarins. NAA (200-300 ppm) have been found useful for deblossoming in Kinnow during "on" year to get optimum crop in "off" year.

vii. Protected Cultivation

Protected cultivation of horticultural crops is production of high quality produce for internal and domestic markets. There are different types of protected structures being adopted by the growers based on the agro-climatic region and the availability of different inputs. Some of the most ideal regions in the country are the western and eastern India. In this region this technology is under maximum adoption for growing export quality strawberry, papaya, *etc.* In northern states, low polytunnel has been used to produce quality strawberries and high value vegetables. Protected cultivation/green-house/low poly tunnels production techniques are now available for growing small fruit plants. Crops like strawberry, papaya, other berries *etc.*, can be successfully grown for high quality under protected cultivation.

viii. Tree Architecture

Accommodation of larger number of trees depend on the designing of orchard architecture which is a combination of main four variables namely variety, rootstock, tree spacing and tree training. Higher yields of superior quality depend on good light distribution within the canopy essential for maintaining the fruit quality. Hence in order to harvest high amount of quality fruits canopy management needs greater attention through manipulation of plant population, plant architect, use of proper scion and rootstock combination and training pruning system. Canopy management also essentially employed for minimization of inside tree shade, increasing temperature in inner part and minimization of shelter points of insect pest and diseases. The tree canopy physically supports the fruit crops and influences the fruit crops and influences the fruit yield and quality. The leaf canopy intercepts light and produces carbohydrates used in growth and metabolic activities. Light distribution within the canopy influences fruit colour and quality, and the rate and timing of canopy development relates to crop quality and quantity, and thus economic yield of an orchard. High yields of good quality fruits produced are attributable to high light interception within canopy which is imperative to ensure the fruit quality. Light is important aspect of canopy studies because of its role in photosynthesis, its function in developmental morphology of leaves and shoots and its role in flower initiation and fruit set and ultimately in fruit development and quality (Rom, 1992). Considering the above points in view, the canopy management deserves to draw the greater attention of researchers by exploiting the proper rootstock and scion combinations, training and pruning system, cultural practices and use of growth retardants. Canopy includes the parts of the tree visible aboveground the trunk, cordon, stems, leaves, flowers, and fruit. The canopy plays a key role in light energy capture via photosynthesis, water use as regulated by transpiration, and microclimate of an orchard. A collective practical methodology to achieve optimum canopy microclimate so that fruit and leaves receive best sunlight exposure and air movement within the canopy is achieved.

The decline of productivity has been attributed to various factors. The most of the problems are due to faulty management *i.e.* unsuitable site and climate, cultivation of intercrops, inadequate nutritions, improper planting, undesirable planting materials, incidence of insect, pest and disease and other biotic and a biotic stresses. The growers do not adopt the proper management practices in terms of plant protection, manuring, irrigation, mulching, pruning *etc.* and the orchards become sick. In general, canopy of fruit crops has irregular shape. Trees of irregular shape and size are difficult to deal with and even culminate into poor yield in the subsequent years as the lower branches of canopy gradually turns inert and infertile as well.

ix. Rejuvenation Strategies

- ☆ Providing technical know-how including plant health coverage and nutritional management programme.
- ☆ Re-plantation of old and uneconomical orchards.
- ☆ Gap filling by providing disease free quality seedlings.
- ☆ The development agencies may prepare comprehensive orchards management programme providing all the necessary inputs like plant nutrient, plant protection chemical, horticultural equipment and periodical trainings.
- ☆ Training is an important component, which improves overall efficiency of the knowledge and skill of field functionaries.
- ☆ Complete technological information on management of decline orchard may be packaged and same may be disseminated in farmer's field.

x. Drip Irrigation/Fertigation

Drip irrigation and fertigation technologies are available for banana, grape, papaya, pomegranate, mandarin, coconut, areca nut and cashew nut. In India, around 88 per cent water is being used in agriculture sector, covering around 85 M ha area under irrigation. Due to liberalization of industrial policies and other developmental activities, the demand for water in industrial and domestic sectors is increasing day by day, which forces to reduce the percentage area under irrigation. The growing demand for the food due to population increasing day by day calls for more efforts to enhance agricultural production. The horticulture sector has emerged as a promising area for diversification in agriculture on account of high income generation for unit area, water and other farm inputs and environmental friendly production systems. The present total irrigated area in the country is about 40 per cent of the total sown area. The conventional system of irrigation employing different methods like flooding, furrow, basin and border irrigation revolve around the concept of replenishing the moisture level to field capacity only after depletion by 50-60 per cent of field capacity (Sivanappan, 1998). Micro irrigation is a low pressure, low volume irrigation system suitable for high-return value crops such as fruit. If managed properly, micro irrigation can increase yields and decrease water, fertilizer and labor requirements. Micro irrigation applies the water only to the plant's root zone and saves water because of the high application efficiency and high water distribution uniformity. Micro irrigation can irrigate sloping or irregularly-shaped land areas that cannot be flood irrigated. Any water-soluble fertilizer may be injected through a micro irrigation system. Micro-irrigation including micro spray, surface drip and subsurface drip irrigation methods can deliver water precisely and efficiently. Micro irrigation is commonly used for irrigation of high value horticultural crops, orchards and vineyards. Subsurface drip irrigation (SDI) is gaining popularity in production of agronomic "row" crops, especially in areas of limited well capacities and where small or irregularly shaped fields give SDI a competitive advantage over other irrigation technologies and methods (Anonymous, 2015).

xi. Mulching

Covering of soil around the root zone of plant with a plastic film is called plastic mulching. It is an effective means to restrict weed growth, conserve moisture and reduce the effect of soil borne diseases. Black plastic prevents the germination and growth of weed seeds. Mulching has been helpful not only in preventing moisture loss through evaporation from the soil and lowering the temperature but also reducing nutrient loss by leaching and weed control, reducing run-off, increasing penetration of rainwater, controlling erosion, correcting the chemical balance of the soil and reducing damage by pests and diseases.

xii. Nutrient Management

Fertilizer schedules have been standardized for a number of crops grown under different agro-climatic regions. Use of micro-nutrient sprays to minimize physiological disorders, improve fruit set and quality, enhanced shelf-life *etc.* are now practiced. Integrated Nutrient Management Strategies (INM) are available and need to be adopted on commercial scale in different perennial crops. Various nutrients such as nitrogen, phosphorus, calcium, magnesium, zinc, potassium and boron have a specific role in maintaining the fruitful behaviour of the plants (Ravishankar *et al.*, 1990). The balance application (through soil and foliar) and availability of these nutrients is very much essential for obtaining good fruit set and yield. Integrated nutrient management (INM) refers to maintenance of soil fertility and plant nutrient supply to an optimum level for sustaining the desired crop productivity through optimization of the benefits from all possible sources of plant nutrients in an integrated manner. Nutrients are essential for productivity and quality of different fruit crops. Among organic inputs, bio-fertilizers are commonly known as microbial inoculants that improve soil fertility and crop productivity. These are the beneficial micro-organisms which help in improved uptake and availability of different essential macro- and micro-nutrients, bio-control properties, enhanced growth, yield and quality (Anonymous, 2007).

xiii. Resource Conservation and Availability

For conservation of soil and water resources, growing perennial fruits offers many advantages when compared with agronomic crops. Fruit trees, vines, and bushes can provide excellent protection for the soil because the persistent roots, leaf litter, and surface cover crops beneath woody plants improve soil structure, nutrient, and water retention. These benefits are especially important for hill side farms, which are ill-suited to tillage and annual crops. Fruit growing and agro- forestry systems are thus widely recommended for fragile soils and flood-prone watersheds (Steppler and Lundgren, 1988). Fruit plantings also can provide a relatively rich habitat for both plant and animal species (including pests), with a variety of niches allowing for greater species diversity than many other agroecosystems. This habitat diversity complicates pest management because it provides many sites for pests such as mites, codling moth, or fruit flies to overwinter or elude natural predators (Croft and Hoyt, 1983). Fortunately, such complex habitats are potentially more suitable for establishing beneficial flora and fauna that might provide biological control of economic pests within integrated pest management (IPM) programs. As pheromone baits and mating disruption techniques are improved, the structural complexity of orchards also should enhance the effectiveness and stability of these low chemical input control tactics. While much is made of soil and water resource limitations, the availability of land or water for fruit growing may be jeopardized more by urban and suburban sprawl than by any direct impacts of conventional farming practices on natural resources. Conversion from flood or sprinkler irrigation to more-efficient drip or sub-irrigation systems may reduce the rural competition for water in arid areas, but, in the present economy, agricultural land usage rarely will preempt commercial or residential development. Growers themselves often oppose protective measures, such as conservation easements or restrictive agricultural zoning, which might deprive them of the lucrative option of selling-out to developers. Once again, we encounter a situation in which a farming system itself may be sustainable, but it is unable to withstand pressures from socioeconomic trends external to the agro-ecosystem.

xiii. Agrichemical Inputs in Fruit Production

Perhaps the greatest perceived threat to the sustainability of modern fruit production is its extensive reliance on synthetic pesticides and fertilizers. Fruit growing often serves as a convenient scapegoat for chemical intensive, high-input agriculture, which is portrayed as exploiting and endangering farmworkers, wildlife, and natural resources (Edwards, 1993). Unquestionably, modern fruit production systems involve relatively high inputs of pesticides per unit of production. Market and regulatory forces discover age agrichemical usage, but also mandate blemish-free fruit. In some comparisons with agronomic crops, fruit growing actually appears more sustainable. For instance, inputs of the herbicides and fertilizers implicated in water pollution are usually lower in fruit production than in field crops (USDA, 1992). This is attributed to the protection of topsoil, use of cover crops, and recycling of nutrients in perennial crop systems. The ratio of energy inputs to crop outputs is higher in fruits than in field crops, but the much greater market value per unit output makes fruits more efficient when energy input costs per crop output value are compared (Pimentel, 1979). Fruit and vegetable crops account for much of the total fungicide usage in the United States. With new horticultural varieties resistant to the major fungal, bacterial and viral diseases are developed for each crop, adopted by growers, and accepted by consumers in the marketplace, fruit growing is now profitable in the long-term.

xiv. Disease and Pest Management

For disease and pest management, use of safe chemicals including bio-control agents has reduced the dependence on chemical pesticides, residues in fresh produce and death of pollinators and useful insects. Several technologies, using bio-control agents for control of insect pests like mealy bug in mango, grape, *etc.* have been developed and need to be adopted by the growers. Plant protection schedules have been established for all commercially cultivated horticultural crops. Biological control methods have been standardized for control of mealy bug in grape, scale insects in citrus, rhinoceros beetle and leaf eating caterpillar in coconut.

Integrated Pest Management (IPM) strategies in crops like Guava, Mango, Papaya, standardized. Apple scab forecasting systems have been developed. Twenty-four IPM packages including those for fruits, plantation and spice crops have also been developed for farmers. Integrated Pest Management (IPM) in agriculture aims at judicious use of cultural, biological, chemical, host plant resistance/tolerance, physical-mechanical control and regulatory control methods. Besides, it also employs bio-pesticides, antagonists *etc.* Different safe chemicals and bio-agents are employed in IPM and can effectively be used to minimize the risk of pesticide residues in fresh and processed produce.

xv. Post-harvest Management

At present, the post harvest losses are about 20-30 per cent in different horticultural crops. This is happening mainly due to mismatch between the production and infrastructure development for post harvest management. Technology development has been in vogue but its adoption is far below. Post-harvest management including precooling, passive evaporative cooling for increasing the shelf-life of fresh fruits, vegetables, floriculture products, processed fruits and vegetables have been standardized. Packing materials like Corrugated Fiberboard boxes (CFBs), perforated punnettes, cling films, sachets, wraps *etc.* have been standardized for fresh horticultural produce. Tetra packs of different products are now house-hold items. Proper handling, packaging, transportation and storage reduce the post harvest losses of fruit. For every one percent reduction in loss will save 5 million tons of fruit per year. Processing and preservation technology helps to save excess fruit during the glut season (off season). The technology has become a necessity to improve the food safety and strengthen nation's food security. The technology helps to boost export of agricultural commodities in the form of preserved and value added products. Presently, mango, pineapple, citrus, grapes, apple and peach are being processed on a large scale.

xvi. Value Addition

Development of new products like dried powder fruit based milk mix, juice punches, banana chips and fingers, mango nectar and fruit kernel derived cocoa substitute, essential oils from citrus, fruit wines, dehydrated products from grape, pomegranate, mango, apricot and coconut, grape wines, value-added coconut products like snowball tender coconut, milk powder and pouched tender coconut water (Cocojal) *etc.* are getting popular day by day. Improved blending/packaging of tea and coffee have opened new markets. Consumer friendly products like dried apricot, raisins, ready to use salad mixes, pineapple, are major retail items.

xvii. Mechanization

To keep pace with improved production and productivity, different machines have been developed for effective cultivation, intercultural operations, harvesting, grading, packaging and value-addition. Development of mango harvester, Kinnow clipper, potato digger, coconut peeler, *etc.* is being adopted by the growers. Machines have also been developed/installed for different specialized uses like cool sterilization (irradiation), dehydration of different produce, vapour heat treatment

(VHT) in major mango growing belts, packaging of coconut water, banana fig and chip making machine, *etc.*

Future Strategies

Fruit plantings are considered by many people to be an aesthetically pleasing part of the rural landscape. In this sense, fruit growing regions actually can be viewed as part of the national heritage a managed natural resource worthy of preservation. In these aspects, fruit growing represents the harmonious, long term integration of agriculture and human civilization with the natural environment. Fruits themselves are powerful symbols in many religions and cultures, but this commonality between fruit growers and the general populace has been forgotten or underused. Fruit growers should pay more attention to these cultural issues in order to maintain public trust and support. They will not succeed by managing their farms like factories in the field, or presenting their fruit to the public as if it were just another commodity.

The demand for fruit is driven by consumers who want to increase the quality and variety of their diet. As countries develop their demand for fruit increases greatly and the traditional home gardens or natural groves that previously supplied them with fruit often are lost to urbanization, deforestation, or replacement by monocultures. Fruit consumption has increased steadily in the industrialized nations, especially with recent reports establishing the importance of fresh fruits for cancer prevention and human health. Certain aspects of current intensive production and marketing systems are probably not sustainable in their present form, but they were developed in response to intensifying market competition and the need for greater short-term profitability. Although, some modern fruit growing techniques enhance sustainability (*e.g.*, micro-propagation, integrated pest management, virus-indexing, clonal rootstocks), most have increased grower reliance on purchased inputs. Today's fruit production thus approaches the goals of sustainability in some respects.

REFERENCES

Anonymous, 2015. Horticulture Training Institute, Uchani (Karnal), Department of Horticulture, Haryana.

Anonymous, 2007. Report of the working group on horticulture, plantation crops and organic farming for the XI Five Year Plan.

Croft, B. A. and Hoyt, S. C. 1983. Integrated management of insect pests of pome and stone fruits. Wiley, New York.

Edwards, C. A. 1993. The impact of pesticides on the environment. In: D. Pimentel and H. Lehman (eds.). The pesticide question, pp. 13-46.

Mai, W. F. and Abawi, G. S. 1981. Controlling replant diseases of pome and stone fruits in north-eastern United States by preplant fumigation. *Plant Disease*, **65**: 859-864.

Merwin, I. A. and Pritts. P. 1993. Are Modern Fruit Production Systems Sustainable. *HortTechnology*, 3: 2.

National Research Council. 1989. Alternative Agriculture. National Academy Press, Washington,D.C.

Pimentel, D., and Pimentel, M. 1979. Food, energy and society. Edward Arnold, New York.

Ravishankar, H. Nalawadi, U. G. and Hulamani, N. C. 1990. Investigations on the use of growth regulators for the control of alternate bearing problem in mango (*Mangifera indica* L.). *South Indian Horticulture*, **38**(5): 234-239.

Rom, C. R. D. 1992. Spur pruning Delicious apple for improved spur quality and yield. *Acta Horticulturae*, **322**: 55-68.

Simmonds, N.W. 1976. Evolution of crop plants. Longman, Essex, U.K.

Sivanappan, R. K. 1998. Status, scope and future prospects of microirrigation in India. Proc. Workshop on microirrigation and sprinkler irrigation system. CBIP New Delhi, 1-7.

Smith, N. J. H., Williams J. T., Plucknett, D. L. and Talbot. J. P. 1992. Tropical forests and their crops. Cornell Univ. Press, Ithaca, N.Y.

Steppler, H.A. and Lundgren. B. O. 1988. Agroforestry: Now and in the future. *Outlook on Agriculture*, **17**: 146-152.

USDA. 1992. Agricultural chemical usage: 1991 fruits and nuts summary. USDA, NASS, ERS, Agr Ch.1(92). Govt. Printing Office, Washington. D.C.

York, E.T. 1991. Agricultural sustainability and its implications to the horticulture profession and the ability to meet global food needs. *HortScience*, **26**: 1252-1257.

Transformation of Indian Agriculture through Innovative Technologies *Pages* ***403–410***
Editor: **Dr. Shahid Ahamad & Dr. Jag Paul Sharma**
Published by: **ASTRAL INTERNATIONAL PVT. LTD., NEW DELHI**

27 Quality Maize for Augmenting Income of the Farmers

Ravinder Singh Sudan and Shahid Ahamad

Maize (*Zea mays* L.) is an adaptable, high-yielding and fast-growing important cereal crop suitable for wider production as grain crop. Recently, it is being grown for diverse uses and specialty purposes. Such maize for specialty and value added purposes are collectively called specialty corn. Compared to field corns, specialty corns possess additional and characteristic features. Their global spread, increasing demand and premium price make them an attractive option for the farmers in many countries including India. Specialty corns are amenable to numerous options pertaining to harvest time and various economical products. Maize with respective quality parameters relating to tender ear characteristics, biochemical components relating to protein, sweetness, starch, oil and popping traits are considered as specialty corn (Ahamad and Narain, 2007; Ahamad, 2009, 2012, 2013, 2006.

Maize is a third prominent cereal food crop in global agriculture as well as in India after rice and wheat. In India maize was cultivated in an area of 8691.2 thousand hectares with a production of 21806.5 thousand tones and an average productivity of 2509 kg/ha during 2015-16. In Jammu and Kashmir Maize was cultivated in an area of 306.1 thousand hectares and the production is 479.2 thousand tones and the yield is 1566 kg/hectare during 2015-16 (AICRP, 2009; Ahamad *et al.*, 2015a,b).

Although much of the world's maize production (approximately 78 per cent) is utilized for animal feed, human consumption in many developing and developed countries is steadily increasing. For example, maize is the most important cereal crop for food in sub-Saharan Africa and Latin America. The growing demand for food consumption in developing countries alone is predicted to increase by around

1.3 per cent per annum until 2020. Between now and by 2050, the demand for maize in the developing world will double.

Maize is emerging as an important *Kharif* rainfed crop in the state of Jammu and Kashmir. Maize grains continue to remain inseparable part of human diet particularly in hilly areas besides its use in animal feed. The potential of Maize crop has been very well recognized in its use in making several products of industrial importance. This crop offers great potential for value addition and diversification in near future.

Figure 27.1. QPM Version of Vivek Hybrid-9.

Apart from the cultivation of normal maize, there is tremendous potential for growing specialty corns like sweet corn, pop corn, baby corn and quality protein maize *etc.* which can ensure additional income to the farmers.

Production Technology of Specialty and other Corn types

Other than grain, maize is also cultivated for various purposes like quality protein maize and other special purposes known as 'Specialty Corn'. The various specialty corn types are quality protein maize (QPM), baby corn, sweet corn, pop corn, waxy corn, high oil corn *etc.* In India, QPM, baby corn and sweet corn are being popularized and cultivated by a large number of farmers. The brief summary of different types of specialty maize is as follows:

i. Quality Protein Maize

More than 85 per cent of the maize is used directly for food and feed, quality has a great role for food and nutritional security in the country. In this respect, discovery of Opaque-2 (O2) and floury-2 (F2) mutant had opened tremendous possibilities for improvement of protein quality of maize which later led to the development of "Quality Protein Maize (QPM). QPM which is nutritionally superior over the normal maize is the new dynamics to signify its importance not only for food and nutritional security but also for quality feed for poultry, piggery and animal sectors as well. Quality Protein Maize has specific features of having balanced amount of amino acids with high content of lysine and tryptophan and low content of leucine and isoleucine. The balanced proportion of all these essential amino acid in Quality

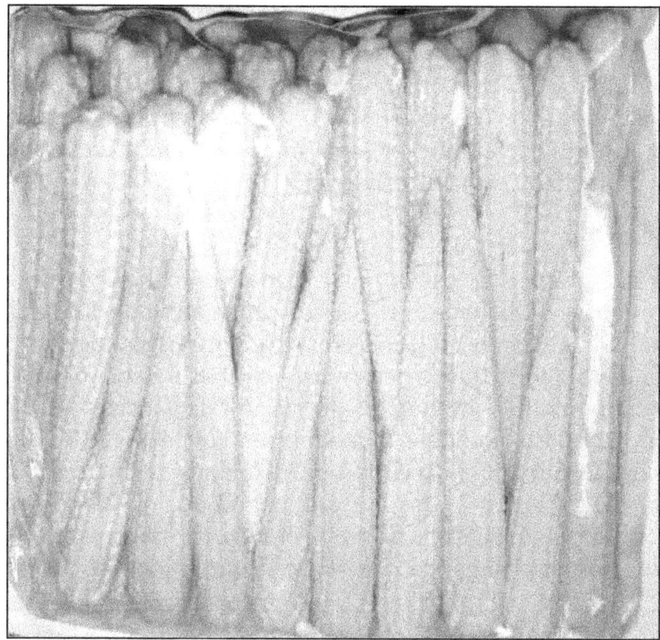

Figure 27.2. Quality Protein Maize.

Protein Maize enhances the biological value of protein. The biological value of protein in QPM is just double than that of normal maize protein which is very close to the milk protein as the biological value of milk and QPM proteins are 90 and 80 per cent respectively. Whereas it is less than 50 per cent in normal maize protein. There are 9 QPM hybrids of different grain colour that have been developed and released in India for their cultivation in different agro-climatic conditions across the country. The production technology of QPM is same as of normal grain maize except isolation as to maintain the purity of QPM.

ii. Baby Corn

Baby corn is a young finger like unfertilized cob with one to three centimeter emerged silks preferably harvested within 1-3 days of silk emergence depending upon the growing season. It can be eaten raw as salad and used in preparation of different recipes such as chutney, pakora, mix vegetables, pickles, candy, murabba, kheer, halwa, raita, Chinese preparations, *etc.* The desirable size of baby corn is 6 to 11 cm length and 1.0 to 1.5 cm diameter with regular row/ovule arrangement. The most preferred colour by the consumers/exporters is generally creamish to very light yellow. Baby corn is nutritive and its nutritional quality is at par or even superior to some of the seasonal vegetables. Besides proteins, vitamins and iron, it is one of the richest sources of phosphorus. It is a good source of fibrous protein and easy to digest. It is almost free from residual effects of pesticides. It can be cultivated round the year therefore, three to four crops of baby corn can be taken in a year. Cost of cultivation of baby corn in India is lowest in the world therefore India can become one of the major baby corn producing countries. It has great potential both for internal consumption and export.

In general, the cultivation practices of baby corn is similar to grain crop except (i) higher plant population (ii) higher dose of nitrogen application because of higher plant population (iii) preference for early maturing single cross hybrids and (iv) harvesting within 1-3 days of silk emergence.

iii. Sweet Corn

Sweet corn is one of the most popular vegetables in the USA, Europe and other developed countries of the world. It is a very delicious and rich source of energy, vitamin C and A. It is eaten as raw, boiled or steamed. It is also used in preparation of soup, salad and other recipes. It is becoming very popular in urban areas of country therefore, its cultivation is remunerative for peri-urban farmers. Besides green cobs, the green fodder is also available to the farmers for their cattle. Generally sweet corn is early in maturity. It is harvested in 70-75 days during *kharif* season. Green cobs are harvested after 18-20 days of pollination during *kharif* but the duration may vary from season to season. At the harvest time the moisture is generally 70 per cent in the grain and sugar content varies from 11 to more than 20 per cent.

Figure 27.3. Sweet Corn.

Color

Sweet corn is generally dull yellow and white but dull yellow colour is preferred.

Precaution

Its picking should be done in the morning or evening time. Green cobs should be immediately transported to the cold storage in refrigerated trucks to avoid the conversion of sugar to starch. It loses flavour if kept in high temperature after picking. Sweet corn with high sugar content should not be planted when temperature is below 16 degree centigrade.

iv. Pop Corn

Popcorn is one of the common snack items in many parts of the world, particularly in cities and is liked because of its light, porous and crunchy texture. The popcorn flour can also be used for preparing many traditional dishes.

Figure 27.4. Pop Corn.

It is consumed fresh, as it has to be protected against moisture absorption from the air. It is hard endosperm flint maize. Kernels of pop corn are very small and oval/round in shape. When heated at about 1700C, the grains swell and burst, turning inside out. Quality of pop corn depends on popping volume and minimum number of non pop corn.

V. Waxy Corn

Waxy corn contains 100 percent amylopectin whereas normal corn contains 75 percent amylopectin and 25 percent amylose. Amylopectin is a form of starch which consists of branched glucose subunits whereas amylose is made up of unbranched glucose molecules. The waxy trait is controlled by a single recessive *wx* gene. Waxy

Figure 27.5.

Figure 27.6.

corn is used by wet-corn millers to produce waxy corn starch which is utilized by the food industry as a stabilizer/thickener.

VI. High Amylose Corn

This corn has amylose content higher than 50 per cent. High amylose corn is grown exclusively for wetmilling. This starch is used in textiles, candies and adhesives.

VII. High Oil Corn

High oil corn has approximately 7 to 8 percent oil. Additionally, protein quality and quantity are increased to some extend in high oil corn. This is because the germ size is larger and it containsprotein of higher quality than the endosperm. The high oil trait is controlled by polygenes. Highly polyunsaturated and high linoleic acid content of corn oil makes it an excellent energy and essential fatty acid source for both humans and livestock.

Table 27.1. Quality Protein Maize, Specialty and other Corn Type Cultivars

Sl.No.	*Corn Type*	*Cultivars*
1	Quality Protein Maize	H*:HQPM 1, HQPM 5, HQPM 7, Vivek QPM 9 Shaktiman 1, 2, 3 and 4
2	Baby corn	H:HM-4, Prakash,PEHM1 and PEHM2
		C: VL Baby Corn1, VL 42
3	Sweet corn	H: HSC-1, Sugar-75
		C:Madhuri, Win orange, Priya
4	Pop corn	C: Jawahar, Amber, Pearl and VL pop corn, Bajaura Pop Corn

Strategies for Increasing the Farmers Income

Production practices for producing maximum grain yield using specialty corn hybrids are similar to those for field corn hybrids. Good fertility, adequate weed control and proper planting date are necessary to produce maximum yields. Care must be taken to avoid cross pollination with normal hybrids. If crosspollination occurs, the cross-pollinated ears of the specialty corn hybrids will produce normal seed and the seed of the hybrid will have the quality trait percentage intermediate between the field and specialty corn hybrid. To avoid cross pollination, specialty hybrids should be grown in an isolated field or the grain from the border rows should be harvested separately from the rest of the field.

Moreover, these specialty hybrids should be grown following crops other than corn to avoid volunteer corn. The potential grain yields of specialty corn hybrids are generally lower than those of field corn. However, quality traits of specialty corn provide competitive advantage over field corn yield. Specialty corn yields vary depending upon location. Test weight can also be lower in QPM than normal maize hybrids so care must be taken to select a hybrid with adequate test weight. To address major limitation of low productivity, initiatives are being made to develop single cross hybrids in specialty corn. Utilizing elite hybrids with higher productivity

Figure 27.7.

and quality would facilitate faster spread and popularization of specialty corn to different stake holders. The huge potentiality of each specialty corn through hybrid technology would facilitate better harnessing multiple benefits. Specialty corn takes advantage of available resources in a more judicious way. It requires special efforts to study value chains and associated market opportunities. Depending upon the actual conditions in a growing season and local market situation, farmers can plan harvesting. Progress in establishing specialty corn research and development would be fostered by transferring technology through public awareness campaigns. Urbanization, rising incomes and convenience foods are enhancing demand of specialty corn. The processing sector can be promoted through streamlined production, market development, focus on production advantages and customer requirements. To strengthen competitiveness, the industry needs to develop quality; reduce post harvest damage by improving on-farm mechanization and labour skills; introducing better storage especially cold storage technologies; enforcing market standards; overcoming inconsistencies in supply (labour availability, seasonal peaks) and providing greater incentives to farmers. In this way, farmers income can be enhanced by growing speciality corn.

REFERENCES

Ahamad, S. 2006. Incidence of maize diseases in two districts of Intermediate zone of Jammu region. *Ann. Agric. Res.*, 40: 231-236.

Ahamad, S. 2009. Plant Disease Management for Sustainable Agriculture Published by Daya Publishing House New Delhi, pp. 373.

Ahamad, S. 2012. Recent Trends in Plant Diseases Management in India, Published by Kalyani Publisher, Ludhiana, India. p. 478.

Ahamad, S. 2013. Hill Agriculture. Published by Astral Publishing House, New Delhi. p. 550.

Ahamad, S. and Narain, U. 2007. Eco-friendly Management of Plant Diseases Published by Daya Publishing House New Delhi, p. 412.

Ahamad, S., Kher, D. and Lal, B. 2015a. Stalk rot of maize diseases in the intermediate zone of Jammu region. *International Journal of Innovative Sciences Engineering & Technology*, pp. 1024-1032.

Ahamad, S., Kher, D. and Lal, B. 2015b. Screening of Maize germs plasms against stalk rot diseases in the intermediate zone of Jammu region. *International Journal of Innovative Sciences Engineering & Technology*, pp. 828-831.

AICRP 2009. AICRP Annual Report, DMR, New Delhi.

Transformation of Indian Agriculture through Innovative Technologies *Pages* **411–419**
Editor: **Dr. Shahid Ahamad & Dr. Jag Paul Sharma**
Published by: **ASTRAL INTERNATIONAL PVT. LTD., NEW DELHI**

28 Hybrid Maize Technology in Hill Agriculture

Ravinder Singh Sudan and Shahid Ahamad

Maize is the third most important crop of country after rice andwheat. Its grain is used as feed, food and industrial raw material.It is cultivated round the year, though more than 80 per cent is grown in rainy or kharif season. In India, it is estimated that maize demand will continue toincrease because of its diversified uses and increasing population (Ahamad, 2006, 2009, 2012, 2013; Ahamad and Narain, 2007). To meet the growing demand, enhancement of maize yield in comingyears across all the growing locations in India is a big challenge in the era of climate change. Meeting such challenge will only bepossible through science-based technological interventions like single cross hybrid technology and application of novel molecular tools and techniques in maize improvement. Globally maize is crop par excellence for food, feed and industrial utilization and is known as queen of cereals because it has the highest genetic yield production among the cereals.

Presently in India, only 10 per cent of maize is used as foodgrain, while remaining is used to meet non-food demand, *viz.* poultry feed, animal feed, brewing alcohol and other industrial uses (AICRP, 2009; Ahamad *et al.*, 2015a,b). It is estimated that one quarter of items found in a modern grocerystore of a developed country contain maize in some form. Toothpaste,dish detergent, paper, clothing dyes, explosives and soaps- there is avast list of products that contain maize constituents. Key ingredients in convenience food like xanthan gum, polyols, artificial sweeteners, fructose *etc.* are all derived from maize. Other products where maizeis used are: food containers and plastic food packaging, disposabledishware and gift cards, batteries, deodorants, hand sanitizers, coughdrops, baby powder, diapers, matchsticks, medicine and vitamin tablets,textile products, colourings and dyes, glue and other adhesives, candies,yoghurt and many more. Maize-based breakfast cereals, cooking oil, snacks, sweetcorn, babycorn, popcorn *etc.* are already gaining much popularity in India.

Further, increasing energy consumption in all walks of life call forfinding alternate sources of energy, in which bio-fuel from maize is adistinct possibility, which might be explored in our country. Maize is themost important natural multiplier and an economical source of starch. Starch comprises of about 68- 74 per cent of maize kernel weight, whichcan easily be converted into glucose and subsequently fermented into ethanol. Presently, US is playing the lead role by using 30 per cent of its maizefor bio-fuel production. Bio-fuel from maize will be need of the houras it will help in energy security along with high procurement prices.The expanding spectrum of demand would necessitate focused researchtailored for specific segments of the myriad value chains.

In Jammu and Kashmir maize is grown in an area of 306.1 ('000 ha) with production of 479.2 ('000 tonnes) and productivity of 1566 (kg/ha) during 2015-16.

Maize is emerging as an important *Kharif* rainfed crop in the state of Jammu and Kashmir. Maize grains continue to remain inseparable part of human diet particularly in hilly areas besides its use in animal feed. The potential of Maize crop has been very well recognized in its use in making several products of industrial importance. This crop offers great potential for value addition and diversification in near future. In the state of Jammu and Kashmir (India), Maize is a second most important crop after rice and is a staple food of some hilly areas people such as Gujjar and Bakarwal (nomadic race). In Himalayan hills extra-early, early and medium maturity maize are grown so as to fit in existing cropping system.

Maize is of special importance in hilly and submountanoius regions of Jammu Division where it forms the stapple diet of the people. It occupies highest area under cultivation in the state. Inspite of the fact that maize occupies of major area during Kharif season, the yield is low. The low yield is because of local seeds and traditional agronomical practices. Hybrids are the high yielding varieties of maize which have the potential of out yielding the local varieties by many folds.

Role of Maize in Jammu and Kashmir

- Used as food, feed and fodder.
- Green cobsas tourist destination.
- Green fodder security to the livestock
- Climatic conditions are highly favourable.
- The organic matter in the soil is high.
- Ensures livelihood security particularly nomadic tribes.
- Tremendous opportunity to generate employment.
- Specialty corns have high export potential and employment generation.

Hybrid Seed

Hybrid variety is first generation (F_1) from crosses between two purelines, inbreds, open-pollinated varieties, clones or other populations that are genetically dissimilar.

Types of hybrid

1. Single cross hybrid:
   Cross between two unrelated inbred lines (A X B)
2. Double cross hybrid: (A X B) X (C X D)
3. Three way hybrid (A X B) X C
4. Top cross hybrid is crossed with an open-pollinated variety

Now, the focus has been shifted to popularize single cross hybrid technology in this country to exploit maximum heterosis to enhance genetic gains in terms of production and productivity target levels of maize to meet envisaged demand in near future. The single cross hybrid technology had already made significant impact in raising productivity levels of maize in states of Andhra Pradesh and Karnataka.

Single cross hybrid maize varieties are the first generation (F_1) from cross between two inbreds that are genetically dissimilar. Seed production of single cross hybrid requires only two parents and one season.

Single Cross Hybrid Maize

- ☆ High acceptability among the farmers.
- ☆ Requires only two parents

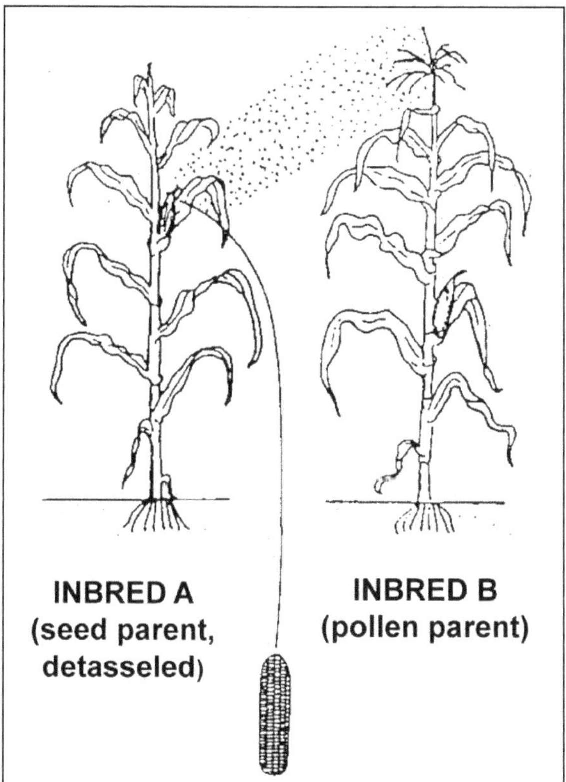

Figure 28.1: Single Cross Hybrid Seed (A x B) Planted by Farmer.

P1
B
P2
Parent 1
F1 Hybrid
Parent 2

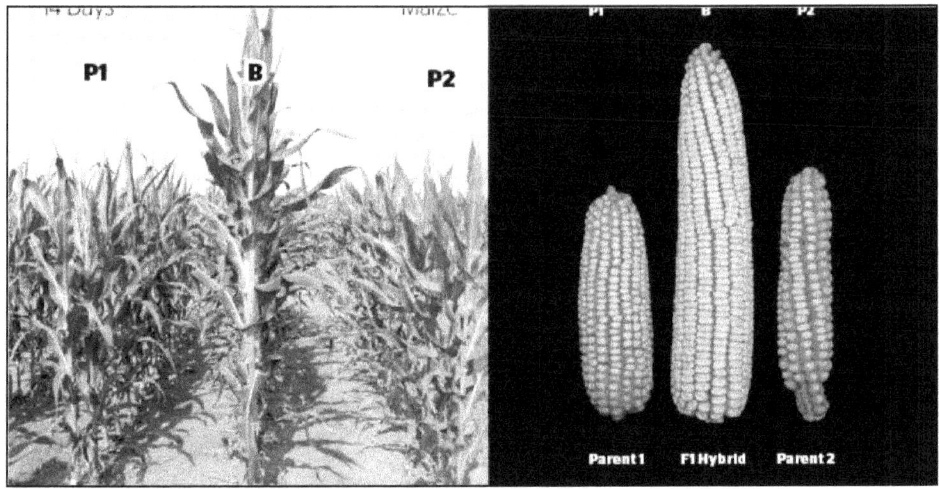

Figure 28.2

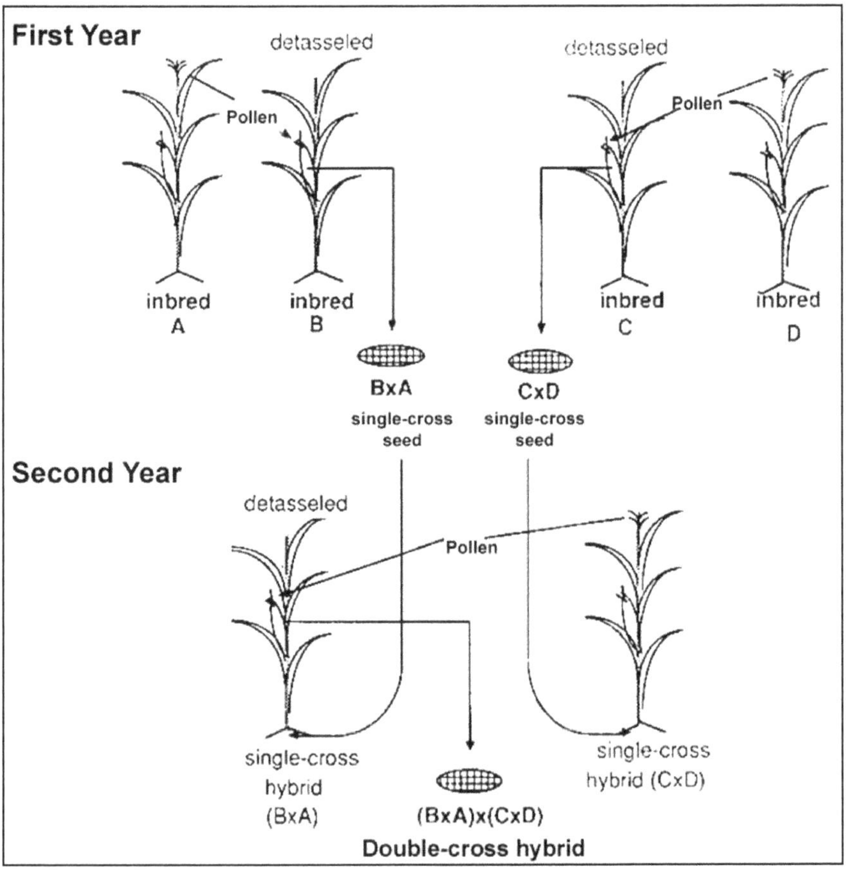
First Year
detasseled
detasseled
Pollen
Pollen
inbred
A
inbred
B
inbred
C
inbred
D
BxA
single-cross
seed
CxD
single-cross
seed
Second Year
detasseled
Pollen
single-cross
hybrid
(BxA)
(BxA)x(CxD)
Double-cross hybrid
single-cross
hybrid (CxD)

Figure 28.3

- ☆ Requires less number of isolations, *i.e.* only three isolations
- ☆ Easy to multiply and involves less labour.
- ☆ Low cost of production.
- ☆ Tolerant to biotic and abiotic stresses.

Prerequisit for Single Cross Hybrid Seed Production

- ☆ Easy to market because of its uniformity and high productivity.
- ☆ Proper site selection- avoid the sites where preceeding crop was maize.
- ☆ Fertile and quality land with good quality and assured irrigation.
- ☆ Productive, uniform and genetically diverse inbred lines.
- ☆ Proper isolation distance.
- ☆ Technically experienced manpower

Characteristics of Good Seed Parent (Female)

- ☆ Productive
- ☆ Strong
- ☆ Long cobs with complete excretion
- ☆ Low cob placement
- ☆ Lax tassel
- ☆ Shorter Anthesis Silking Interval (ASI)
- ☆ Nutrient responsive
- ☆ Stay green trait
- ☆ Erect leaves
- ☆ Resistant/tolerant to biotic and abiotic stresses
- ☆ Strong root system.

Characteristics of Good Pollen Parent (Male)

- ☆ Lax tassel, long main branch with few secondary branches.
- ☆ Long duration of pollen shedding.
- ☆ Taller than female parent.
- ☆ Attractive grain colour.

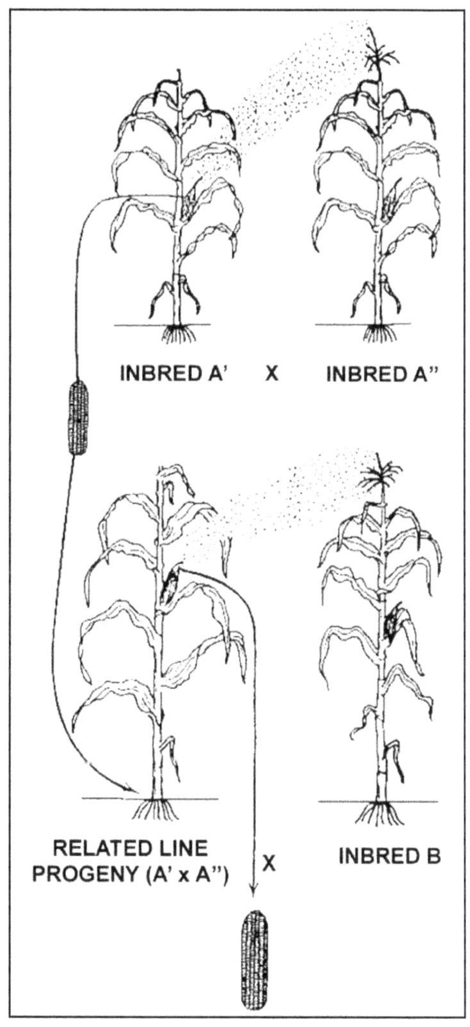

Figure 28.4: Three Way Cross Hybrid.

- ☆ Strong plant resistant to lodging with better root system.
- ☆ High yield potential.
- ☆ Resistant/tolerant to biotic and abiotic stresses.

Importance Considerations for Hybrid Seed Production

The Hybrid seed production should be taken either in the area where no other maize variety is planted near by the seed production plot or at least 400-500 metre distance is required between two maize genotypes to maintain the genetic purity.

Male : Female Ratio

The male: female ratio depends on

(a) Pollen shedding potential of male parent.

(b) Male: female synchrony

(c) For better seed setting flowering of female should be earlier than male.

In general the male female ratio should be 1: 2: 1: 2: 1 or 1: 3: 1: 3: 1 or 1: 4: 1: 4: 1

How to Bring Male : Female Synchrony

Male: female synchrony can be brought about by:

- ☆ Staggered planting of male and female.
- ☆ Manipulation in plant distance by spaced and narrowplanting
- ☆ Irrigation along with fertilizer application.
- ☆ Application of FYM in either male or female to induce earliness and vigor

Technology for Hybrid Seed Production Seed Production Site

Seed production should be taken in well drained, weeds and disease free soil and preferably the fields where preceding crop was not maize to minimize rouging and maintain the genetic purity.

Time of Sowing

In Jammu region the sowing of maize ranges from April to 1st week of July.

Spacing

Row to Row spacing 60cm and plant to plant 20cm

Seed Rate

The Seed rate is 15kg/ha for female and 10kg/ha for male with at least 80 per cent germination.

Seed Treatment

Seed Should be treated with fungicides and insecticides before sowing to protect it from seed and soil borne diseases and some insect pests.

Table 28.1. Notified Single Cross Maize Hybrids and Open Pollinated Varieties for J&K

Hybrid/Composite	*Average Yield (Q/h)*	*Maturity*	*Grain Colour and Type*	*Year of Release*	*Source*
Vivek Hybrid 33	60	Early	Yellow, dent	2008	VPKAS, Almora
Vivek Hybrid 21	50	Extra Early	Yellow, semi flint	2007	VPKAS, Almora
Vivek Hybrid 25	55	Extra early	Yellow, semi dent	2007	VPKAS, Almora
Vivek Hybrid 15	50	Extra early	Yellow, flint	2005	VPKAS, Almora
Pusa Extra Early Hybrid 5	50	Extra early	Yellow orange, flint	2004	IARI, New Delhi
Vivek QPM 9	55	Extra early	Yellow orange, dent	2008	VPKAS, Almora
HQPM 1	62	Late	Yellow, dent	2006	HAU, Karnal
Shalimar Hybrid Maize-1	55-70	135-145 Higher belts	White, flint	2009	SKUAST-K, state release
PMH-12	55-60	120-130 Mid hills	Yellow semi dent/flint	2015	SKUAST-J
Shalimar Maize composite	40	135-50	Yellow semi flint	2013	SKUAST-K
KDM - 438	40	135-145	White flint	2013	SKUAST-K

Nutrient Management

Inbreds requires high fertility as compare to hybrids. It is desirable to apply FYM @ 15t/ha 15days prior to seeding. The fertilizer nutrients required for inbreds are

a) N-180-200kg/ha.
b) P_2O_5 80kg/ha.
c) K_2O 80 kg/ha.
d) Zn SO_4 25kg/ha.

Full dose of phosphorus, Potash and half N should be applied as basal. The remaining dose of nitrogen should be applied in four splits as per details given below:

1. 20 per cent N at 4 leaf stage
2. 30 per cent N at 8 leaf stage
3. 30 per cent N at flowering stage
4. 10 per cent N grain

Water Management

- First irrigation should be applied very carefully. Water should not overflow on the ridges. The irrigation should be applied in furrows upto 2/3rd height of the ridges.
- Irrigation should be given as and when required by the crop depending upon the rains and moisture holding capacity of the soil.
- Young seedlings, knee high stage, flowering, grain filling and 10days after grain filling are the most sensitive stages for water stress for inbreds and irrigation should be ensured at these stages.

Weed Control

- Two hoeings to the crop should be give one at 15days and other at 30days after sowing.

Chemical Method

- Atrazine being the selective and broad spectrum herbicides in maize checks the emergence of both broad leaves and most of the grasses. Preemergence application of atrazine @ 1.0 to 1.5kg a.i/ha in 600 liter water is effective way for control of weeds.
- Insect Pest and Disease Management in Maize crop should be done as desired.

Removal of Off-Type Plants and Thinning

1. At early stage *i.e.* after 12-15 days of sowing, off-type plants and excess plants should be removed and proper plant to plant distance of 20-25cm

should be maintained to provide an equal opportunity to each plant to grow,
2. At knee high stage
3. At flowering *i.e.* before anthesis.

Dissimilar plants should be removed from male and female lines to maintain the genetic purity of seed. Dissimilar tassel bearing male plant should also be removed.

Earthing Up

Third dose of nitrogen should be applied followed by earthing up. This operation should be done before tasseling stage to save the crop from lodging.

Detasseling

Detasseling in female should be done before anthesis. It should be practiced row-wise. One person should be follow to monitor the each row to check that no part of tassel is left inside. The process of detasseling should continue for 8-10 days. While detasseling, leaf should not be removed which will other wise reduce the photosynthesis. The removed tassel should not be thrown in the field but feed to cattle as it is nutritive fodder.

Harvesting

Male parents should be harvested first than the female and should be kept separately. Optimum moisture content in grain at harvesting should be around 20 per cent.

REFERENCES

Ahamad, S. 2006. Incidence of maize diseases in two districts of Intermediate zone of Jammu region. *Ann. Agric. Res.*, 40: 231-236.

Ahamad, S. 2009. Plant Disease Management for Sustainable Agriculture Published by Daya Publishing House, New Delhi, p. 373.

Ahamad, S. 2012. Recent Trends in Plant Diseases Management in India, Published by Kalyani Publisher, Ludhiana, India. pp. 478.

Ahamad, S. 2013. Hill Agriculture. Published by Astral Publishing House, New Delhi. pp. 550.

Ahamad, S. and Narain, U. 2007. Eco-friendly Management of Plant Diseases Published by Daya Publishing House New Delhi, pp. 412.

Ahamad, S., Kher, D. and Lal, B. 2015a. Stalk rot of maize diseases in the intermediate zone of Jammu region. *International Journal of Innovative Sciences Engineering & Technology*, pp. 1024-1032.

Ahamad, S., Kher, D. and Lal, B. 2015b. Screening of Maize germs plasms against stalk rot diseases in the intermediate zone of Jammu region. *International Journal of Innovative Sciences Engineering & Technology*, pp. 828-831.

AICRP 2009. AICRP Annual Report, DMR, New Delhi.

Transformation of Indian Agriculture through Innovative Technologies *Pages* **421–448**
Editor: **Dr. Shahid Ahamad & Dr. Jag Paul Sharma**
Published by: **ASTRAL INTERNATIONAL PVT. LTD., NEW DELHI**

29 Entrepreneurship Development in Beekeeping

Devidner Sharma, D.P. Abrol, Hafeez Ahmad and Mahesh Kumar

Indian Agriculture has been a "gamble of monsoons" as described by Pt. Jawahar Lal Nehru. India Occupies 2.4 per cent of the land of the world and has 21 per cent population of the countries of the world, because of its contribution to GDP. Agriculture and allied sector will remain the life-life Indians with Over 85 per cent of the rural population still dependent on agriculture for their livelihood. With the increasing population over the last five decades, the per capita share of land and water resources has reduced substantially. As a result, rural people are faced with the problem of unemployment. Although Indian economy transformation from a predominantly agriculture based to an industrialized economy is still a continue process. The Indian agriculture is facing emerging threats and problems on many front and how effectively it will sustain is a real challenge. To maintain pace with current industrialization in every sector, there is need to develop entrepreneurships in agriculture and allied sector. A logical solution to employment is for people to engage in activities that increase their earnings and one of those activities is beekeeping. Beekeeping is the management of colonies of bees for the production of honey and other bee products. Bee keeping is of very low capital investment but with high rate of turnover. Beekeeping is a business opportunity with social, economic, and ecological benefits that requires minimal time, labor, and resources. Crop farmers and skilled people in their communities also increased their earnings through the production and sale of beekeeping equipment. Entrepreneurs earn additional income through the production and sale of bee products as well as making and selling other secondary products. These products are sold to provide more income for farmers. Other crop farmers also benefit from bee pollination, as many crops are improved if bees are kept nearby, as larger and more profitable fruits can be grown. When beekeepers plant trees that provide a source of nectar for bees, they also help to prevent soil erosion and landslides. Keeping bees also encourages people to value and preserve the forest. Entrepreneurship plays an important role in the process of economic growth and technological change. In the absence of local entrepreneurship, the opportunities in agriculture are highjacked by outsiders, particularly the urban businessmen and traders, leading to exploitation and deprivation of employment for the farmers. Considering the growing unemployment in rural areas and slow growth of the agricultural sector, it is necessary to tap the opportunities for promoting entrepreneurship in agriculture, which in turn can address the present problems related to agricultural production and profitability.

Introduction

The words entrepreneur, intrapreneur and entrepreneurship have acquired special significance in the context of economic growth in a rapidly changing socioeconomic and socio-cultural climates, particularly in industry, both in developed and developing countries. Entrepreneurial development is a complex phenomenon. Productive activity undertaken by him/her and constant endeavor to sustain and improve it are the outward expression of this process of development of his personality (Ducker, 1998). An entrepreneur is a person who either creates new combinations of production factors such as new methods of production, new products, new markets, finds new sources of supply and new organizational forms or as a person who is willing to take risks or a person who by exploiting market opportunities, eliminates disequilibrium between aggregate supply and aggregate demand or as one who owns and operates a business. The entrepreneurial motivation is one of the most important factors which accelerate the pace of economic development by bringing the people to undertake risk bearing activities. In many of the developing countries a lot of attention is being paid to the development of entrepreneurship because it is not the proprietary quality of any caste and community. The entrepreneurship is usually understood with reference to individual business. Entrepreneurship has rightly been identified with the individual, as success of enterprise depends upon imagination, vision, innovativeness and risk taking. The production is possible due to the cooperation of the various factors of production, popularly known as land, labour, capital, market, management and of course entrepreneurship. The entrepreneurship is a risk-taking factor, which is responsible for the end result in the form of profit or loss. According to A Schumpeter "The entrepreneurship is essentially a creative activity or it is an innovative function". The economic activity with a profit motive can only be generated by promoting an attitude towards entrepreneurship. The renewed interest in the development of entrepreneurship to take up new venture should emphasize on the integrated approach. The developments of entrepreneurship will optimize the use of the unexploited resources; generate self-employment and a self sufficient economy. The young entrepreneur should be motivated to come out with determination to do something of their own and also to contribute to the national income and wealth in the economy. If the country wants to achieve the growth at the grass root level, through social justice and the crimination of poverty, it will have to provide institutional support and structural changes in organization of financial institutions to promote entrepreneurship development. Industrial development in any region is the outcome of purposeful human activity and entrepreneurial thrust. David Melelland emphasized the importance of achievement motivation as the basis of entrepreneurial personality and a cause of economic and social development through entrepreneurship by fulfilling the following needs such as 1) Need for power 2) Need for affiliation and 3) Need for achievement. Another school of thought says "entrepreneurship is a function of several factors *i.e.* individual socio cultural environment and support system".

An entrepreneur is typically in control of a commercial undertaking, directing the factors of production–the human, financial and material resources–that are required to exploit a business opportunity. They act as the manager and oversee the launch and growth of an enterprise. Entrepreneurship is the process by which an individual (or team) identifies a business opportunity and acquires and deploys the necessary resources required for its exploitation. The exploitation of entrepreneurial opportunities may include actions such as developing a business plan, hiring the human resources, acquiring financial and material resources, providing leadership, and being responsible for the venture's success or failure. Entrepreneurial behaviour can accordingly be found in all kinds of enterprises, regardless of their size, age or profit-orientation. Entrepreneurship describes the process of value creation through the identification and exploitation of opportunities, *e.g.* through developing new products, seeking new markets, or both. It focuses on innovation by identifying market opportunities and by building a unique set of resources through which the opportunities can be exploited, and is usually connected with growth. Reynolds *et al.* (1999) *e.g.* proposed that 15 percent of the highest growth enterprises created 94 percent of all new jobs. One of the key challenges for entrepreneurs is to deal with the strategic changes required with the growth of their enterprise. Many scholars have thus decided to separate (growth-oriented) entrepreneurship from small business management (*i.e.* mom and pop enterprises or lifestyle businesses), describing growth as "the essence of entrepreneurship." Entrepreneurial enterprises identify and exploit opportunities that their competitors have not yet observed or have underexploited. An appropriate set of resources is required to exploit entrepreneurial opportunities with the greatest potential returns. Entrepreneurial enterprise resources are often intangible, such as unique knowledge or proprietary technology. According to Ireland *et al.* (2001), entrepreneurial behaviour arises through the "...concentration on innovative, proactive, and risk-taking behaviour." In real business life, though, there is not yet a cogent, direct treatise to formally recognize entrepreneurial behaviour as a new "dominant logic" in enterprises. Farmers have been entrepreneurs since the beginning of agricultural education and research. The Farmer/growers plan, implements, operates and assumes financial risks in a farming activity or agricultural business. In entrepreneurship programmes, the student owns the materials and other required inputs and keeps financial records to determine return to investment. Today, agriculturists are involved in many different types of entrepreneurial activities. Entrepreneurship in agriculture can still be raising livestock and growing crops, but it can be much, much more than that. Examples of Agriculture entrepreneurship are: Floriculture nursery, Lawn maintenance service, Custom Crop Harvesting, Tractor and Farm Equipment, Operating a Small Engine Repair service, Mushroom Growing, Apiculture and many more.

Scope in India

India is a land of enterprises, where almost 70 per cent of the population is still self-employed, and some place thi estimate as high as 80 per cent. (Srilatha 2013).

Entrepreneurship, a form of human behaviour, is indispensable for the growth and development of any society. Generally, the entrepreneur is considered as a person who initiates, organizes the activities, manages and controls the affairs of business unit combining the actors of production to supply goods and services. Farmers deciding to take particular activity also exhibit entrepreneurial behaviour (Rao and De, 2009;). The economic development of a country is correlated with the population, employment, literacy and effective utilization process, which accelerates the growth including social change. To accelerate economic development, it is necessary to increase the supply of entrepreneurs. The economic development of a country is correlated with the population, employment, literacy and effective utilization process, which accelerates the growth including social change. To accelerate economic development, it is necessary to increase the supply of entrepreneurs. Entrepreneurs have to do conscious decision making with great deal of thought concerning starting of an enterprise, finalizing different products, expansion of enterprise, borrowing money, hiring workers and considering new ways o marketing. So, entrepreneurs have to make decisions under certain uncertainties. In order to know, whether entrepreneurs under such uncertain conditions take decisions at their own or they do in consultation with somebody else.

Based on area under a 12 selected crops in the country, the classical estimate made out by Dr N.P. Goyal in 1994 mentions of the requirement of about 73.7 million colonies for only pollination services while currently, it is estimated that there are only 8.5 million colonies including 1.9 million colonies of hive honey bee species, in the country. Dr Goyal, based on the prices of these plant commodities existing at that time, reported that if the pollination requirement w.r.t. the required number of colonies are met, then it would result in additional income of Rs. 30 billion. According to an expert estimate, 10,000 honey bee colonies require 4,62,900 man days, equivalent to full time job for 1,543 persons annually. Accordingly, India has potential of providing employment opportunities for about 11 million people. The employment generation through development of allied industry is tremendous. According to recent estimate by National Bee Board, for optimum pollination of all the agricultural crops, India needs nearly 200 million honey bee colonies, and on meeting this requirement, employment opportunities will swell to nearly 30 million. Currently, India is producing a total of about 72,000 tonnes of honey including 37,000 tonnes of apiary honey the major chunk of which is exported. Based on the recommended consumption of two spoon of honey a day, India will need 0.67 million tonnes of honey annually. Thus, there is a wide scope in the further promotion and expansion of beekeeping in India. There is also a vast potential and scope from diversification in apiculture itself *i.e.* besides honey, it offers scope for production and marketing of other bee products *e.g.* pollen, propolis, royal jelly, bees wax and bee venom. Besides, sale of bee packages and rearing and sale of pedigree queen bees offers a great scope. Besides, sale of bee packages and rearing and sale of pedigree queen bees offers a tremendous scope. According to a recent estimate, based on the current price status of inputs, an apiary unit of 100 colonies put under diversification plan can earn a profit of about Rs. 7 lakh per year. Demand of honey bees for crop pollination is also increasing in some parts of the country. This profession offers a great scope in generating employment besides a good livelihood. Apart from direct employment to the beekeepers, there would be need for good artisans, hive manufactures, apicultural equipment and machinery manufactures, transport system for irrigation of colonies, traders, product quality experts, packers, sellers, raw material dealers *etc.* and allied industries. The value addition, so far, has remained little explored and offers tremendous scope. According to expert estimates, based on 15 g consumption of honey per capita per day would result in creating demand of 0.5 million tonnes of honey in the country while the current level is only 1/10th of the ideal. The low honey consumption in the country, besides it being promoted as luxury item and being largely promoted as medicine, is owing to very few value added products of honey developed and available in Indian market. Apart from direct employment to the beekeepers, there would be need for good artisans, hive manufactures, apicultural equipment and machinery manufactures, transport system for irrigation of colonies, traders, product quality experts, packers, sellers, raw material dealers *etc.* and allied industries.

Beekeeping in India

Honey and bee keeping have a long history in India. Honey was the first sweet food tasted by the ancient Indians inhabiting rock shelters and forests. They hunted bee hives for this gift of god. India has some of the oldest records of bee keeping in the form of paintings by prehistoric man in the rock shelters. With the development of civilization, honey acquired a unique status in the lives of the ancient Indians. They regarded honey as a magical substance and was given credit for maintaining health, vigour, fertility of not only humans but also of cattle and plants. In India bee keeping has been mainly forest based. Several natural plant species provide nectar and pollen to honey bees. Thus, the raw material for production of honey is available free from nature. Tribal populations and forest dwellers in several parts of India have honey collection from wild honey bee nests as their traditional profession. The methods of collection of honey and beeswax from these nests have changed only slightly over the millennia. The major regions for production of this honey are the forests and farms along the sub-Himalayan tracts and adjacent foothills, tropical forest and cultivated vegetation in Rajasthan, Uttar Pradesh, Madhya Pradesh, Maharashtra and Eastern Ghats in Odisha and Andhra Pradesh.

Beekeeping is emerging as very successful agricultural enterprise for local people in rural areas of the less developed, developing and developed countries alike. The practice has intrinsic health benefits through provision of food rich in nutritional value lacking in most rural diets, requires little capital investment and capitalizes on the existing natural environment.. Beekeeping systems range from the primitive rock and tree cavity shelters from which honey hunters harvested honey using smoke and fire to chase away the bees, to the modern beehives provided by beekeepers to protect bees from heat, cold, rain and pests, which are comfortable for the bees and suitable for the beekeepers (Shade, 1999). Honey bees produce a variety of useful products such as honey, beeswax, pollen, royal jelly and propolis, all of which are highly valuable and marketable. the introduction of commercial beekeeping as a vehicle through which the Sinendet – Chepkatet Youth Group in Kapseret, Kenya, will achieve its stated goal: "to reduce unemployment and increase incomes so as to improve the living standards of the youth in India.

Beekeeping Systems Beekeeping systems refer to the nature of the bee colony management. These range from the primitive rock and tree cavity shelters from which honey hunters harvested honey using smoke and fire to chase away the bees, to the modern beehives provided by beekeepers to protect bees from heat, cold, rain and pests, which are comfortable for the bees and suitable for the beekeepers (Shade, 1999). According to Shade, there are two types of beehives: the fixed comb hives and the movable comb hives. The former represent any kind of man-made container which can serve bees as a breeding place. Combs are fixed to the roofs and sides of the hives and are cut out and removed during harvesting. These include the traditional log hives. The latter hives are the modern types, including the KBTH, the mud hive and the Langstroth frame hive. this microenterprise activity, to be undertaken by the youth for youth, will create wealth and employment, thereby contributing to the fight against poverty in line with the UN Millenium Development Goals.

The Bee-keeping is a village industry and listed in the Government artisan sector. This indigenous industry has been taken by Khadi And Village Industries Commission to develop in the rural areas to create employment and generate production to use as food and medicine.

With the development of apiculture (bee keeping) using the indigenous bee (Apis cerana), use of the European bee (Apis mellifera) also gained popularity in Jammu and Kashmir, Punjab, Himachal Pradesh, Haryana, Uttar Pradesh, Bihar and West Bengal. Wild honey bee colonies of the giant honey bee and the oriental hive bee have also been exploited for collection of honey. 24 Bee hives neither demand additional land space nor do they compete with agriculture or animal husbandry for any input. The beekeeper needs only to spare a few hours in a week to look after his bee colonies. Beekeeping is therefore ideally suited to him as a part-time occupation. Beekeeping constitutes a resource of sustainable income generation to the rural and tribal farmers. It provides them valuable nutrition in the form of honey, protein rich pollen and brood. Bee products also constitute important ingredients of folk and traditional medicine. Resources and Potential The raw materials for the beekeeping industry are mainly pollen and nectar that come from flowering plants. Both the natural and cultivated vegetation in India constitute an immense potential for development of beekeeping. About 500 flowering plant species, both wild and cultivated, are useful as major or minor sources of nectar and pollen. There are at least four species of true honey bees and three species of the stingless bees. Several sub-species and races of these are known to exist. In recent years the exotic honey bee has been introduced. Together these represent a wide variety of bee fauna that can be utilized for the development of honey industry in the country. There are several types of indigenous and traditional hives including logs, clay pots, wall niches, baskets and boxes of different sizes and shapes. In modern beekeeping, the combs are built on wooden frames that are moveable.

Beekeeping Industry

Beekeeping or apiculture is the maintenance of honey bee colonies, commonly in hives by beekeepers. A hive is a simple rectangular box. A beekeeper or apiarist keeps bees in order to collect honey and other products of the hive (including beeswax, pollen and royal jelly), to pollinate crops, or to produce bees for sale to other beekeepers. Honey bees convert nectar of flower into honey and store them in combs of hive. A location where the bees are kept is called an apiary or bee yard. Bee colonies consist of three types of bee: A Queen bee which is normally the only breeding femalev in the colony. A large number of female worker bees typically 30,000-v 50,000 in number. A number of male drones ranging from thousands in av hive. The queen is the only sexually mature female in the hive and all of the female worker bees and male drones are her offspring. A queen lives three to four years and capable of laying more than half a million eggs in her lifetime.

There are four well known species of true honey bees in the world: 1. Rock bee, *Apis dorsata* 2. Little bee, Apis florea 3. Asian bee, Apis cerana 4. European bee, Apis mellifera Out of these four bees species, Apis dorsata and Apis florea are wild and are open nesting and cannot be domesticated in modern hives. Each colony of A.dorsata yields 30-40 kg of honey per year whereas A.florea yields only about 500g per colony. A.cerana and A.mellifera are domesticated in modern hives and are cavity nesting *i.e.* live in enclosures. The wild bees construct single comb whereas hive bees construct multiple parallel combs. The body size as well as the comb cell size of these bee species varies to a great extent. Among the two domestic bee species, each has many subspecies in different parts of the world *e.g.* A. cerana has three subspecies in India like A.cerana in 25 Himachal Pradesh and Jammu Kashmir (North India), A.cerana indica in Kerala, Tamilnadu and Karnataka (South India) and A.cerana Himalaya in Nagaland, Manipur, Mizoram, Assam and Meghalaya (Eastern Parts of India). In addition to above three subspecies, A.cerana japonica has been identified from Japan. Honey yield of A.cerana per colony per year is 3-5kg

There are four species of honey bees in India. Out of which the following two species are of commercially important (a) Indian bee (Apis cerana indica) (b) European bees or Italian bee (Apis Mellifera). The structure of a domesticated bee colony is normally rectangular box called hive within which there are eight to ten parallel frames vertically fitted with artificial honey comb made of wax which contain the eggs, larvae, pupae and food for the colony. The bees store surplus honey in combs above the brood nest. In modern time, beekeepers uses separate boxes called supers above the main box. In these boxes several shallow combs made of wax are provided for storage of honey. This enables the beekeeper to remove the supers to extract honey without damaging the colony and its brood nest. The annual cycle of a bee colony is location specific. It is completely dependent upon how the flora of the region blooms. In the plains of Assam, this starts with the flowering of many wild flowers after the break of monsoon season *i.e.* in the months of November which is followed by nectar flow from mustard, silk cotton (simul), mandarine oranges, litchi, pears cashew and other wild flowers. It continues upto the month of June depending upon the location. Formation of new colonies is very important as all colonies are dependant on their queens, who are the only egg layer. Beekeepers use the ability of the bees to produce new queens to increase their colonies. This procedure is called splitting a colony. Production of honey has been the major aim of bee industry. Modern beekeeping also includes production of beeswax, bee collected pollen, bee venom, royal jelly as also of package bees, queen bees and nucleus colonies. Modern beekeeping makes use of beekeeping equipment and honey processing plant. Seasonal management of bee colonies varies with the location of the project. There are special management techniques like queen rearing, migration of honey production or for colony multiplication which the commercial beekeeper adopts to enhance productivity and income generation.

Suitability of Beekeeping as an Agro-based Enterprise

i. Beekeeping involves low cost technologies. To begin with, basically only hives and bees are required which following training, can be acquired even on subsidy or the loans may be obtained for initiating beekeeping. Majority of the apicultural equipment required for bee husbandry are simpler and can be fabricated by skilled rural artisans and are cheaper. So initial heavy expenditure is not required.

ii. Beekeeping does not have any special land or elaborated structural requirements, as are required for diary, poultry, piggery, *etc.* The beekeepers do not need their own land for keeping the colonies or for growing the flora. The common bee flora already being grown in the area or even in nearby areas can be used for beekeeping. So it is also highly suitable for small and marginal farmers.

iii. Recurring expenditure on the activity is also not very high.

iv. Beekeeping does not compete with any other agricultural enterprises for resources.

v. It also does not require continuous labour and any heavy physical work, thus, is very ideal as a part-time occupation, even for women and children.

vi. Since beekeeping requires simple equipment and encourages the rural artisans to undertake the jobs for their fabrication, it helps in generating employment opportunities.

vii. Beekeeping is very hygienic activity; it does not lead to any filthy conditions and any breeding of flies and mosquitoes, and does not produce any offensive smell. Rather, it is fragrance emanating particularly during nectar flow.

viii. Honey itself is a very hygienic food, tonic and medicine and makes the diet more balanced.

ix. Beekeeping is a multiple source of income. Income from honey would supplement the main income from the crops. Beeswax is another bee product which has great commercial and industrial value. Sale of queen bees and nucleus colonies by division of parent colonies are other sources of income. If managed properly, honey bee colonies are self-perpetuating and very soon accounting for the cost, can start yielding profits in terms of increase in honey bee population. Production of other bee-products like royal jelly, bee venom, pollen and propolis can further be new sources of income. It leads to generation of several other avenues of beekeeping related trades and employment.

x. Above all, pollination by bees improves the quantity and quality of the crop produce and benefits the community as a whole rather than only the individual beekeepers through sustenance of food security. Beekeeper can further augment his income by renting out his bee colonies for pollination services.

Beekepring as Entrepreneurial Activity

Beekeeping is a vehicle to sustain the environment and improve the earning capacity of poor rural people. Honey production per hive increases steadily as apiary management improves. Agrolod (1991) stated that bee keeping is the manipulation of a bee colony by man based on the understanding of the science of tapping the economic benefits of the honey bees. Bee keeping can also be viewed as an agricultural activity whereby bee farmers employ their understanding of bee biology to provide good housing, appropriate feeding in addition to needed management practices to bees for the purpose of harvesting the honey. In the view of Chamber (2007) bees can be reared as it is adaptable to many climatic and laboratory conditions Ntega and Muogongo (1990) stated that honey from bees are valued, because of its food, medicinal and industrial uses. The authors further explained that in some part of the world, honey is extracted from the bees reared by farmers for generating income. Other bye products from the bee that provide the farmer income include wax, propolis and venom for medical uses. In order to produce bee bye products efficiently and profitably, the individuals require entrepreneurial skills. Hull (1991) said that skill is the habit of doing something well. Drucker (1970) stated that entrepreneur refers to a person who is ready to take risk. The author stated further that the behaviour of the entrepreneur reflects a kind of person willing to put his or her career and financial security on the line and take risks in the name of an idea, spending much time as well as capital on the venture. Therefore, an entrepreneurial requires mastery of entrepreneurial skills in order to succeed in a business. Application of the entrepreneurial skills to bee keeping can increase the income of the individuals in the enterprise such as farmers, business men, teachers and retirees.

Beekeeping is suitable for a wider arena of people including well-to-do farmers, landless labourers, small and marginal farmers, employed persons, ex-servicemen and retired persons, house wives/farm women, students and unemployed youth. Thus, beekeeping enterprise is suitable for people from all walks of life as a hobby, subsidiary occupation for supplementing income or as a whole time job for self employment. It is particularly suitable for underemployed/unemployed youth residing in or around rural areas. After investing once on honey bees, hives and other equipment (non recurring expenditure), regular/annual (recurring) expenditure is very low which can even be further reduced with better management practices. Beekeeping has a lot of employment generation potential (Figure 29.1).

Beekeeping is an important livelihood activity providing sources of food and income for many people around the world. Beekeeping as an activity spans forestry, horticulture, agriculture, natural environment, animal husbandry, entomology *etc.* without precisely fitting into any one of these sectors per se. The same assertion

Figure 29.1. Beekeeping: A Multipronged Source of Income and Employment Generation.

goes with pollination which is an important aggregate of crop farming/orchard; still bee management is often considered part of animal husbandry. The historical evolution of ideas about the entrepreneur is a wide-ranging subject and one that can be organized in different ways – theorist by theorist, period by period, issue by issue and on... Entrepreneurship in simplest term can be defined as the discovery and evaluation of opportunities as well as the creation of new opportunities and possibilities with calculated risk.

Entrepreneurship fosters sustainability as the bees are essential to biodiversity. Moreover, through entrepreneurship, beekeepers can increase their revenues and enrich their social capital. The innovative entrepreneurial behavior determines improved production conditions in order to maximize the production, together with the reduction of costs.

Essentially, sustainable entrepreneurship should innovate, adopt appropriate environmentally friendly technologies, develop skills and human resources, and enhance productivity to remain competitive in national and international markets. Entrepreneurship influences sustainable development as it entails a better access to certain markets and opportunity recognition. The entrepreneurial behavior influences sustainable development through the maximization of the rural beekeepers' income and by reducing poverty, maintaining the biological systems through pollination. Entrepreneurship as a strategy for the improvement of sustainability combines the interests of beekeepers and of the environment.

Therefore, entrepreneurship in the beekeeping sector is considered to be a factor that influences sustainable development and sustainable development represents a development that meets the needs of the present without compromising the ability of future generations to meet their own needs.

According to UN Conference on the Human Environment (1972), sustainable development is a concept that describes the social goal of improving and maintaining human well-being over a long-term time horizon within the critical limits of life-sustaining ecosystems. Researchers that analyze sustainable development from an entrepreneurial orientation start from the concept of ***creative destruction*** arguing that new sustainability pressures create various types of market failure, opening up opportunities for new. In essence, entrepreneurship is a means by which economic, environmental and social disruptions can be improved. It is a well-known fact that opportunities abound for business development through beekeeping. For instance; most small-scale beekeepers sell their honey only in bulk which may not be profitable at the end of the day. They do this because of their immediate need for money, but more income can be earned by establishing honey packaging for the retail market. At the highest end of the retail market is the gift size, which typically is the smallest size. The honey processor must also tailor products for natural-food consumers, restaurants, and for those who use honey to make other products. There exist other bee products that are money-spinning apart from adding values to them. Another important consideration for a prospective commercial honey entrepreneur is to provide incentives to the beekeeper through training and re-training programs focused on improving bee products. The long term overall objective is to contribute to economic growth and creation of more and better jobs.

Beekeeping is dynamic hence the need to integrate a number of interventions, which are important to the growth and expansion of businesses, cannot be over-emphasized. These include both training, non-training and extension services interventions. Beekeepers must be taught not only how to keep bees using good apicultural practices for quality honey production, but also how to render wax properly into a pure cake and how to harvest and handle other bee products.

Bee Conservation and Food Security

According to Albert Einstein *"If the bee disappeared off the surface of the globe, then man would only have four years of life left. No more bees, no more pollination, no more plants, no more animals, no more man"*. No doubt, bees are considered to be essential to biodiversity, a main indicator of the health of an ecosystem. In fact, it is estimated that one-third of all plants or plant products eaten by humans depend directly or indirectly on bee pollination. Moreover, crops pollinated by bees give higher yields and better quality products. No wonder farmers in Europe, America, Australia and the rest of advance countries pay beekeepers to bring their hives to the fields to ensure good yields and better product quality.

The important role played by honeybee in maintaining ecosystems and biodiversity is so great to the extent that without bees there would be no flowering plants, and without flowering plants there would be no bees. Without bees biodiversity would not be so great whereas biodiversity is measured as the number of different plant and animal species found in a certain unit area.

Pollinated flowers produce seeds that give life to the next generation of plants; fruits and seeds provide food for other animals; and through cross pollinations, genetic diversity is increased. Considering the role of bees and other pollinators in food production and the maintenance of biodiversity, it is important to create awareness among farmers, especially in developing countries, on the role of the pollinators in agricultural production and to manage them correctly.

Bees are the only livestock capable of harvesting nectar and pollen without any other animals competing with them and without bees, these valuable resources would have been wasted. Unfortunately, in spite of the overwhelming facts supporting the indispensability of honeybees to optimum crop production, crop farmers and agricultural experts in India, with the exception of South Africa have not seen the need to incorporate beekeeping development as an integral part of agriculture to boost crop production for more income

Begining Beekeeping: Important Considerations

To be a successful beekeeper one should acquire full knowledge of bee behaviour, hive operations and acquaintance with nectar and pollen plants. Sincere efforts and promptness in meeting the needs of bee colonies is the backbone of the profession. Bees face many hazards of diseases, enemies and poisoning and beekeeper should be alert to safeguard bees against these. Beekeeping should be started with clear-cut objectives as a hobby, subsidiary enterprise or a whole time job. Thorough planning, interest, zeal, confidence and enthusiasm in starting beekeeping and timely adoption of various management operations play a great role in its success. For starting beekeeping and for its subsequent success, following prelims need be considered.

Basic Requirement

- Acquire knowledge of bee behaviour and management from books and bulletins.
- Gain experience by working with a successful beekeeper.
- It is highly recommended to attend short training courses in apiculture.
- It is advisable to start with fewer colonies and multiply and increase number later as you acquire experience.
- Bee colonies can be purchased in hives from beekeepers/organization.

Training

A beginner must acquire the critical knowledge and practical skill by undergoing beekeeping training from some recognized institution or at least by working with a well experienced beekeeper for some time. The study of farm magazines, bulletins, books, *etc.* regarding beekeeping also help to refresh and update the knowledge of the beekeeper about the latest beekeeping technologies. Become fully conversant in handling of the colonies, in seasonal management practices and for common specific beekeeping situations.

Purchase of Essential Equipment

The essential bee equipment must be arranged well before starting beekeeping. The other beekeeping equipment required from time to time should also be arranged as and when required.

When to Start

There are regional climatic variations and, therefore, it is difficult to specify the exact time but as a thumb rule, enough blossoms should be available for supply of food requirements for brood rearing. Well established colonies can survive and face the problems but poorly started colonies in bad floral and climatic conditions are difficult to establish. A beginner is not conversant of needs of colonies. Beekeeping can even be started in any season, provided the beginner knows the needs of the colonies at that time and manages accordingly.

Purchase of Nucleus Colonies

For starting bee keeping, nucleus colonies should be purchased in the beginning of suitable season. A nucleus colony must be of four or five frame bee strength with a young, newly mated quality queen bee; should have sufficient amount of rightly laid eggs, worker brood (sealed and unsealed) and food reserves (honey and pollen stores) and should not have excessive amount of drone brood area. Unsealed honey during transportation should be avoided n the colonies.

Area/Site Requirement

An ideal location should provide enough spring forage for repid build up and later avail major nectar flow for long duration. Nectar producing plants of wide variety and seasonal distribution provide a base for profitable beekeeping. The locality should provide economic surplus of nectar and pollen. To start beekeeping, the knowledge of local flora is essential. For hobbyist any location can be good but for commercial venture the assessment of the flora should be very exact. Commercial beekeepers generally migrate their bees from one place to another and my produce more than one honey crop of different types. Large number of amenity, highway, forest and social forestry trees are good bee flora. Many crop plants also surve as good bee forage. Irrigated cropped areas present ensured flora. An easily approachable site with minimum disturbance, plenty of fresh water, air, shade, sun light and availability of bee flora the year round is well suitable for setting up an apiary. The land selected should be well drained and should not be flood prone.

Climatic Protection

The bee colonies should be located in shade in areas where the temperatures go above 95°F in summer and fall. This shade can be provided by boards laid on top of the hives, by harbors, by shrubbery or under thick shade giving trees. Bees have greater difficulty in lowering down hive temperature during summers. During winter in north India, the colonies are better if these hives are not shaded and located in sun. Suitable season for starting beekeeping coincides with moderate climatic conditions and availability of bee flora in plenty. Normally, spring season

(February-April) and post monsoon (September-November) are the best periods to start beekeeping under the North Indian conditions. These are generally the periods when colonies would rear drone brood.

Good Air Drainage

Air with high moisture should be regularly drained away from the apiary. Dampness is bad for bee flight and effects the colony's capacity to regulate nest temperature. Air drainage is more important in rainy season in India.

Wind Breaks

Wind breaks of trees or artificial structures give protection against cold winds. This is especially during build up period when colonies are weak and weather is unsettled in cooler areas.

Water

Artificial source of water should be managed in the apiary if natural sources are not available. Adding some common salt makes water more attractive.

Hive Arrangement and Density

Spread and stagger the colonies in both directions, space them as wide as possible. Bees get advantage of land marks. Economic foraging distance of Indian honeybee is about one km and that of *A. mellifera* is 2 km, though, they go in search of nectar and pollen and fly farther if necessary but longer foraging distances are not advantageous in terms of energy balances. Assessment of suitable density of hives in a locality is a difficult job. Overstocking of a locality means reduced honey production. Under tick floral conditions the apiaries can be set apart at about one kilometer. One hive per 2.5 hectare area should be appropriate. A radius of 0.6 km would present a foraging area of about 115 hectare and this area can economically support about 50 bee hives. The number of bee colonies in an apiary would be lower under poorer floral conditions. As a principle each new location should be tested with fewer colonies then regular the number according to seasonal conditions affecting plant growth.

Bee Products

Honey bee colony productivity (including honey, other bee products and benefits) in India is dismally low as compared with other apiculturally developed countries. *Inter alia,* the reason for low productivity is non-adoption of apicultural diversification in the country. Increasing colony productivity by adopting apicultural diversification will help making our beekeeping internationally competitive and pave way for the country to enter into the global market of other bee products too, thus, enhancing the income to the beekeepers. The beekeepers in India domesticate bees only to produce honey and some beeswax. However, apart from honey and bees wax, bees provide us pollen, propolis, royal jelly, bee venom, queen bees, package bees, brood *etc.* Honey bees can also be managed as and when required for pollination of field and horticultural crops and for hybrid seed production in vegetables and other bee pollinated crops. A beekeeper can get at least Rs. 1000 to 1300 per colony for pollination of fruit crops like apple in the state of Himachal Pradesh. Thus, renting out bee colonies for pollination can also be another source of income to the beekeepers. As such, all these hive products present a great scope for increasing colony productivity and, thus, the beekeepers income, as these products have a great demand in some other countries. Thus, in addition to honey, production of bees wax, propolis, pollen, royal jelly, pedigree queen bees, package bees and renting out honey bee colonies for crop pollination service are potential areas of apicultural diversification in the State. These potential areas for diversification in beekeeping are briefly discussed below:

Honey

Honey is defined as "the sweets substance produced by honeybees from the nectar of blossoms or from secretions on living plants, which the bees collect transforms and store in honey combs". Honey bee produces dense and stable energy food called as nectar which ripened into honey. Honey is a complex natural product, containing more than 400 different substances, *e.g.* various carbohydrates, organic acids, proteins, amino acids, enzymes, aroma substances, mineral substances, pigments, waxes, *etc.* The composition and quantity of these components vary widely according to the floral and geographic origin of the honey. Honey is a functional food and has different biological properties such as antibacterial (bacteriostatic properties), anti-inflammatory, wound and sunburn healing, antioxidant, radical scavenging, antidiabetic and antimicrobial activities. India produces around 70,000 tonnes of honey annually out of which 25,000- 27,000 tonnes is exported to more than 42 countries. Honey exports are estimated around USD 3.2 mn in 2007-08. United States, EU and Middle East are key exports destination for Indian honey. Presently there are more than one million bee colonies in India but considering the market demand, the Country has the potential for more than 120 million bee colonies to achieved what is called 'Amber Revolution'. This would result in the production of around 1.2 mn tonnes of honey and 15,000 tonnes of beeswax. Further it can provide employment to over 6 million rural and tribal families.

Bee Venom

In the Honeybee, the stinging apparatus and the venom, originally evolved as devices to paralyse preys, became arms to defend the colony mainly from the attacks of vertebrate predators. The venom gland of worker bee is located in posterior portion of the abdomen, between the worker's rectum and ovaries. The two glands (Dufours and Venom gland) associated with sting apparatus of the worker produce venom. The venom gland is a thin long, distally bifurcated integumentary gland with cuticular lining. It consists of a secretary filamentous region, connected to a reservoir at its proximal portion, in which the venom is stored. The small flat cells also bearing canaliculi form the distal region of the reservoir where their products contributes to venom composition. The workers sting only once, which leads to their death. Bee venom contains 88 per cent water. At least 18 pharmacologically active components have been described so far; including various enzymes, peptides and amines (Dotimas and Hider, 1987). A newly emerged bee has very little venom content, but the amount gradually accumulates with age, to about 0.3 mg in a 15 day old *A. mellifera* worker bee, after the age of 18 days no additional venom is produced. Subsequently, the weight of the venom in the venom sac remains unchanged. Bee venom has interesting pharmacological properties and is used in the treatment of various health conditions such as arthritis, rheumatism, pain, cancerous tumors and skin diseases. Bee venom is synthesized in venom glands of worker bees (150-300 µg) and queen bees (700 µg), however, only the worker bees are exploited for venom production. Normal recovery of dry venom per bee is 0.5 – 0.1 µg. One million stings result in the production of 1 g dry venom. For venom collection, the bees are exposed to low current (33-44 volts) under which effect the bees sting a taffeta spread under the current carrying element wires (spaces 6 mm) and deposit their venom on the glass plank under the taffeta. The deposited venom is scraped off and stored in dark container under refrigerated condition. Twenty colonies would yield about one gram of venom. LD50 of bee venom is 2.8 mg per kg, thus about 600 stings could be lethal for human being. The main venom producer is USA. It produced only 3000 gm of venom over 30 years and all the venom needed in the USA is produced essentially by one beekeeper operating part time. The price of bee venom on the U.S. open market during 1999 varied form $ 100-200/g (Rs 6700-1,25,00/g).

Royal Jelly Production

Royal jelly is a viscous substance secreted by the hypopharyngeal and mandibular glands of worker honeybees as an essential food for the queen bee larva and for the queen herself. If fed to an ordinary female bee in the larval stage, royal jelly will transform her into the queen bee. As a queen, she will grow 1½ times normal size, become extremely fertile and lay over a thousand eggs each day. Incredibly, she may live over five years while all the other bees live only a few weeks. The only difference is that she receives royal jelly while the others don't. The chemistry of royal jelly Royal jelly is a complex mixture of proteins (12 per cent), sugar (12 per cent), fats (6 per cent) and variable amounts of minerals vitamins and pheromones. About 15 per cent of royal jelly is 10-hydroxy-trans-(2)-decanoic acid (HDA), which is probably the substance that causes the queen bee to grow so large. Royal jelly is particularly rich in B vitamins, with pantothenic acid dominating. For royal jelly production, 90 primed plastic queen cell cups are to be grafted with one worker larva each of less than 24 h of age and given to a 20 BF queen-right cell builder colony. The royal jelly is extracted 72 hours following larval grafting and is stored in refrigerated condition in dark coloured vials. It should not be exposed to sun and any metallic container. Average yield of royal jelly is 300 mg/queen cell/72 h. Well managed colony can, however, produce more than 800 gm of it in a year. Royal jelly is used in several medicines, cosmetics and as dietary supplement. Price of royal jelly in different countries varies from US$ 50-120 (Rs.2500-6000)/kg for raw and over US$ 3000/kg for processed or freeze dried royal jelly. Royal jelly folk use Royal jelly has a history of folk use as a skin tonic and hair growth stimulant. The skin benefits are supposed to include a nourishing process that reduces wrinkles, although there is no actual scientific research that supports these claims (or the hair growth claims). Royal jelly has also been considered to be a general tonic that has a general systemic action rather than any specific biological function, which benefits menopause and sexual performance. Perhaps it's most significant use has been as an aid for increasing energy.

Bees Wax

Beeswax is produced by a set of eight wax glands on the underside of the worker bee. A bee secreting wax will pull a tiny scale of wax from her gland, and add it to the wax comb. Using her mouth and antennae, she then shapes the wax scales into the hexagon pattern of the comb. A strong, healthy hive is a much better wax producer and comb builder than a weak one. Over time, the wax will change from white to yellow. The wax will eventually become very dark where the bees are rearing brood. It is a consequence of the growth of the larvae, and pigments in pollen carried by bees walking over the comb. Old, black comb should be removed and replaced with new foundation. If you want to collect wax for crafts or candles, use new, light-coloured wax. The 49 cappings from the honey frames are best. The cappings will be left over when you have extracted your honey. Wax is cleaned first with warm water to remove any honey. It should be melted only under low heat. Hot wax is flammable.

Bees wax is the secretion of paired epidermal glands situated on the underside of 4th to 7th abdominal segments of bees of 14-18 days age and is used for comb construction. The sources of bees wax include honey cells cappings, burrs, braces, old combs and combs of wild bees. This wax has to be purified by melting it in water and then through its fine filtration. The wax being lighter than water, floats on the surface of water and the solid disc of wax is removed after it has cooled down. It is then bleached either chemically or through keeping small pieces of it in sun. One colony can yield about 300g wax in a year. This wax is used in candle making, shoe polish, vehicle and floor polishes, varnish, gum, carbon paper, electrical appliances, fabric industry, wax crayons, metal casting and modelling, food processing and packing and in cosmetics industry. Bees wax is used in over 300 items. In churches, candles of bees wax are lighted. In beekeeping industry, it is used for preparation of comb foundations. Generally the comb foundations are prepared manually though recently, imported German machines have been installed in the Punjab for continuous comb foundations fabrication. In India, its whole sale price is Rs. 250-300 per kg and its comn foundations are sold at Rs. 350 per kg.

Pollen Production

Pollen is the male gametophyte of flowers. Bee pollen, which is also called "bee bread" and sometimes "ambrosia" is a staple food for honey bees and their young. Commercially traded pollen is mainly pollen collected by the honey bee *Apis spp.* for the purpose of feeding its larvae in the early stages of development. Collected flower pollen is accumulated as corbicular pellets in pollen baskets on the rear legs of the honey bee and it is a mixture of these pellets that comprises bee pollen. When visiting flowers, bees touch the stamens and their bodies become covered with pollen dust. The bees use their hind legs to compress the pollen into the pollen baskets. The bees moisten the pollen with mouth secretions which help the pollen to cling together and to the basket hairs. These secretions contains different enzymes, *e.g.* amylase, catalase *etc.* Bees require pollen for growth and development. Immature (larval) bees are fed a mixture of brood food and bee bread. Newly emerged bees consume bee bread so that their bodies can complete development. The amount of pollen required to rear a single worker larva has been estimated at 124-145 mg, this containing about 30 mg of protein. The minimum level of protein required for honey bees has been estimated to be between 20-25 per cent crude protein. Pollens with protein levels in this range are more useful to colonies and allow them to meet their protein requirements readily. A diet of high protein pollen increases worker bee longevity, while brood rearing is reduced when supported by pollens low in protein. A pollen load contains up to 10 per cent nectar, which is necessary for packing. One pollen granule contains from one hundred thousand to five million pollen spores each capable of reproducing its entire species. Bee pollen is often referred to as nature's most complete food. Human consumption of bee pollen is praised in the Bible, other religious books, and ancient Chinese and Egyptian texts. Pollen is the source of proteins, enzymes, vitamins, minerals and fat for bees for brood rearing or tissue building. The pollen protein contains from 15 to 19 amino acids, including all essential amino acids. Pollen was found to be particularly rich in glutamic acid, aspartic acid, proline, leucine, lysine, arginine and in serine. The authors found substantial differences in the content of individual amino acids depending on the plant origin. Serine, cysteine and histidine was found to show the most extensive variation, as well as valine, leucine, isoleucine and lysine. Many people who use bee pollen as a supplement or remedy feel that it is a natural superfood. Bee pollen is created to feed growing bees as well as adult bees and it believed to contain unparalleled nutritional complexity. "Nutritional complexity" means that the food has many different kinds of vitamins, minerals, and nutrients. Bee pollen enthusiasts generally believe that bee pollen is incredibly well nutritionally balanced. Excessive pollen trapping adversely affects brood rearing and colony development. Studies have shown that 25 per cent pollen can be safely trapped without adversely affecting the colony strength. Different types of fabricated traps for pollen collection can be used and the trapped pollen should be stored either afresh in refrigerated condition or as shade dried. The pollen can be collected during the period of its excess availability and can be fed to honey bee colonies during pollen dearth to maintain brood rearing and thus the growth the colonies to best exploit the ensuing honey flow. One 10 bee frame strength colony can consume about 250 g of pollen in about a fortnight. This hive product finds its use as highly nutritive dietary supplement in health clubs for muscle building, in several medicines and cosmetics industry. Its wholesale price in the international market varies from US$ 5 -13 (about Rs 225 -575)/kg and retail price from US$ 11 -30 (about Rs 550 -1500)/kg. The tablets and capsules made from the pollen are sold at US$ 900 (about Rs. 45,000)/kg.

Propolis Production

Propolis (bee glue) is a sticky dark coloured material that honeybees collect from living plants, mix with wax and use in construction and adaptation of their nests. The term 'propolis' was used by authors in Ancient Greece: *pro* (for, in front of, *e.g.*, at the entrance to) and *polis* (city or community); a substance that is for or in defence of the city or hive. Bees apply propolis in a thin layer on the internal walls of their hive or other cavity they inhabit. It is used to block holes and cracks, to repair combs, to strengthen the thin borders of the comb, and for making the entrance of the hive weather tight or easier to defend. Propolis also is used as an "embalming" substance to cover hive invaders which bees have killed but cannot transport out of the hive. Bees make use of the mechanical properties of propolis and of its biological action: bee glue contains the putrefaction of the "embalmed" intruders, it is responsible for the lower incidence of bacteria and moulds within the hive than in the atmosphere outside. Propolis is produced only by *Apis mellifera* bees. Even in *Apis mellifera,* some races collect more propolis than others. Propolis is sticky and resinous material, collected by foragers as exudates of buds, bark and wounds of plant/trees. Propolis is used by bees to plug the cracks and crevices in the hive, varnishing the comb surface and as a repellent against ants at the hive entrance. It is the only antimicrobial arsenal in the honey bee colony. Its production potential is 300 g per colony. Propolis can be collected from colony by scraping it. However, the refined technique of pollen collection includes using a propolis screen, which has perforations and is spread on the top bars of combs under inner cover. After the perforations are filled up with propolis, the screens are taken out and kept under deep freeze condition for overnight. After this exposure, the propolis screens are twisted to make the dried pollen to fall down in small pieces. The same is then purified by dissolving it in suitable solvent and filtration and solvent evaporation. The Propolis is used in human and veterinary medicine, especially in skin and dental ailments. The price of propolis in international market varies between US$ 4 - 12 (about Rs. 200-600) though in some countries, it is as high as US$ 52 (about Rs. 2600)/kg. Beeswax is used commercially to make cosmetics and pharmaceuticals, including bone wax these (cosmetics and pharmaceuticals) account for 60 per cent of total consumption, in polishing materials particularly shoe and furniture polish and as a component of modelling waxes. It is commonly used during the assembly of pool tables to fill the screw holes and the seams between the slates. It is a good adhesive when blended with pine rosin to attach reed plates to the structure inside a squeezebox. Beeswax is used by the cosmetics industry; the bee-keeping industry, for the production of honeycombs; the pharmaceutical and cosmetics industry; in candle making; and for the production of polishes and varnishes (FAO 1996). Compared to the market for honey, beeswax is a minor market.

Queen Bees and Brood

A colony of honeybees contains a queen, drones and workers in which the queen controls productivity and behavior of the colony. These castes are developed as a result of preferential feeding during their larval stages. Ones the queen started egg laying, there will be no longer mating and consequently a queen remains economically productive for only one to two years (Akyol *et al.*, 2008). While colony of bees rear queen naturally to replace a lost queen, replace the queen when swarming or to supersede a failing queen, beekeepers manipulate a colony to duplicate one or more of these natural circumstances (Johnstone, 2008). Generally, queen rearing is done to allow beekeepers reproduce good stock for replacing old or undesirable queens in their colonies or to start new colonies. While bee colony rears queen naturally to replace lost queen, replace the queen when swarming or to supersede a failing one, beekeepers manipulate their colonies to duplicate one or more of these natural circumstances (Johnstone, 2008). Queen rearing is done to allow beekeepers reproduce good stock for replacing old or undesirable queens or to build up new colonies. Beekeepers can get replacement queens either by purchasing or by rearing (Somerville, 2009) but it is preferable to rear queens at self apiary because of cost, time, availability, quality, diseases and mites (Bush, 2007). Queen rearing enables beekeepers to select and reproduce colonies out of their stock. Queens can be reared out of colonies that have demonstrated better performances among the stock although such outstanding traits are less likely to be maintained and inherited for generations under natural mating, where there is no control of the several drones that met with a queen. About 150 to 200 queens/ colony can be produced in 5-6 months (during breeding season) in northern India especially in the Punjab. The price of single newly mated quality queen bee varies from $ 9 to 11 (about Rs. 450-550) in USA. In India a good quality/pedigree queen bee can be sold for Rs 250. Nucleus colonies with brood frames and package bees (bees by weight without brood and frames) can also be sold by the beekeepers to supplement their apicultural income. Brood (larvae and pupae) of honey bees is very hygienic food for human beings and livestock. The bee larvae/pupae provide 15.4 per cent proteins and 3.7 per cent fat, Vit. A and B *etc.* In many countries, bee brood is relished for human consumption and is also used as apitherapy.

Renting out Bee Colonies for Pollination Service

The importance of honeybees to agriculture – more specifically, the role that the tiny-winged insects play in pollination – is too well-known to require elaboration. Not as well recognised a reality, however, is the destruction of beehives that the indiscriminate usage of pesticides have wrought and, in turn, contributed to a not-insignificant agricultural crisis. There's no better proof of that than farmers like Takle today having to hire beehive boxes to "aid" in pollination. Highly cross-pollinated crops like onion, cotton, oilseeds and most fruits and vegetables depend entirely on winged insects that, during their flights to collect nectar, also transport pollen from one flower to another. "Honeybees are best suited to perform this function. Our experiments have shown an average 30 per cent increase in crop yields – from 17-19 per cent for cotton to 48 per cent in sunflower and 150-170 per cent in lychees – if honeybees are artificially introduced in the fields during the flowering stage even in normal conditions. Promotion and growth of the pollination enterprise can help improve livelihoods for several mountain farming families by generating employment and income as well as boosting the production and quality of crop. Managed pollination of apples as practiced in Himachal Pradesh, India is an excellent example of enhancing income and food security of not only apple farmers but also beekeepers. In recent years pollinator abundance and diversity are declining worldwide, due to habitat loss, excessive use of pesticides, climate change and other factors. Many varieties of the cash crops currently cultivated in mountain areas are self sterile and require cross pollination for producing fruit or seeds. A diversity of pollinators including bees, flies, butterflies, moths, beetles help in cross pollination of these crops, thereby helping in maintaining (or improving) their yield and quality. Pollinators, thus, play a crucial role in improving food security and livelihoods of mountain households through provisioning of the pollination services. Pollinators also play an often unrecognized role in combating soil degradation by enhancing the replenishment cycle *i.e.* more pollination, more seed, more plants, retuning more biomass to the soil, more food for birds, insects and other animals. Globally the annual contribution of pollinators to the agricultural crops has been estimated at about 153 billion Euros.Honey bees are excellent agents of crop pollination because they can be easily managed at any time and in any number. The body of honey bees is extremely hairy and thus they collect higher amount of pollen which results in better pollination of crops. Moreover, pollen being the food of larvae and adult bees, the foragers purposely visit the flowers for pollen collection and, hence, prove to be efficient pollinators. Bees also have a behaviour to constantly visit blossoms of a crop (floral fidelity) until it is exhausted of pollen/nectar resources. Major emphasis of beekeeping in apiculturally developed countries is on the use of honey bees for pollination services. In India also, there is a demand of honey bee colonies for pollination of apple, other temperate fruit plants and for hybrid seed production of various vegetable, oilseed and other field crops. Thus, beekeepers can earn additional income by renting out their honey bee colonies to the farmers for pollination service to their crops.

Economics of Beekeeping

After going through different aspects of beekeeping including seasonal management, it is important to work out the economics of this enterprise if one goes for commercial beekeeping. It requires information of expenditure and income from a unit of an apiary. For commercial beekeeping it is recommended that one should start with minimum of 100 bee colonies. Details of expenditure and income are given below:

Table 29.1. Expenditure for 100 *Apis mellifera* Bee Colonies

Expenditure Head	*Number*	*Rate/unit*	*Total Amount (Rs.)*
Bee hives	100	1500	150000
Bee colonies	100	2000	200000
Honey extractor	1	5000	5000
Smoker bee veil, hive tools *etc.*	1 set	1000	1000
Miscellaneous	-	5000	5000
Total			361000 (A)

Table 29.2. Recurring (per year)

	For stationary beekeeping		
Labour (Full time)	1	Rs.5000/month	60000
Comb foundation sheets	1000	Rs20/sheet	20000
Sugar feeding	500	Rs.40/kg	20000
Chemicals	-	Rs. 20/colony	2000
Miscellaneous	5000		5000
		Total	127000 (B)

For migratory beekeeping			
Transportation	No. of trips	Cost per trip	16000
By truck	4	Rs 4000	
Total			143000 (C)

Table 29.3. Details of Expanses

Stationary beekeeping	
Interest on non-recurring cost @15 per cent	34200
Recurring cost	73000
Interest on recurring for 6 months @15 per cent	5475
Depreciation on permanent articles except bees @ 10 per cent	10800
	1,23,475 (D)
Migratory beekeeping	
Interest on non-recurring cost @15 per cent	34200
Recurring cost	73000
Interest on recurring for 6 months @15 per cent	5475
Depreciation on permanent articles except bees @ 10 per cent	10800
	1,23,475 (E)

Income from 100 Colonies

Table 29.4. Stationary Beekeeping

Commodity	*No./ Average Yield*	*Quantity*	*Rate/item (Rs.)*	*Amount (Rs.)*
Honey	15 kg/colony	1500	150	225000
Sale of divided bee colonies	30 per cent of	30 bee colonies	2500	75000
Beeswax produced	2 per cent of honey produced	75	300	22500
Colonies for pollination	10 per cent bee colonies	10 colonies	800	8000
Commercial queen production (one breeding season) from selected stock	From 10 per cent colonies	100	500	50000
Total				380500 (F)

Table 29.4. For Migratory beekeeping

Commodity	*No./ Average Yield*	*Quantity*	*Rate/item (Rs.)*	*Amount (Rs.)*
Honey	60 kg/colony	6000	150	900000
Sale of divided bee colonies	40 per cent of	40 bee colonies	2500	100000

Commodity	No./ Average Yield	Quantity	Rate/item (Rs.)	Amount (Rs.)
Beeswax produced	2 per cent of honey produced	75	300	22500
Colonies for pollination	10 per cent bee colonies	10 colonies	800	8000
Commercial queen production (one breeding season) from selected stock	From 10 per cent colonies	100	500	50000
Total				**1080500 (G)**

Net Income Stationary beekeeping (F-D) 380500-1,23,475 = Rs

Migratory beekeeping (G-E) 1080500-123475 = Rs. 2,68,925

Additional Benefits of bee keeping are increase in crop yield and it boosts employment in bee equipment related industry.

Factors Affecting Entreprenerushіp Development in Beekeeping

The present research examines some of the factors that determine the intention to start a business in the beekeeping sector. Factors such as motivation, knowledge and social capital are seen to explain entrepreneur's capacity to discover opportunities and decide to start a business. Entrepreneurship in general terms is concerned with the "discovery and exploitation of profitable opportunities" (Shane and Venkataraman, 2000). All human action is the result of both motivational and cognitive factors, the latter including ability, intelligence, and skills (Locke, 2000). Entrepreneurial activity "can be conceptualized as a function of opportunity structures and motivated entrepreneurs with access to resources" (Aldrich and Zimmer, 1986). According to Baum *et al.* (2007) motivated entrepreneurs are important to the entrepreneurial process, and so the inclusion of human motivation in theories of the entrepreneurial process is crucial. Hisrich (1985) found that one of the prime motivations for starting a business was the desire for independence. Passion and the desire for independence were found to be one of the main factors that determine beekeepers to practice apiculture. Surprisingly, there have been virtually no quantitative studies of the role of passion in entrepreneurship. One exception is the study by Baum *et al.* (2001) who entered passion for the work as a separate variable along with 29 other variables from five domains (personality, situational motivation, skills, strategy, and environment). In 288 Baum's research passion had a direct significant effect on firm growth. The human capital literature in entrepreneurship has begun to show the effect of certain types of knowledge and skills on the start-up (Shane *et al.*, 2003). According to Shane and Venkataraman (2000) some people and not others discover particular entrepreneurial opportunities because they possess prior information necessary to identify an opportunity and the cognitive properties necessary to value it. Knowledge is one of the aspects that particularly matter for the performance of an entrepreneur. According to Penrose's (1959) theory of the growth of the firm, firm's knowledge to productively integrate resources into higher-level capabilities determines its growth. According to Spender

(1994), competitive advantage is not merely explained by individual resources, but by a different type of resource – knowledge, referring to the coordination of resources, which inheres in the activity itself and therefore the firm (rather than its individual members). Competitive advantage has come to be seen as based on knowledge, not raw materials. In this way, firms create and sustain competitive advantage by protecting valuable knowledge through preventing its migration and reducing its imitability (Prashantham, S., 2008). Knight and Cavusgil (2004) sustain that "knowledge is the most important resource, and the integration of individuals' specialized knowledge is the essence of organizational capabilities". Closely related is the aspect of absorptive capability, which is the ability of a firm to recognize, assimilate and apply information from the external environment (Prashantham, S., 2008). Aldrich and Zimmer (1986) show that stronger social ties to resource providers facilitate the acquisition of resources and enhance the probability of opportunity exploitation. According to Bourdieu (1986), the social capital is "the aggregate of the actual or potential resources which are linked to possession of a durable network of more or less institutionalized relationships of mutual acquaintance or recognition". Because very often entrepreneurs do not have access to extensive information sources, they are backed up by actors in their environment who influence their decision-making process. Social capital contributes to the availability of information, and it has a positive impact on the innovative performance of small and medium-sized enterprises (Gibcus *et al.*, 2008). Social capital fosters trust and decreases barriers to the exchange and combination of new knowledge (Nahapiet and Ghoshal 1998).

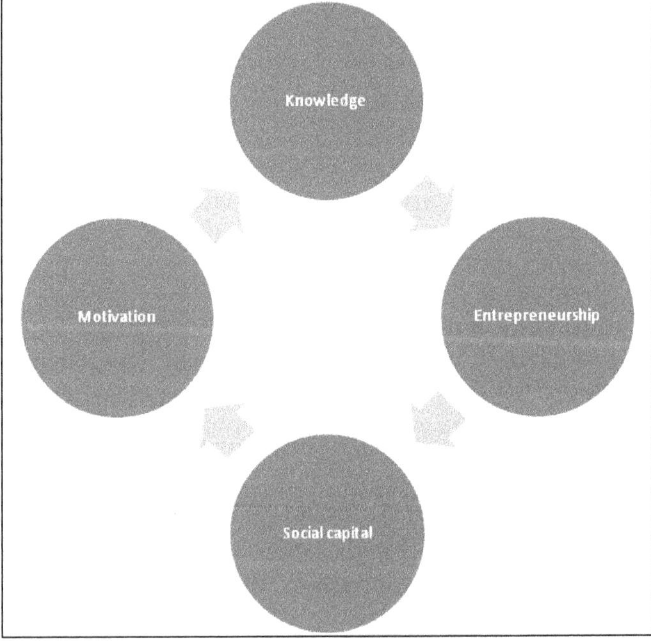

Figure 29.1. Factors that Influence Entrepreneurship.

Beekeepers' motivation, desire to gain knowledge and social capital are connected to the beekeepers' willingness to start a business. Experience in the beekeeping sector was the only factor that did not influence the decision to start a business. Therefore, motivation, knowledge and social capital can facilitate the creation and acquisition of market and technological knowledge, thereby leading to useful innovation. Motivation determines beekeepers to acquire the necessary knowledge for the design of the business, development of the product that will be provided to customers and assembly of human and financial resources. Therefore, the strengths required for successful entrepreneurship depend to a great extent on the individual's motivation, knowledge and social capital. In order to sustain entrepreneurship in the beekeeping sector, networks of collaboration have to be developed, as well as management and marketing training programs, distance learning technologies and business support systems. The benefits that the entrepreneur can obtain from the social network are often related to motivation and access to valuable resources like information, customers and suppliers.

Problems of Entrepreneurship Development in Beekeeping

Entrepreneurship in agriculture is not only an opportunity but also a necessity for improving the production and profitability. However, the rate of success is very low in India, because of the following reasons.

- For most of the farmers, agriculture is mainly a means of survival. In the absence of adequate knowledge, resources, technology and connectivity with the market, it is difficult for the illiterate small holders to turn their agriculture into an enterprise.
- Before promoting various services by self employed persons, there is a need to create awareness among the farmers, who are the users, about the benefits of these services.
- For popularisation of services, the present practise of providing free service by the Government agencies should be discontinued. In fact, many farmers, particularly the politically connected leaders are of the impression that the government is responsible for providing extension and technical advisory services to the farmers. However, over the years, the credibility has eroded and the services of these agencies are not available to small farmers, particularly those living in remote areas. Nevertheless, the concept of free service makes the farmers reluctant to avail of paid services, offered by the local self-employed technicians.
- The self-employed technicians need regular back up services in the form of technical and business information, contact with the marketing agencies, suppliers of critical inputs and equipment and research stations who are involved in the development of modern technologies.
- There are several legal restrictions and obstacles, which come in the progress of agri-business, promoted by the People's Organisations and Cooperatives. Private traders engaged in such business tend to ignore these rules and disturb the fair trade environment.

- ☆ People's Organisations often hesitate in taking the risk of making heavy investments and adoption of modern technologies, which in turn affect the profitability. With low profitability and outdated technologies, farmer members lose interest in their own enterprises as well as in that of their leaders.

Strategy for Promoting Beekeeping as an Enterprises

Indian honey offers tremendous export potential. For tapping its potential, there is need to chalk out suitable export strategy. Considering the present problems faced by the entrepreneurs engaged in agri-business, it is necessary to create a congenial atmosphere in the field. Some of the important conditions necessary for successful agri-business are presented below:

- ☆ There should be a unanimous option among government officials and farmers about the need and benefits of promoting self-employed youth or private entrepreneurs to facilitate the farmers to enhance agricultural production and profitability.
- ☆ The Government should discontinue the practice of providing free services in those sectors where the work has been assigned to private entrepreneurs.
- ☆ The technical skills and ability of the entrepreneurs should be evaluated to ensure high standards. There should be a monitoring agency to check the quality of the services and the charges collected from the farmers to avoid exploitation.
- ☆ To popularize the services of the entrepreneurs, the line department *viz.* Agriculture deparment, Horticulture department, Agricultural universities, NGOs and other agencies should give wider publicity about the services available to the farmers/beekeepers/unemployed youths. Such publicity can enhance the credibility of the services provided by the entrepreneurs.
- ☆ The Government should encourage the entrepreneurs by introducing various concessions and incentives.
- ☆ Networks of beepreneurs may be established to share their experiences. These networks can also establish a close link with Research Institutions and Universities to become acquainted with the latest research findings and seek solutions for their field problems.

Conclusions

Beekeeping industry in the country can well become a major foreign exchange earner if international standards are met. Beekeeping is an age-old tradition in India but it is considered a no-investment profit giving venture in most areas. Of late it has been recognised that it has the potential to develop as a prime agri-horticultural and forest-based industry. Honey production is a lucrative business and it generates employment. There is a need for efficient support organisations to monitor the activities of small enterprises. Prediction of the future demand, introduction of modern technologies, cost control and business expansion are the important areas, where entrepreneurs need regular support. Suitable legal support may also be required to protect the traders engaged in unfair trade practices. It is better to promote agro-based enterprises in rural areas, as the local people have the required skills and most of the businesses help the entrepreneurs to ensure food security. The outputs of such business have ready demand even in rural areas and hence the market opportunities are better. With a strong agro-based programme, non-farm activities can also be initiated when the entrepreneurs are more experienced and capable of taking risk and can manage the programme better. Promotion of direct marketing by establishing close interaction between producers and consumers will further enhance the benefits, while encouraging a large number of unemployed rural youth to turn into micro-entrepreneurs and traders. Recognizing the importance of apiculture in improving the rural economy and its role in rural upliftment, and all the more in enhancing crop productivity through cross pollination services towards food security, the concerted and earnest efforts are required in promoting scientific beekeeping in holistic way and in doing away the above problems which shall boost the Gandhian dreamt cottage industry, ensure higher food production and bring rural prosperity. India is an emerging economy. It is not a hidden fact that out farmers who still do not get the much needed assistance of some well professed people. Entrepreneurs are the ones who can take this initiative and give their contribution in this area.

REFERENCES

Aldrich, H., C. Zimmer, 1986. Entrepreneurship through social networks. In D. Sexton and R. Smilor (Eds.), The art and science of entrepreneurship, Cambridge, MA: Ballinger, 3-23

Baum, J. R., E. A. Locke, K. G. Smith, 2001. A multi-dimensional model of venture growth. *Academy of Management Journal*, 44(2), 292–303.

Baum, J. R., M. Frese, R. A Baron, 2007. The psychology of entrepreneurship, Lawrence Erlbaum Associates, New Jersey, 110-173.

Bourdieu, P., 1986, 'The forms of capital', in J.E. Richardson (ed.) Handbook of Theory of Research for the Sociology of Education, New York: Greenwood, 46-53.

Gibcus, P., P.A.M. Vermeulen and J.P.J. de Jong (2008). 'Strategic decisionmaking in small firms: a taxonomy of small business owners', forthcoming in *International Journal of Entrepreneurship and Small Business*, 14-33.

Hisrich, R. D., 1985,. The woman entrepreneur in the United States and Puerto Rico: a comparative study. *Leadership and Organizational Development Journal*, 5, 3–8.

Knight, G.A. and Cavusgil, S.T. (2004). 'Innovation, organizational capabilities, and the born-global firm', *Journal of International Business Studies*, 35: 124–141.

Locke, E. A. (2000). Motivation, cognition and action: an analysis of studies of task goals and knowledge. *Applied Psychology: An International Review*, 49, 408–429.

Nahapiet, J. and Ghoshal, S. (1998). 'Social capital, intellectual capital, and the organizational advantage'. *Academy of Management Review,* 23: 242–266

Penrose, E., 1959, The Theory of the Growth of the Firm, New York: Wiley.

Prashantham, S. (2008). The internationalization of small firms, A strategic entrepreneurship perspective, Routlege, 112-122.

Shane, S., S. Venkataraman, 2000. The Promise of entrepreneurship as a field of research. *Academy of Management Review* 2000, Vol. 25, No. 1. 217-226.

Spender, J.-C. (1994). 'Organizational knowledge, collective practice and Penrose rents'. *International Business Review,* 3: 353–367.

Shane, S., E.A. Locke and C.J. Collins (2003). 'Entrepreneurial motivation', *Human Resource Management Review,* 13 (2), 257–79.

Transformation of Indian Agriculture through Innovative Technologies *Pages* **449–464**
Editor: **Dr. Shahid Ahamad & Dr. Jag Paul Sharma**
Published by: **ASTRAL INTERNATIONAL PVT. LTD., NEW DELHI**

30 Farmer Motivation: A Tool for Improving Sericulture in J&K State

R.K. Bali and Mokshe Sajgotra

Sericulture is an age old agro industry of J&K State. The quality of silk produced is ranked with the best type known. It ranks at number three in revenue generation after tourism and horticulture. Due to climatic compulsions it is practiced only as a subsidiary occupation with single cocoon crop production per year in spring season. Industrial aspects of sericulture *i.e.* reeling of cocoons and silk weaving are least developed inspite of good quality bivoltine cocoon production. Most of 9.73 lac kg of cocoon produced is purchased by reelers from as far as West Bengal. Farmers are deprived of benefits accruing from value addition.

Stiff competition from other crops and vagaries of nature which determines the cocoon output has brought this trade to a cross road. Farmer is fast getting disillusioned with it and is gradually shifting to other trades as well as daily labour job. The traditional farmer is highly frustrated on account of above mentioned facts. The motivation to continue the occupation is slowly dying. The day is not far when he will abandon the trade completely. It is high time the motivation is brought to high speed for good results and increase in trade.

Motivation Factors

During the recent past, technologies in sericulture have been developed at a very high pace and has already reached a plateau. The advantage of these technologies can be taken if these are totally adopted by the farmer. However the adoption level is far below the technology development level. The gap is widening day by day. Factors motivating farmer towards adaption of new technology are briefly enumerated.

Appropriate Technology

In J&K State bivoltine silkworm rearing technology has been well worked out. Techniques for incubation, brushing, chawki rearing, late age rearing, disease control using bed disinfectant, seriposition, post harvest care *etc.* have been developed for increasing cocoon output to the international standards. However, farmers are still using some of the age old practices.

New technology can itself be a motivating factor. Depending upon specific requirements, farmers are always looking forward for technology which may alienate their problems and fulfill their needs under given sources. Farmer no doubt is ready to adopt a new technology but feels shy to invest. He needs to be made to believe by live demonstration that returns are going to be much higher than the inputs made on new technology. This aspect becomes all the more important in the present case where status of occupation is purely subsidiary. Also the strata involved in this occupation belong to lowest level being mostly landless farmers who are ready to work for very small profits as their investment are also negligible. In this scenario technology involving large inputs would be impossible for adoption. Although the gap between technology generated and technology adopted is large primarily because some of the technology may not be worth the efforts but major factor of gap in sericulture remain the state of practicing farmers. This necessitates vigorous efforts by extension worker.

Now again the role of extension worker has an important status. He should be well versed with the process of technology and must also know the impact it can create. Mastering of technology should be first step toward its transfer to field. A skeptical extension worker may not be able to produce desirable results. To attain refinement in this, researcher (technology generator) and extension worker should think on the same wave length. Extension worker too needs updating and support from research. Simple technology transfer for production to spread may not have desirable result. The finest quality of an extension worker should be his persistence. He should be sensitive to the prevailing status of farmer and his needs. Subtle effort can result in change getting incorporated in the psyche of user. Sudden and abrupt means may be hard to penetrate and may become counterproductive.

In the recent times, due to changing scenario in agriculture, new prospects and problem are emerging at a fast rate. Some of these challenges are manmade where as some are natural. After the onset of WTO, liberalization, diversification and commercialization on global level is the order of day. These manmade challenges can be met by increasing productivity and betterment in quality to make the produce competitive. Here improved technology plays an important role. Natural challenges including rapid and frequent changes in weather cycles are having major effect on productivity of main crops. These challenges offer an opportunity to try some other avenues including subsidiary occupation which can withstand fluctuation in climatic condition and have technology ready for extrigencies.

In brief, technology becomes a motivation only when it produces economic gains in shortest possible time. Output is higher than input and the difference is evident from the beginning itself.

Cultivation Technologies for Sericulture Development

The agro-climate conditions of Jammu division are most suitable for bivoltine silkworm rearing during spring (Feb-March) and autumn (Sept.-Oct.) season. Quality cocoon production at commercial' level suffer due to non adoption of proper rearing technology. In view of prevailing conditions and facilities available at rearers level following cultivation technology for bivoltine silkworm rearing is recommended for sub-tropical areas of the country.

a) Mulberry Cultivation

Selection of land: Mulberry flourishes well in slightly acidic (PH 6.2 to 6.8), flat and well drained soil. Sloppy soil too can be used for its cultivation after proper contouring terraces. It prefers clay loam to loam soil.

Preparation of Land

For raising bush mulberry garden the land should be deeply dug during December for winter plantation or with the onset of pre-monsoon showers for monsoon plantation followed by preparation of 35 x 35x 35 cm pits. However, for raising tree plantation pits of 45 x 45 x 45 cm could be directly dug and irrespec-tive of the type of plantation to be raised the pits should be filled back with one part of decomposed FYM and 2 parts of soil before taking up plantation.

Mulberry Variety

Chinese white, S_{1635}, V_1, Chak majra, TR-10 and Vishala are well suited for subtropical areas.

Planting Material

Cutting of about 15-20 cm length with a thickness 1.25 cm having 4 healthy buds should be prepared from 9-12 month old disease and insect, pests free shoots. The middle position of the shoot should be used giving a sloping cut at the base and a flat cut at the top. Cutting should be planted immediately. In case of a delay cuttings should be stored properly within extra care for maintenance of proper humidity.

Planting of Cuttings

Care should be taken to plant the cutting without damaging the buds. The proximal bud should be positioned out of the soil facing upwards. The cutting needs to be planted in slanting position. A spacing of 15 x 10 cm may be maintained between the cuttings. Before planting lower end of the cutting may be dipped in BHC and then planted to prevent any soil borne insect attack. Initial watering at the time' of planting is essential for proper establishment. The mulberry beds should be regularly watered depending upon the available prescription. Beds should be kept devoid of weeds and first round of weeding must be carried out after 35-40 days of planting. The 2nd weeding may be done after 60 days of planting of SOS. The unwanted sprouts arising from the cutting other than main should be trimmed off to reduce inter twig compe-tition for nutrients.

Figure 30.1a-d

Transportation and Transplantation

Cutting planted in monsoon will be ready as saplings for transplantation in ensuing winter and vice-versa. The saplings should be planted with care, one or two days prior to uprooting the nursery beds should be watered to loosen the soil for easy removal of saplings without damaging the roots. The uprooted saplings may be kept in shad covered with wet gunny cloth with regular water sprinkling. The uprooted saplings should be transported to the trans-planting sites immediately and this longer storage should be avoided. Before transportation/transplantation, the leaves on the saplings should be trimmed off. This practice is essentially important in winter raised saplings for monsoon plantation. The length of the saplings should be maintained depending upon the type of the plant to be trained. It *is,* however, essential to plant healthy and vigorous saplings with well deve-loped root system. On planting the planting media shou-ld be pressed firmly all around the sapling and not watered. However water could be skipped during rainy day.

Training of Plants

Saplings after planting can be trained either bush, high bush or even as trees depen-ding upon the land topography and available irrigation. The culture of raising tree type plantation has air avocation under rainfed conditions while bush planta-tion is recommended under irrigation conditions in view of its higher leaf production. For raising bush type of garden a low cut at 10-20 cm above the ground level is advised after planting and different pruning schedules after one year of plantation are adopted. For high bush plants, an apical cut at the height of 60-75 and at 180-200 cm for tree is advocated.

Pruning

Under subtropical conditions mulberry sprouts almost throughout the year except in dormancy during winter. The grown up trees can suitably be pruned for

Figure 30.2

maintaining desired shape. No pruning is advised during the first year of plantation except the one given for training of the plants. The first pruning is to be taken up with the onset of monsoon at the crown height (Bottom pruning) whereas 2nd pruning should be carried out in the 2nd or 3rd week of December at the height of 125 cm from

Figure 30.3a-b

level irrespective of type of plantation. During dry spells of summer *i.e.*, after April rearing is over, no pruning should be carried out till the onset of monsoon rains.

Application of FYM and Chemical Fertilizers

Application of farmyard manure and chemical fertilizers is necessary for overall growth of the plant and leaf harvest thereof. FYM is advocated to be applied in the pit during first year of plantation itself and from next year onward it should be applied and properly mixed with soil at the rate of 10 and 20 tons per hectare per year at the time of digging during winter as pre--monsoon season.

The application of chemical fertilizers, N,P,K in the ration of 150: 75: 75 kg/ hectare/year under rainfed condition and in the ration 300: 120: 120 kg/hectare/ year under irrigated cond-itions could be applied. The first doze of fertilizers (NPK) can be applied in the month of February without splitting the phosphorus and Potash. The 2nd close fertilizer (N) can be applied in the month of August. However application 'of chemical fertilizers would vary with chemical status of the soil, moisture availability *etc.* As such it is suggested to get the soil analyzed.

Inter-cultivation

In mulberry plantation the first digging in advocated only after about two months of plantation. Thereafter it is advisable to go for at least three diggings per year under irrigated conditions and two diggings under rainfed conditions. In both cases annual digging should be carried out after bottom pruning of plants during monsoon season followed' by another digging in the month of January after carrying out middle pruning. Under irrigated condition, third digging could be carried out in the month of April. Fortnightly to monthly weeding, depending upon season, is required to keep the field free of weeds which normally compete with mulberry for soil nutrients and lead to biomass losses.

Management of Diseases

Leaf spot and powdery mildew are common prevalent during autumn. Both the diseases could be managed by use of 0.05 to 0.1 Carbendozim or 0.210 suffix after 20-25 days of bottom pruning.

b) Silkworm Rearing

Disinfection

The rearing equipments must be thoroughly washed and disinfected with 1 per cent bleaching powder in 0.3 per cent slaked lime or 2 per cent formalin solution at 24-25°C. The surroundings of rearing house should be sprinkled with 10 per cent high grade bleaching powder mixed with slaked lime (200 g/sq m.) every 6-7 days interval.

Incubation and Brushing

The following incubation schedule must be adopted for obtaining more than 95 per cent hatching in single day.

Figure 30.4a-b

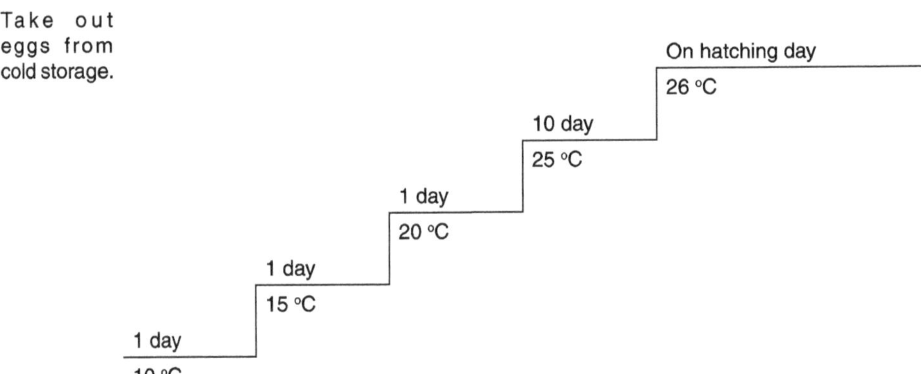

☆ Note: - Provide 18 hours light of 300 LUX and maintain 75-80 per cent RH during incubation.

Brushing should be completed by 11 AM on brushing day with finely chopped succulent leaf.

Chawki Rearing

Chawki rearing is recommen-ded for bringing overall improvement in cocoon production. Chawki reared worms must be distributed to rearers and worms should be transported from chawki rearing centers to rearers house after resum-ption for third stage during cool morning hours. Alternate would be to conduct chawki rearing on Co-operative basis for a group of rearers maintaining most optimal conditions till distribution of worms to individual rearers post second moult.

For chawki rearing mulberry leaves should be individually plucked at early morning or late evening hours. The moisture content of Chawki leaf must be

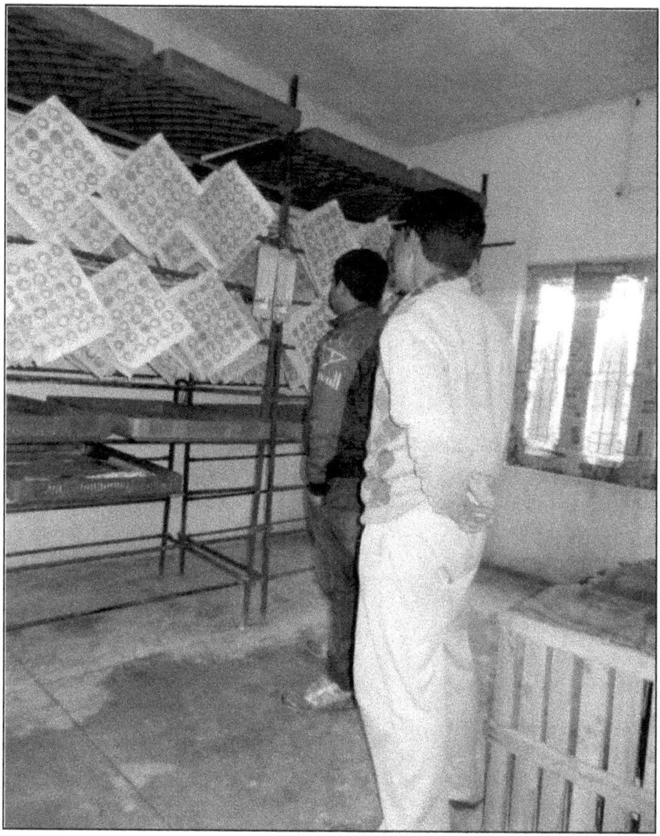

Figure 30.5

Figure 30.6a-b

maintained above 75 per cent for which regular irrigation of chawki garden every third day is recommended. Wet foam pads and maintained high temperature (27°C-28°C) is recommended for robust growth of worms during Chawki stages.

Factor	INSTAR	
	I	II
Temperature (°C)	27-28	27-28
Relative Humidity (per cent)	80-90	80-90
Leaf size	Finely chopped	Finely chopped
Bed cleaning	At resumption for IInd stage	At resumption for IIIrd stage

Precautions during Chawki Rearing

Chawki silk-worms should be reared on a sheet of paper. From the beginning of IInd instar RKO should be dusted @ 120 gm/100. Dfls. RKO dusting should be done before resum-ing the worms for next instar. The RKO dusted worms should be covered with paraffin/newspaper for 30 minutes. At the time of resumption, finely chopped leaf should be fed. During moulting period, relative humidity should not be more than 70 per cent.

Figure 30.7a-b

Late Age Rearing

The IIrd, IVth and Vth age rearing in silkworm is termed as late age rearing. The recommended temperature and relative humidity for late age rearing is as under

Factor	INSTAR		
	III	IV	V
Temperature (°C)	25-26	24-25	23-24
Relative Humidity (per cent)	75-80	70-75	65-70
Leaf size	One cut daily	Entire daily	Entire daily

Figure 30.8

Precaution during Late Age Rearing

Soiled, diseased, dusty and over mature leaves should be avoided during late age feeding. The undersized and lethargic worms should be removed carefully to avoid further contamination. These worms should be collected in 20 per cent formalin or lime solution and pro-perly disposed. The silkworm refuse must be collected in refuse bags and disposed of properly in pits. Good aeration and dry conditions in rearing room should be ensured during moulting period. RKO should be applied uniformly 30 minutes prior to resumption from every moult. In late age rearing bed should be cleaned once in a day and floor moped with 5 per cent bleaching powder solution after every feed. The feeding quant-ity and leaf size must be reduced with onset of spinning.

Spinning and Harvesting

Clean/disinfected mountages material/plastic mountages should be used for, spinning of cocoons. About 30-40 larvae should be mounted per sq-ft. old newspaper must be placed under the mounting frame/material for absorption of urine and excessive moisture. The diseased and dead worms should be removed carefully while removing the news-paper placed under mounting frame/material on 3rd day of mounting. A temperature of 25°C and RH of 60-65 per cent is recommended during spinning. During spring season (Feb.-March) harvesting of cocoons must be done on 7th/8th day of mounting while in autumn season (Sept.-Oct.) cocoons should be harvested on 6th/7th day. The stifling of cocoons should be done after eliminating the deformed, thin ended and double cocoons. Preferably stifling should be done in Hot Air oven or in Sun under black cloth cover.

Adopting the recommended technologies for integrated development for sericulture the rearers can harvest quality grade cocoon yield which can fetch them more price in the cocoon market arranged in each district by the State Sericulture Development Department.

Pricing Policy

It is one of the major discouraging factors in sericulture in J&K State. It is a well known fact that increased production results in glut, leading to crash in prices. This is one of the reason that the farmers are shy to adopt or discard the technology for fear of losses. Another aspect in today's scenario is monopoly of buyer, especially in case of sericulture produce in J&K State. It becomes worst when buyer cartel steps in. They choke the natural progression of prices leading to static prices. Farmers are forced to sell their produce at lower rates.

This problem can be averted by announcing floor prices of a commodity commensurate with input cost, so that the producer is able to get profitable margins. The floor price be reviewed periodically so as to accommodate losses due to inflation. Floor prices of raw material should also commensurate with the finished product so that the gains of value addition are distributed at all levels. As has been calculated farmer gets only 18.5 per cent of gain while doing the bulk of work in cocoon production. Traders share of market margin is above 50 percent. The floor price of cocoon has not increased for the last about ten years where as the prices of finished product, silk is soaring. Farmer is not getting any benefit of increasing silk rates. This needs immediate attention so that farmer is attracted back with higher floor price of silk raw material.

Movement of Goods

In the event of surplus production, better price can be obtained by moving the goods to scarce areas. It requires a perfect transport chain. This problem is very acute in remote areas where due to lack of transportation farmers do not get the appropriate return. Most of the profits

Figure 30.9a-b

are made by middlemen. The poor farmers have to bear the brunt. Technology gains are negated. Bivoltine cocoons seldom fetch above Rs. 300 per kg whereas same material could sell at 50 per cent premium in south India.

Income Guarantee

The advantages of technology by way of higher productivity may not get transferred to greater income due to various reasons. Farmer's prime motive is to get more money. Technology or productivity are meaningless for this. Thus a guaranteed high income brings farmer closer to adoption of new technology irrespective of any costs involved. Lure of higher income and its guarantee act as great motivator.

Insurance of produce is an excellent guarantee for losses in unforeseen circumstances. It needs to be publicized widely and should cover all aspects of a particular crop. There should be complete transparency in this scheme with least beaurocratic hic ups in its implementation. Govt. intervention in market during glut situation can bring relief to farmer to some extent.

Marketing

One of the weakest link in the development of sericulture in Jammu and Kashmir is marketing of raw material (cocoons) produced by farmers. It has been estimated that out of 9.73 lac Kg of cocoons produced only about 15-30 per cent processed locally for reeling. Rest of the produce is sold in govt. controlled cocoon markets where open bidding takes place. Most of the bidders come from outside state mainly from west Bengal and have formed a sort of cartel. It is completely a buyer's market, seller is left at their mercy resulting in vide variation in sale rates which seldom commensurate with the produce. Delayed marketing forces farmers to shed their produce so as to salvage some income. Moreover since the farmer is landless or marginal one, he does not have infrastructure for storage of cocoon for a long period. Storage losses too are to be borne by the hapless farmer. In this direction, farmers need to be provided help about when to sell and to whom to sell. Government or Co-operative agency could develop storage facility to help farmer to sell when prices are good.

Cocoon Banks

Computerization and internet facility be made available at selected spots providing information on prevailing prices. This shall help the farmer to regulate his ware.

Steps for Successful Motivation

Motivation needs verification of baseline information about the person to be motivated. The prerequisite include knowledge of requirements and interests of subjects. Creation of a sense of dissatisfaction with the existing and old thought will definitely create zest for the new thought and would lead to change in shortest possible time.

Before embarking on the introduction of new thought, it is important to have an insight on existing problems, inputs, constraints *etc.* This would help in formulating a flow chart of action for motivating the subject. Some of the suggestion for motivating farmers is given in brief.

- ☆ Demonstration is the best tool for exhibiting innovation effects. It is very successful a tool as it involves all the three modes; hearing, seeing and doing. Farmer needs to be involved fully in demonstration of a new technology. It makes transfer very easy. It would be appropriate to quote Dr. S.A. Knapp, "What a man hears, he may doubt; what he sees, he may possible doubt, but what he does himself, he cannot doubt".
- ☆ Progressive farmers can be the best tool in demonstration. A progressive farmer already using some technology can be adopted. Visit by other farmers be organized to the progressive farmer. This will help in speedy transfer of technology.
- ☆ Mass media coverage of field demonstration helps in wider dissemination of success of a particular technology.
- ☆ Viability of technology is very important for being put forth for adoption. It should take care of existing facilities available at field level, extra input involved and finally productivity. Lower input cost and higher productivity is sure to succeed as a motivating factor.
- ☆ Persistence of extension work is very important a factor in motivation. A worker should never feel heartbroken on initial negative reaction of farmer which normally is a rule. Our farmers are in general orthodox in their thinking and resist change. Farmer's aggressive behavior towards new technology needs analysis. Comparative advantage of new technology needs high lighting. The presenter should have pleasing manner in communication at the farmer level using the local dialect.
- ☆ Drawback of a new technology, if any, should not be hidden from farmer. These should be described in details vis-à-vis benefit. Farmer will himself make the profit/loss estimation on the basis of his expression keeping in mind the disadvantages if any.
- ☆ The benefits may never be exaggerated. It is wise to communicate a few degrees less than the estimated benefits. On practical demonstration the result will show higher value than as advocated. This will add to moral of farmer and convince him about its benefits.
- ☆ Involvement of local leader, religious head, panchayat *etc.* will have better effect. If convinced, these leaders can pass on the message to the farmers who normally have full faith on them and will not feel disappointed in the eventually of some failure.
- ☆ The motivation of younger generation with some educational background yields, better result. They act as agent for convincing elders and can successfully pass on the technology.

☆ Incentive is the most important motivation factor. However it should not be in the form of govt. subsidy or input costs. This leads to pilferage of funds by unscrupulous farmers. Needy farmer fails to get anything. Moreover in the present competitive world incentive kills the genuine entrepreneurship. Incentive should be productivity based. Govt. should devise ways to assist those farmers who are able to produce above the set benchmark for the crop. This will install competition to work hard and produce more. Prospects of added incentives will definitely lead to high product as well as productivity.

REFERENCES

A report on the diseases of silkworm in India. 1984. Jameson, A. P. International Books and periodicals supply service, New Delhi.

An introduction to sericulture. 1971. Ganga,G. and Chetty, J.S. Oxford and IBH Publishing Co. Pvt. Ltd. New Delhi.

Hand Book of Practical Sericulture. 1987. Ullal, S.R. and Narasimahanna, M.N. Central Silk Board, Bangalore.

Lectures on sericulture. 1994. Boraiah, G. Super Graphics Rajaji Nagar, Bangalore.

Mulberry and silkworm crop protection (Vol. III) 1989. A.K. Dhote, NCERT, New Delhi.

Appropriate Sericulture Techniques. 1987. M.S. Jolly, CSR and TI, Mysore.

Mulberry cultivation. 1988. Zheng Ting, Oxford and JBN Pub.

Mulberry Silk Industry. 1984. Chowdhury, S.N. Dibrugarh, Assam, India.

Principles of Sericulture 1994. Aruga, H. Oxford and IBH Publishing Co. Pvt. Ltd. New Delhi.

Principles of Temperate Sericulture. 2000. Kamili, A.S. and Masoodi, M.A. Kalyani publishers.

Proceedings of Training Course in Sericulture. 1995. Agarwal and Seth, H.P. University, Shimla.

Sericultural Manual - I. Mulberry cultivation. 1976. Rangaswami, G. Narasimhanna, M.N., Kasiviswanathan, K., Sastry, C.R. and Jolly, M.S. Agricultural Service Bulletin, FAO, Rome.

Sericulture Manual - 2 Silkworm rearing. 1973. Krishnaswami, S. Narasimhanna, M.N., Suryanarayan, K. and Kumararaj, S. FAO, Rome.

Silkworm Biology, Genetics and Breeding. 1998. Sarkar, D. Vikas Publication House, New Delhi.

Silkworm rearing, 1998. Pang Chuan, Oxford.

Silkworm seed production Technology (Vol. IV) 1989. A.K. Dhote, NCERT, New Delhi.

Synthesized Science of Sericulture. 1962. Yokoyama, T. Central Silk Board, Bombay-2.

Text Book of Tropical Sericulture. 1975. Japan overseas Co-operation volunteers Agency, Tokyo, Japan.

Transformation of Indian Agriculture through Innovative Technologies *Pages* **465–471**
Editor: **Dr. Shahid Ahamad & Dr. Jag Paul Sharma**
Published by: **ASTRAL INTERNATIONAL PVT. LTD., NEW DELHI**

31 Role of Markets and Market Led Extension in Doubling the Farmers' Income

Pawan Kumar Sharma, R.K. Arora, Vishal Mahajan and Banarsi Lal

Introduction

Income of a businessman is actually a factor of many complex circumstances existing on the ground and income of a farmer is much more complex than a normal business man due to the intervention of nature and of course man made effects. When nature is not in favour of farmers, production decreases and ultimately income of farmers' decreases and if nature is conducive to agricultural production, productivity increases but price of the commodity is bound to decrease because of higher production and then again the income of farmers' decreases. Therefore, doubling the farmers' income in a span of five years would not be easy and if it is to be achieved, the effectiveness of marketing stakeholders in agriculture have to play a very crucial role. Chandrasekhar and Mehrotra, 2016 suggested measures to achieve the objective of doubling the farmers' income. In this era of globalization and liberalization a farmer will have to transform himself from mere 'producer' in the domestic market to 'producer-cum-seller' in order to best realize the returns for his investments. The focus should be turned towards high quality produce, low cost of production and high productivity. In this context the extension system may face crisis in the form of knowledge, efficacy, credibility, reorganizational structure *etc.* Extension personnel will have to be ready to perform some additional roles like SWOT analysis of the market, organizing Farmers' Commodity Interest Groups, supporting and enhancing the capacities of locally established groups like Water Users Associations, Self-Help Groups, Watershed Committees *etc.*, creation of websites of successful farmers besides establishing marketing agro-processing linkages and market intelligence for motivating our farmers to shift from 'supply driven' to 'market driven'.

Agricultural Marketing System

Market size in India is large and continuously expanding, but marketing system not kept pace with it. The unregulated and private trade constitutes more than 80 per cent of marketed surplus. Direct marketing "farmer–consumer" is negligible. Out of 85 per cent of the 27,294 rural periodic markets, facilities for efficient trade are still almost absent. 7161 market yards/sub yards is inadequate, ill equipped and mismanaged. Due to lack of proper handling at farm gate lead to 30 per cent F and V, 7 per cent grains, 10 per cent spices loss before reaching market.

Traditionally agricultural outputs were seasonal in their production and supply. The modern technology and cultural practices mean that food manufacturers need not have their production schedules dictated by the seasons. Indeed the capital-intensive food industry cannot afford to incur the high costs of under utilizing its capacity. This means that farmers will have to complete in terms of reducing seasonality or fitting into a pattern of social competitiveness.

An efficient marketing system is essential for the development of agricultural sector. The marketing system contributes greatly to the commercialization of subsistence farmers. Failure to develop the agricultural marketing system is likely to negate most (if not all) the efforts to increase agricultural production (Anon 2000). In the changing scenario of Indian agriculture the extension system is likely to undergo following series of crisis: Knowledge (skill impact crisis): The extension educationists will now have to be well-equipped with the latest market information besides the knowledge of production technologies. This requires their further training and additional funding. Efficacy crisis: The extension system has already been under criticism. With the increased and enriched role the extension functionaries they will have to perform multiple activities to prove their efficacy. Credibility crisis: Despite market knowledge and efficacy in performing their role the extension system may face the credibility crisis due to rapid and unexpected changes in the market. The producers of agricultural produce will be increasingly judged on their reliability in all of these respects. Ease of processing will become an increasingly important expectation of the food industry. Like all industries, reductions in the costs of capital equipment, wages and inventories are important objectives. For example, farmers who can deliver on the 'justin-time' principle will contribute towards reducing a manufacturer's working capital and space requirements. Farmers who can do part of the secondary processing and/or performing functions such as the post harvest treatment of the crop or transporting will be adding another advantage. Crops that are specially bred or designed to facilitate processing are another type of advantage that the food industry could expect from agriculture. In short, the competitive advantage will rest with those able to add most value and can differentiate what they are offering from that of other suppliers. In competitive brand marketing, the food industry has to innovate continuously to create new products that are different from and superior to existing ones of their own or competitors. The scope of innovation has traditionally been at the processing stage. Whilst this will continue to be an important area for innovation, manufacturers will increasingly tend to look for innovative changes in the agricultural produce itself. This may be in terms of novel tastes, improved texture, more attractive shapes, *etc.*

New Options in Agricultural Marketing

In the more sophisticated food markets, healthy eating can become a priority among consumers. Therefore, farmers will have to consider the health connotations of what they choose to grow. The consumers may be interested in the food itself *i.e.* low/no sugar or low/no salt. Nutrition is important in all segments of the market. Thus farmers have to be concerned about the nutritional value of the produce they grow. Further, the consumers are may be more, or equally, concerned about the food production methods *i.e.* the avoidance of chemicals like herbicides, pesticides *etc.* This may mean a change to the farmer's cultivation practices with implications for the costs of production. The consumer and the food industry will expect the farmer to produce without potentially dangerous chemicals, but at no extra cost to them. This will be another challenge for agriculture.

Steps to Empower Farmers through Market

The availability of assured market is the most crucial aspect of increasing the farmers' income. The following points need to be considered for empowering farmers to enhance their options for getting remunerative prices for their produce and thus increased incomes.

Enhancing Risk bearing Ability of Farmers

In both the production and marketing of produce, the possibility of incurring losses is always present. Market risks are those of adverse change in the value of the produce between the processes of production and consumption. The provision of warehouse receipts and price information can considerably enhance risk bearing ability of farmers. Whether storage takes place on the farm or in silos off the farm, increases in the value of products due to their time utility must be sufficient to compensate for costs at this stage, or else storage will not be profitable. These costs will include heating, lighting, chemical treatments, store management and labour, capital investment in storage and handling equipment, interest charges and opportunity costs relating to the capital tied up in stocks.

Grading and Standardization

Grading system accurately describes products in a uniform and meaningful manner. Grades and standards contribute to operational and pricing efficiency by providing buyers and sellers with a system of communicating price and product information. By definition, commodities are indistinguishable from one another. However, there are differences between grades and this has to be communicated to the market. By the same measure, buyers require a mechanism to signal which grades they are willing to purchase and at what premium or discount. Prices vary among the grades depending upon the relative supply of and demand for each grade. Since the value of a commodity is directly by its grade, disputes can and do arise. The absence of grades and standards restricts the development of effective and efficient marketing systems.Standardization is concerned with the establishment and maintenance of uniform measurements of produce quality and/or quantity. This function simplifies buying and selling as well as reducing marketing costs by enabling buyers to specify precisely what they want and suppliers to communicate what they are able and willing to supply with respect to both quantity and quality of product. In the absence of standard weights and measures trade either becomes more expensive to conduct or impossible altogether.

Processing

Most agricultural produce is not in a form suitable for direct delivery to the consumer when it is first harvested. Rather it needs to be changed in some way before it can be used. Of course, processing is not the only way of adding value to a product. Storing products until such time as they are needed add utility and therefore adds value. Similarly, transporting commodities to purchasing points convenient to the consumer adds value.

Quality Products

Quality differences in agricultural products arise for several reasons. Quality differences may be due to production methods and/or because of the quality of inputs used. Technological innovation can also give rise to quality differences. In addition, a buyer's assessment of a product's quality is often an expression of personal preference. Thus, for example, in some markets a small banana is judged to be in some sense 'better' than a large banana; white sugar is considered 'superior' to yellow sugar; long stemmed carnations are of 'higher quality' than short stemmed carnations. It matters not whether the criteria used in making such assessments are objective or subjective since they have the same effect in the marketplace. What does matter in marketing is to understand how the buyer assesses 'quality'.

Use of Information and Communication Technology

Mass media has an important role to play in dissemination of price signals, especially in cash crops like cotton (Dagwal, *et al.*, 2009; Helen *et al.*, 2008 and Lwoga*et. al.*, 2010). Internet technology has percolated down up to block level and in some states up to village level. Search engines and the present websites furnish general information presently. Agricultural Market related information on the internet is inadequate. Hence, a whole network of skilled personnel need to be engaged in collection of current information and creation of relevant websites pertaining to/serving specific needs of farmers. Creation of websites should be mandatory in different languages to equip the farmers with information. These websites should contain information like market networks, likely price trends, current prices, demand status options for sale, storage facilities *etc.* Information technology should be able to provide answers to questions like what and how much to produce, when to produce, in what form to sell, at what price to sell, when to sell and where to sell.

Market Intelligence

Marketing decisions should be based on sound information. The process of collecting, interpreting, and disseminating information relevant to marketing decisions is known as market intelligence. The role of market intelligence is to reduce the level of risk in decision-making. Through market intelligence the seller finds out what the customer needs and wants. The alternative is to find out through sales, or the lack of them. Marketing research helps establish what products are right for the market, which channels of distribution are most appropriate, how best promote products and what process are acceptable to the market.Generation of data on the market intelligence would be a huge task by itself. Departments of market already possess much of the data. Hence, establishment of linkages between agriculture line departments and departments of market strengthens the market-led extension.

Financing

In almost any production system there are inevitable lags between investing in the necessary raw materials (*e.g.* machinery, seeds, fertilizers, packaging, flavorings, stocks *etc.*) and receiving payment for the sale of produce. During these lag periods some individual or institution must finance the investment. The question of where the funding of the investment is to come from, at all points between production and consumption, is one that marketing must address.

e-National Agriculture Market

National Agriculture Market (NAM) is a pan-India electronic trading portal which networks the existing APMC mandis to create a unified national market for agricultural commodities.

The NAM Portal provides a single window service for all APMC related information and services. This includes commodity arrivals and prices, buy and sell trade offers, provision to respond to trade offers, among other services. While material flow (agriculture produce) continue to happen through mandis, an online market reduces transaction costs and information asymmetry.

Agriculture marketing is administered by the States as per their agri-marketing regulations, under which, the State is divided into several market areas, each of which is administered by a separate Agricultural Produce Marketing Committee (APMC) which imposes its own marketing regulation (including fees). This fragmentation of markets, even within the State, hinders free flow of agri commodities from one market area to another and multiple handling of agri-produce and multiple levels of mandi charges ends up escalating the prices for the consumers without commensurate benefit to the farmer.

NAM addresses these challenges by creating a unified market through online trading platform, both, at State and National level and promotes uniformity, streamlining of procedures across the integrated markets, removes information asymmetry between buyers and sellers and promotes real time price discovery, based on actual demand and supply, promotes transparency in auction process, and access to a nationwide market for the farmer, with prices commensurate with quality of his produce and online payment and availability of better quality produce and at more reasonable prices to the consumer. 455 mandis have been enrolled under e-NAM within one year of its launching, covering 35 commodities, including vegetables, fruits, pulses and cereals.

Shifting towards Market Led Extension

The agricultural extension system in the country is heavily burdened with performance of multi-farming activities in the field. During the last 50 years, major emphasis has been given on Production-led Extension (Duraisamy, 2007). The system acts as liaison between the researchers and the farmers. Extension personnel are assigned the responsibility of conveying research findings from the scientists to the farmers and feeding back the impressions from the farmers to the scientists. The new dimensions of marketing may overburden them and become an agenda beyond their comprehension and capability. In the light of this scenario the extension personnel are required to be motivated to learn the new knowledge and skills of marketing before assigning them marketing extension jobs to establish their credibility and facilitate significant profit for the farming community. Today advanced and scientific agriculture is considered as professional industry and enterprise. So crop should be grown to earn as much profit as possible. Emphasis should be given on climate-based extension education by devising training programmes,Market-led extension, demonstrations *etc.* to the farmers empowering them to adopt improved technologies for higher yields and other high-tech ventures. For this purpose the farmers need to know the answers to questions like what to produce, when to produce, how much to produce, when and where to sell at what price and in what form to sell their produce. In responding to such questions the extension system should be oriented with knowledge and skills related to the market with the objective to improve the quality of agricultural production to compete in the market particularly in the global market. We have to accept the concept and paradigm of market-led extension and find out the new thrust areas in market-led extension in the interest and welfare of farming community.

The public extension system is already under severe criticism for its inability to deliver the services. In the light of this, the challenge remains to motivate the extension personnel to learn the new knowledge and skills of marketing before assigning them marketing extension jobs to establish their credibility and facilitate significant profits for the farming community. Extension cadre development poses a new challenge to the newly designed role. The present extension system suffers from several limitations of stationery, mobility, travel allowances, personnel development, *etc.* There is a dire need to upgrade these basic facilities and free the extension cadres from the shackles of the hygiene factors and enthuse them to look forward for the motivating factors like achievement, job satisfaction, recognition *etc.* There is a need to innovate approaches for the transfer of technology and information to farmers so as to empower them to face the challenges of market liberalization and globalization (Birthal and Ganesh, 2009). The strategies towards adoption of market led extension include:

- ☆ Promoting Groups, including Farmers Interest Groups, Women Interest Groups, Commodity Interest Groups for strengthening Farmers'-led extension.
- ☆ Capacity building of groups for advanced agricultural production technologies.
- ☆ Promotion of farmers' participatory research and extension programmes for location specific technology development, refinement and dissemination.
- ☆ Strengthening of market intelligence and Information and Communication Technology (ICT) for access to market information.

Conclusion

Agricultural marketing is the most crucial part of agricultural system and enhancing the efficiency of this system can play a huge role in achieving the objective of doubling the farmers' income in a shorter period of time. Capacity building of farmers is needed not only in adoption of production related technologies but also in terms of market related technologies which would enhance farmers' ability to earn more percentage in the price paid by the ultimate consumer. The wider usage of Information Communication Technology and market intelligence can boost the chances of farmers for obtaining better price and thus better income for their produce.

REFERENCES

Birthal, A. and Kumar, G. 2009. Strengthening pluralistic agricultural information delivery systems in India", *Agricultural Economics Research Review*, 22: 71-79.

Chandrasekhar, S., Mehrotra, N. 2016. Doubling Farmers' Incomes by 2022. *Economic and Political Weekly*, 51(18): 30 Apr, 2016.

Dagwal, G.R, Gohad, V.V., Chorey, A. and Dhapate, S.M. 2009. Mass media utilization by cotton growers", *Agriculture Update* 4(1 and 2): 136-137.

Duraisumy, D.G. 2007. Market led extension: Emerging perspectives. In F.M.H. Kaleel, Jayagree Krishnonkutty and K. Satheesh Babu (eds). *Market Led Extension: Dimensions and Tools*. Agrotech Publishing Academy, Udaipur: 42-51.

Helen, M., Ikoja-Odongo, R. Nasinyama, G.W. and Lwasa, S. 2008. Information seeking and use among urban farmers in Kampala District, *Uganda AALD AITA WCCA*.

Kokate, K.D., Kharde, P.B.,Patil S.S. and Deshmukh, B.A. 2009. Farmers'- Led Extension: Experiences and Road Ahead. *Indian Res. J Ext. Edu.* 9(2): 18-21.

Lwoga, E.T., Ngulube, P. and Stilwell, C. 2010. Information needs and information seeking behaviour of small-scale farmers in Tanzania", *Innovation*, 40: 82-103.

NAM Portal: http: //enam.gov.in/NAM/home/index.html, accessed on 10th August, 2017.

Index

A

B

C

D

J

K

L

M

N

O

P

www.ingramcontent.com/pod-product-compliance
Ingram Content Group UK Ltd.
Pitfield, Milton Keynes, MK11 3LW, UK
UKHW021529300726
14060UKWH00011B/132

9 789390 384655